2015 U.S. Higher Education Faculty Awards: Fine Arts, Humanities, Liberal Arts and Social Sciences

2015 U.S. Higher Education Faculty Awards: Fine Arts, Humanities, Liberal Arts and Social Sciences

FACULTY AWARDS

River Publishers

ISBN: 978-87-93379-00-8 (Paperback)

Published, sold and distributed by:
River Publishers
Niels Jernes Vej 10
9220 Aalborg Ø
Denmark

River Publishers
Lange Geer 44
2611 PW Delft
The Netherlands

Tel.: +45369953197
www.riverpublishers.com

Contents

Introduction

FacultyAwards.org is the first and only university awards program in the United States based on faculty peer evaluation. Faculty Awards was created to recognize outstanding faculty members (as viewed by their Faculty peers) at colleges and universities across the United States. Faculty members voted through the 2014–2015 academic year for their peers at their academic departments and schools within a number of categories.

Access to FacultyAwards.org to nominate and vote for Faculty was limited to university professors or faculty members at accredited U.S. institution of higher education.

Faculty members were nominated and voted for by other faculty members in their own academic departments and schools. We strove to maintain an accurate peer-review process. Voting was not open to students or the public at large. In addition, faculty members voted for educators only at their own college or university.

Winners for the 2014–2015 academic year, in all departments and colleges across U.S. institutions of higher education were announced in March 2015 and are permanently archived at FacultyAwards.org, as well as recognized in this 2015 print edition of the Faculty Awards Compendium.

For the academic year 2014–2015; thousands of votes were cast to nominate and vote for Faculty members, and no self-voting was allowed, to assure the integrity of the entire process. A total of 4106 winning Faculty members were determined after tallying the votes.

The awards are not intended to carry merit towards a typical faculty tenure or promotion process. However, as Faculty peer review of teaching, scholarship, research and service is a crucial component of the evaluation processes in U.S. institutions of higher education; the results provide a valuable resource to acknowledge the stellar perception of a particular set of distinguished Professors by their peers and, hopefully, will continue in the years to come to provide a further incentive for Faculty members to continue on their varying paths of academic and scholarly excellence so they would be recognized by their institutional peers.

FacultyAwards.org fully recognizes that peer evaluation is only one of many aspects that define a distinguished educator, scholar and researcher; which also include: student evaluations, scholarly production and publication records, innovative curricula development, distinguished teaching outcomes, outreach and service activities, external funding records, collegiality, research eminence and outcomes, and several other factors. The intention of this compendium is to highlight and reward *only* the peer-evaluation factor for Faculty members and none of the other factors involved in what typically would define an eminent and well-rounded Faculty member at a U.S. institution of higher education.

This volume of the Faculty Awards Compendium includes Faculty awardees within the *Fine Arts, Humanities, Liberal Arts and Social Sciences* disciplines for the 2014–2015 academic year. We would like to thank all Faculty members who participated in the voting process and to wish all the Faculty awardees continued success in their academic endeavors. We look forward to resuming the voting process for the 2015–2016 academic year awards.

Biographies

Professor Tarek M. Sobh received the B.Sc. in Engineering degree with honors in Computer Science and Automatic Control from the Faculty of Engineering, Alexandria University, Egypt in 1988, and M.S. and Ph.D. degrees in Computer and Information Science from the School of Engineering, University of Pennsylvania in 1989 and 1991, respectively. He is currently the Senior Vice President for Graduate Studies and Research, Dean of the School of Engineering and Distinguished Professor of Engineering and Computer Science at the University of Bridgeport (UB), Connecticut; the Founding Director of the Interdisciplinary Robotics, Intelligent Sensing, and Control (RISC) laboratory; the Founder of the High-Tech Business Incubator at UB (CTech IncUBator), and a Professor of Computer Engineering, Computer Science, Electrical Engineering and Mechanical Engineering.

He was Vice President from 2008–2014, Vice Provost from 2006–2008, Interim Dean of the School of Business, Director of External Engineering Programs, Interim Chairman of Computer Science and Computer Engineering, and Chairman of the Department of Technology Management at the University of Bridgeport. He was an Associate Professor of Computer Science and Computer Engineering at the University of Bridgeport from 1995–1999, a Research Assistant Professor of Computer Science at the Department of Computer Science, University of Utah from 1992–1995, and a Research Fellow at the General Robotics and Active Sensory Perception (GRASP) Laboratory of the University of Pennsylvania from 1989–1991. He was the Founding Chairman of the Discrete Event and Hybrid Systems Technical Committee of the IEEE Robotics and Automation Society from 1992–1999, and the Founding Chairman of the Prototyping Technical Committee of the IEEE Robotics and Automation Society from 1999–2001. His background is in the fields of computer science and engineering, control theory, robotics, automation, manufacturing, AI, computer vision and signal processing.

Research Interests and Activities:

Dr. Sobh's current research interests include reverse engineering and industrial inspection, CAD/CAM and active sensing under uncertainty, robots and electromechanical systems prototyping, sensor-based distributed control schemes, unifying tolerances across sensing, design, and manufacturing, hybrid and discrete event control, modeling, and applications, and mobile robotic manipulation. He has published over 200 refereed journal and conference papers, and book chapters in these and other areas, in addition to 18 books. Professor Sobh is also interested in developing theoretical and experimental tools to aid performing adaptive goal-directed robotic sensing for modeling, observing and controlling interactive agents in unstructured environments.

Dr. Sobh serves on the editorial boards of 15 journals, and has served as Chair, Technical Program Chair and on the program committees of over 150 international conferences and workshops in the Robotics, Automation, Sensing, Computing, Systems, Control, Online Engineering and Engineering Education areas. Dr. Sobh has presented more than 100 keynote speeches, invited talks and lectures, colloquia and seminars at research meetings, University departments, research centers, and companies.

Some of the current research and development activities that he leads at the RISC laboratory include work on tolerance representation and determination for inspection and manufacturing, hybrid controllers for robotics and automation, service robots, prototyping and synthesis of controllers, simulators, and monitors for manipulators, algorithms for uncertainty computation from sense data, and web-based prototyping, control synthesis, and simulation of robots.

Dr. Sobh has supervised more than 50 award-winning graduate and undergraduate students working on different projects within robotics, prototyping, computer vision, control, and manufacturing; in addition to more than 300 undergraduate and graduate students working on their B.S. projects, Master's thesis or Ph.D. dissertations. Dr. Sobh is active in consulting and providing service to many industrial organizations and companies. He has consulted for several companies in the U.S., Switzerland, India, Malaysia, England, the United Arab Emirates, Kazakhstan and Egypt, to support projects in robotics, automation, manufacturing, sensing, numerical analysis, and control. He has also worked at Philips Laboratories in New York, and a number of companies in Egypt. Dr. Sobh has been awarded over 45 research grants to pursue his work in robotics, automation, manufacturing, and sensing. Dr. Sobh is a Fellow of the African Academy of Sciences and a member of the Connecticut Academy of Science and Engineering. He has received many awards and merits in recognition of his research and scholarly activities in engineering, education and computing and his services to the academic community.

Dr. Sobh is a Licensed Professional Electrical Engineer (P.E.) in the State of Utah, a Certified Manufacturing Engineer (CMfgE) by the Society of Manufacturing Engineers, a Certified Professional Manager (C.M.) by the Institute of Certified Professional Managers at James Madison University, a Certified Reliability Engineer (C.R.E.) by the American Society for Quality, a member of Tau Beta Pi (The Engineering Honor Society), Sigma Xi (The Scientific Research Society), Phi Beta Delta (The International Honor Society), Upsilon Pi Epsilon (The National Honor Society for the Computing Sciences), Phi Kappa Phi (The Academic Honor Society) and an honorary member of Delta Mu Delta (The National Honor Society for Business Administration).

Dr. Sobh is a member, senior member, founding or board member of several professional organizations including; ACM, IEEE, the International Society for Optical Engineering (SPIE), the National Society of Professional Engineers (NSPE), the New York Academy of Sciences, the American Society of Engineering Education (ASEE), the American Society of Quality (ASQ), the American Association for the Advancement of Science (AAAS), the Society of Manufacturing Engineers (SME), the International Association of Online Engineering (IAOE), the Discovery Museum, the Connecticut Pre-Engineering Program (CPEP), the Northeast Center for Computers and Information Systems Security (NECCISS), the International E-Learning Association (IELA), and the Society for Industrial Computing. Dr. Sobh is a graduate of Victoria College, Alexandria, Egypt, in 1983 and a life member of the Old Victorians Association. He is also a life Member of the Egyptian Engineering Syndicate, a licensed engineer of the Egyptian Engineers Association and the Alexandria Engineering Organization.

Professor Khaled Elleithy is the Associate Vice President for Graduate Studies and Research at the University of Bridgeport. He is a professor of Computer Science and Engineering. He has research interests include wireless sensor networks, mobile communications, network security, quantum computing, and formal approaches for design and verification. He has published more than three hundred fifty research papers in national/international journals and conferences in his areas of expertise. Dr. Elleithy is the editor or co-editor for 12 books published by Springer.

Dr. Elleithy received the B.Sc. degree in computer science and automatic control from Alexandria University in 1983, the MS Degree in computer networks from the same university in 1986, and the MS and Ph.D. degrees in computer science from The Center for Advanced Computer Studies at the University of Louisiana – Lafayette in 1988 and 1990, respectively.

Dr. Elleithy has more than 20 years of teaching experience. His teaching evaluations are distinguished in all the universities he joined. He is the recipient of the "Distinguished Professor of the Year", University of Bridgeport, academic year 2006–2007. He supervised hundreds of senior projects, MS theses and Ph.D. dissertations. He developed and introduced many new undergraduate/graduate courses. He also developed new teaching/research laboratories in his area of expertise. His students have won more than twenty prestigious national/international awards from IEEE, ACM, and ASEE.

Dr. Elleithy is a member of the technical program committees of many international conferences as recognition of his research qualifications. He served as a guest editor for several International Journals. He was the chairperson for the International Conference on Industrial Electronics, Technology & Automation. Furthermore, he is the co-Chair and co-founder of the Annual International Joint Conferences on Computer, Information, and Systems Sciences, and Engineering virtual conferences 2005–2014.

Dr. Elleithy is a member of several technical and honorary societies. He is a Senior Member of the IEEE computer society. He is a member of the Member of the Association of Computing Machinery (ACM) since 1990, member of ACM SIGARCH (Special Interest Group on Computer Architecture) since 1990, member of the honor society of Phi Kappa Phi University of South Western Louisiana Chapter since April 1989, member of circuits & systems society since 1988, member of the IEEE computer society since 1988, and a lifetime member of the Egyptian Engineering Syndicate since June 1983.

ALLEN UNIVERSITY

Gregory Grooms

Most Helpful to Students, Liberal Arts, Social Sciences, and Humanities

AMERICAN RIVER COLLEGE

Jiayi Ling

Best Researcher/Scholar, Fine Arts

Ken Magri

Best Researcher/Scholar, Fine Arts

Diane Richey-Ward

Best Teacher, Fine Arts
Most Helpful to Students, Fine Arts
Best Overall Faculty Member, Fine Arts

Michael Sheldon

Best Overall Faculty Member, Fine Arts
Best Teacher, Fine Arts
Most Helpful to Students, Fine Arts

Mick Sheldon, the director of the James Kaneko Gallery at ARC, had his artworks exhibited at The Painting Center in New York's Soho district in a show titled "Artful Jesters" from the book of the same name two years ago, a one person exhibition at the South Dakota Art Museum in 2007 and a solo exhibition of war protest paintings at Sacramento's Brickhouse Gallery in 2008, titled "Give War a Chance."

ANDREWS UNIVERSITY

Susan Zork

Best Teacher, Religion

"Compassionate."

Susan Pervier Zork, MA Religious Education, is an Assistant Professor of Religion. She serves the department in teaching General Education Religion to more than 400 undergraduate students each year and was recently awarded the AUSA 2006 Teacher of Year. She is the faculty sponsor of Andrews Ministerial Association; a student association for Religion and Theology majors. She is also an active academic advisor in the department. In addition to her busy schedule teaching and preaching, Susan is a published composer and recording artist. She regularly speaks and gives sacred concerts throughout the States and beyond. Additionally, she is a member of University Singers, the Andrews University select touring choir under the direction of her husband, Stephen Zork. She is currently taking her course work toward a PhD in Theological Curriculum and Spiritual Formation. Susan would, undoubtedly, claim her four grown children; Nicholas (Noelia), Benjamin (Miroslava), Timothy and Jasmine as her greatest claim to accomplishment and joy.

ANNE ARUNDEL COMMUNITY COLLEGE

Steve Canaday

Best Researcher/Scholar, Liberal Arts, Social Sciences, and Humanities

B.A. (English), University of Virginia, 1989 M.A. (English Language and Literature), University of Maryland, College Park, 1995 Ph.D. (English Language and Literature), University of Maryland College Park, 2002 (Dissertation: Underground Celebrity: Sam Shepard and the Paradox of Off-Off Broadway Success)

Scott Cooper

Best Overall Faculty Member, Languages

Professor Scott Cooper teaches Spanish, Portuguese and beginning Italian at AACC and is currently the chair of the Department of World Languages. He holds a BA in music and Spanish from Heidelberg University (Tiffin, OH) and an MA in Francophone studies and Spanish from the University of Louisiana (Lafayette, LA), and did doctoral work in Latin American literature at Tulane University (New Orleans, LA). His research area is contemporary Latin American narrative.

Professor Cooper came to AACC in 2006 and plays bassoon in the AACC Symphony Orchestra and Opera AACC. In his free time, he enjoys the literature, cooking, films and music of Spain, Portugal and Latin America. He is a world traveler and especially enjoys going by train wherever possible.

Harold Counihan

Most Helpful to Students, History

Miyoko Dickerson

Most Helpful to Students, Languages

Suzanne Spoor

Most Helpful to Students, English
Best Teacher, Liberal Arts, Social Sciences, and Humanities
Most Helpful to Students, Liberal Arts, Social Sciences, and Humanities
Best Overall Faculty Member, Liberal Arts, Social Sciences, and Humanities

APPALACHIAN STATE UNIVERSITY

Vicki Clift

Best Teacher, Fine Arts
Most Helpful to Students, Fine Arts

Andrew Ferguson

Best Teacher, Sociology

Gordon Hensley

Best Overall Faculty Member, Theater

Gordon Hensley holds a BS in Teaching Theatre K–12, and a MFA in Theatre Education/Theatre for Youth. He has been a college professor since 1999. He currently serves as faculty and theatre education coordinator at Appalachian State University in Boone, North Carolina.

Gordon has an extensive background in theatre performance, directing, conference presentations, and service to the field of theatre and education.

Gordon holds a North Carolina Teaching License for grades K–12, and as an arts specialist he has had the privilege of working with children and adults of all ages. Additionally, he has authored the popular teacher resource www.cre8tivedrama.com, and has written several other published works on topics such as playwriting, hybrid learning, and puppetry.

With ample experience in curriculum development and program design, Gordon has been chosen as the state representative for the American Alliance for Theatre and Education.

His hobbies include volunteering, home improvement, landscaping, web design, and motorcycles. Learn more about Gordon at www.gordonhensley.com.

Eric Koontz

Most Helpful to Students, Music

Eric Koontz has led the viola sections of the Barcelona Symphony Orchestra (Orquestra Simfònica de Barcelona i Nacional de Catalunya) and the Jerusalem Symphony Orchestra before returning to the United States to pursue studies towards the Doctor of Musical Arts at the University of North Carolina at Greensboro. In Barcelona, he premiered Max Bruch's Double Concerto for Viola, Clarinet and Orchestra on the Iberian Peninsula under the baton of Lawrence Foster. An active chamber musician, he was a founding member of the Nayades Trio (flute, harp and viola); the Reinecke Trio (clarinet, viola and piano); and the Quartet Glinka, with whom he participated in performances of the complete string quartets of Dmitri Shostakovich at the Juan March Foundation in Madrid. In Israel, Koontz has performed with Music in the Desert, a festival for new music at Kibbutz Sde Boker, directed by Michael Wolpe.

As a chamber and symphonic violist, Koontz has recorded with the RCA, EMI, Ars Harmonica, Columna, Koch and Decca labels. He earned the Bachelor of Music at the Cincinnati College-Conservatory of Music, and holds the Master of Music and the Master of Musical Arts degrees from Yale University. He has studied with Donald McInnes, Scott Rawls, Msao Kawasaki, Jesse Levine, Hatto Beyerle, Gerald Stanick, Jaap Schroder, and Lawrence Dreyfus.

Koontz joined the string faculty at the Hayes School of Music at Appalachian State University in 2005 and the faculty of the Brevard Music Center in 2004. Examples of his writing in print include the articles "Els músics i els reptes de l'OBC" (4 April 2002, Avui newspaper, Barcelona) and "Una orquesta eficaz y plural" (November 2002, Scherzo magazine, Madrid)

Susan Mills

Best Researcher/Scholar, Music
Best Researcher/Scholar, Liberal Arts, Social Sciences, and Humanities
Most Helpful to Students, Liberal Arts, Social Sciences, and Humanities

Dr. Susan W. Mills is the Music Education Coordinator in the Hayes School of Music at Appalachian State University. Her career is focused on the development of music education that is relevant, balanced, and inclusive of both the classical and folk music traditions. In 2010, Dr. Mills was awarded a Fulbright Fellowship for her work with the Freedom and Folk Song traditions in South Africa, where she lectured and conducted research and community service in partnership with the University of Zululand and the University of the Free State. She is a 2007 recipient of the Digital Sound Archives/Appalachian Music Fellowship of Berea College.

In addition to multicultural music education, Dr. Mills has woven service learning and teacher education research into her career as a K–12 general and choral music specialist since 1989. She continues in her quest for quality folk music by obtaining experience for herself and her students with several master musicians from world cultures including Azerbaijan, The British Isles, Canada, Cuba, Haiti, Hungary, Ghana, New Zealand, Puerto Rico, South Africa, Southern Appalachia, and the U.S. Gulf Coast.

Gary Nemcosky

Best Overall Faculty Member, Fine Arts

Amy Page

Best Overall Faculty Member, Sociology
Best Researcher/Scholar, Sociology

Eric Reichard

Best Overall Faculty Member, Graphic Arts

"Patience with students."

Rodney Reynerson

Best Teacher, Music
Best Teacher, Liberal Arts, Social Sciences, and Humanities

Rodney Reynerson is a graduate of Murray State University and the Indiana University. Among his principal teachers are Jorge Bolet and Menaham Pressler. Dr. Reynerson, formerly a member of the Indiana State University music faculty, joined the Appalachian faculty in 1980 where he serves as the coordinator of class piano. Dr. Reynerson has performed solo literature, chamber music, and piano concerti throughout the eastern United States.

Ed Rosenburg

Most Helpful to Students, Sociology

Reeves Shulstad

Best Overall Faculty Member, Music
Best Overall Faculty Member, Liberal Arts, Social Sciences, and Humanities

Reeves Shulstad has been a member of the Appalachian faculty since 2009. With earned degrees that include a Masters and Ph.D. in Historical Musicology, Shulstad teaches music history and appreciation, world music and music and gender courses at the Hayes School. In addition to teaching, Shulstad is working on a book about microtonalist composer Tui St. George Tucker, a contemporary of John Cage who split her time between Greenwich Village and Camp Catawba, a boys camp outside of Blowing Rock, NC. Shulstad is also working on a critical edition of Tucker's music. Other research interests include the Scholarship of Teaching and Learning with her most recent project on student engagement. Shulstad is also active in curriculum development and assessment, serving as the chair of the Hayes School of Music's curriculum committee and on university-wide general education assessment committees.

ARKANSAS STATE UNIVERSITY

Gretchen Hill

Best Teacher, Sociology

Larry Salinger (Posthumously)

Best Overall Faculty Member, Sociology
Most Helpful to Students, Sociology

ARKANSAS TECH UNIVERSITY

Tom Deblack

Best Teacher, History

Jeff Woods

Best Researcher/Scholar, History

ART CENTER COLLEGE OF DESIGN

David Bailey

Best Overall Faculty Member, Fine Arts

Victoria Hochberg

Most Helpful to Students, Film

"indispensable core of the graduate film dept"

ATHENS TECHNICAL COLLEGE

Carol Myers

Best Overall Faculty Member, English
Best Overall Faculty Member, Liberal Arts, Social Sciences, and Humanities

Betty Watts

Best Overall Faculty Member, Social Work

"Mrs. Watts is an incredible instructor. She is innovative in her presentation of information. She brings a wealth of knowledge to the classroom from her educational background, work experience, and the tenacity to keep up with current trends. She also is very caring and concerned about the students in her class. As a colleague, she is a delight to work with."

ATLANTIC CAPE COMMUNITY COLLEGE

Joy Jones

Best Researcher/Scholar, Communication

"She is passionate about communication apprehension and related-research and can present in an engaging fashion for a variety of audiences."

AUBURN UNIVERSITY

Cynthia Kristan-Graham

Best Overall Faculty Member, Fine Arts
Best Researcher/Scholar, Fine Arts
Best Teacher, Fine Arts
Most Helpful to Students, Fine Arts
Best Researcher/Scholar, Fine Arts
Best Teacher, Fine Arts
Most Helpful to Students, Fine Arts

Joshua Pifer

Most Helpful to Students, Music

Dr. Joshua Kelley Pifer, Lecturer in Piano at Auburn University where he teaches Class Piano, Applied Piano, and Music Skills. Born and raised in Tokyo, Japan, Joshua made his piano debut playing Haydn's Concerto in A Major with the Kotani Chamber Ensemble. At age 10, Joshua performed as a vocalist for the MGM Tokyo Red Carpet premier of the movie Mr. Mom. Joshua has an exciting solo and collaborative career that has taken him to four continents. He regularly appears as guest lecturer and clinician at other universities and National and State Music Teachers Conferences, and has adjudicated at over 25 festivals and competitions. The summers find Joshua as faculty and artist at the Orfeo Music Festival in the Italian Alps. Joshua has also been sought out by many college and graduate level pianists who have suffered injuries from excessive tension in their playing, to help them overcome their problems and learn to play with much more individualistic freedom and ease. Joshua has also served as a choral conductor and organist for several churches, and taught general music, keyboard, and choir at the elementary school levels. Joshua completed his bachelor's degree in pre-medicine, piano, and organ from Wittenberg University, his Master's degree from Miami University, and in 2009 he graduated with honors and a Doctorate of Musical Arts from the University of Southern California. His major teachers include Antoinette Perry, Stewart Gordon, Dennis Thurmond, Robert Thomas, Donald Busarow, Magen Solomon, Sharon Lavery, and Robert Howat. Joshua has previously served as faculty at The Florida State University, Wittenberg University and Miami University. He is a founding member of the Oto Trio, including internationally acclaimed artist Chikuen Kato, renown for exploring the interconnections between the Japanese arts of Sumie (black ink painting), shodo (calligraphy), and ikebana (flower arrangement), and the music of Japanese composers. Joshua has commissioned and premiered works by 21st Century composers Andre Cormier, Justin Aftab, Gary Shields, Douglas Townsend, Andrea Clearfield, and is an advocate for the piano music of Alexander Tcherepnin. Aside from his musical and teaching activities, Joshua enjoys spending time with his wife and son playing tennis, golfing, swimming, cooking, and making music together.

Rosephanye Powell

Best Overall Faculty Member, Music
Best Teacher, Music

Dr. Rosephanye Dunn Powell serves as Professor of Voice at Auburn University, where she teaches applied voice, vocal pedagogy, art song literature, and conducts the women's chorus. She holds degrees from The Florida State University (D.M. in vocal performance, University Fellow), Westminster Choir College (M.M. in vocal performance and pedagogy, with distinction), and Alabama State University (B.M.E., summa cum laude. Dr. Powell served on the faculties of Philander Smith College (AR) and Georgia Southern University prior to Auburn University. Dr. Powell has been hailed as one of America's premier women composers of choral music. She has an impressive catalogue of works published by some of the nation's leading publishers, including the Hal Leonard Corporation, the Fred Bock Music Company/Gentry Publications, Oxford University Press, Alliance Music Publications, and Shawnee. Dr. Powell is commissioned yearly to compose for university choruses, professional, community and church choirs, as well as secondary school choruses. Dr. Powell's works have been conducted and premiered by nationally and internationally-renowned choral conductors, including, but not limited to, Anton Armstrong, Philip Brunelle, Bob Chilcott, Rodney Eichenberger, Tom Hall, Albert McNeil, Tim Seelig, André Thomas and Judith Willoughby. Her work has been auctioned by Chorus America and her compositions are in great demand at choral festivals around the country and internationally, frequently appearing on the regional and national conventions of the American Choral Directors Association, as well as Honor Choir festivals. Dr. Powell's compositions include sacred and secular works for mixed chorus, women's chorus, men's chorus, and children's voices.

William Powell

Most Helpful to Students, Music

AUSTIN PEAY STATE UNIVERSITY

Stuart Arkovitz

Best Teacher, English

Mike Gotcher

Most Helpful to Students, Liberal Arts, Social Sciences, and Humanities

"I have never met a department chair who is always available to his students, who makes things work for the benefit of his students and his faculty. Mike Gotcher schedules classes to meet both faculty and student needs. He helps students achieve their goals, even altering their curriculum with substitutions to make essential, emergency, or career-oriented learning better for them. He is a wonderful student advisor and makes decisions that are in the best interest of both his students and faculty. He has a vision and a mission that has the very best interest of students, faculty, and administration in mind for the growth of this Department and Austin Peay State University."

Interim Graduate Dean

Christina Hicks-Goldston

Best Researcher/Scholar, Communication

"Dr. Hicks-Goldston employs service learning in her classes and has become an expert in this region regarding service learning. She travels extensively presenting papers, panels, and training in service learning at most of the professional organization's conventions, conferences, and meetings."

Dr. Hicks-Goldston received her Master's degree from Emerson College in Boston, and her doctorate in mass communication from the University of Southern Mississippi. Her concentration in political communication at Emerson College resulted in internships at Bill Clinton's campaign headquarters in Little Rock, where she reported to George Stephanopoulos, and the office of Congressman Joseph P. Kennedy II.

While completing her doctorate, Hicks-Goldston continued research in the area of critical-cultural studies, particularly gender studies, and focused her dissertation on film portrayals of women in World War II. She expanded such interests to popular culture, and examined gender in series television, such as Buffy the Vampire Slayer. Dr. Hicks-Goldston examined the press coverage of women's suffrage in the Tenenessee press prior to the ratification of the 19th Amendment for the Eastern Communication Association in 2007. She has presented a student workshop on teaching intercultural communication and adapting to the cultural diversity of the student population in the 21st Century. She is presently working on a project about Generation X and the mass media.

Dr. Hicks-Goldston has published articles in scholarly journals and made conference and convention presentations at professional meetings. She teaches Applied Communication Theory graduate course.

Barry Kitterman

Best Researcher/Scholar, Liberal Arts, Social Sciences, and Humanities

"Barry Kitterman, teaching creative writing in the APSU English Department, is a prolific writer and the professional author of two good-selling novels. He is a role model for his discipline and interacts effectively with his students by encouraging their creative writing and modeling the behavior they should hope to achieve professionally. Kitterman is a scholar and a wonderful research asset to Austin Peay State University and his discipline."

Barry Kitterman's novel, The Baker's Boy was published in 2008, and his collection of stories From the San Joaquin followed in 2011. Barry Kitterman grew up in the San Joaquin Valley of central California. He attended the University of California at Berkeley from 1971–1975, receiving his A.B. in English. In the mid 70's he served as a Peace Corps volunteer in Belize (Central America) where he taught in a boys reform school. After completing the MFA program at the University of Montana, he traveled to China with his wife Jill Eichhorn, where they spent two years teaching at Chinese universities, and later at a children's school in Taiwan. He was Hudson Walker Fellow at the Fine Arts Work center in Provincetown, Mass. in 89–90. He has taught at Indiana University East, Miami University (Ohio) and, since 1995, at Austin Peay State University in Clarksville, Tennessee. He has published his short fiction in a number of literary magazines, including Cutbank, Flyway, The Chariton Review and The Green Hills Literary Lantern, and has served as fiction editor of Zone 3 Magazine, Shankpainter and The Green Hills Literary Lantern. In 2001, he received an Individual Artist Grant from the Tennessee Humanities Council.

John Moseley (Posthumously)

Best Overall Faculty Member, Communication
Best Teacher, Communication
Most Helpful to Students, Communication
Best Teacher, Liberal Arts, Social Sciences, and Humanities

David Rands

Best Researcher/Scholar, History
Most Helpful to Students, History

Timothy Winters

Best Overall Faculty Member, Liberal Arts, Social Sciences, and Humanities

"Timothy Winters is a classicist at Austin Peay teaching Latin. He is loved by his students, participates in shared governance of the University and has genuine care and concern for his peer faculty as he lobbies for their rights. Dr. Winters is a musician, a classical historian, semanticist, scholar, and wonderful teacher at our regional institution. He is a virtual Renaissance man in the 21st Century where he serves as a Deacon in the Catholic Church, deputy sheriff in Montgomery County, TN, and a full time faculty member at Austin Peay."

BAPTIST COLLEGE OF FLORIDA

Rich Elligson

Most Helpful to Students, Religion

BARD COLLEGE

Celia Bland

Best Teacher, Liberal Arts, Social Sciences, and Humanities

Stephanie Kufner

Best Teacher, Languages

Teaching Diploma, Certification for English Language and Business Administration, University of Munich; Diplom-Handelslehrer, University of Munich; M.A. and Ph.D. in German, SUNY Albany. Articles, papers, and talks on Schiller, language acquisition, technology, and language pedagogy. Academic Director, Center for Foreign Languages and Cultures, Bard College (1995–). At Bard since 1990.

Melanie Nicholson

Most Helpful to Students, Liberal Arts, Social Sciences, and Humanities

Philip Pardi

Best Overall Faculty Member, Liberal Arts, Social Sciences, and Humanities

B.A., Tufts University; M.F.A., Michener Center for Writers, University of Texas. Poet and translator. Author, Meditations on Rising and Falling (University of Wisconsin Press, 2008); has published poems and translations in Gettysburg Review, Best New Poets 2006, Exile Quarterly, Marlboro Review, Mid-American Review, New Orleans Review, Nimrod, Seneca Review, others. Editor, Borderlands: Texas Poetry Review (2003–05). Has worked as human rights activist in El Salvador and as labor organizer in the Hudson Valley; taught writing at Marist College; led poetry workshops at University of California–Los Angeles Writers' Program. Recipient, Brittingham Poetry Prize; American Literary Translators Association Conference Fellowship; Adele Steiner Burleson Poetry Award. Codirector, Center for Faculty and Curricular Development (2012–). At Bard since 2005.

Eric Trudel

Best Researcher/Scholar, Liberal Arts, Social Sciences, and Humanities

B.A., Concordia University, Montreal; M.A., French literature, McGill University; Ph.D. in Romance languages, Princeton University. Author, La terreur à l'oeuvre: Théorie, poétique et éthique chez Jean Paulhan (Paris, Presses Universitaires de Vincennes, coll. "L'imaginaire du texte," 2007). Coeditor of "Tout peut servir": Pratiques et enjeux du détournement dans le discours littéraire des XXe et XXIe siècles (Presses de l'Université du Québec, 2011); coeditor of a special issue of L'Esprit Créateur, "Old and New, Avant-garde and Arrière-garde in Modernist Literature" (Fall 2013); coeditor and translator, Jean Paulhan on Poetry and Politics (University of Illinois Press, 2008); coeditor, Figures et Frictions: La littérature au contact du visuel (Etudes Françaises, Vol. 42, No. 2, 2006). Numerous articles on Jean Paulhan, Chris Marker, Remy de Gourmont, Tanguy Viel, and modern and contemporary French and Francophone poetry. Managing editor of Formes Poétiques Contemporaines (SUNY Buffalo). At Bard since 2002.

BAYLOR UNIVERSITY

Louis Mazé

Most Helpful to Students, Languages

"Professor Maz goes beyond course requirements and the expectations we have of our students. He gets to know students personally, and bends over backwards to help them–not only with the course material, but with important 'life lessons.'"

Yuko Prefume

Best Teacher, Languages

"Yuko is a real asset to our department and to the study of modern languages in the State of Texas. What an advocate for language study and for the effective use of technology in our discipline! She is both intelligent and dedicated–highly respected by her peers and loved by her students."

Michael Thomas

Best Researcher/Scholar, Languages

"Mike has always loved the Spanish language and Spanish literature. His books and articles are of the highest quality. Throughout his tenure at Baylor, he has helped and encouraged junior scholars and professors with wise and perceptive counsel."

BELLARMINE UNIVERSITY

Robert Pfaadt

Most Helpful to Students, History

"Bob Pfaadt spends the most time helping students with problems and plans."

Eric Roorda

Best Researcher/Scholar, History

"Eric Roorda has 3 books published and his second one received national recognition. He has also received several grants to aid him in travel for his research on the Dominican Republic."

BELLEVUE COLLEGE

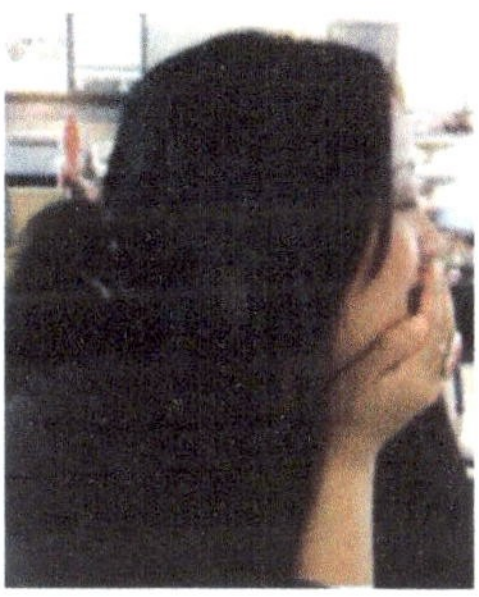

Donna Miguel

Best Overall Faculty Member, Liberal Arts, Social Sciences, and Humanities

Kumusta! My name is Donna Miguel, and I teach a gamut of classes in the English department: pre-college, college and advanced composition, research writing, technical writing, literature classes, recently an Accelerated 101 pilot, and an Interdisciplinary Class (English and Ethnic and Cultural Studies). My passion for reading started in high school when my mom gave me a romance novel and with a play I read in my AP English course: Sophocles' Antigone. It was only in the first year of community college, in a writing course taught by my first instructor of color, that I started to understand how reading and writing could become a power tool, especially for people of color. As a Filipina-American born to immigrant parents, my experiences and learned knowledge of diversity/multicultural issues allows me to integrate issues of gender and race inequalities in my writing classes. I'm still continuing to learn more about the resiliency of oppressed groups, issues of privilege and power, and intersectionality, which drives my efforts to also learn best practices that break through barriers students may have in their classes. I hope to continually put my philosophy into action by providing mentorship for underrepresented populations along with collaboration from colleagues in college programs (that assist in their success). In essence, I'm blessed to be in a career that I love, feel passionate about, and with colleagues who are dedicated in providing a holistic environment that puts the entire student at the center of the educational process. What do I do in my free time? Of course I spend time with my family and friends, and as a new homeowner, seem to be forever organizing the garage and yanking weeds from the yard.

BENEDICTINE UNIVERSITY

Jane Boumgarden

Most Helpful to Students, Sociology

BENTLEY UNIVERSITY

W. Tim Anderson

Most Helpful to Students, Liberal Arts, Social Sciences, and Humanities

Sociology teaching interests include drugs and society, deviance, medical sociology and interaction analysis. Research interests include Native American culture, medical sociology, ethnomethodology and conversational analysis. Instituted Bentley summer program for students who want to learn about Crow Indian culture by living with a Crow Indian tribe in Montana. Published in Discourse and Institutional Authority: Medicine, Education and Law, New Direction in the Study of the Social Order, The New England Sociologist, Patients, Physicians & Illness, and Perspectives in Sociology. Has taught at Wheaton College, Massachusetts College of Pharmacy & Allied Health Services, Boston College and Boston University. Formerly a post-doctoral fellow at Harvard Medical School.

Alexander Gribanov

Most Helpful to Students, Liberal Arts, Social Sciences, and Humanities

"brilliant and attentive to students"

Diane Moul

Best Teacher, English

"most attentive to students"

K Tippo

Most Helpful to Students, Liberal Arts, Social Sciences, and Humanities

Christine Williams

Best Overall Faculty Member, Liberal Arts, Social Sciences, and Humanities

Teaching interests include American national elections, parties and interest groups and the media. Publications and research interests include political socialization, women and politics, media and politics, teaching with the World Wide Web, and curriculum issues (reform, process, content, assessment). Formerly taught at Mount Holyoke College and State University of New York, College at Fredonia. Recipient of the Joseph M. Cronin Award for Excellence in Academic Advising. Recipient of a research grant from the Joan Shorenstein Center on the Press, Politics and Public Policy at Harvard University.

BERGEN COMMUNITY COLLEGE

Stacey Balkan

Best Teacher, Literature Most Helpful to Students, Literature

BERKELEY CITY COLLEGE

Elizabeth Wing-Paz

Most Helpful to Students, Philosophy

BLACK HILLS STATE UNIVERSITY

Robert Campbell

Best Overall Faculty Member, History
Best Researcher/Scholar, History

BLINN COLLEGE

Jeffery Scott

Best Overall Faculty Member, English

BOROUGH OF MANHATTAN COMMUNITY COLLEGE

Mabel Y Asante

Best Teacher, English

Mila Brisbon

Best Teacher, Liberal Arts, Social Sciences, and Humanities

Rafael Corbalan

Best Teacher, Languages

Cheryl J. Fish

Best Researcher/Scholar, English

Cheryl J. Fish is a scholar, fiction writer, and poet, who teaches English at BMCC/City University of New York, and has been appointed docent lecturer at University of Helsinki, Finland. She has written about Sami film and cultural identity, including Liselotte's Wajstedt's Sami Nieda Joik, for the collections Rivers to Cross: Sami Land Use and the Human Dimension (Umea University 2012) and Critical Norths: Space, Nature, Theory (forthcoming, University of Alaska Press). Fish is the author of Black and White Women's Travel Narratives: Antebellum Explorations and has published essays on environmental justice and literature, film, and architecture in journals and anthologies including MELUS and Discourse. Her most recent poetry chapbook is Make It Funny, Make it Last (#171, Belladonna Chaplets, 2014). Her poetry has appeared in The Bloomsbury Anthology of Contemporary Jewish American Poetry; Far from the Centers of Ambition: The Legacy of Black Mountain College; Terrain.org; New American Writing; Talisman; The Village Voice, and Volt. Her fiction was featured in Liars League NYC, and she was a finalist for L Magazine's Literary Upstart search for pocket fiction. A short story from the innovative Lower East Side journal Between C&D was featured in an exhibit at the Fales Library, New York University. Fish was a Fulbright professor in Finland, writer in residence at Mount St. Helens National Volcanic Monument, and was recently awarded a Grimes Scholar at the NYU Faculty Resource Network.

Jose Haro

Best Researcher/Scholar, Philosophy

"He is young, energetic and completed his dissertation in a matter of a few years moving to the Assistant Professor tenure track."

Sandra Poster

Best Overall Faculty Member, Communication

"It gives me great pleasure to be able to nominate Dr. Sandra Poster. Dr. Sandra Poster is the best at what she does. She is one of the most energetic, creative instructors in the field of Communications. In today's connected economy she is at the forefront of new ideas and insights that take her colleagues and students to heights beyond imagination. She has created two educational departments within our institution with her intelligence, compatibility, dedication and professionalism. Thank you for allowing me the opportunity to nominate this outstanding educator. Kerry Ruff of the Speech and Communications Department at the Borough of Manhattan Community College in New York City."

BOSTON COLLEGE

Lisa Cahill

Best Researcher/Scholar, Theology

"Her scholarship expands the horizons of a critical field, yet she keeps it readable for lay people and makes herself generously available to all students and fellow scholars."

Lisa Sowle Cahill received her BA from the Santa Clara University and her MA and Ph.D. from the University of Chicago Divinity School. She has taught at Boston College since 1976, and has been a visiting professor at Georgetown and Yale Universities. She and her husband Larry are the parents of five children.

John McDargh

Most Helpful to Students, Theology

"Enables students to probe depths of meaning and of selfhood with endless energy and creativity as a teacher."

BOWDOIN COLLEGE

Elena Cueto Asin

Best Overall Faculty Member, Languages

Enrique Yepes

Best Teacher, Languages
Most Helpful to Students, Liberal Arts, Social Sciences, and Humanities

Ph.D. Spanish-American Literature, Rutgers University, May 1998. Dissertation: Los poderes del goce: La revisión cultural en poesía hispanoamericana 1960–1995. (The Powers of Enjoyment: Cultural Revision in Spanish American Poetry 1960–1995). Dissertation director: Dr. Carlos R. Narváez. Licenciatura (B.A. equivalent) in Education-Modern Languages, Univ. Pontificia Bolivariana (Medellín-Colombia), 1989.

BRADLEY UNIVERSITY

Kyle Dzapo

Best Researcher/Scholar, Music

Kyle Dzapo enjoys a diverse career as teacher, scholar, and performing artist. She teaches flute students and music history classes in the Department of Music and also leads seminars through the Honors Program and participates in Study Abroad programs in Vienna, London, and Salzburg. Dr. Dzapo is currently under contract with Oxford University Press for Notes for Flutists: A Guide to Selected Works, the flagship book in Oxford's "Notes for Performers" series for which she serves as

Series Editor. She is also a pre-concert lecturer for the Chicago Symphony and has given presentations for the Cleveland Orchestra and the New York Philharmonic. As the leading authority on Danish flutist, composer, and conductor Joachim Andersen, she has published a compact disc recording, Joachim Andersen: Etudes and Salon Music (Naxos, 2010); the book Joachim Andersen: A Bio-Bibliography (Greenwood Press, 1999); articles for journals of the American, British, French, Dutch, and Finnish flute associations and the Lexikon der Flöte (ed. Andràs Adorjàn, Laaber, 2009); and new editions of Andersen's Fünf leichtere Stücke, Op. 56, and Salonstücke, Op. 52, with his original publisher, Zimmermann of Frankfurt. Dr. Dzapo served as Principal Flutist of the Peoria Symphony Orchestra for twenty years, performing as soloist in works by Bach, Ciardi, Ibert, Martin, Mozart, and Nielsen with reviews praising her "elegant" performances, her "full, lovely tone and her expressive, masterful phrasing," and the way "she executed dazzlingly complex lines with seeming ease." She has presented recitals in London and Denmark, on live broadcasts for WFMT's Dame Myra Hess Memorial Concert Series and Wisconsin Public Radio's "Sunday Afternoon Live," and at Lincoln Center's Bruno Walter Auditorium. Bradley University has recognized her work as a dedicated teacher and scholar with awards including the Caterpillar, Inc. New Faculty Achievement Award for Excellence in Teaching and the Samuel Rothberg Professional Excellence Award. In 2010, she was awarded a Caterpillar Professorship. Dr. Dzapo earned the Doctor of Music degree from Northwestern University, Master of Music degree with Distinction in Performance from New England Conservatory, and a Bachelor of Music Education degree with High Distinction from the University of Michigan.

Steve Heinemann

Best Teacher, Music

Dr. Stephen Heinemann teaches second-year and upper-division music theory courses and private lessons in composition and clarinet, and is music director of the new-music ensemble Peoria Lunaire. In 2009, he received the Slane College Faculty Achievement Award for Excellence in Teaching. He is a member of the Peoria Symphony Orchestra with which he plays clarinet and E-flat clarinet. An ardent jazz musician, he plays alto and soprano saxophones with the Todd Kelly Quintet, for which he has composed a number of pieces, and is lead alto saxophonist with the Central Illinois Jazz Orchestra. He has performed at the jazz festivals of Vienne (France), Montreux and Brienz (Switzerland), and Umbria and Tuscany (Italy). Dr. Heinemann's compositions have been performed in North America and Europe; a notable recent premiere was in Minneapolis (November 2009) of Metropassacaglia, composed for the Metropolitan Symphony Orchestra and its conductor, William Schrickel. (Metropassacaglia received its Illinois premiere on Jan. 14, 2012 by the Peoria Symphony Orchestra.) His music-theoretical writings have been published in Music Theory Spectrum, Mitteilungen der Paul Sacher Stiftung, and other journals, and he wrote the program booklets for two CDs released by Cedille Records (Early Chamber Music of Elliott Carter and Clarinet Chamber Music by Hindemith). In the winter and spring of 2000, he did research in the Pierre Boulez and Elliott Carter Collections at the Paul Sacher Foundation in Basel, Switzerland, which awarded him a fellowship for this work. He is a contributing author of Elliott Carter Studies (Marguerite Boland and John Link, editors), forthcoming in the summer of 2012 from Cambridge University Press. He is on the editorial board of iSCI, the online journal of the Society of Composers, Inc., and is an editor of CUR Quarterly, the journal of the Council on Undergraduate Research.

He was among the first councilors with the Council on Undergraduate Research to represent a background in the arts. He received the Bachelor of Music and Master of Arts degrees in composition from San Francisco State University and the Doctor of Musical Arts degree in composition from the University of Washington. His principal teachers in San Francisco were composer Roger Nixon, clarinetist Donald Carroll, and saxophonists John Handy and James Matheson, and in Seattle included composer/clarinetist William O. Smith and music theorists Jonathan W. Bernard and John Rahn. His own students have continued their studies in highly regarded graduate programs and distinguished themselves in performance and academia.

John Jost

Best Overall Faculty Member, Music

John Jost, Professor of Music, has been Director of Choral Activities at Bradley since 1989. He also serves as director of the Ecole Sainte Trinité summer music program for Haitian youth in Léogâne, Haiti. He has held both principal viola and principal second violin positions with the Peoria Symphony Orchestra and has served as president of the Illinois Chapter of the American Choral Directors Association. He is co-founder and co-director of the Peoria Bach Festival which presents vocal and instrumental works of Bach each year during the first week of June. A native of California, Dr. Jost received his bachelor's, master's, and doctorate degrees from Stanford University where he was assistant concertmaster of the Stanford Symphony and principal violist of the Stanford Chamber Orchestra. He spent four years teaching music in Haiti, after which he returned to the San Francisco area, serving as Director of Creative Ministries at Peninsula Christian Center, teaching in the Palo Alto and Redwood City schools, and serving as assistant concertmaster and assistant conductor of the Peninsula Symphony Orchestra. After receiving his doctorate in choral conducting he became Director of Choral Activities at Houghton College in New York, where his select choir was chosen to perform for state and regional music educators conventions. Dr. Jost's duties at Bradley include directing the Chorale, Chamber Singers, and Community Chorus, and teaching private voice and conducting. He has led the Bradley Chorale in over two dozen concert tours to locations including Poland, the Czech Republic, Italy, Austria, Germany, Hungary, England, Ireland, Scotland, Denmark, the National Cathedral in Washington D.C. and the Cathedral of St. Louis, King of France in New Orleans. He has performed with the Chorale for Illinois Music Educators Association Conferences and for state and regional conventions of the American Choral Directors Association. He also oversees Bradley's efforts to serve the central Illinois choral community, which include skills clinics for area choral directors, an honor choir for area high schools, a festival audition preparation workshop for district high schools, and rehearsal clinics for area school choirs. He received Bradley's First Year Teacher Award in 1990, the Charles Putnam Award for Excellence in Teaching in 2004, a service award from the Haitian Ministry of Culture in 2004, and the Bradley Parents Association Award of Excellence in 2005.

Todd Kelly

Most Helpful to Students, Music

Dr. Todd Kelly is Professor of Trumpet and Director of Jazz Ensembles at Bradley University in Peoria, Illinois. Dr. Kelly teaches trumpet, directs the Bradley Jazz Ensembles and Jazz Combos, teaches Jazz Improvisation, and serves as the advisor for the Music Business program. Under his direction, the jazz program includes two big bands, one combo, and a year - long Jazz Improvisation sequence. The Bradley Jazz Ensemble has performed with Mike Vax, Mike Lee, Matt Wilson, Rachael Lee, Allen Vizzutti, Ron Jones, Tom Garling, Byron Stripling, Chris Vadala, the Jeff Hamilton Trio, Wayne Bergeron, Denis DiBlasio, Michael Davis, Allen Vizzutti, Wycliffe Gordon, Mark Colby. and Bobby Shew. During the summer of 2003, the group performed at the Montreux Jazz Festival in Switzerland, and they returned in the summer of 2006, performing to enthusiastic audiences at the Vienne Jazz Festival, the Montreux Jazz Festival, and the Umbria Jazz Festival. In the summer of 2009, the band toured Italy for ten days, performing at Lake Como, Serremaggiore, Valdarno, Viterbo, and the Umbria Jazz Festival in Perugia, and they returned in 2012. Dr. Kelly holds the Doctor of Arts degree in trumpet performance from Ball State University in Muncie, Indiana. While at Ball State, he was the Assistant Director of the Jazz Studies program, directing the Studio Jazz Ensemble, coordinating the Jazz Combo program, and coordinating the "Dimensions in Jazz" high school jazz festival. He also served as Adjunct Instructor of Trumpet and Jazz at Anderson University in Anderson, Indiana. Kelly received a Bachelor of Music Education degree with high honors from the University of Montana, and a Master of Music degree in Trumpet Performance from the University of Northern Colorado. His principal teachers include Lido Vizzutti, Dr. Thomas Cook, William Pfund, Paul Everett, and Bill Adam. He spent three years working as a musician for Carnival and Royal Caribbean cruise lines. From 1992–1994, Kelly taught grades 6–12 bands in Gilcrest, Colorado. Dr. Kelly maintains a busy performing schedule in the Central Illinois area as a member of the Todd Kelly Jazz Quintet, Four Other Brothers, JC and the Redemption, the Brazilionaires, and as a freelance jazz and classical musician. He is the musical director of the Central Illinois Jazz Orchestra, a big band comprised of seventeen of the finest musicians from the central Illinois area. He has backed such performers as Wayne Newton, Jerry Lewis, Charo, the Fifth Dimension, Lesley Gore, and Mannheim Steamroller, and the national touring companies of Hairspray, Wicked, The Rat Pack is Back, Mary Poppins, Billy Elliot, and Chicago. His professional memberships include the Jazz Education Network, Music Educators National Conference, and Phi Mu Alpha Sinfonia. He is also listed in "Who's Who Among American Teachers." In the spring of 2004, he was given the Bradley Parents Association's "Award of Excellence" for outstanding service to students, and in the spring of 2013, he was given the Slane College of Fine Arts Faculty Achievement Award for Excellence in Teaching. He is in demand throughout the Midwest as a clinician, adjudicator, and guest artist. Dr. Kelly has been published in Band World, Downbeat, JAZZed, and The Instrumentalist. His book, "Chet Baker: 21 Selected Solos" is available through Jamey Aebersold Jazz and ejazzlines.com. He is a contributing author to "The Jazzer's Cookbook," published by Meredith Music. Todd Kelly is an artist/clinician for Conn/Selmer in Elkhart, Indiana. He and his wife Kirsten are the proud parents of Connor and Maren.

BRANDEIS UNIVERSITY

Michael Randall

Best Overall Faculty Member, French

"Michael Randall is an outstanding leader in the ROMS department. He has a vision about planning for the future of Romance Studies. Michael brings faculty and administration to work together in harmony and in the most effective and successful way possible."

Michael Randall's primary research has been devoted to late medieval and Renaissance culture in France. He has published two books, as well as several articles, on subjects related to this area of study. He teaches courses in many different areas of French and comparative literature. These courses cover subjects such as literature and politics, literature and urbanism, the relationship of necessity and freedom, the notion of love, and comparative political systems in the modern Francophone world. He is currently chair of the Department of Romance Studies.

Andreas Teuber

Best Researcher/Scholar, Liberal Arts, Social Sciences, and Humanities

His scholarship encompasses the internationalization of human rights, issues of social inequality, risk analysis, the nature and limits of democratic forms of deliberation, civic participation and renewal, cosmopolitanism, and the relation of U. S. Constitutional Law to international covenants, understandings and agreements as well as the relation of decisions of the U.S. Supreme Court to the opinions of international courts and tribunals.

BRAZOSPORT COLLEGE

Joy Kennedy

Best Teacher, Liberal Arts, Social Sciences, and Humanities

I grew up in Sweeny, Texas and began my studies here at Brazosport College. I also studied in San Marcos, Texas; Canterbury, England; and Indiana, Pennsylvania

Gail Steele

Most Helpful to Students, Liberal Arts, Social Sciences, and Humanities

BRENAU UNIVERSITY

Heather Casey

Best Teacher, Liberal Arts, Social Sciences, and Humanities

BRIGHAM YOUNG UNIVERSITY

Orlando Alba

Best Researcher/Scholar, Spanish

Orlando Alba is a well-known Dominican linguist. He completed his Masters Degree in Linguistics at the University of Puerto Rico, Río Piedras, and his PhD in Hispanic Linguistics at the Universidad Complutense de Madrid. At present, he is a Full Professor in the Department of Spanish and Portuguese, Brigham Young University, Provo, UT. His academic career begun as a professor in the Department of Filosofía y Letras, at the Pontificia Universidad Católica Madre y Maestra, Santiago, Dominican Republic, where he was the Director of the Master Program in Linguistics (1988–1991). From 1987 until 1991 he was the Regional Delegate for the Dominican Republic of the Asociación de Lingüística y Filología de América Latina (ALFAL), and Treasurer of the same institution from 1999 through 2005. Dr. Alba is a member of the Academia de Ciencias de la República Dominicana, and in 2006 was elected to become Académico Correspondiente of the Academia Dominicana de la Lengua. In 2005 he was distinguished by the Dominican Government with the Orden de Duarte, Sánchez y Mella, grado de Comendador. He has written numerous articles in academic journals, and more than twelve books: Variación fonética y diversidad social en el español dominicano de Santiago (1990); Estudios sobre el español dominicano (1990); Enriquecimiento del vocabulario (1993); El léxico disponible de la República Dominicana (1995); El español dominicano dentro del contexto americano (1995); Vocabulario Básico del Español (1996); Nuevos aspectos del español en Santo Domingo (2000); Manual de fonética hispánica (2001); Cómo hablamos los dominicanos (2004); Lengua y béisbol en la República Dominicana (2006); La identidad lingüística de los dominicanos (2009); Nombres propios de persona en la República Dominicana (2013); Observación del cambio lingüístico en tiempo real: el nuevo léxico disponible de los dominicanos (2014).

Brian Cannon

Best Researcher/Scholar, History

Brian Q. Cannon graduated from BYU with a BA in American Studies in 1984. He completed an MA in History at Utah State University in 1986 and a PhD in History at the University of Wisconsin-Madison in 1992. Since that time he has been teaching at BYU. He teaches upper division courses in the American West in the Twentieth Century, Utah History, and U.S. History from 1890 to 1945. Much of his research focuses upon agricultural settlement, rural community development and federal rural policy in the twentieth century. He has received fellowships or other awards from the Western History Association, the Agricultural History Society, the Charles Redd Center for Western Studies, the Ford Foundation and the Aspen Institute for Humanistic Studies. He and his wife, Anna Lea, are the parents of one daughter and three sons.

David Laraway

Best Teacher, Spanish

David Laraway received his PhD in Romance Studies (Hispanic Literature) from Cornell University in 1998, writing his doctoral dissertation on the notion of place in the works of Jorge Luis Borges. In addition to Árbol de imágenes: nueva historia de la poesía hispanoamericana (with Merlin H. Forster), he is the author of numerous articles and book chapters on a wide range of topics in Hispanic literature and culture, from modernista poetry to contemporary Basque narrative to Chilean cyberpunk. He was an erstwhile doctoral student in Philosophy at the University of Utah (completing several seminars and passing the PhD qualifying exam before logistical complications prevented him from continuing) and is currently pursuing another PhD in Media and Communication at the European Graduate School (EGS) in Saas-Fee, Switzerland, where he has taken seminars from Simon Critchley, Slavoj Zizek, Michael Hardt, Peter Singer, Manuel de Landa, and Lev Manovich, among others. In addition to his interests in Hispanic studies, he takes an active interest in philosophy and new media. He is working intermittently on two book projects, one on Borges, digital culture, and the new materialism and another on outsider art and music. The latter will also serve as his dissertation at EGS, which he is writing under the direction of Simon Critchley.

Dr. Laraway currently serves as the chair of the Department of Spanish and Portuguese at Brigham Young University, where he teaches undergraduate courses on Hispanic literature and culture and graduate seminars on topics such as modernismo, contemporary Latin American poetry, and Borges and philosophy. The product of an unlikely convergence of Basque and Mormon stock in Idaho, he takes a keen personal interest in Basque and Basque-American literature and culture and hopes to expand his teaching repertoire to include these fields as well.

BRIGHAM YOUNG UNIVERSITY-IDAHO

Wade Huntsman

Most Helpful to Students, Art

Wade Huntsman was raised in Shelley, Idaho, a small farming community in Southeast Idaho. He spent most of his time as a child exploring the outdoors, playing sports, and drawing anything and everything. Wade's formal art education includes an Associate Degree in Illustration from Ricks College, a Bachelor of Fine Art in Illustration from Art Center College of Design in Pasadena, CA and a Master of Education from A.I.U. in Boston. Bro. Huntsman has created illustrations in both the publishing market and the entertainment arena. Clients include Hartcourt, Brace Publishing, Field and Stream, Peterson's Hunting, Walt Disney, Sony, Warner Brothers, NBC, Cartoon Network, and Dreamworks SKG. Bro. Huntsman spent several years working on full feature films with Dreamworks. He contributed to such films as Prince of Egypt, The Road to El Dorado, Spirit, Sinbad, Over the Hedge, and Flushed Away. He is currently a full time illustration professor at Brigham Young University-Idaho. Wade embraces both the traditional artmaking mediums and the digital art programs. Wade is enjoying the southern Idaho lifestyle with his wife Mary Jo and his three kids, Trey, Lauren, and Paige.

Randall Kempton

Best Overall Faculty Member, Music

Dr. Randall Kempton (M.M. Brigham Young University, D.M.A. Arizona State University) came to BYU-Idaho in 1999 and currently serves as Director of Choral Activities and conductor of the Collegiate Singers. He teaches private voice, diction, conducting, arranging, and choral methods courses, and has directed the University Choir, Men's Choir, and Women's Choir. Prior to his arrival in Rexburg, he conducted ensembles at Arizona State University, Brigham Young University, Utah Valley State College, and the Meridian School in Provo, Utah. His professional experience includes work as a pianist, vocal coach, and vocalist in classical, jazz, and musical theater settings. He studied conducting and arranging with Mack Wilberg and Ron Staheli.

Daniel Kerr

Best Teacher, Music

Dr. Daniel Kerr directs organ studies and oversees the Musicianship program at BYU-Idaho. He also teaches orchestration and works with composition students. He received his bachelor's, master's and Doctor of Musical Arts degrees from the University of Southern California, Thornton School of Music, where he studied organ with Dr. Ladd Thomas and Cherry Rhodes. Prior to his appointment at BYU-Idaho, he taught at USC and California State University, Los Angeles, was accompanist for the Claremont Chorale, musical director of the Foothill Summer Theatre, and organist and hand bell choir director at Pasadena Presbyterian Church. He has performed with the Los Angeles Philharmonic, at the American Guild of Organists' 2004 National Convention, at the 2008 Eccles Organ Series at Salt Lake's Cathedral of the Madeline, and is a regular recitalist on the organs of Temple Square in Salt Lake City.

Ryan Nielsen

Most Helpful to Students, Music

Ryan Nielsen, trumpet and jazz: Praised as a musician of "high intelligence, tremendous instrumental skill, and outstanding creativity" by jazz icon, John McNeil, Dr. Ryan Nielsen has been playing the trumpet since the age of 11. His love of music and trumpeting has grown ever since, leading to a Masters Degree in classical trumpet performance from Arizona State University, and a doctorate in jazz performance from the New England Conservatory of Music in Boston. Dr. Nielsen grew up in a musical home in Rexburg, Idaho, and has gone on to record with such artists as Corey Christiansen, Brent Jensen, Jon Bishop, Jay Laurence, Justin Nielsen, Rakalam Bob Moses, John Lockwood, and Clark Sommers. He has performed with such icons as Harvey Mason, the New York Voices, Eric Marienthal, Kevin Mahogany, Brian Bromberg, Byron Stripling, Wycliffe Gordon, Conrad Herwig, Delfeayo Marsalis, Kobie Watkins, John McNeil, Tiger Okoshi, and the Summit Brass (which included such brass legends as David Hickman, Ronald Romm, Alan Dean, Robert Tucci, and Dan Perantoni). Dr. Nielsen's recent album "Weeds in the Wall," recorded with the band Sister and Brothers, is an eclectic blend of folk, bluegrass, americana, and jazz music (available on itunes). He also recently recorded an album of experimental world music with vocalist Bisan Toron, and pianist Justin Nielsen (to be released 2014). Dr. Nielsen's love of teaching has garnered him a reputation as an effective and inspiring clinician throughout the intermountain west. He has received many awards for his playing, including the Doc Severinsen award for outstanding classical and jazz trumpeting and first place at the Lionel Hampton International Jazz Festival. Dr. Nielsen is an alumnus of BYU-Idaho.

Leon Parson

Best Overall Faculty Member, Art
Best Teacher, Art

Andrea Radke-Moss

Best Teacher, History

Dale Sturm

Best Teacher, Religion

Mark Watkins

Best Researcher/Scholar, Music

Mark Watkins received his degrees from Brigham Young University and Indiana University and has studied with Eugene Rousseau, Daniel Deffayet, Ray Smith, and Yushi Ishiwata and jazz with David Baker. Watkins has performed in more than 20 countries and with artists ranging from Bob Mintzer (jazz saxophone) to Marlene Pauley (classical clarinet). In 1994 he was offered a position in the premier Washington DC Army

Field Band but was unable to enlist due to a visual disability. Dr. Watkins has lectured at conferences world wide including the 12th, 14th, 15th, 16th World Saxophone Congresses and published in periodicals including The Saxophone Symposium and JAZZed Magazine. He has several book projects nearing completion with titles: Studies with the Bird: Foundational Jazz Articulation and Saxophone Sound: A View from the Inside, culminating over 15 year's of research and pedagogical application, as well as his ebook, Fundamental Jazz Improvisation: What Everybody Thinks You Already Know. Many positive reviews regarding Dr. Watkins' performance and composition are extant from Jazz Times, Saxophone Journal, Cadence Magazine, and other sources. Ed Calle comments about the CD, FOUR: On Warm Summer's Evenin' (2010 Jazz Hang), "Two of the most appealing dimensions of FOUR's offering are Watkins' pen and the virtuosity of each individual. Watkins' diversified, fresh, and engaging writing results in new discoveries with every listen." Noah Peterson, KPSU Radio, states, "...You never know what's coming next and, like Beethoven, it's always the perfect thing that follows." Dr. Watkins has over 100 compositions and arrangements, 38 are available from Walrus Music, Roncorp, and Dorn and he has produced six CDs with a seventh scheduled for June 2013. He has been Director of Woodwind Studies at North Dakota State University and is currently Director of Jazz Studies at Brigham Young University-Idaho.

Greg Williams

Most Helpful to Students, Religion

BRONX COMMUNITY COLLEGE

Mary Jo Viola

Best Overall Faculty Member, Art & Music
Best Researcher/Scholar, Art & Music
Best Teacher, Art & Music
Most Helpful to Students, Art & Music
Best Researcher/Scholar, Fine Arts
Best Teacher, Fine Arts
Most Helpful to Students, Fine Arts
Best Overall Faculty Member, Fine Arts

BROOKHAVEN COLLEGE

Aaron Clark

Most Helpful to Students, English
Best Teacher, Liberal Arts, Social Sciences, and Humanities

Nita Drescher

Best Overall Faculty Member, Liberal Arts, Social Sciences, and Humanities

Jim Dudlo

Best Overall Faculty Member, Liberal Arts, Social Sciences, and Humanities

Diane Plotkin

Best Researcher/Scholar, English
Best Researcher/Scholar, Liberal Arts, Social Sciences, and Humanities

Margaret Riley

Best Teacher, Developmental Studies
Best Overall Faculty Member, Developmental Studies

Bill Sigsbee

Most Helpful to Students, Liberal Arts, Social Sciences, and Humanities
Best Teacher, Liberal Arts, Social Sciences, and Humanities
Best Overall Faculty Member, Liberal Arts, Social Sciences, and Humanities

"A wonderful mentor to many students, even long after they have left his classes. He is also a great team player."

Stephanie Venza

Best Overall Faculty Member, English
Best Researcher/Scholar, English
Most Helpful to Students, English

"Professor Venza receives many accolades from her students. She has also been my mentor for many years. I have learned to be a better teacher because of her."

Stephanie Venza

Best Researcher/Scholar, Liberal Arts, Social Sciences, and Humanities

Marla Ziegler

Best Researcher/Scholar, Fine Arts
Best Teacher, Fine Arts
Most Helpful to Students, Fine Arts
Best Overall Faculty Member, Fine Arts

BROOKS INSTITUTE

Chris Broughton

Best Researcher/Scholar, Fine Arts
Most Helpful to Students, Fine Arts

Christopher Broughton earned his BA in Photography-Industrial/Color from Brooks Institute of Photography in 1985 and his MS in Photography from Brooks Institute in 1998. Christopher has been a faculty member of Brooks Institute since 1996, teaching courses in both the MFA and BFA programs and specializing in the History of Photography, Optics in Fine Art, Digital Photography, the View Camera, Introduction to the Studio, Advanced Studio Applications, and the Zone System.

He has recently co-authored a new textbook on photography called Capture: Digital Photography Essentials from publisher Rocky Nook. Christopher's photography is represented by the Four Friends Gallery in Thousand Oaks, CA. You can see his work at www.christopherbroughton.com.

Bruce Burkhardt

Most Helpful to Students, Fine Arts

Bruce Burkhardt earned his BA Still Photography-Color and Technology from Brooks Institute of Photography in 1988. Bruce has also studied art and photography at Ohio University in Athens, Ohio, and at The American College of Switzerland in Leysin, Switzerland.

He is a noted photographer, educator and consultant to the photographic industry and also an internationally published free-lance photographer. His images are marketed through the www.Corbis.com image agency. He works as a freelance photographer and also owns and manages a small greeting card company featuring his photography located in Santa Barbara, Calif.

Prior to coming to Brooks Institute, Bruce was a consultant to Tamron U.S.A., testing new lenses in the field as well as being a guest lecturer on various international expeditions as a photographic expert. Bruce also spent time lecturing in Kyoto, Japan. He has been active in many professional organizations, including P.P.A. and A.S.M.P. Specializing in location photography, he is equipped to work in most all environments

Ralph Clevenger

Best Overall Faculty Member, Fine Arts
Best Teacher, Fine Arts

Ralph Clevenger earned his BS in Zoology from San Diego State University in California in 1974 and a BA in Industrial Photography from Brooks Institute of Photography in 1979. He owns a photography company specializing in location photography and video projects of eco-travel, environmental portraits, natural landscapes, wildlife and undersea subjects.

Ralph has traveled throughout the world on assignment, including Australia, New Zealand, the Caribbean, Central America, Brazil, Antarctica, Alaska, Ireland, Mexico, Kenya, Tanzania, Zimbabwe, Zambia, Botswana and South Africa. Selected publication credits include Audubon, Conde Naste Traveler, Forbes, Islands, National Geographic Magazine, National Geographic Traveler, Nature Conservancy, Oceans, Sea, Outside, Orion Nature Quarterly, Terre Sauvage, National Geographic Books, Smithsonian Books, Sierra Club Books, Reader's Digest Books, and many others.

Ralph is the author of the book "Photographing Nature," published by New Riders/Peachpit Press and his stock images are represented worldwide by Corbis Images. For over 29 years, Ralph has taught courses in Lighting Theory, Digital Asset Management, Digital Workflow, Natural History Photography, Video Production, Stock Photography, and Underwater Photography. www.ralphclevenger.com

Paul Meyer

Best Teacher, Fine Arts
Best Overall Faculty Member, Fine Arts

Paul Meyer earned his BA in Art from the University of Hawaii-Manoa in 1977, his BA in Industrial Photography from Brooks Institute of Photography in 1981, and an MS in Photography from Brooks in 2000.

From 1982 to 1990 Paul worked with some of the top commercial and advertising photographers in New York and Los Angeles before joining the faculty at Brooks Institute. He has traveled extensively doing commercial and personal assignments, including going around the world twice with the Semester-at-Sea program.

Marcia Rubenstein

Best Researcher/Scholar, Fine Arts

Marcia Rubenstein earned her AB in Art from UC Berkeley in 1973 and a MFA from The School of the Art Institute of Chicago in Painting in 1977. Marcia has been an educator in the photographic field for nearly 25 years at Brooks Institute and is currently the Program Chair for the MFA in Photography, teaching advanced seminar courses and guiding graduate students through their culminating exhibitions.

With a background in painting, she began teaching traditional color photographic printing while working with photographic artists to develop their print portfolios. As technology changed, she transitioned into digital image production. For over a decade, she taught Photoshop.

Assuming various leadership positions throughout her career, Marcia has served as a leader in the development of new curriculum reflecting changing industry standards. In addition to her contributions to Brooks Institute, she has served as president of the Photo Imaging Education Association (PIEA), an educational organization.

BROOME COMMUNITY COLLEGE

Mindi Townsend

Best Teacher, Sociology

"Has represented the adjunct counsel as a founding member, served on CAI and goes above and beyond advising students and working with them to better their class instruction and grades."

BROWARD COLLEGE (ALL CAMPUSES)

Holley Gilley

Best Teacher, Languages

"She cares deeply for her students both academically and personally."

Edward Harris

Best Teacher, Reading

It is my personal belief that all pupils have the ability to learn. My philosophy of education is simply this. Students are not "sponges" that soak in knowledge, but they have special gifts and talents that are waiting to be discovered. It is the educator's job to facilitate that creativity and to tap into their genius. My mission as an educator is to go the distance for every learner in my classroom. I will do all that I can possibly do for each student. Some students may be talented in more academic fields, whereas others may be gifted in a more artistic field such as drawing, painting, or music. I think that as a teacher, my first objective should be to identify the different learning styles and then to educate; thereby simplifying the communication process between student and teacher. This means I will give feedback to students morning, noon and night. Moreover, I feel I can be an asset in the classroom. I know what it is like to overcome educational handicaps. I also believe that it is my duty to ensure that I am meeting the needs of the students that I come in contact with. I can motivate and inspire young people to surpass mediocrity and to strive for excellence in the 21st century. As a teacher, it is my vision to make a tremendous impact on the lives of the students I teach. Hands on activities will be included in each subject area (because individuals learn best by doing). Lastly, I will use the constructivist method and strategies, because many students learn better when you use different kinds of teaching methods. My education beliefs are directly related to all of the NBPTS core propositions

Thea Knight

Best Teacher, Reading

Quakish Liner

Best Overall Faculty Member, Reading

Bindu Ranaut

Best Overall Faculty Member, English

BUFFALO STATE COLLEGE

Stephen Caruana

Most Helpful to Students, Fine Arts

I graduated from Buffalo State College with a BFA in painting and a minor in art history. In 1992, I received my MFA in painting with a minor in printmaking from the State University of New York at Albany. I have been showing my artwork since elementary school.

My paintings, woodburnings, and woodcarvings are highly representational and carefully crafted. I also do plein air painting of various rural and urban scenes throughout Western New York. Subject matter ranges from figures to scenes and still life images.

I have an exhibition record which includes international, national, and regional gallery shows. My work has been exhibited in the Salmagundi Club In NY City, Triumph of Academic Realism Exhibition, International Cook Forest Woodburning Exhibition, and national woodcarving shows. In addition to art shows, I illustrate books for Queenship Publishing Companies.

Louis Colca

Best Overall Faculty Member, Social Work
Best Teacher, Social Work
Most Helpful to Students, Social Work

Gerhard Falk

Best Researcher/Scholar, Sociology

Ronnie Mahler

Best Researcher/Scholar, Social Work

Ronnie Mahler has earned a Ph.D. in sociology, an M.S.W., and a B.A. in psychology with a double major in sociology, all from SUNY at Buffalo. Areas of special interest and/or research include evaluation research, quality assurance, mental health, self-help, and social action groups. Mahler is a member of the Board of Directors of the Mental Health Association of Erie County and a mediator in-training at the Center for Dispute Resolution. She has published in the Journal of Social Work Education and Continuum: Developments in Ambulatory Mental Health Care, a Jossey-Bass quarterly. Mahler currently teaches Social Work practice (SWK 423), Research Methods in Social Work (SWK 317), and Field Instruction (SWK 493–494).

Janet Reed

Best Researcher/Scholar, Fine Arts
Best Teacher, Fine Arts

BUTLER UNIVERSITY

Liliana Goens

Best Overall Faculty Member, Languages

Gail Lewis

Best Teacher, Music

"enthusiastic, supportive - always helpful to all students"

Dr. Gail Lewis joined the faculty of the Jordan College of Fine Arts at Butler University in 1997, as Assistant Professor of Horn and Music Theory. She has also served as Assistant Professor of Horn and Music Theory at the Columbus State University (Georgia), and as Instructor of Horn at the University of Wisconsin-Oshkosh. She received the Doctor of Musical Arts degree in performance from the University of Wisconsin, Madison, the Master of Music degree in performance from the Eastman School of Music, and the Bachelor of Music in performance and music theory from Capital University in Columbus, Ohio. Her principal teachers have included Douglas Hill, Verne Reynolds, Froydis Ree Wekre, and Nicholas Perrini.

She has maintained an active schedule as a freelance performer with orchestras and chamber groups throughout the Southeast and the Midwest, including the Indianapolis Chamber Orchestra, the Symphony Orchestras of Evansville, LaFayette and Kokomo Indiana, Owensboro Kentucky, Columbus, Macon, Albany, and Valdosta Georgia, Greenville and Charleston South Carolina, Oshkosh and Madison Wisconsin, and Dubuque Iowa. She performs extensively with many chamber groups such as the Resurrection Brass Quintet, Monument City Brass Quintet, White River Brass Quintet, the Indianapolis Brass Choir, and is a former member of the Southwind Woodwind Quintet and the Fountain City Horn Quartet. She is also in demand as a studio musician at various recording studios in Indianapolis. She has presented solo and chamber recitals in Indiana, Georgia, Missouri, New Jersey, Alabama, Tennessee, Florida, Ohio, and Wisconsin.

Dr. Lewis conducts clinics and master classes at many high schools in Indiana, presents programs through the auspices of the Young Audiences of Indiana organization, and participates in clinics presented for high school students at Huntington College and Butler University. She has also presented clinics and concerts at the Georgia Music Educators Association annual conference in Savannah. Dr. Lewis is active with students of all ages through summer programs which have included the Southeastern Music Center and the University of Wisconsin Summer Music Clinic, and has participated in teaching courses in the Elderhostel music program for senior citizens. She maintains a private teaching studio, which currently includes horn students from the fifth grade through age 65.

She is currently pursuing a new interest in playing and researching the alphorn. Additionally, she explores the art of home-brewing beer, and in the summers enjoys backpacking, canoeing, and hiking with her dog Hershey.

BUTTE COLLEGE

Cristina Dahl

Best Researcher/Scholar, English

Cristina has a B.A. and M.A. in English Literature from CSU, Chico and an M.A. and Ph.D. in Comparative Literature with a Certificate in Gender and Sexuality Studies from Cornell University. She has also studied at the University of Guadalajara, Jalisco, Mexico and through Bryn Mawr University at Avignon, France. Her M.A. thesis focused on literary realism in the work of James Joyce and Gustave Flaubert, and her dissertation focused on gender roles and discursive violence in the work of Virginia Woolf, Jacques Derrida and Elena Garro. Currently, she is working on a multi-genre book project that examines the transgenerational trauma of slavery and some of its effects on her family in Cuba and in the U.S. and a research project on eastern and western feminisms.

Frances Gibbons

Best Overall Faculty Member, English
Best Teacher, English

Frances Vargas Gibbons teaches language, literature, and ideas as things that should be apprehended intellectually, as well as experienced emotionally, psychologically, and aesthetically. She believes that education transforms us into more ample, receptive, discerning, and embracing individuals. A good education sharpens our analytical skills, increases our awareness of ourselves and others, and makes us more empathic and constructive members of the species. Education humanizes us.

Gibbons, a native of Venezuela, has an MA from Columbia University in New York City, and a doctorate in literature from The University of Florida in Gainesville, Florida. In her previous career as a university teacher, she published research using psychological approaches to literature, and continues to employ these methods today in the writing she does for professional growth and pleasure. Gibbons cherishes her profession and delights in the challenges of assisting her students to become compassionate and enlightened citizens of the world.

CALIFORNIA STATE POLYTECHNIC UNIVERSITY

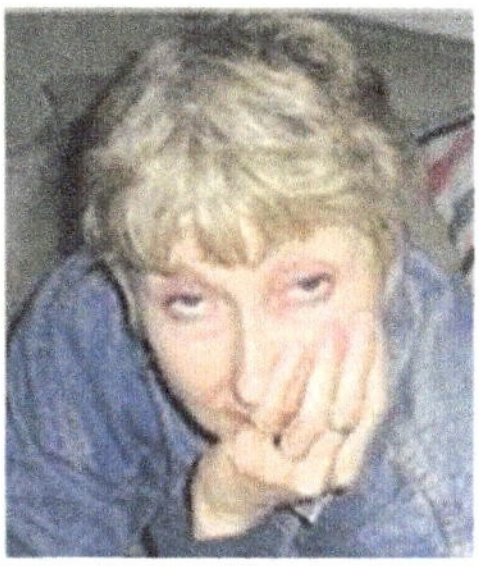

Barbara Thomason

Best Teacher, Fine Arts

Barbara A. Thomason is a native of Southern California and lives in Los Angeles. She received a Masters Degree in printmaking from California State University, Long Beach, After graduation she worked as a master printer in lithography at Gemini G.E.L. where she printed for Robert Rauschenberg, Claus Oldenberg, Jasper Johns, Frank Stella, Ed Ruscha, Ellsworh Kelly and many others

Her Paintings can be found in both public and private collections. During the last thirty years she has shown her work in exhibitions at galleries, museums and universities.

Her artwork includes public site specific works as well as paintings, drawings and prints. The work has moved from image based through abstraction and back again. The focus has mainly been on the urban environment but also includes more personal explorations and portraiture. The mediums used are watercolor, oil paint, cel vinyl, wood cut and etching.

Barbara currently teaches printmaking, sculpture and painting at California Polytechnic University at Pomona. She has been on the Art faculty at the University of California, Santa Cruz, University of Redlands, Otis College of art and Design to mention a few.

CALIFORNIA STATE UNIVERSITY

Kevin Elstob

Most Helpful to Students, Languages

Kevin Elstob teaches French. He is an avid traveler who puts biking and hiking in California and Provence at the top of his activities list. His most recent journey was to the Alpes Maritimes in the south of France. Elstob has also spent a lot of time in Canada (from coast to coast). If you would like to see more, see Elstob's Facebook and blog pages.

Since working on the theatre of French playwright Michel Vinaver for his PhD, Elstob has continued to examine ways in which the visual arts, particularly theatre and film, represent

French and Francophone culture. He has led a number of post-film discussions and question and answers sessions at the Crest movie theater.

He has a profound interest in multilingualism and how it changes one's appreciation of life. One of the projects he is presently working on is the multi-dimensional and multilingual theatrical work of Robert Lepage.

Elstob has also collaborated on a number of translations for organizations in Northern California. Recently, he was lead translator of an Avian Flu workshop for UC Davis Wildlife Center's programs in Djibouti and West Africa.

Elstob is board president of the Sacramento French Film Festival and former president and founding member of the Alliance Française de Sacramento. He is vice president of the Sacramento French Club where he plays Pétanque. He is also a qualified Pétanque umpire.

He likes to write fiction and poetry when he has time, but he mostly is successful with film and theater criticism. Elstob loves to laugh; he adores palindromes, word plays and a good joke.

Timothy Fong

Best Overall Faculty Member, Liberal Arts, Social Sciences, and Humanities

Timothy P. Fong Is Professor of Ethnic Studies at California State University, Sacramento. He is also the Project Director and Principal Investigator for the Full Circle Project (FCP) a project funded by the U.S. Department of Education. FCP is a comprehensive approach to implement a strategically focused, campus-wide effort to improve retention and graduation rates of underrepresented Asian American and Pacific Islander (AAPI) students. FCP aims to assist AAPI students throughout his or her entire college careers, and provide ample opportunities to engage in service both on and off campus to enhance their university experience.

In addition, Dr. Fong is a Steering Committee member of the California State University Asian American & Pacific Islander Initiative. The steering committee advises the Chancellor's Office on Asian American & Pacific Islander issues, provides direction to the CSU AAPI Initiative's programs on campus, and develops long-term strategies designed to increase the number of students who go to and succeed in college.

Dr. Fong received his doctorate in Ethnic Studies from the University of California at Berkeley in 1992. His research specialty areas include comparative race and ethnic relations, contemporary immigration, politics and public policy, community studies, higher education equity and student engagement, and qualitative methodology (ethnography and oral history).

Dr. Fong has written and edited several books (see Publications below) and has presented at numerous professional conferences across the nation and internationally. He was also the editor of the Critical Perspectives on Asian Pacific Americans Series published by AltaMira Press (a division of Roman & Littlefield). The Series presented books that are theoretically engaging, comparative, and multidisciplinary. Special emphasis is placed on works that reflect contemporary concerns critically important to understanding and empowering Asian Pacific Americans.

A recognized authority on multiculturalism and social change, Dr. Fong has been interviewed in stories by the New York Times, Los Angeles Times, San Francisco Chronicle, Sacramento Bee, the World Service of the British Broadcasting Corporation, and on National Public Radio, among others.

Kazue Masuyama

Best Overall Faculty Member, Languages

CALIFORNIA STATE UNIVERSITY BAKERSFIELD (CSUB)

Monica Ayuso

Most Helpful to Students, English

"Her students flock to her for help; she has letters to show it."

Mustafah Dhada

Best Researcher/Scholar, Liberal Arts, Social Sciences, and Humanities

Mustafah Dhada was born and brought up in Mozambique. He spent nearly eleven years in Asia before studying sculpting under abstract and neo-representationalist masters. He got his doctorate from the University of Oxford in 1987. In 1984 he was elected Fellow of the Royal Society of Arts, London. Dhada's works are to be found in collections internationally: The United States, Spain, Portugal, Brazil, Switzerland, and Mozambique.

Winner of several juried exhibition awards, Dhada was a Fulbrighter in 1995 to Mozambique where he sketched and held exhibitions of his work in Maputo.

The nineteen eighties proved to be years of artistic uncertainty for Dhada, a period of turmoil and despair in search of a creative identity uniquely his. Dhada's sculptures have emerged stronger, powerfully reflective, teasingly subtle in manipulating empty spaces as an integral part of experiencing sculpture as a three dimensional art form. His works compel you to touch and caress them. They are impregnated with inner organic compactness but express this externally with tensile irruptive vitality. They seek the chalice of life, the liberation of the self from within. Dhada calls his work Sensuous Abstractionism because it attempts to fuse his life's temporal trilogy – the raw carvings of his African childhood, the sinuous rotundity of Vedic expressions in stone and the poise and grace of European representational forms.

Brett Schmoll

Best Teacher, Liberal Arts, Social Sciences, and Humanities

Influences from bolh disciplines infuse his nascent research and publication in the scholarship of learning and American history. A similar trend evinces in the service portion of his record, which is highly suggestive of an interdisciplinary approach to the profession.

Students described the class as interactive, exploratory, not intimidating, inviting and creative. Several students noted that he is passionate about teaching. One student commented, "the classroom atmosphere was perfect, attempting to get students out of their bubbles", which is a testimony to Brett's success at getting students to engage with the material and participate in discussion and other classroom activities.

Many students noted that they normally hated history, but that Brett had changed their minds. One student comment sums it up nicely: "went from hating history to loving history, that is a pretty big improvement." One observed. "He made it so that everyone wanted to come to class." Another wrote, "everyone is smiling & laughing but LEARNING at the same time."

Although many students commented that the course was fun. it also appears that Brett maintains high, but reasonable, standards and that he expects students to write according to the standards of the discipline, developing a clear argument supported by evidence and communicating it in good English.

Me has shared his experience and ideas with his colleagues in History and is a member of the department's committee on on-line/hybrid teaching, which hopes to find a way to offer students in on-line courses the rich experience they get in face-to face courses. Clearly Brett is, quite simply, a gifted teacher and the committee commends him for his teaching excellence and his willingness to teach the service courses so important to the department and the university.

CALIFORNIA STATE UNIVERSITY CHANNEL ISLANDS

Catherine Bae

Best Teacher, History

Rainer Buschmann

Best Researcher/Scholar, History

Rainer F. Buschmann is Professor of History at the California State University Channel Islands (CI). His training in anthropology awoke a passion for oceanic settings that took him to Hawai'i (where he obtained his M.A. in anthropology and his Ph.D. in history), and lastly the California Channel Islands. He has formerly taught at Hawai'i Pacific University and Purdue University. At CI, he was a founding faculty member of the history program contributing to a curriculum that actively emphasizes world history.

His publications explore European perceptions of the Pacific Ocean in an attempt to recover this body of water's global history, a largely understudied area in contemporary scholarship. He is author of Anthropology's Global Histories: The Ethnographic Frontier in German New Guinea, 1870–1935 (U of Hawai'i Press, 2009) and Oceans in World History (McGraw-Hill, 2007). He has collaborated with Edward Slack and James Tueller on Navigating the Spanish Lake: The Pacific in the Iberian World, 1521–1898 (U of Hawai'i Press, 2014). His most current monograph is Iberian Visions of the Pacific Ocean, 1507–1899 (Palgrave Macmillan, 2014). He has held visiting fellowships at the Institute of Social Studies at the University of Lisbon and the European Institute of Global Studies at the University of Basel. He is section editor for world history at the History Compass and co-edits a new publication series entitled Studies in Pacific Worlds issued by the University of Nebraska Press.

His new research topics include a textbook on the Pacific Ocean in global perspective and a co-written monograph on Portugal and the Pacific between the years of 1511 and 1999.

Amy Caldwell

Most Helpful to Students, History

James Meriwether

Best Overall Faculty Member, History

"Decolonization & Desegregation: Africa, African Americans, and the Johnson Administration." Paper delivered at American Politics, World Politics: International Dimensions of U.S. History. Clare College, Cambridge University, UK, April 2011.

"To Defeat Racialism Whenever it Raises it's Ugly Head': Africa and American Political Life in the 1960's." Paper delivered at Invited Talk, Princeton University, NJ, October 2011.

Recent Publications:

Review of: Secular Missionaries: Americans and African Development in the 1960's. Journal of American History. 98:2, September 2011, 583–83.

Review of: The United States and the End of British Colonial Rule in Africa, 1941–1968. H-Diplo Reviews. XIII: Vol. 9, October 2011, 2–4

Review of: Cold War in Southern Africa: White Power, Black Liberation. H-Diplo Reviews. XIII: Vol. 10, November 2011, 15–18

Andrea Reilly

Best Teacher, English

CALIFORNIA STATE UNIVERSITY CHICO

Victoria Junco

Best Overall Faculty Member, Languages

Daun Weiss

Most Helpful to Students, Music

"Daun Wiess is friendly, knowledgeable, helpful to her students, and an outstanding musical talent."

Daun Weiss holds a B.S. in Music Education from Western Baptist College and an M.Mus. in Vocal Performance from the University of Oregon . She teaches singing lessons at Simpson University in Redding and at California State University, Chico. In the past, she has taught on the faculties of Western Baptist College, Salem, Oregon and the University of Oregon in

Eugene. She is currently the president of the Sierra North Chapter of the National Association of Teachers of Singing. A frequent regional soloist and a founding member of the Chico Early Music Ensemble, Daun specializes in Baroque music. She is a featured soloist at Chico's Bach Festival each Spring. She has performed the role of the Queens in Handel's Solomon and of Rodisette in Telemann's The Patience of Socrates. Daun also enjoys work as an oratorio soloist, having recently sung Orff's Carmina Burana, Beethoven's Ninth Symphony, and Mozart's Requiem. Daun has performed with the North State Symphony, the Sacramento Early Music Consort and with the Sacramento Baroque Soloists. She has just returned from the Bay Area where she sang the role of the Angel in Stradella's L'Anime del Purgatorio with the California Bach Society.

Rocky Winslow

Best Teacher, Music

"He is a rare combination of a top professional musician and master teacher."

Professor Rocky Winslow has been the director of Jazz Studies at California State University, Chico since joining our music faculty in the fall of 2002. Mr. Winslow holds bachelor and master's degrees from the University of North Texas where he was a featured soloist with the One O'Clock lab band and can be heard on five of their CDs. Jazz Studies at CSU, Chico boasts two big bands and several jazz combos. The flagship ensemble, Jazz X-Press has been proud to host concerts with guest artists such as Grammy Award singer Terry Richards, Bobby Shew, Greg Gisbert, Sinatra's jazz pianist and conductor Vincent Falcone, trombonist Mark Patterson, drummer and recipient of the 2006 Presidential Teaching Award Paul Romaine, along with many others. The X-Press was a featured group in summer 2008 at the Telluride Jazz Festival in Colorado. Rocky is also a popular clinician having served as the all-state guest conductor and a clinician for Music Education Associations in Texas, Nevada, California, Arkansas, Hawaii, Australia, Japan, and Scotland. As a performer, Rocky has shared the stage with, recorded, and toured with artists such as Maynard Ferguson, Tony Bennett, Vincent Falcone Orchestra, Chaka Khan, Peabo Bryson, The Temptations, Don Menza Big Band, Natalie Cole, Drew Carey, Paul Anka, The Four Tops, Bob Newhart, and many others. Rocky's last two CD recordings, both on Troppe Note Records, "Simple Complications, and "Out of the Blue" feature many of his original compositions and boast some terrific side men including former members of the famed Bill Evans trio Eddie Gomez and Elliot Zigmund.

CALIFORNIA STATE UNIVERSITY FRESNO (FRESNO STATE)

Peter English

Most Helpful to Students, Criminology

Peter English is a psychologist and Associate Professor in the Department of Criminology at California State University, Fresno where he has taught courses such as Psychology of Crime, Principles and Applications of Forensic Behavioral Sciences, Ethics in Forensic Behavioral Sciences, Juvenile Delinquency, and Law and Criminal Justice. He graduated from the University of Arizona with a degree in psychology. Professor English has published and continues to conduct research on topics related to psychology, public policy, and law. His most recent research interests concern the adjudication of mentally ill offenders, comprehension of legal language, and the use of behavioral forensic science in legal decision making. He is co-author of a book called More Than the Law: Behavioral and Social Facts in Legal Decision Making (2005) with Bruce D. Sales, PhD, JD, published by the American Psychological Association. Professor English worked previously at NASA – Ames Research Center at Moffett Field Naval Air Station in the Office of Space Human Factors and Perception and Cognition Lab. He has published and presented nationally on topics related to attention, visual search and information processing. Professor English has served as an editorial assistant on the journal Psychology, Public Policy, and Law where he serves as an ad hoc reviewer. He is a member of the American Psychological Association, American Psychology-Law Society, and the Association for Psychological Science.

Ofelia Gamez

Most Helpful to Students, University Studies

Emma Hughes

Best Teacher, Criminology

Dr. Emma Hughes is an Associate Professor in the Criminology Department at California State University, Fresno. She joined the Department in 2007. Dr. Hughes earned her Ph.D. in Criminology from Birmingham City University in the UK, where she also taught for five years before moving to Fresno. Dr. Hughes earned her M.Phil. in Criminology from the University of Cambridge. She has an M.A. in History from Columbia University and earned her B.A. in English from Stanford University. Dr. Hughes is the coordinator for the Corrections option within the Criminology major and she regularly teaches the community-based and institutional corrections courses as well as courses in the graduate program. She is the director of the undergraduate Criminology Honors Program. Her research interests focus on jails and prisons, with a particular emphasis on rehabilitation programs. She is the author of the book Education in Prison: Studying through Distance Learning, published by Ashgate in 2012. She contributed a chapter on prison education to the book How Offenders Transform Their Lives (eds: Veysey, Christian & Martinez, 2009, Willan Publishing) and she has also contributed chapters to edited volumes on prison education. Dr. Hughes serves as an advisor to the Fresno County Local Criminal Justice Coordinating Council for whom she conducts research on the dynamics of the jail population. Whilst living in the UK Dr. Hughes conducted research for British charities involved in educational and arts provision in prisons. She regularly presents papers at conferences such as the American Society of Criminology's annual meeting. She is currently undertaking research on the role and experiences of community volunteers who provide programming in prisons and jails.

George Kikuchi

Best Researcher/Scholar, Criminology

"Is constantly conducting pioneering research in the field of crime mapping and its related inpacts. Is engaged with providing students with research skills and collaborates with scholars in the US and in Japan."

Dr. George Kikuchi is an Assistant Professor in the Department of Criminology, California State University, Fresno. He has received his Ph.D. in Sociology from Purdue University in 2008. He is a certified crime and intelligence analyst by the California Department of Justice and has worked as a crime analyst intern at several law enforcement agencies in the U.S. Additionally, he has worked as a postdoctoral researcher at the National Research Institute of Police Science, which is affiliated with the National Police Agency of Japan (roughly equivalent to the F.B.I. in the U.S.). His current research interests include spatio-temporal analyses of crime, data-visualization, and data mining in crime analysis.

Loredana Lo Bianco

Best Researcher/Scholar, Liberal Arts, Social Sciences, and Humanities

Maria Lopes

Most Helpful to Students, Liberal Arts, Social Sciences, and Humanities

"Interacts with students and faculty at an exceptional level. Is always warm and friendly and is an intense scholar."

Gary Rice

Best Overall Faculty Member, Liberal Arts, Social Sciences, and Humanities

"This is the consumate scholar who does it all. Is great with his students, knows his material very well, has a great national reputation and is engaged with the community."

Director, Community Journalism program; Director, Central California War Veterans Oral History Project; Director, San Joaquin Valley Scholastic Journalism Conference

Education: Ph.D., M.A., American History, University of Texas; B.A. Applied Arts and Sciences, Southwest Texas State University

Teaching areas: Community Journalism, International Journalism, Investigative Reporting, Public Affairs Reporting Research Interests: War and Media; Community Journalism, International Journalism

Community Journalism: Under the direction of Dr. Gary Rice, a 21-year reporting and editing veteran, students produce stories for area newspapers and magazines including the Fresno Bee, Vida en al Valle; Reedley Exponent, Sanger Herald, Fowler Ensign, Madera Tribune, Visalia Times-Delta, Fresno magazine and other publications. Since the program's inception in 2004, students have published more than 1,600 stories professionally, won four professional awards for public service journalism (Gruner Awards) and finished as runners-up for three others. Student work as been cited by the Fresno County grand jury and led to criminal investigations.

Yoshiko Takahashi

Best Overall Faculty Member, Criminology

"Yoshiko Takahashi does it all: conducts empirical research, is a scholar, is loyal to her students, knows her subjects, volunteers her time to an NGO and is a great fellow faculty member."

Dr. Yoshiko Takahashi began her career at Fresno State in the fall of 2008. Previously, she was a management analyst at the Mecklenburg County Sheriff's Office where she conducted jail-related studies. She has an M.S. in Criminal Justice and a Ph.D. in Public Policy with a Justice Concentration from the University of North Carolina at Charlotte.

Her research interests are jail inmates, comparative studies of criminal justice, and recovery of victims.

CALIFORNIA STATE UNIVERSITY FULLERTON

Heather Battaly

Best Overall Faculty Member, Philosophy
Best Overall Faculty Member, Liberal Arts, Social Sciences, and Humanities

"Heather works harder than several of us put together, and is always focused on making our students' and her colleagues' lives and educational experiences better."

Randolph Baxter

Best Researcher/Scholar, American Studies

Randy Baxter received his BA degree in European History from UC Berkeley, his MA in International Affairs from Columbia University (SIPA), and his PhD in US History from UC Irvine. Since 1999, he has taught American Studies and US History at Cal State Fullerton.

Amy Coplan

Most Helpful to Students, Philosophy
Best Overall Faculty Member, Philosophy
Best Researcher/Scholar, Philosophy
Best Teacher, Philosophy

"Publishes in multiple areas of philosophy and develops many cross-discplinary areas of research.

Inspires students to work harder and take more courses.

Prof. Coplan cares deeply about her students, makes herself available to them in unprecedented ways and to a degree that is well above and beyond the call of duty. She also runs the coolest classes and has some of the most fun ideas for assessment and participation of any professor I've ever encountered. She provides students with remarkable freedom to express themselves and to engage personally with complex philosophical material.

Covers all the bases (research, teaching, student interaction and even extended links to the philosophical community nationally and internationally in an exemplary way!"

Jessica K. Grimes

Most Helpful to Students, Sociology

"Jessica Grimes engages students with the community to apply sociology to real life learning."

Brady Heiner

Best Researcher/Scholar, Philosophy

"Prof. Heiner's work on race, gender and the prison industrial complex in California is some of the most genuinely important and radical research and activism I have yet encountered in our discipline."

Melanie Horn-Mallers

Best Overall Faculty Member, Human Services
Best Teacher, Human Services
Most Helpful to Students, Human Services

Dr. Melanie Mallers is an assistant professor in the Department of Human Services at Cal State Fullerton. She was trained to be a life span developmental researcher with a focus on stress, health and coping across adulthood. Her educational background also includes intensive emphasis on family systems theory, social ecology and biopsychosocial models of human development. Currently, she is exploring the relationship between disrupted parent-child attachments during childhood and daily reactivity (stressor response, mood and physical health symptoms) during middle adulthood. A related thread of research includes the unique role of father-son relationships. She is also currently working on a grant to examine the role of technology as a function of health adherence and maintenance among older adults. Dr. Mallers's research has been published in Developmental Psychology and Psychology and Health, as well as in several books.

John Koegel

Best Researcher/Scholar, Music

John Koegel, Professor of Musicology, serves as Graduate Advisor for the School of Music, and Coordinator of Music History. He teaches courses and conducts research in American, Mexican, and European musical topics, particularly musical theater and opera, and music in the context of immigration. His book Music in German Immigrant Theater: New York City, 1840–1940 (University of Rochester Press, 2009) was given the Irving Lowens Book Award of the Society for American Music in 2011, and he was a Finalist for the Theater Library Association's 2010 Freedley Award. Opera News (Metropolitan Opera Guild) described it as "Deep-delving. . . . Readable and entertaining. . . . A tour guide down a forgotten byway of the American immigrant experience."

In 2013–14, Koegel was the recipient of a year-long Research Fellowship given by the National Endowment for the Humanities for work on his book Mexican Musical Theater in Los Angeles, 1910–1940. In 2013, he served as Chair of the Cátedra Jesús C. Romero, of the Centro Nacional de Investigación, Documentación e Información Musical Carlos Chávez, in Mexico City. His articles and reviews appear in journals and dictionaries in the United States, Mexico, Spain, and Britain, such as the Journal of the American Musicological Society, American Music, Latin-American Music Review, Journal of the Royal Musical Association, Historia Mexicana, The New Grove Dictionary of Music and Musicians, and Diccionario de la Música Española e Hispanoamericana.

Koegel holds degrees from Claremont Graduate University (Ph.D. musicology), Cambridge University (M.Phil. ethnomusicology), and Cal State Northridge (B.A. music education and Spanish). He has served as Book Review Editor for the Journal of the Society for American Music (2010–14), has been a member of the Council of the American Musicological Society, and was Board Member for Musicology for the College Music Society (2011–13). He is a member of the Board of Trustees of the Society for American Music (2013–16), and was a contributing editor for the Grove Dictionary of American Music, 2nd ed. (2013).

Patricia Literte

Best Teacher, Sociology

"Patricia fills classes because students know she cares about their learning and she sets high expectations while mentoring them."

Patricia Literte received her Ph.D. in Sociology from the University of Southern California. Her current research interests include race and ethnicity, higher education, racial politics, and urban sociology. More specifically, Dr. Literte has conducted research on Black-Latino/a relations in higher education, Black-Latino/a relations in low-income housing developments, Black immigrants in Southern California, student affairs' interaction with Hapa students and multiracial students, and historically Black sorority members' attitudes about gays and lesbians. She teaches social inequality, race and ethnic relations, sociology of families, sociology of city life, and gender, sex, and society. In addition to her teaching and research, Dr. Literte serves as the Academic Coordinator for the Sociology Online Degree Completion Program and the Academic Coordinator for the Ronald E. McNair Post-Baccalaureate Scholars.

JeeLoo Liu

Best Teacher, Philosophy

"Prof. Liu works harder to provide helpful feedback on her students' written work than anyone else, pushes them to succeed, and cares deeply about the welfare and success of our students, however closely or distantly they are involved with our programs."

I am currently working on constructing analytic Chinese philosophy, in topics such as Chinese metaphysics, Confucian moral psychology and Neo-Confucian virtue ethics. My work-in-progress is Neo-Confucianism: Metaphysics, Mind and Morality, contracted with Wiley-Blackwell.

Laura Lohman

Best Overall Faculty Member, Music

Laura Lohman combines her backgrounds in musicology and ethnomusicology to teach courses in music history, world music, and music research in the School of Music and the Honors Program. With funding from the National Endowment for the Humanities and the American Musicological Society, her own research has focused on American and Middle Eastern music. Her first book, Umm Kulthum: Artistic Agency and the Shaping of an Arab Legend, 1967–2007 (Wesleyan University Press, 2010), explores how the legendary Egyptian singer expanded her international fame and shaped her legacy following the Six-Day War, and the conflicting ways that her image and music have been interpreted since her death in 1975. Dr. Lohman is completing a second book on American music and politics in the early national period. In addition to her scholarship and teaching, she has been continuously involved in collegial governance at CSUF, where she has provided leadership and coordination in diverse areas such as faculty affairs, facilities, curriculum, advising, and student recruitment. She earned the Bachelor of Music in flute performance from Indiana University and Doctor of Philosophy in musicology from the University of Pennsylvania, in addition to graduate study in flute performance and ethnomusicology from the Royal Northern College of Music and the University of California at Santa Barbara.

Amanda Perez

Best Teacher, Liberal Arts, Social Sciences, and Humanities

Michael Perez

Best Researcher/Scholar, Sociology
Best Overall Faculty Member, Sociology
Best Teacher, Sociology
Most Helpful to Students, Sociology

"Dr. Perez produces engaging scholarship that is readable and accessible."

Michael P. Perez is a native diasporic Chamorro/Chamoru ancestrally rooted in Guam, and born in Long Beach, CA. He is Professor of Sociology and the Faculty Athletics Representative at Cal State Fullerton. He was previously an Assistant Professor of Sociology at the University of Guam. He received his Ph.D. in sociology at the University of California at Riverside, after conducting fieldwork and completing his dissertation on indigenous identity in Guam. His teaching and scholarly interests include critical race studies, race, ethnic and indigenous relations, Native Pacific studies, sociology of sports, social justice, education and critical pedagogy, deviance and delinquency. Perez's published work has appeared in Amerasia Journal, Critical Criminology, Deviant Behavior, Ethnic Studies Review, Faculty of Color Teaching at Predominantly White Institutions (Anker Publishing), Global Processes, Local Impacts: The Effects of Globalization in the Asia Pacific Region (University Press of America), Pacific Studies, Social Identities: Journal of the Study of Race, Nation and Culture, Social Science Quarterly, Sociological Spectrum, and Sovereignty Matters: Locations of Contestation and Possibility in Indigenous Struggles for Self-Determination (University of Nebraska Press).

Chris Ruiz-Velasco

Best Teacher, English

Stella Ting-Toomey

Best Researcher/Scholar, Human Communication Studies

Dr. Stella Ting-Toomey is a Professor of Human Communication Studies at California State University, Fullerton. Her teaching passions include intercultural communication theory, intercultural communication training and design, and intercultural and interpersonal conflict management.

Her research interests have focused on testing and fine-tuning the conflict face-negotiation theory and the cultural/ethnic identity negotiation theory. She is the author and editor of 17 books (plus 2 Instructor's Manuals and 2 Interactive Student Study Guides).

Stella's most recent books include: Understanding Intercultural Communication, Second Edition (with Leeva C. Chung; Roxbury/Oxford University Press); text details at: www.oup.com/us/tingtoomey) and The Sage Handbook of Conflict Communication, Second Edition (co-edited with John G. Oetzel; Sage). Other book titles include: Managing Intercultural Conflict Effectively (with John G. Oetzel; Sage), Communicating Effectively with the Chinese (with Ge Gao; Sage), and Communicating Across Cultures (Guilford Publications). She has also published more than 100 journal articles and book chapters in academic journals such as the International Journal of Intercultural Relations, Human Communication Research, Communication Monographs, International Journal of Conflict Management, and Communication Research, among others.

Stella has held major leadership roles in international communication associations and has served on more than fifteen editorial boards. She has delivered major keynote speeches on the theme of mindful intercultural communication in South Africa, Germany, Ireland, Norway, Portugal, China, Hong Kong, Japan, Canada, and different regions of the U.S. As a trainer, consultant, and certified mediator, she has designed and conducted a variety of intercultural conflict competence workshops for corporations, universities, and non-profit centers and institutes. Stella is the 2008 recipient of the 23-campus wide California State University Wang Family Excellence Award. She is also the 2007–08 recipient of the CSU-Fullerton Outstanding Professor Award in recognition for superlative teaching, research, and service.

Berna Torr

Best Overall Faculty Member, Sociology

"Dr. Torr teaches a core course, Statistics, in a way that students can understand. She is a committed department citizen–always volunteering and suggesting improvements in policies, curricular issues, and practices."

As a social demographer, my research explores how changing demographic context shapes family life, health, and well-being. I examine how changes in demographic context—attributable to both historical trends and individual events, such as the experience of migration and being an ethnic minority—influence family formation, health, and well-being. I also continue to be active in applied research and demography. I love research and working with data. I try to share this enthusiasm by regularly teaching undergraduate and graduate classes on statistics and research methods. In addition, I teach substantive classes on the family and am currently developing a class on migration. My approach to teaching stresses the value of hands-on learning and application of the research methods of sociology through individual and group research projects. In all classes, I assign a hands-on project that (1) helps students understand how sociological research works in practice, (2) develops practical research skills, and (3) provides a venue for critical thinking and interpretation.

Fureya Unal

Best Overall Faculty Member, Music
Best Researcher/Scholar, Music
Best Teacher, Music
Most Helpful to Students, Music
Best Researcher/Scholar, Liberal Arts, Social Sciences, and Humanities
Best Teacher, Liberal Arts, Social Sciences, and Humanities
Most Helpful to Students, Liberal Arts, Social Sciences, and Humanities
Best Overall Faculty Member, Liberal Arts, Social Sciences, and Humanities

Ünal is a Faculty member at California State University Fullerton. She has been teaching for over 20 years at schools in Istanbul, Germany and the United States. Her students have won numerous competitions and festivals. Ünal has given a number of concerts and recitals in Turkey, Germany, Monaco, France, Bulgaria, Italy, Portugal, Switzerland, and the United States, both as a solo artist, accompanist, and ensemble player. She was a recipient of Cultural Exchange Fellowship in 2012 given by the Turkish Cultural Foundation to give concerts and master classes in The People's Republic of China. She has been awarded "the Woman of Distinction 2013 from 65th District of California" by Assemblywoman Sharon Quirk-Silva. Ünal is the founder, co-artistic director and pianist of the new music ensemble Divan Consort. Her ensemble has given numerous concerts and master classes since their beginning in 2009 in the US and Europe. The group has released their first album "Refuge" (Albany Records), which has been awarded the Gold Medal in Album category and the Silver Medal in composition category by Global Music Awards.

Toya Wyatt

Best Overall Faculty Member, Human Communication Studies

"Dr. Wyatt is a very dedicated faculty who contributed at all levels. I have great respect for her and feel very fortunate to work alongside with her!"

My primary goal as a faculty member here at CSUF is to: a) provide our students with the types of educational, and clinical training that will enable them to work effectively with children and families from diverse cultural and linguistic backgrounds, no matter where in the state, country or world their future professional lives might take them; b) help our students develop the type of critical thinking skills that would enable them to pursue future research opportunities in our field at the Ph.D. level and beyond, if they should ever decide to choose that path, and c) help our students learn the types of professional advocacy skills that are important for becoming fully functioning members and powerful voices in our national, state and local associations on issues of relevance to the field and the diverse communities that we serve.

CALIFORNIA STATE UNIVERSITY LONG BEACH

Carolyn Bremer

Most Helpful to Students, Music

Carolyn Bremer has been dubbed a composer "driven by hobgoblins of post modernist cant." Bremer came to composition on the heels of intensive training as an orchestral bassist. Her catalogue contains works based on feminist symbolism (Athene), baseball (Early Light), and popular culture (It Makes Me Nervewracking). Recently, Bremer has incorporated her photography and music into multimedia works.

Bremer has had recent performances of her works at Carnegie Hall; in Germany, Norway, and Sweden; and for the gala 150th anniversary concert at West Point. Her commissions include the Symphony for WindBand, premiered by Ray Cramer at Indiana University; Returns of the Day, premiered by Thomas Dvorak at University of Wisconsin-Milwaukee, Pieces of Eight premiered by the California State Honor band, and Spark, premiered by Adam Brennan at Mansfield University. CDs released since Spring 2002 include the El Paso Wind Symphony on Summit Records, the Heritage of American Band of the US Air Force, the Towson University Symphonic Band, and the Monarch Brass Ensemble.

Her work Early Light is a mainstay in the wind ensemble repertoire, receiving hundreds of performances each year. The original version for orchestra has been performed by professional orchestras including the Houston Symphony, California Philharmonic, Chattanooga Symphony, Oklahoma City Philharmonic, and Waco Symphony.

Bremer studied at the Eastman School of Music, CalArts, and received a Ph.D. in composition from UCSB. She was Chair of Composition at the University of Oklahoma from 1991–2000 where she held the O'Brien Presidential Professorship. Currently, she is a Professor of Composition/Theory and Chair of the Bob Cole Conservatory of Music at California State University Long Beach.

Joanie Conley

Best Overall Faculty Member, Recreation
Best Teacher, Recreation

Jane Dabel

Best Overall Faculty Member, History
Best Researcher/Scholar, History
Best Teacher, History
Most Helpful to Students, History

Alicia Doyle

Best Overall Faculty Member, Music

Alicia Doyle is an Associate Professor of musicology, graduate adviser and associate chair in the Bob Cole Conservatory of Music. Specializing in medieval liturgical music, 20th-century Latin American popular and art music, and music appreciation, Dr. Doyle teaches courses in music history, world music, and music appreciation at California State University, Long Beach. Dr. Doyle received her Bachelors degree in Music Performance (Horn) from the University of Southern California where she studied with James Decker. She received her Masters degree in Musicology from the University of California at Santa Barbara under the advising of Dr. William Prizer. Her thesis focused on Florentine carnival music from 1500–1510. Dr. Doyle's Ph.D. was also from the University of California at Santa Barbara. Her dissertation, a study of a tenth-century Aquitanian troper, was completed under the direction of Dr. Alejandro Planchart.

Kristine Forney

Best Researcher/Scholar, Music

Hiroko Kataoka

Best Overall Faculty Member, Languages
Best Researcher/Scholar, Languages
Best Teacher, Languages
Most Helpful to Students, Languages

Pat Kearney

Best Overall Faculty Member, Communication
Best Teacher, Communication
Best Overall Faculty Member, Liberal Arts, Social Sciences, and Humanities

Nancy Matthews

Most Helpful to Students, Recreation
Best Teacher, Recreation

Katherine McLoone

Best Overall Faculty Member, Comparative Literature
Best Teacher, Comparative Literature

Gary Metzker

Best Teacher, Journalism

Gary Metzker has been a lecturer at Cal State Long Beach since 2008. He is design adviser for the Daily 49er and Dig Magazine's content and design adviser.

Metzker started in the newspaper business more than 40 years ago by delivering the Sunday Philadelphia Bulletin.

He received his Bachelor of Journalism degree in 1976 at the University of Missouri and his Master's in Communications in 2013 from California State University Fullerton. Metzker has worked at newspapers in Delaware, Missouri, Florida and California. For almost 25 years, he worked at the Los Angeles Times in positions ranging from sports news editor, metro news editor, A1 editor and Senior Editor.

During that time, Metzker was a part of four different Pulitzer Prize-winning staffs for breaking and spot news and was the designer for stories that earned a photographer the Pulitzer for photography and a writer the Pulitzer for best feature. Metzker has also won two Los Angeles Press Club awards and two Medals of Excellence from the Society for News Design.

Metzker teaches Media Design, Introduction to Mass Communications, Writing Across the Media, Introduction to Sports Journalism and Magazine Production at Cal State Long Beach.

Metzker has presented seminars and workshops for the Journalism Association of Community Colleges, College Media Association, National High School Journalism Association and the Associated Collegiate Press. He is the outreach chairman for the Department of Journalism and Mass Communication and he instructs at newspapers and high school journalism boot camps during the summer. Metzker was editor-in-residence at the Salem (Ore.) Statesman-Journal in 2013.

Nancy Meyer-Adams

Most Helpful to Students, Liberal Arts, Social Sciences, and Humanities

Dr. Nancy Meyer-Adams began as Director of the School of Social Work in the Fall, 2014. She received her Ph.D. in social work with a specialty in school social work at The University of Tennessee College of Social Work in 2002. The focus of her dissertation research was the effects of school bullying on the culture and climate of middle schools in urban communities. Dr. Meyer-Adams joined the faculty of California State University, Long Beach School of Social Work in the summer of 2005 where she has been recognized with the Most Valuable Professor award for four years (2008, 2009, 2011, and 2012) and with a College of Health and Human Services Community Service Award in

2010. Prior to joining CSULB's faculty, she taught at the University of Southern California, School of Social Work where she received the Jane Addams Faculty Award for academic, administrative and moral support for students in 2005. Her professional experience includes over ten years of direct practice experience with diverse family and child client populations as a school social worker, mental health professional, and researcher. She was the founding project coordinator of the Pasadena Unified School District's School-based Mental Health Program in 2002 where she served as a consultant to the district on bullying and school violence as well as a direct practitioner. She has also served as a school social work consultant to the Montebello Unified School District. Dr. Meyer-Adams also teaches in both the graduate and undergraduate programs.

Kirran Moss

Most Helpful to Students, Communication

Aparna Nayak-Guercio

Best Teacher, French

Rachelle Okawa

Best Researcher/Scholar, Comparative Literature

Tim Plax

Most Helpful to Students, Liberal Arts, Social Sciences, and Humanities

Nancy Quam-Wickham

Best Researcher/Scholar, Liberal Arts, Social Sciences, and Humanities
Best Teacher, Liberal Arts, Social Sciences, and Humanities
Most Helpful to Students, Liberal Arts, Social Sciences, and Humanities
Best Overall Faculty Member, Liberal Arts, Social Sciences, and Humanities

Nancy Quam-Wickham is a Professor of History, and was Chair of the Department from 2005–2015. Her research interests include environmental, gender, and social history, oral history, history pedagogy, and assessment practices in higher education. Prior to becoming department Chair, she was Editor of The History Teacher journal. She is a participant in the American Historical Association's Tuning project, and has presented her work on Tuning in History at many professional meetings. In October 2014, Dr. Quam-Wickham was one of twelve scholars nationwide to be named as a faculty coach for the Degree Qualifications Profile project, a collaborative initiative endorsed by the American Association of Colleges and Universities, with origins in the Bologna Process.

Professor Quam-Wickham is particularly interested in the impact of natural resource development and exploitation in the American West, including the Pacific Basin. She is currently working on a study of whaling in the Pacific Ocean from the late 1700s to the mid-1900s. She is also finishing an article on transforming the lower-division United States history survey course by embracing active learning pedagogies; see the Daily 49er story here and the links below for more information about Dr. Quam-Wickham's "U.S. History Lab" class, in which students produced curriculum materials for at-risk high school history students in the Long Beach Unified School District. In her spare time, Professor Quam-Wickham enjoys traveling, sea-kayaking, hiking, and backpacking in the Sierras.

James Sauceda

Best Teacher, Liberal Arts, Social Sciences, and Humanities

James Steven Manseau Sauceda, a third generation Mexican American, is a native son of Los Angeles. Born in 1952, he grew up and attended schools in Compton, California. He received his B.A. Degree in Speech Communication from CSULB, being honored as his department's top graduate in 1974. His Masters Degree, also from CSULB, included international study in London, Dublin, and Wales. He graduated Summa Cum Laude in 1976. For his doctorate in Communication Arts and Sciences, he attended the University of Southern California. He graduated Summa Cum Laude in 1980. Currently, Dr. Manseau Sauceda is a Professor in Communication Studies at CSULB.

Dr. Manseau Sauceda's background in multiculturalism, intercultural communication, and international education is far reaching and varied. He has taught courses in Ethnic Studies at USC, the California School of Professional Psychology, and CSULB. His published work has appeared in the prestigious text Intercultural Communication: A Reader (edited by Samovar and Porter), and he has written a second monograph for the Reader's 9th edition. Dr. Manseau Sauceda is completing a major work, Communicating Across The Cultural Divide: Understanding and Experiencing a Multicultural America. Additionally, Dr. Manseau Sauceda has presented many scholarly papers and professional workshops, including those for the annual conventions of the International Communication Association, the Western States Communication Association, the East meets West Conference, the Association of International Educators and the Speech Communication Association, among others.

Dr. Manseau Sauceda facilitated and created the curriculum for the pilot program Latino Leadership Academy in the fall of 2010. The program will continue with various faculty and student facilitators as it moves forward.

Dr. Manseau Sauceda continues to pursue his creative writing projects. His play, Magic Kit, took a top prize in the National Chicano Literary Competition. Furthermore, he has completed writing 'Hey Zeus' his second original screenplay, (a project in which he also composed the musical score.) In addition, Dr. Manseau Sauceda is a poet who has performed his work throughout the greater Los Angeles area. His poetry has been translated into Spanish and is featured in an anthology published in Mexico; Nueva Poesia de Los Angeles, Universidad De Guadalajara, 1994.

Finally, Dr. Manseau Sauceda is in demand nationally as a speaker. He has been Keynote and Plenary speaker for the United Nations Association, the Association of International Educators (NAFSA), the California Association for Counseling and Development, the California Cooperative Education Association, TESOL, the Honor Society for International Scholars (Phi Beta Delta), Summer Bridge and Minority Engineering Programs, Hispanic Youth Leadership Conferences to name but a few engagements. Dr. Manseau Sauceda is also available for Multicultural/Intercultural Communication Consulting and Workshops tailored to the needs of students, faculty and staff, as well as companies in the private sector.

Dr. Manseau Sauceda is the founding director of the Multicultural Center at California State University, Long Beach. He is also the founder and director of "The Rainbow Voices," a multicultural theater troupe which brought issues of diversity to elementary, middle, high school, and college campuses. The Rainbow Voices have brought their message of acceptance to over 60,000 students nationwide.

Nhora Serrano

Most Helpful to Students, Comparative Literature

Craig Smith

Best Researcher/Scholar, Communication
Best Researcher/Scholar, Liberal Arts, Social Sciences, and Humanities

Craig R. Smith, is professor of communication studies at CSULB and the founder of the Center for First Amendment Studies. In addition to publishing more than 15 books and 60 articles, he served as a full-time speechwriter to President Gerald Ford and Chrysler CEO Lee Iacocca, and was a consultant to George H. W. Bush during his 1980 run for the presidency. In 2010, he received the Douglas W. Ehninger Award for lifetime achievement in rhetorical scholarship by the National Communication Association.

Jordan Smith

Best Overall Faculty Member, Liberal Arts, Social Sciences, and Humanities

Michiko Takeuchi

Best Teacher, Liberal Arts, Social Sciences, and Humanities
Most Helpful to Students, Liberal Arts, Social Sciences, and Humanities

Jonathon Talberg

Best Teacher, Music

Dr. Jonathan Talberg, Director of Choral, Vocal, and Opera Studies and Associate Director of the Conservatory, is conductor of the nationally renowned CSULB Bob Cole Conservatory Chamber Choir. He has twice conducted at the Music Educator's National Conference regional honor choir, at the 2008 and 2012 American Choral Directors Western Convention and 2009 and 2013 National Collegiate Choral Organization Conferences, at numerous All-State choir concerts, and in various venues throughout Europe and Asia, including the Sistine Chapel, St. Peter's and St. Mark's Basilicas in Italy, the Karlskirche in Vienna, the Matyas Templom in Budapest, and at the Great Hall of the People in China. In constant demand as a guest conductor, he has worked with all levels of singers—from elementary to professional—throughout the United States and Europe. He has prepared choirs for the Cincinnati Symphony, the Long Beach Symphony Orchestra, the Los Angeles Master Chorale, the Pacific Symphony and the Pasadena Pops. Dr. Talberg is Music Director at First Congregational Church of Los Angeles, where he conducts both the Cathedral Choir and the Cathedral Singers, a 16-voice professional chamber choir that sings weekly in service. He also serves as director of the Los Angeles Bach Festival, which celebrated its 80th anniversary in October. Prior to his appointment at CSULB in 2000, Dr. Talberg served as Conducting Assistant to the Cincinnati Symphony and the Cincinnati Pops and as principal choral conductor at Arrowbear Music Camp. Before graduate school, he taught high school choir in Orange County and at the Los Angeles County High School for the Arts. A ten-year member of the California ACDA board, he is currently serving as Past-President. He is also an editor at Pavane Music Publishing, where a choral series is published under his name. Of the many hats he wears each day, the one he is most proud of is as mentor to the next generation of choral musicians. Alums of the Bob Cole Conservatory Choral Studies program are teaching at elementary, middle and high schools, as well as community and four-year colleges throughout the country. Recent Bob Cole Conservatory graduates are currently earning their doctorates in choral music at the University of Michigan, the College-Conservatory of Music at the University of Cincinnati, Indiana University, the University of Kentucky, the University of Iowa, and the University of Southern California. Dr. Talberg received his BM in Choral Conducting from Chapman University and his MM and DMA in Choral Conducting from the University of Cincinnati's College-Conservatory of Music. He completed a post-doctoral fellowship, graciously underwritten by the Oliver Family Foundation, with the Cincinnati Symphony Orchestra, Cincinnati Pops, and the May Festival Chorus. His teachers include Roger Wagner, William Hall, Earl Rivers, John Leman and Elmer Thomas.

Oliver Wang

Best Researcher/Scholar, Liberal Arts, Social Sciences, and Humanities

Steve Wilson

Best Teacher, Social Work

Dr. Wilson teaches research courses including Social Work Research (Parts I and II) and Thesis Advising. In addition, he also teaches direct practice courses including Social Work Practice with Death, Dying, and Loss (micro practice) and Social Work Administration (macro practice). Prior to his work in academia, Dr. Wilson had extensive experience in human resource management with various organizations and worked for several years as a hospice social worker. Dr. Wilson's research interests and areas of specialization are in medical and geriatric social work with older adults including death, dying, and bereavement (e.g., terminal care bio-ethics, cross-cultural bereavement), hospice care, and family caregiver issues. Previous research experience has also included workplace diversity issues, AIDS and HIV, as well as gerontological social work labor force demand. In 2012, Dr. Wilson was the recipient of the University's Distinguished Faculty Teaching Award for his teaching accomplishments in the school of social work.

CALIFORNIA STATE UNIVERSITY SAN MARCOS

Ibrahim Al-Marashi

Best Researcher/Scholar, History
Best Researcher/Scholar, Liberal Arts, Social Sciences, and Humanities

Ibrahim Al-Marashi is Assistant Professor of Middle East History at California State University San Marcos. His research deals with the modern history of Iraq. He is the co-author of Iraq's Armed Forces: An Analytical History (Routledge, 2008). He obtained his D.Phil. at the University of Oxford, completing a thesis on the Iraqi invasion of Kuwait. He is an Iraqi-American who lived at various times in Saudi Arabia, Yemen, Egypt, Morocco, and Turkey, and has traveled extensively through the Middle East.

Catherine Armas-Matsumoto

Most Helpful to Students, Communication

Catherine Armas-Matsumoto earned her B.A. in Communication at CSUSM in 2000, and her M.A. in Communication at San Diego State University in 2003. With interest in Organizational and Intercultural Communication, Ms. Armas-Matsumoto worked as a communication consultant assisting organizations with communication audits, diversity training and small group development. Since 2005 she has taught as an adjunct instructor for CSUSM and Palomar College. Currently in the dissertation phase of her Ph.D. in Education at Capella University, Professor Armas-Matsumoto's doctoral research examines the use of effective communication skills during internships as a means for students to gain social capital.

Katherine Brown

Best Teacher, Communication

Katherine Brown earned her Ph.D. in (1996) from the Communication Department at U.C. San Diego. She joined the faculty of CSUSM in 2001, and teaches Organizational Communication; Communication Research Methods and Designs, Communication and Collaboration, and Communication Portfolio, as well as the Mass Media core course Mass Media and Technology.

In 2013, Dr. Brown was appointed Faculty Director of the College of Humanities, Arts, Behavioral and Social Sciences Career Readiness Initiative.

The CRI involves faculty, students, college and Career Center staff and community partners in expanding the number and quality of internship opportunities for students in our college, is creating a mentoring network, and is bringing speakers to campus for a wide variety of career-themed panels. The overarching emphasis of CRI is helping CHABSS students prepare for next steps after college.

In her teaching and research, Dr. Brown draws upon cultural, historical and other systems level theoretical frameworks, theories of collaboration, organizational communication theory, technology studies and qualitative research methods. She has studied the use of communication technologies to support collaboration in educational, professional and research settings. Several of her projects arose from work with members a large, distributed international and interdisciplinary research collaborative based at UC San Diego called the Laboratory of Comparative Human Cognition, with whom she is continues to work with collaborators to develop an electronic archive and wiki of the lab's several decades of work.

Pat Hall

Best Teacher, Communication

Mrs. Hall holds M.A. and B.S. degrees in Communication from Arizona State University with emphases in intercultural and interpersonal communication. Her current areas of interest include persuasion/critical thinking, interpersonal interactions and research, and oral communication as it relates to first year students. In addition, she has worked in conflict resolution as a mediator in California and Arizona, both in private practice and for the Arizona State Attorney General's Office. She has handled a variety of cases including civil disputes, child protective custody cases and has mediated high school youth issues.

Melissa McGuire

Best Teacher, History
Best Teacher, Liberal Arts, Social Sciences, and Humanities

Terri Metzger

Best Overall Faculty Member, Communication

Terri Metzger (MA. San Diego State University) began teaching at CSUSM in the Spring of 1998. She coordinates the oral communication program and teaches Oral Communication as well as Argumentation and Dialogue, Research Methods, and Scientific Communication in Biotechnology. Her interests include faculty, instructional and curriculum development and public communication in a variety of contexts.

Kimber Quinney

Most Helpful to Students, History
Most Helpful to Students, Liberal Arts, Social Sciences, and Humanities

Lilianacastaneda Rossmann

Best Researcher/Scholar, Communication

Dr. Liliana Castañeda Rossmann holds a Ph.D. in Communication from the University of Massachusetts, Amherst, and an M.A. in International Peace Studies from the University of Notre Dame. She is a mediator and works facilitating public dialogue with community groups. She teaches courses in Communication Theory; Language and Social Interaction; Interpersonal and Intercultural Communication; and Conflict, Mediation and

Dialogue. Her teaching approach derives from the empowerment model in transformative mediation. Her research interests include storytelling by Mexican migrants, especially as it pertains to the social construction of identity. Her academic work has been published in national (American Behavioral Scientist) and international (Text; Human Systems) journals. Her recent book "Transcending Gangs: Latinas Story Their Experience," was published by Hampton Press in 2013. She is a member of the Mayor's Gang Prevention Task Force in San José, California. She is a native of Tampico, México.

Andy Strathman

Best Overall Faculty Member, History
Best Overall Faculty Member, Liberal Arts, Social Sciences, and Humanities

CALIFORNIA STATE UNIVERSITY STANISLAUS

Stephanie Paterson

Best Teacher, English

Stephanie is a Professor of English who teaches courses in Rhetoric and Teaching Writing M.A. concentration. She leads teacher research work at GVWP.

CALIFORNIA UNIVERSITY OF PENNSYLVANIA

Greg Harrison

Best Researcher/Scholar, Fine Arts
Most Helpful to Students, Fine Arts
Best Overall Faculty Member, Fine Arts

Scott Lloyd

Best Overall Faculty Member, Art
Best Teacher, Art
Most Helpful to Students, Art

Richard Miecznikowski

Best Teacher, Fine Arts

Cynthia Persinger

Best Researcher/Scholar, Art

Cindy Persinger's research focuses on the development of the discipline of art history in the United States and its relationship to the political events of the 20th century. In addition to historiography, her areas of interest include European modernism, American art, and the history of photography. In her teaching, she emphasizes the various perspectives from which we understand art objects. She enjoys introducing the students in her large appreciation and survey courses to visual arts and architecture from around the globe. Prior to her arrival at California University of Pennsylvania in Fall 2010, she taught art history courses at West Virginia University, the University of Pittsburgh, and Duquesne University.

CALVIN COLLEGE

William Romanowski

Best Teacher, Communication

BA - Indiana University of Pennsylvania, English, 1976
MA - Youngstown State University, English, 1981
PhD - Bowling Green State University, American culture studies, 1990

CAMERON UNIVERSITY

Justin Walton

Best Overall Faculty Member, Communication
Best Researcher/Scholar, Communication
Best Teacher, Communication

Dr. Justin Walton is an Associate Professor of Communication Studies at Cameron University in Lawton, Oklahoma. He completed his undergraduate education at Oklahoma State University and received both an M.A. and Ph.D. from the University of Oklahoma. He frequently teaches coursework in public speaking, small group communication, research methods, communication theory, and multicultural communication. Walton also holds an Oklahoma teaching certificate in the subject areas of speech/drama/debate and psychology/sociology.

Walton's scholarship focuses on communication pedagogy, citizenship education, and critical social theory. His specific research interests include:

Critical Approaches to Communication and Citizenship Symbolic Interaction and the Sociology of Knowledge Qualitative Research Approaches and Designs Western Philosophy In addition to his teaching and research, Walton serves as the department's mentor for students concentrating in communication education.

Dr. Walton is a two-time president of the Oklahoma Speech-Theatre-Communication Association (2004–2005 & 2005–2006) and holds professional membership with the National Communication Association, the Central States Communication Association, and National Council for the Social Studies. He was honored with the "Outstanding College Communication

Educator" Award in 2001 and the "Outstanding Young College Educator" Award in 2002 from the Oklahoma Speech-Theatre-Communication Association.

In his spare time, Walton enjoys reading, movies, and spending time with friends and family.

CAMPBELLSVILLE UNIVERSITY

Lisa McArthur

Best Overall Faculty Member, Music

Dr. Lisa McArthur teaches flute, music theory, and graduate courses in music history and music education. She has a Bachelor's degree in Music Education from the Crane School of Music (SUNY Potsdam College), Masters degrees in Performance and Music Theory from Kent State University, and a Ph.D. in Music Theory from the University of Kentucky. She performs, teaches, and adjudicates regularly throughout KY and surrounding states and is the director of the award-winning Quarteto Gracioso (flute quartet). Her arrangements for flute ensemble are published by Falls House Press.

CARNEGIE MELLON UNIVERSITY

Alex London

Best Teacher, Philosophy

Alex John London is Professor of Philosophy and Director of The Center for Ethics and Policy at Carnegie Mellon University. Professor London is an elected Fellow of the Hastings Center and recipient of the Distinguished Service Award from the American Society of Bioethics and Humanities (ASBH) and a New Directions Fellowship from the Andrew W. Mellon Foundation.

Professor London's research focuses on foundational ethical issues in human-subjects research, issues of social justice in the trans-national context, and on methodological issues in theoretical and applied ethics. His papers have appeared in Mind, Science, The Lancet, PLoS Medicine, Statistics In Medicine,

The Hastings Center Report, and numerous other journals and collections. He is co-editor of Ethical Issues in Modern Medicine, one of the most widely used textbooks in medical ethics.

In 2012 he joined the Working Group on the Revision of the CIOMS 2002 International Ethical Guidelines for Biomedical Research Involving Human Subjects and in 2011 he was appointed to the Steering Committee on Forensic Science Programs for the International Commission on Missing Persons (ICMP). Since 2007 he has served as a member of the Ethics Working Group of the HIV Prevention Trials Network. He has testified before the Presidential Commission for the Study of Bioethical Issues and has been commissioned to write papers for the Centers for Disease Control and Prevention (CDC) and the Institute of Medicine (IOM). He has served as an ethics expert in consultations with numerous national and international organizations including U.S. National Institutes of Health, the World Health Organization, the World Medical Association and the World Bank.

Professor London is an affiliated faculty member of the Center for Bioethics and Health Law at the University of Pittsburgh and was previously a Visiting Scholar at the Harvard University Program in Ethics and Health. Before joining Carnegie Mellon in 2000, he was a post-doctoral fellow at the University of Minnesota's Center for Bioethics and he received his PhD in Philosophy from the University of Virginia.

Peter Spirtes

Best Researcher/Scholar, Philosophy

CENTRAL CHRISTIAN COLLEGE OF THE BIBLE

Paul Axton

Best Overall Faculty Member, Liberal Arts, Social Sciences, and Humanities

CENTRAL CONNECTICUT STATE UNIVERSITY

Julie Ribchinsky

Best Teacher, Music
Most Helpful to Students, Liberal Arts, Social Sciences, and Humanities

CENTRAL MICHIGAN UNIVERSITY

Betsy Richard

Best Teacher, Theater

Sue Woods

Best Overall Faculty Member, Communication Disorders

"Dr. Sue Woods is an amazing teacher, mentor and advisor to our undergraduate and graduate students. She is friendly, kind, knowledgeable and passionate about communication disorders."

CENTRAL NEW MEXICO COMMUNITY COLLEGE

Terri Brown-Davidson

Best Teacher, English

Jean Silesky

Best Overall Faculty Member, Languages

"She is the faculty advisor for the Spanish Club. As advisor she planned and participated in many wonderful activities with the students last year. She has also shared many of her great Power Point activities with the faculty members through workshops and the Community section of Blackboard."

CENTURY COLLEGE

Cullen Bailey Burns

Best Overall Faculty Member, English

Cullen Bailey Burns lives in Minneapolis and Sturgeon Lake, Minnesota. Her first book, Paper Boat, was a finalist for a Minnesota Book Award in poetry. Her poems have appeared in many journals, including The Denver Quarterly, Rattle, The Laurel Review and Court Green. She is the recipient of a Minnesota State Arts Board Artist Fellowship, the Neil Postman Award for Metaphor from Rattle, and has twice been nominated for a Pushcart Prize. She teaches English at Century College and keeps bees on her farm in Northern Minnesota.

Dana Bruhn

Best Researcher/Scholar, English
Best Teacher, English

Scott Guenthner

Most Helpful to Students, English

CERRITOS COLLEGE

John Swanson

Best Teacher, English

CHAFFEY COLLEGE (ALL CAMPUSES)

Eva Rose

Best Overall Faculty Member, Communication Studies

"She oversees all operations for two large departments, Communication Studies and Modern Languages. She speaks fluently in five languages, a rarity in the United States. She has chaired the Chaffey College Honors Program, held stewardship of the Faculty Development Center, has done keynotes and other presentations at leadership conferences around the country and the world. She is a relentlessly kind and great communicator in any and all contexts and is liked, revered, and emulated wherever she goes. Prof. Rose sets a high bar and a great example as she models daily, weekly, and annually, the important communication precepts that she advocates, wherever she goes, in all settings and contexts. She is the jewel of the faculty at Chaffey College and has been for nearly thirty years. I recommend Professor Eva Rose for this earned distinction highly and without reservation."

CHAMPLAIN COLLEGE

Francine Page

Best Overall Faculty Member, English

A lifelong Vermonter, I earned my Bachelor of Arts in English with Honors from Middlebury College and my Master of Arts in English from the University of Vermont. At UVM, I got a taste of teaching college students through a two-year teaching fellowship.

Here at Champlain, I have taught many courses-from interpersonal communication and public speaking to literature and writing. For several years, I directed the Champlain Writing Center and served as co-chair of the College's very successful Community Book Program. I taught rhetoric courses in the first year of Champlain's unique Core Curriculum.

My professional interests include helping students to read more actively and working with colleagues to improve instruction in writing. I have served both as vice president and as secretary of the Faculty Senate. Individually and with colleagues, I have presented half a dozen workshops at regional and national conferences.

I live in Essex but spend most weekends in Garfield, Vermont, a family meeting place. I enjoy reading, walking, gardening, snow shoeing, skiing, and visiting with our two daughters and their families, including two grandsons (four and seven years old) who live with their parents in northern New Mexico, near Taos. Our younger daughter and her husband live in Salem, Massachusetts.

CHAPMAN UNIVERSITY

Emily Carman

Best Researcher/Scholar, Film

Emily Carman earned her PhD from the Cinema and Media Studies program at UCLA's Department of Film, Television, and Digital Media, where she channeled her passion for film history, especially classic Hollywood cinema in the 1930s, into various scholarly and professional pursuits in film. While her primary research expertise focus is on film stardom, the Hollywood studio system and media industries, other scholarly interests include postwar European cinemas (especially Italy) and moving image archive theory and practice. Carman is currently preparing her book for publication at University of Texas Press, titled Independent Stardom: Female Stars and Freelance Labor in 1930's Hollywood, in which she documents how female stars like Carole Lombard, Irene Dunne, Barbara Stanwyck (and more) became freelance artists in the 1930's, and by doing so, they proactively took ownership of their career and star image decades before it became standard practice in Hollywood. She is also currently co-editing the volume Hollywood and the Law for BFI/Palgrave press, which explores how legal infrastructures influence and regulate the content, distribution, and consumption of film and media.

In addition to her scholarly pursuits, she has worked for various motion picture archives and cultural institutions, including the Warner Bros. Archives of the School of Cinematic Arts at the University of Southern California, where she was the curator responsible for the preservation of the collection and its public access for research, the Academy Film Archive, L'Immagine Ritrovata at the Cineteca di Bologna, and the UCLA Film and Television Archive. She helped curate the UCLA Film and Television Archive centennial Barbara Stanwyck Retrospective, "A Lady to Talk About" in Spring 2007 and coordinated the 17th Annual Thinking Gender International Graduate Student Conference for the UCLA Center of the Study of Women.

Amy Graziano

Best Overall Faculty Member, Music
Most Helpful to Students, Music

Amy Graziano is Associate Professor and Chair of the Conservatory of Music at Chapman University. She is also the Director of Music History for the Conservatory. Dr. Graziano received her Ph.D. in musicology (1996) and M.M. (1990) from the University of Texas at Austin and her B.A. (music and psychology, 1985) from Vassar College. She also completed a two-year Post-Doctoral Fellowship in Music Cognition at the University of California, Irvine. At Chapman she teaches introductory and advanced courses in music history, The Psychology of Music, and Film Music. Her research focuses on the history of music psychology, in particular the relationship between music and neurology in the nineteenth century. She has published articles on the history of music and brain studies (with Julene Johnson) in Music Perception, Brain and Cognition, Journal of the History of the Neurosciences, and other journals, as well as a chapter in Neurology of Music (F.C. Rose, Ed., 2010). She has published on the general history of music psychology in Psychomusicology and has co-edited a double issue of Psychomusicology (History of Music Psychology in Autobiography). Dr. Graziano has received several Chapman University awards for excellence in teaching and scholarship.

Marilyn Harran

Best Researcher/Scholar, Liberal Arts, Social Sciences, and Humanities

Marilyn J. Harran is professor of history and religious studies, holder of the Stern Chair in Holocaust Education, and the founding director of the Rodgers Center for Holocaust Education and the Sala and Aron Samueli Holocaust Memorial Library at Chapman University where she teaches courses on Nazi Germany and the Holocaust. In her teaching and research, she draws upon a strong background in European history, particularly that of the Reformation, for which she has contributed articles and chapters in several volumes, as well as the monographs, Luther on Conversion: The Early Years (Cornell University Press) and Martin Luther: Learning for Life (Concordia Publishing House) and the edited volume Luther and Learning: The Wittenberg University Luther Symposium (Susquehanna University Press). Dr. Harran was a contributing writer and consultant to The Holocaust Chronicle. As founding director of the Rodgers Center for Holocaust Education, she has established a multi-faceted program with academic minor, distinguished lecture series, library, and community outreach, including a contest reaching 5,000 middle and high school students each year. She is the recipient of fellowships from the National Endowment for the Humanities, the Mellon Foundation, and the International Research & Exchanges Board. She has been a Visiting Scholar at the Hoover Institution at Stanford University and a Mellon Fellow at the Aspen Institute for Humanistic Studies and is the recipient of numerous awards and honors, including the Spirit of Anne Frank Award in 2008. She is currently working on several projects, including co-authoring a Holocaust survivor memoir, and is a member of the governing board of the Association of Holocaust Organizations.

Sean Heim

Best Researcher/Scholar, Music

Born in Philadelphia in 1967, Sean Heim began his first serious musical training in secondary school and soon after began studies in composition with Harold Oliver at Rowan University. He then worked with Native American composer Louis W. Ballard, completed his Masters degree in composition with Chinary Ung at Arizona State University, and holds a Ph.D. from The University of Queensland in Brisbane, Australia where he worked with Philip Bracanin. Sean has taught at Rowan University, Rutgers University, and the Settlement Music School. He currently resides in Southern California and is Associate Professor of Music at Chapman University. The primary focus of Sean Heim's work as a composer has been to develop an imaginative personal language that strongly reflects the compositional techniques and aesthetic of his own western tradition as well as the distillation and infusion of philosophical ideas and musical elements found in numerous cultures. His work also reflects a deep interest in the visual arts, physics, and the natural world, out of which he has created unique fusions that continually evolve through aesthetic and technical abstraction.

Sean Heim has received numerous prizes, awards, honors, and commissions from such prestigious institutions as the Fromm Foundation at Harvard, the National Endowment for the Arts, Meet the Composer, the American Music Center, the Atlantic Center for the Arts, The American Composers Forum, The ASCAP Foundation - including the Rudolf Nissim Prize and the Morton Gould Young Composers Award, Chapman University, Arizona State University, and The University of Queensland. Sean's music has been performed to critical acclaim throughout the United States and abroad by such distinguished performers as the California E.A.R. Unit, Piano Spheres, Ensemble Green, the Focus Festival at Juilliard, the

Thailand International Composition Festival, inauthentica, Topology, the Auros Group for New Music, Perihelion, and the New York Miniaturist Ensemble.

His discography includes releases on Sono Luminus, Eine Klang, and Capstone Records.

Gerald Hicks

Most Helpful to Students, English

"Always goes above and beyond to give students individual care and attention. Jerry Hicks is truly a credit to Chapman's English program."

David Kost

Best Teacher, Film
Best Overall Faculty Member, Fine Arts

Dave Kost is an Assistant Professor in the Dodge College of Film and Media Arts at Chapman University. He has an M.F.A. in film production from New York University and is an award-winning maker of short films. His many films have been seen in festivals and on television around the world. He has worked in all areas of writing, production and post-production in film, television, music videos, and commercials. He has also created an instructional video about safety on film sets that is internationally distributed through First Light Media and has written a revolutionary book used in directing and acting classes that is published by Focal Press. At Chapman University, Professor Kost has help to build both the graduate and undergraduate film production programs, designing curriculum, revising degree requirements, creating new courses, writing syllabi used in hundreds of course sections, and creating a unique, inter disciplinary degree in screen acting.

Michael Kowalski

Best Overall Faculty Member, Film

As a documentarian, Michael Kowalski says, you start with many views of reality, but you have to find an angle or focus. It's the creative process of structuring the story that he finds so fascinating, as a documentary filmmaker and as a sound designer. "Reality doesn't work dramatically," he says, "and we don't care." When he shows a sound class the opening of Star Wars, the students are not concerned with the fact that there is no sound in space, rather, they respond to the dramatic use of sound in the moment.

Kowalski's background—growing up in Europe, earning a Masters in Visual Anthropology at USC, teaching documentary techniques to students in architecture as a way of understanding how the city affects people—provided multiple ways to see the world. His own work, either as sound designer or documentarian, covers a wide range of subjects, from the Alamo and medieval knights to a comedic view of a Britney Spears impersonator and an examination of an African-American community as a microcosm of underdeveloped neighborhoods in urban American today.

Kowalski applies his interest in storytelling to teaching as well: "The more boring the subject, say sound physics, the greater the storytelling challenge."

Michael Nehring

Best Overall Faculty Member, Theater

Professor Michael Nehring is an accomplished actor, director, teacher, and choreographer. Michael is a founding member of several successful theatre companies, most recently Son of Semele Ensemble in Los Angeles. For Son of Semele, he played Napoleon in the Ovation Award winning musical Animal Farm. As a founding member of Shakespeare Orange County, Michael has been seen as Iago in Othello, Touchstone in As You Like It, Horatio in Hamlet (Dramalogue Award), Don Pedro in Much Ado About Nothing, the Fool in King Lear, Autolycus in Winter's Tale, Mark Anthony in Julius Caesar, Feste in Twelfth Night, Gratiano in The Merchant of Venice, a witch in Macbeth and as Caliban in The Tempest Michael was also a founding member of the award-winning Friends and Artists Theatre Ensemble (FATE)in Los Angeles. For FATE he appeared as the Herald in Marat-Sade (L.A. Weekly Award, Dramalogue Award), Semyon in The Suicide (Dramalogue Award), and Precious in Dolores and Her Loved Ones. Other roles include La Flech in The Miser and Pippin in Pippin at the Gem Theatre. He spent four seasons as a leading player with the Hollywood Theatre Ensemble in Pennsylvania and performed in over twenty musicals. He can be seen in the lead role of the 1998 feature film release Celestial Rhapsody, which was featured at the Berlin Film Festival.

Directing and choreography credits include If I Should Die Before I Wake, La Mirada Civic; West Side Story, San Jose Civic Light Opera; Looking Glass, Laughing Horse Repertory - Washington; Mame, the Gem Theatre; The Threepenny Opera, FATE and A Shakespearean Christmas for Shakespeare Orange County. Michael has served as an on-camera acting coach for The Disney Channel and for Sal Romeo Workshops in Los Angeles during the last fifteen years. He was awarded Chapman University Faculty of the Year in 1990 and a Chapman University Excellence in Creative Activity Award in 2000.

Directing credits at Chapman include: The Crucible, She Loves Me, The Creation, Cloud Nine, The Grapes of Wrath, Hair, The Madwoman of Chaillot, Playboy of the Western World, The Bacchae, Animal Farm, The Fantasticks, Brigadoon, Oklahoma and Die Fledermaus.

Janice Park

Best Teacher, Music

Janice Park, DMA, is an instructor of music and Coordinator of Secondary Piano at the Hall-Musco Conservatory of Music at Chapman University. She is also 1st Vice President and Program Chair of the Music Teachers' Association of California Orange County Branch. She received both a Masters degree and Doctorate of Musical Arts in Piano Performance from the University of Southern California and a Bachelor's degree in Piano Performance from Chapman University. She taught music theory and was the pianist for Women's Choir at Chapman. She has also been a faculty member at the Irvine Valley College, the Pasadena Conservatory of Music and an Assistant Lecturer at USC. Dr. Park has received numerous honors and awards.

Dr. Park has been active in solo, ensemble and collaborative recitals, including: performances with the concert master of the Pacific Symphony, as a soloist with Chapman Orchestra, Piano Trio recital at USC and chamber recital at the Colburn School in Los Angeles. This year, she was invited by Detroit Chamber Winds & Strings to perform with the Principal Trumpet of Los Angeles Philharmonic. In the past year she was the featured pianist for two separate world premiere concerts, one locally and one in South America. She is frequently sought to act as a clinician and as an adjudicator for competitions and piano festivals.

She is a member of Music Teachers National Association (MTNA), Music Teachers' Association of California (MTAC), and Pi Kappa Lambda National Music Honor Society. She has served as a member of the Board of Directors of the Musical Arts Club of Orange County. Her article on how to teach piano students was published in the piano magazine, Keyboard Companion.

Dan Pavelin

Best Teacher, Film
Most Helpful to Students, Film

If Dan Pavelin hears voices where others don't, it would be no surprise. After all, as an audio designer and engineer, Pavelin's training keeps him well attuned to the slightest nuance of pitch, the faint stumble at the beginning of a sentence, the hesitant breath that leaves just enough room for an edit. With a master's touch, he makes hundreds of corporate and commercial projects sound better than they could in real time by working his digital magic. An on-air personality as well as a composer (with an M.F.A. in Music, Composition and Technology emphasis), and a behind-the-scenes director, mixer and sound designer, Pavelin brings that same sensitivity to listening in the classroom. He teaches his students to hear what they never heard before and to find the power in sound that galvanizes or changes the shape of their film projects. But Pavelin also practices what he preaches, listening closely to the sounds of his students' interest—heading off boredom should it ever think to enter his classroom and putting real time to shame as his students are caught up in the sounds of Pavelin's magic.

Wendy Salmond

Best Overall Faculty Member, Fine Arts

Wendy Salmond is a scholar of Russian and early Soviet art, architecture, and design. She is particularly interested in exploring the intersection of diverse cultural traditions in Russia and in the formation of national identity. She has written and lectured extensively on the Arts and Crafts movement, on Art Nouveau, and on Russian modernism. Her current project is a book tracing transformations in the perception and function of icons in Russia, from objects of devotion to works of art.

Professor Salmond has been a visiting curator at Hillwood Museum and Gardens in Washington DC and a guest curator of exhibitions at Hillwood (Tradition in Transition: Russian Icons in the Age of the Romanovs, 2004) and The New York Public Library (Russia Imagined, 1825–1925: The Art and Impact of Fedor Solntsev, 2006). She is a prolific translator of texts on Russian art and culture, and has edited volumes on the sculptor Sergei Konenkov, the Bolshevik sales of Russian art in the 1920s and 1930s, and the reception of Art Nouveau in Russia.

Professor Salmond teaches courses on European modernism, graphic design history, gender and Western art, art history methods, and Russian art, design, and visual culture of the Imperial and early Soviet era. Her classes emphasize close attention to visual analysis, sustained discussion, and the development of research skills.

Pat See

Most Helpful to Students, Liberal Arts, Social Sciences, and Humanities

Pat See (Ph.D., Mississippi State University) is a Professor of Sociology & Director of the University Honors Program. She teaches HONORS 303: In Search of Relationships and oversees the Honors Forum class and all Honors activities, such as the fall retreat, social events and attendance at the National and Regional Honors Council conferences. See also teaches courses in microsociology, marriage and the family, sociology of consciousness, and holistic health. Both students and faculty bodies have recognized her with outstanding teacher awards. She has served as both department and division chair and been active on faculty, department, and academic committees, in addition to holding positions with community organizations such as Planned Parenthood, YWCA, Orange County Wellness Coalition and the Sierra Club. She has also served for the past twelve years as a volunteer at the California Institute for Women.

Myron Yeager

Best Teacher, English
Best Teacher, Liberal Arts, Social Sciences, and Humanities
Best Overall Faculty Member, Liberal Arts, Social Sciences, and Humanities

Myron Yeager, whose primary research interests are Samuel Johnson, Philip Dormer Stanhope (the Fourth Earl of Chesterfield), and Oscar Wilde, has published articles and reviews on eighteenth-century British literature and business communication and is co-author of the Hutton Story. He has presented papers at numerous international and national conferences, has served as president of the Samuel Johnson Society of Southern California and co-president of the Western Society of Eighteenth-Century Studies, and is currently secretary of the Samuel Johnson Society of Southern California and executive board member of the Western Society for Eighteenth-Century Studies. Additionally, he has served on editorial review boards for journals and texts in literature and business communication. Dr. Yeager was the recipient of a William Andrews Clark Postdoctoral Fellowship (UCLA), and a David Ross Doctoral Fellowship (Purdue University). Dr. Yeager teaches courses in eighteenth and nineteenth-century British literature and Gay, Lesbian, Bisexual, Transgendered (GLBT) Studies.

At Chapman, Dr. Yeager has served as dean of the School of Arts and Communication (formerly the School of Communication Arts), chair of the Department of English and Comparative Literature, chair of the faculty, and chair of numerous faculty committees. He was the founding director of Chapman's program in GLBT Studies. Dr. Yeager has received the Valerie Scudder Award for Excellence in Teaching and the Chapman University Alumni Faculty of the Year Award as well as Chapman's Excellence in Collaborative Teaching Award and Merit Award for Scholarly/Creative Activity.

CHARLESTON SOUTHERN UNIVERSITY

Nancy Barendse

Best Researcher/Scholar, English

Nancy Barendse earned her PhD from the University of South Carolina and her MA from Clemson University. She joined the English department in 1996. She is a regular reviewer at Sandwich-It-In and co-chair with Scott Yarbrough since fall 2009. She has presented papers at literature, composition, linguistic and popular culture conferences; her articles and reviews have appeared in Journal of Teaching Writing, The SECOL Review and the Concise Dictionary of Literary Biography. She holds professional memberships in the National Council of Teachers of English, Assembly for Literature for Adolescents, Council of Writing Program Administrators and Carolinas Writing Program Administrators. One of her favorite activities at CSU is tutoring in the Writing Center.

Nancy Canavera

Best Overall Faculty Member, Liberal Arts, Social Sciences, and Humanities

Nancy Canavera earned her MA from the Auburn University and completed additional graduate work at Michigan State University and The Citadel. Her professional interests revolve around improving skills in composition, grammar and critical thinking among high-risk, nontraditional and first generation college students. As a Peace Corps volunteer, she developed a strong interest in global cultures and issues, world religions and peace. She is a member the National Association of Developmental English, the South Carolina Association of Developmental English and the National Council of Teachers of English. She cherishes the time spent with the women of Psi Delta Phi, a Christian women's sorority, as she serves as faculty advisor. Likewise, her work on the Cultural Diversity Committee helps to bring awareness of other cultures to students at CSU.

CHATTANOOGA STATE COMMUNITY COLLEGE

Juan Alonso Santillana

Most Helpful to Students, Languages

"As a new instructor for our college Juan Alonso has shown dedication, enthusiasm and passion for his work, and for the success of his students."

Julie Barcroft

Best Teacher, Liberal Arts, Social Sciences, and Humanities

Becky Cantrell

Most Helpful to Students, Liberal Arts, Social Sciences, and Humanities

Rachael Falu

Most Helpful to Students, English

Brian Hale

Best Overall Faculty Member, English
Most Helpful to Students, Liberal Arts, Social Sciences, and Humanities

Joel Henderson

Best Researcher/Scholar, Liberal Arts, Social Sciences, and Humanities
Best Overall Faculty Member, Liberal Arts, Social Sciences, and Humanities

Kathy Long

Best Overall Faculty Member, History
Best Researcher/Scholar, History
Best Teacher, History
Most Helpful to Students, History

Erica Lux

Best Researcher/Scholar, English

Linda Miller

Best Teacher, Liberal Arts, Social Sciences, and Humanities
Best Overall Faculty Member, Liberal Arts, Social Sciences, and Humanities

Bill Stifler

Best Teacher, English
Best Researcher/Scholar, Liberal Arts, Social Sciences, and Humanities

I was born in northeast Maryland, and grew up in York County, Pennsylvania, attending Red Lion Kindergarten, Canadochly Elementary, and graduating in 1972 from Eastern York High School. While a senior in high school, I attended Lancaster Bible College part-time in the evenings. I came to Chattanooga, Tennessee, in 1972 to attend Tennessee Temple University, where I received a Bachelor of Arts degree in 1976 (Major: Biblical Studies, Minor: Psychology). I received a Master of Religious Education: Christian School Administration in 1981 from Temple Baptist Theological Seminary, and in 1991, a Master of Arts in English: Writing from the University of Tennessee at Chattanooga. I began teaching at Chattanooga State as an adjunct in 1989 and full-time in

1992 in Developmental Studies. I now teach mythology, composition, and, occasionally, Intro to Humanities and creative writing in the Humanities & Fine Arts division. I am the division webmaster and also serve as webmaster for the Meacham Writers' Workshop.

Katheryn Thompson

Best Overall Faculty Member, Humanities
Best Researcher/Scholar, Humanities
Best Teacher, Humanities
Most Helpful to Students, Humanities

BA and Secondary Credential, History and Spanish, California State University, Chico. MA and PhD, Spanish, University of California, Chico.

Josh Tucker

Most Helpful to Students, Liberal Arts, Social Sciences, and Humanities
Best Overall Faculty Member, Liberal Arts, Social Sciences, and Humanities

CHRISTOPHER NEWPORT UNIVERSITY

Lynn Shollen

Best Teacher, Leadership
Most Helpful to Students, Leadership

"Many students say she is the best."

CITRUS COLLEGE

Theresa Villeneuve

Best Overall Faculty Member, Communication

"She is the most awesome."

CITY COLLEGE OF NEW YORK

Matt Mead

Best Overall Faculty Member, English
Best Researcher/Scholar, English
Best Teacher, English
Most Helpful to Students, English

Harriet F. Senie

Best Overall Faculty Member, Art History
Best Researcher/Scholar, Art History
Best Teacher, Art History

Harriet F. Senie is director of the MA program in museum studies and teaches the required museum studies seminars in that area. She is the author of The 'Tilted Arc' Controversy: Dangerous Precedent? (2002), Contemporary Public Sculpture (1992), and co-editor of Critical Issues in Public Art (1992; 1998), as well as numerous articles and essays on public art, memorials, and audience response. She is currently working on a book tentatively titled Transforming American Memorials: From Symbolic Cemetery to a Light in the Sky: Vietnam to 9/11.

Nicole Treska

Best Overall Faculty Member, English
Best Overall Faculty Member, Liberal Arts, Social Sciences, and Humanities

"Nicole Treska is creative, good humored, extremely competent, and dedicated."

Christian Volkmann

Most Helpful to Students, Fine Arts

"Prof. Volkmann is a tireless educator whose efforts have benefited our students in receiving an exceptional education."

CLAREMONT GRADUATE UNIVERSITY

Ingolf U. Dalferth

Best Researcher/Scholar, Religion

Ingolf U. Dalferth has studied theology, philosophy, and linguistics in Tübingen, Edinburgh, Vienna and Cambridge. His major areas of research and teaching are philosophical and theological hermeneutics in the 20th century; systematic theology in the 19th and 20th century; ecumenical theology (Lutheranism and Anglicanism); analytic and phenomenological philosophy of religion in the 20th century; subjectivity theory; religion and emotion; altruism; trust; and evil. From 1998 to 2012 he directed the Institute of Hermeneutics and Philosophy of Religion, located in Zurich where he is also Professor Emeritus of Systematic Theology, Symbolism and Philosophy of Religion.

Tammi Schneider

Best Overall Faculty Member, Religion
Best Teacher, Religion
Most Helpful to Students, Religion

Tammi J. Schneider teaches ancient Near Eastern History, literature, archaeology and religion and also women in the Hebrew Bible. Her research draws together the varied fields of archaeology, Assyriology and biblical studies in an effort to understand the ancient Near East, especially the interactions among various peoples. She received her doctorate in Ancient History from the University of Pennsylvania. Her books include; Judges in the Berit Olam series, Sarah: Mother of Nations, Mothers of Promise: Women in the Book of Genesis, and An Introduction to Ancient

Near Eastern Religion. She was the editor for the ancient Near East section of the journal Religious Studies Review, Vice President of the American Schools of Oriental Research and edits the series People of the Ancient World for the Society of Biblical Literature. She has worked on numerous archaeological excavations in Israel and is presently the head of the educational program at the Tel Akko Excavations where she excavates every year with students from CGU and the Claremont Colleges. Along with her teaching responsibilities, she presently serves as Dean of the School of Arts and Humanities.

CLARION UNIVERSITY OF PENNSYLVANIA

Mary Pat McCarthy

Best Overall Faculty Member, Communication Disorders

"An outstanding colleague who is flexible, focused on the needs of the department and university and works well with students."

Mary Pat McCarthy, D.Ed., CCC/SLP is an associate professor in the Department of Communication Sciences and Disorders at Clarion University. She has also served as a clinical supervisor in the Department of Communication Sciences and Disorders since 1994. She has presented at both regional state and national conferences on topics pertaining to clinical pedagogy/supervision, and clinical practice issues in a variety of areas including adult and child language and speech disorders. She has published articles in the ASHA SIG 11 Perspectives in Administration and Supervision and served as a column coordinator from 2003–2011. She is a past Vice President of Publications for the Pennsylvania Speech language Hearing Association. She is a member of ASHA, PSHA, and ASHA SIG 11 Administration and Supervision, SIG 16 Schools and SIG 14 Culturally Linguistic Diverse Populations.

CLAYTON STATE UNIVERSITY

Ron Jackson

Best Teacher, Philosophy

CLEMSON UNIVERSITY

Alton Brant

Most Helpful to Students, Languages

Toshiko Kishimoto

Best Overall Faculty Member, Languages
Best Teacher, Languages

"Excellent in teaching, service, and section coordination."

Toshiko Kishimoto started the Japanese program at Clemson in 1988. She has been coordinating the program since then. She teaches all levels of Japanese, but mainly 300 and 400 levels. Her primary research interests are Japanese pedagogy, Japanese business culture and heritage Japanese language. Her recent publications include: Chapter 3 of Japanese Language for Professional Purpose (2010) and Chapter 7 of Japanese Children in the United States: Light and Shadow of Bilingualism (2008). Her contribution to Curriculum Development of Business Japanese (2012), Chapter 4, is upcoming. She has served as CLIP director since 2006.

She received the Excellence Award in Teaching by Clemson University Student Government in 2008. She was also recognized by the Consulate General of Japan in Atlanta for her contribution to US-Japan relations, especially for her efforts as founder and principal of the Japanese Saturday School in Greenville, SC.

Jeff Love

Best Researcher/Scholar, Languages

Jeff Love is Professor of German and Russian. He has a Ph.D. in Russian literature from Yale University and a law degree from Queen's University in Kingston, Canada. Jeff is interested in the classic Russian Novel, German Idealist philosophy and literary theory. He has written two books on Tolstoy, Tolstoy: A Guide for the Perplexed (Continuum, 2008), and The Overcoming of History in War and Peace (Rodopi, 2004). He has also published an annotated translation of F.W. J. Schelling's Philosophical Investigations into the Essence of Human Freedom (State University of New York Press, 2006) with Johannes Schmidt as well as articles on Pushkin, Dostoevsky and the French philosopher Alain Badiou, among others.

Lee Morrissey

Best Overall Faculty Member, Liberal Arts, Social Sciences, and Humanities

"Teaching and research"

Lee Morrissey's work focuses on relationships between literature and intellectual history, political philosophy, and theory, largely in the seventeenth and eighteenth centuries. His main areas of interest include: the history and theory of literary criticism, Restoration and eighteenth-century English literature, the Enlightenment, the early English novel, early democratic political theory, and relationships among the arts.

Graciela Tissera

Best Overall Faculty Member, Languages
Best Overall Faculty Member, Liberal Arts, Social Sciences, and Humanities

"I have the pleasure to nominate Dr. Graciaela Tissera for the effort and dedication in the Department of Languages at Clemson University.

I've never worked with such an intelligent and dedicated person at work. I am proud to work with her and learn as much I can."

Graciela Tissera completed her graduate studies in Spanish, Literature, and Latin, with honors, at Universidad Nacional de Córdoba, Argentina, and received her Ph. D. in Romance Languages and Literature from the University of Pennsylvania. Dr. Tissera's research and teaching interests include Hispanic and comparative literature and film, literary and critical theory, and Spanish for the professions. Her Creative Inquiry projects focus on international health, global business, film studies, and service learning activities. Students in her classes present their research at national and international professional conferences. As Director of the L&IH Program, she coordinates the L&IH Symposiums and the study abroad program in the Dominican Republic and Spain. Dr. Tissera is the General Editor of Textos, Works & Criticism and also serves as a Contributing Editor for national and international literary journals. She has been recognized as an outstanding professor by Clemson University Student Government, and her "Introduction to Hispanic Literary Forms" course was selected as one of the ten best courses in the national study of Spanish literature courses conducted by the Education Policy Improvement Center (EPIC). Dr. Tissera was inducted into the Order of the Discoverers, one of the National Collegiate Hispanic Honor Society's highest honors. She was also awarded the Certificate of Service by the State of South Carolina.

CLEVELAND STATE UNIVERSITY

Sanda Kaufman

Best Researcher/Scholar, Urban Studies

"She is widely recognized internationally as a prolific researcher on conflict resolution and has presented her research at many academic conferences."

Education: Ph.D., Public Policy Analysis, Heinz School at Carnegie Mellon University, 1985 B.A., Architecture, Technion, Israel Institute of Technology, 1975 M.S., City and regional planning, Technion, 1977

Sanda Kaufman is Professor of Planning, Public Policy and Administration. She holds degrees in Architecture and in City and Regional Planning from the Technion. She received her Ph.D. in Public Policy Analysis from the Heinz School at Carnegie Mellon University. Her research field is public decision making. Her research focus is negotiation and intervention in the urban, environmental, and organizational. She has mediated small claims disputes, designed dispute resolution systems and mediation training materials, worked on organizational conflict management, and facilitated community consensus building. Her articles have been published in the Journal of Planning, Education and Research; the Journal of Architectural Planning and Research; the International Journal of Conflict Management; the Negotiation Journal; Evaluation and Program Planning; Environmental Practice; Conflict Resolution Quarterly; Revue Négociations; International Journal of Economic Development; Public Works Management and Policy; and Fractals.

Professor Kaufman is Director of the Master of Arts in Environmental Studies Program and she chairs the Quantitative Methods Group at the Levin College. She teaches courses in quantitative reasoning, negotiation and conflict management, environmental policy, environmental dispute resolution, and strategic planning. She has built and is maintaining web resource pages for Planning, Public Administration, Conflict Management, Environment, Decision Making, Research Tools, and Teaching. Her architectural photography – assembled into a web-based Image Collection for the Levin College - has been requested internationally for journals, posters, books and websites.

Wendy Kellogg

Most Helpful to Students, Urban Studies

Wendy A. Kellogg is an Professor of Urban Planning and Environmental Studies. Dr. Kellogg earned her Ph.D. (City and Regional Planning) at Cornell University in 1993. Her major fields of research interest are local government environmental protection, urban and regional sustainability, neighborhood sustainability, and Great Lakes water quality and land use issues. She has published analyses of Great Lakes water quality planning programs and citizen participation in neighborhood-based environmental planning. Her research projects have included an environmental history and inventory of a neighborhood in central Cleveland, Ohio, the role of local decision makers in coastal and watershed protection, and the role of training programs in shaping local decision maker behavior toward coastal management. She currently is conducting research on collaborative environmental planning processes, including governance systems in the Chagrin River Valley in Northeast Ohio. Dr. Kellogg was a USEPA-funded Fellow at the CSU Program on Risk Analysis in 2000–2001. Dr. Kellogg was a member of the Ohio Lake Erie Commission Blue Ribbon Taskforce on Balanced Growth in the Lake Erie Basin in 2001–2004 and has continued her work with the Balanced Growth program in a research and technical advisory role. She currently represents Cleveland State on the Cleveland Water Alliance. Dr. Kellogg teaches courses in urban planning, environmental planning, environmental policy, and urban studies.

Robert Simons

Best Overall Faculty Member, Urban Studies

"Rob teaches at the graduate and undergrad levels. He coordinates the MUPDD Program's Real Estate Development and Finance specialization. He conducts two competitions annually with cash prizes from the real estate industry for his best students. He and his family established a student scholarship. He has hired many students to work on various projects. He served as the president of ARES. He was a Fulbright Scholar in South Africa and a Lady Davis Scholar in Israel."

Robert A. Simons is a Professor and former director of the Master of Urban Planning, Design and Development program at the Levin College of Urban Affairs at Cleveland State University in Cleveland, Ohio. He is also the faculty advisor for the Certificate Program in Real Estate Development and Finance, offered in conjunction with the Nance College of Business at CSU. During Fall 2005, Dr. Simons was a Fulbright Scholar at Wits University in Johannesburg, South Africa.

Dr. Simons received his Ph.D. from the University of North Carolina at Chapel Hill in City and Regional Planning, with an emphasis in real estate. He also holds a Master of Regional Planning and a Master of Science in Economics, both from U.N.C. His undergraduate degree in anthropology was earned at Colorado State University. He has been a member of the American Institute of Certified Planners (AICP) since 1983. Dr. Simons is in the inner leadership group of the American Real Estate Society (ARES), and will be program chair in 2009–2010 and President in 2011–2012.

At the Levin College of Urban Affairs, Dr. Simons teaches courses in real estate development, market analysis and finance, public economics, Ph.D. research methods and environmental finance. Dr. Simons has published over 45 articles and book chapters on real estate, urban redevelopment, environmental damages, housing policy and brownfields redevelopment. He authored a book entitled Turning Brownfields into Greenbacks, (published by Urban Land Institute), and When Bad Things Happen to Good Property, (published by Environmental Law Institute in 2006), and was the lead editor for an international research monograph on Indigenous Property and Valuation (2008, ARES). Another Adaptive Reuse book by ULI on New USes for Religious Buildings and Schools is due out in 2010. Dr. Simons has an active consulting practice, and has served as an expert witness in over 50 matters related to real estate, housing markets, and environmental contamination.

Winifred Weizer

Most Helpful to Students, Urban Studies

"Win has long helped students deal with statistics and research methods, which many find daunting. She has gone out of her way to tutor them, as well as teaching courses on these subjects."

After a successful career in private industry, Ms. Weizer returned to Cleveland State and completed her Master in Public Administration with a concentration in Public Works Management in 1997. She is currently working towards a Phd in Urban Studies and Public Affairs with a concentration of public administration and public policy. Teaching Areas: Statistics, Urban Studies, Budgeting, Management.

Professional Affiliations: American Society for Public Administration Government Finance Officers Association American Public Works Association.

University Service: Cleveland State University CERT, Treasurer Academic Technology Committee.

COASTAL CAROLINA UNIVERSITY

Jose Luis Mireles

Most Helpful to Students, Languages

"This professor truly cares about his students and their success. Jose Luis Mireles always encourages student participation in the Foreign Language Instructional Center and other language centered campus events. Students of his who have received help in the center always comment positively on his teaching style and appreciate the efforts he makes to help them succeed in language study. I have had the opportunity to take his Spanish 211 course and know first hand that his classroom activities and teaching style are a reflection of his compassion and enthusiasm for teaching. He is always prepared, organized and provides a learning environment that is positive for student learning. He is also an exceptional colleague and faculty member."

Jose Luis Mireles was born and raised in Ensenada, Baja California, Mexico. He attended the University of California, Davis where he double majored in Political Science and Spanish. He continued his career in the Spanish language and earned an MA in Spanish also from UCD. Jose Luis joined the CCU faculty as an Instructor of Spanish in 1999.

COLLEGE OF CHARLESTON

Vincent Benigni

Best Overall Faculty Member, Communication

Vince Benigni has more than two decades of experience in public relations and mass media. He regularly teaches Addressing Problems in Context (Comm 336) and Media in the Digital Age (Comm 214), as well as courses in the graduate program. From 2009–2013, Dr. Benigni was the director of the master's degree program in Communication at the College. He was also appointed by the President in 2009 to be the College's Faculty Athletics Representative, just the second in the school's history. In 2013–14, he will serve as a Faculty Administrative Fellow, helping the College's transition to the Colonial Athletic Association and its Colonial Academic Alliance.

Dr. Benigni received his Ph.D. in Mass Communication at the University of Georgia (UGA) in 1999. While at UGA, he served a three-year research assistantship with the prestigious George Foster Peabody Awards, considered the world's premier broadcast award. A summary of his dissertation, a study of external communication strategies in sports heritage venues, was presented at the International Association of Sports Museums and Halls of Fame Conference in Fall 1999. In Summer 1998, Dr. Benigni co-produced a video for the Ty Cobb Museum (Royston, GA) that was a finalist for the prestigious Vision Award. His previous degrees came from Indiana University of Pennsylvania (Journalism) in 1983 and Slippery Rock University of Pennsylvania (English) in 1990.

For 10 years, Dr. Benigni was a sports information director (SID), beginning with a two-year stint as an assistant SID at the U.S. Naval Academy from 1986–88. Among his duties was being the primary media contact for National Player of the Year David Robinson in men's basketball. From 1988–96, he was SID for a 22-sport program at Mary Washington College (VA).

Charles Calvert

Best Overall Faculty Member, Theater
Best Teacher, Theater
Most Helpful to Students, Theater

Charlie Calvert teaches Scene Design, Drafting, Rendering and Scene Painting. He has recently designed The Love of the Nightingale and Letters to Sala for the College. He also mentors student designers on productions for the Department of Theatre and Dance. Charlie, a member of United Scenic Artists, has designed scenery professionally for many theatre companies in New York City and across the country, including multiple projects at The Shakespeare Theatre of New Jersey. He also spent years working in NYC as an Assistant Scenic Designer for Broadway, Off Broadway, and International Opera productions.

Steve Della Lana

Best Teacher, Languages

Born in New Jersey and with an undergraduate degree from Rutgers University, Mr. Della Lana's association with CDS has been over 20 years in the making. He was a participant in the prestigious Congress-Bundestag Exchange Program for Young Professionals (1989–1990) in which he did two months of intensive German language training in Cologne, followed by a semester at the University in Erlangen and then a six month internship at a small, private language school and translation service in Nürnberg. There, he taught Business English, coordinated the English language instructors and did translations for a wide variety of clients. He was hired on full time and remained in Germany until 1994, when he returned to the U.S. to pursue other opportunities and a M.A.T. degree from the University of South Carolina.

Upon graduation from the University of South Carolina in 2000, he was hired by the Department of Classics, German, Italian, Japanese and Russian at the College of Charleston as a visiting instructor and then as an Instructor until he was promoted to rank of Senior Instructor in 2006. His specialty is Business German, which he teaches every other year. He set up the only testing center in the state for the Goethe Institute's battery of German proficiency examinations, serves as the faculty advisor of the German Club and, in his role as the coordinator for the Summer Internship Program in Germany, recruits, promotes and visits student participants. Mr. Della Lana works closely with the School of International Business to make business contacts from German companies in South Carolina, which has produced several internships for students at the German American Chamber of Commerce, Weber AG, Ifa Rotorion, among several others. Mr. Della Lana has written several book reviews in the field of Business German which appeared in the journal of the American Association of Teachers of German (AATG) die Unterrichtspraxis. He is the current president of the AATG South Carolina Chapter and represents the interests of South Carolina German teachers in other Foreign Language Teaching Associations, both on the state and national levels.

Mr. Della Lana was recently inducted into the College of Charleston's Global Scholars program. He has participated in the First Year Experience program, the Faculty Technology Institute as well as numerous professional development seminars and workshops through the years. He has a particular interest in the division and reunification of Germany, and helped organize a 20th anniversary commemoration of the fall of the Berlin wall, and has guest lectured on this topic.

Mr. Della Lana's other area of interest is German-English Business translation, for which he earned a supplemental certification. He has monitored students doing translation projects for Motley Rice LLC, worked as an independent auditor for BMW's German language program in Spartanburg, and does translations when time permits. He has twice taught an upper level special topics course "German Business Translation", to which he invited guest speakers from various industries employed at German companies throughout the state to speak to students about unique translation issues.

Sara Frankel

Best Overall Faculty Member, Fine Arts
Best Researcher/Scholar, Fine Arts
Best Teacher, Fine Arts
Most Helpful to Students, Fine Arts
Best Researcher/Scholar, Fine Arts
Best Teacher, Fine Arts
Most Helpful to Students, Fine Arts
Best Overall Faculty Member, Fine Arts

Sara Frankel was born in Peoria, IL and holds a Bachelor of Fine Arts degree from the University of Iowa (1980) and a Masters Degree of Fine Arts in Painting from Yale University (1983).

She is currently an Assistant Professor of Drawing at the College of Charleston. Her teaching experience also includes posts at Dickinson College, Gettysburg College, University of Hawaii, Wesleyan University, and University of Connecticut.

Her exhibition background includes solo exibitions at the Honolulu Contemporary Museum, College of Charleston's Halsey Gallery, Franklin and Marshall University, Gettysburg College and several invitational and group shows in NYC and Washington D.C. Sara lectures nationally on her work.

Susan Kattwinkel

Best Researcher/Scholar, Theater

Amanda Ruth

Most Helpful to Students, Communication

Dr. Ruth-McSwain came to the College of Charleston from the University of Florida, where she earned her Ph.D. in Agricultural Communication. During her graduate work, Dr. Ruth-McSwain received the College of Agricultural and Life Sciences Jack Fry Teaching Excellence Award and the University of Florida's 2005 Graduate Student Teacher of the Year Award for her role as a graduate instructor.

She has received several recognitions, including Outstanding Graduate Student Research Paper for two consecutive years from The Association for Communication Excellence, as well as Outstanding Faculty Research Paper in 2005 and 2006. Her dissertation, titled "Sources of Agricultural News: The Media Relations Environment of Agricultural Communication Professionals" won the 2006 Outstanding Dissertation Award from the Association for Communication Excellence. In 2009, Dr. Ruth-McSwain received the article of the year distinction from the Journal of Applied Communication.

Dr. Ruth-McSwain has a co-authored chapter in The Practice of Government Public Relations as well Learning to Teach: What You Need to Know to Develop a Successful Career as a Public Relations Educator. She also co-authored the instructor's manual and test bank for the popular Public Relations Strategies and Tactics text.

Under her direction, Dr. Ruth-McSwain's student won awards from both the American Marketing Association and the Public Relations Society of America for their campaign work.

Three fun facts: two kids, dog lover, loyal chai drinker

Christopher Warnick

Best Researcher/Scholar, English

Robert Westerfelhaus

Best Teacher, Communication

Robert Westerfelhaus has been awarded a Fulbright Lectureship for 2009–2010. He will serve as the Distinguished Fulbright Chair at the Instytut Anglistyki of the Maria Curie-Sklodowska University, located in Lublin, Poland. He will teach courses in film, mass media, and American popular culture.

Dr. Westerfelhaus came to the College of Charleston from the University of Houston-Downtown (UHD), where he taught as an assistant professor for three years. He received his Ph.D. and M.A. from Ohio University (OU), from which he earned two awards for rhetorical scholarship and was named Graduate Student of the Year (1998). Dr. Westerfelhaus earned his B.A. from Ohio Dominican University (ODU).

COLLEGE OF SOUTHERN NEVADA

Duprey-Smith

Best Teacher, English

Kathy Baker

Best Overall Faculty Member, English

Jennifer Basquiat

Best Researcher/Scholar, Liberal Arts, Social Sciences, and Humanities

CSN Communication Professor Dr. Jennifer Basquiat has done extensive field research on a wide range of ethnographic, religious and anthropology related topics, including Haiti, Vodo and Mormonism. She is currently exploring the controversial issue of polygamy as practiced by one of many fundamentalist Mormon groups in Utah and Arizona. Dr. Jennifer Basquiat received her bachelor's and masters from California State University, Los Angeles in Communication Studies and a second masters as well as a Ph.D. from Claremont Graduate School in Cultural Studies. Her doctoral dissertation received the "Dissertation of the Year" award from the Religious Communication Association and portions of her work have also been published in Dialogue: A Journal of Mormon Thought. In addition, she has had her research regarding Mormon feminism published in The Harvard Divinity School's Journal of Feminist Studies in Religion for which she was awarded the journal's prestigious New Scholar Award.

Levia Dinardo Hayes

Best Overall Faculty Member, English

"Levia is both a teacher and department chair. I've worked with her closely and know her concern for educational quality and for her students as individuals is unsurpassable. In the two years she has chaired the department, she has completely revitalized it. Her support of each faculty member's professional efforts is incomparable, and her personal concern for each faculty member is so outstanding that whenever I speak of her to people outside the college, they ask how they can get her as a boss. She is extremely active in college committees of all types and on all levels. Faculty is convinced she never sleeps, yet she is the most energetic, up-beat person I know. She is the perfect role model, and I strive to be as much like her as I can. In my 15 years teaching at community colleges, I've never met her peer, both in the classroom and outside of it. I urge you to give her the recognition she as earned by presenting her with this award."

Shelley Fischer

Most Helpful to Students, Liberal Arts, Social Sciences, and Humanities

Shelley Fischer, a tenured professor of English at CSN, has been the subject of a number of news articles for her knowledge about the world of astrology.

A member of the American Federation of Astrologers since 1978, Shelley has been practicing astrology for more than 40 years. She has a local and international astrology business and previously had her own talk radio show on KDWN on astrology in the 1990s. She has taught astrology classes throughout Las Vegas, including CSN non-credit courses.

She has appeared on several radio and television programs as a guest astrologer and lectured nationally to organizations about the subject. She has been honored by the Distinguished Women of Southern Nevada for her work in education and astrology.

She has taught and worked at the University of Nevada, Las Vegas, including a position as academic advisor to the Rebels Men's Basketball team. She is also the former Director of Planned Parenthood of Southern Nevada.

Linda Gannon

Best Overall Faculty Member, Liberal Arts, Social Sciences, and Humanities

Kevin Mitchell

Best Overall Faculty Member, Communication
Best Teacher, Liberal Arts, Social Sciences, and Humanities

Kevin has a Masters degree from Northern Illinois University and completed doctoral coursework from Arizona State University. His interests include Interpersonal Communication, Intercultural Communication and Rhetoric. Additional academic specialty in humor studies.

Jennifer Nelson

Best Researcher/Scholar, English

Deborah Okey

Most Helpful to Students, English

Patricia Vasquez

Best Teacher, Liberal Arts, Social Sciences, and Humanities

COLLEGE OF STATEN ISLAND (CUNY - STATEN ISLAND)

Jeffrey Bussolini

Most Helpful to Students, Liberal Arts, Social Sciences, and Humanities

Jeffrey Bussolini was educated at Georgetown, CUNY Graduate Center, the Sorbonne (Université de Paris I) and the École des Hautes Études en Sciences Sociales, and is a sociologist, philosopher, and historian of technology by training. His areas of inquiry are: ailourography (etho-ethnography), ethnography of national security institutions, ancient and contemporary philosophy (including presocratics and critical philosophies of violence, war, and state power), and television studies. He has conducted ethnographic and historical study of Los Alamos and related nuclear security sites since 1991 and etho-ethnographic study of feline-human interactions since 1995. He has translated French and Italian works for English publication. He has been visiting professor at Macquarie University (CRSI, CISAB) and University of New South Wales (Environmental Humanities).

Gerry Milligan

Best Teacher, Languages

Gerry Milligan authored and co-directed the passage of CSI's new Italian Studies Major, officially launched in 2010. His research interests are Italian Renaissance literature, gender studies and women writers of late-medieval/Renaissance Europe. In addition to his classes on Italian language, he teaches courses on Medieval and Renaissance literature and culture. He encourages students to investigate literature as productions of cultures that are distant in time, yet have informed contemporary life in interesting and unexpected ways. Professor Milligan often focuses on social histories of the past in tandem with literary representations in order to question our assumptions of pre-modern life. Students have specifically enjoyed his teaching modules on women's makeup, Renaissance dress, and sexuality. Taking advantage of New York's cultural richness, he organizes trips to the Cloisters Museum, the Frick collection, and Italian cinema events. Additionally, he encourages students to travel beyond New York, especially to study abroad programs in Italy. While he was a student himself, Professor Milligan participated in several study abroad programs in both Italy and France and then later went to work in Italy as a coordinator for several universities' programs. Because of his sincere interest in study abroad, he has been an active participant in the development and continuation of CSI's "Isola Scholarship," which has sent several deserving students to Italy.

Ananya Mukherjea

Best Overall Faculty Member, Liberal Arts, Social Sciences, and Humanities

Ananya Mukherjea received her PhD in Sociology from the City University of New York Graduate Center and her BA from New College, Florida. She joined CSI's program in Women's, Gender, and Sexuality Studies and the Department of Sociology, Anthropology, and Social Work in the fall of 2004. Her research interests include: the study of gender and sexuality; medical sociology (the social politics of infectious disease epidemics and their management, particularly viral epidemics); urban sociology; the sociology of culture and popular culture; and the study of animals in society. Dr. Mukherjea teaches courses on gender studies, urban sociology, community studies, and the sociology of culture at CSI, and she has co-taught the Introduction to LGBTQ Studies at the CUNY Graduate Center with Dr. Paisley Currah. Dr. Mukherjea also is a member of the faculty for the doctoral program in public health at the CUNY Graduate Center.

Peter Simpson

Best Researcher/Scholar, Philosophy

Dr. Peter Simpson is a Full Professor for the Philosophy Department. Dr. Simpson is also a Professor of Philosophy and Classics at the CUNY Graduate Center. He has taught at the Catholic University of America, Washington DC; University College, Dublin, Ireland; and Manchester Polytechnic, United Kingdom.

Saadia Toor

Best Teacher, Liberal Arts, Social Sciences, and Humanities

Dr. Toor's scholarship revolves around issues of culture, nationalism, gender/sexuality, state formation, and international political economy. Her book, The State of Islam: Culture and Cold War Politics in Pakistan was published by Pluto Press in 2011. A special issue of Women's Studies Quarterly on the theme of Solidarity co-edited by Dr. Toor was published in November 2014.

COLLEGE OF THE HOLY CROSS

Lynn Kremer

Best Overall Faculty Member, Theater

Ms. Kremer's works have been performed in Minneapolis, Boston, Fort Worth, Chicago, Washington D.C. and Ireland. As a director and librettist she is particularly known for the development of new plays. Representative productions include a work titled Journeys Through Imaginary Landscapes that was inspired by the sculpture of Isamu Noguchi, the prints of Hiroshige, and by Haiku poetry. Journeys was granted two month-long residencies at the Smithsonian Institute and several performances at The Walker Art Center in Minneapolis. She directed Crawl Space, a performance piece, for the Boston Center for the Arts (BCA) at the Mills Gallery and The Separate Prison: A Jazz Opera at the BCA's Cyclorama.

Ms. Kremer has received grants and awards for her work from the National Endowment for the Arts, The New England Foundation for the Arts, the Massachusetts Cultural Commission, the Worcester Arts Lottery, the LEF Foundation, the Cambridge Multicultural Arts Center and Holy Cross College. With funds provided by the National Endowment for the Arts/Dance

Program, Ms. Kremer lived in Micronesia where she produced and directed a video program on the chant/dance traditions of Palau. She has conducted research on Javanese classical dance and Balinese topeng (mask/dance/drama), where she studied with acclaimed mask dance I Gusti Ngurah Windia. She has also traveled extensively and studied the performing arts traditions of Nepal and India.

Ms. Kremer has collaborated with composer, Shirish Korde as director of the video for his piece for solo flute Tenderness of Cranes, performed by Jean DeMart. They have collaborated on the chamber opera Rasa, which has been funded by the NEA Opera program and has been performed in the Boston area and in Fort Worth and is recorded on Neuma compact discs; Bhima's Journey, a contemporary music theatre piece based on an episode from the Mahabharata, and on The Separate Prison: A Jazz Opera.

Scott Malia

Best Teacher, Theater

Scott Malia is an Assistant Professor of Theatre who specializes in GLBTQIA Theatre, Comedy, Television, and Italian Theatre. His book, Giorgio Strehler Directs Carlo Goldoni was published by Lexington Books in 2013. Other articles and writings have appeared in Theatre Journal and New England Theatre Journal. His play, The Interview, was performed as part of the Boston Theatre Marathon in 2009 and published by Smith & Krauss. The play was also performed in Ireland as part of the Cork Theatre's 10x10 Festival in 2013.

At Holy Cross, Scott has taught Basic Acting, Theatre History I & II, Gay Theatre & Film, Comedy on Stage and Screen, 17th-Century Italian Theatre, and Commedia in Performance. His productions for the College include The Servant of Two Masters (which he also translated into rhyming verse from Italian), Hay Fever, and The London Merchant. Scott is a member of the Association for Theatre in Higher Education, the American Society for Theatre Research, the New England Theatre Conference and the Popular Culture Association.

COLLEGE OF THE REDWOODS

Paul Geck

Best Teacher, History

COLLEGE OF WESTERN IDAHO - NAMPA

Michelle Bennett

Most Helpful to Students, Communication

Michelle Bennett has been teaching at the College of Western Idaho in Communication since 2008 and as a full-time instructor since January 2010. She received her B.A. in Communication and a Master of Arts in Communication from Boise State University. Michelle is completing her PhD from the University of Idaho in Educational Leadership and is currently ABD.

During Bennett's tenure at CWI she has led the Speech and Debate team to three Pi Kappa Delta national championships. Through her work, PKD honored Bennett with the 2012 Derryberry Forensics Educator award. This is presented to an educator directing forensics for less than five years. At the 2013 PKD National Convention Bennett was also awarded the Rising Star Award. Bennett has received recognition from the Phi Theta Kappa CWI student organization and CWI as Professor of the Year in 2011. Michelle has served on numerous college-wide committees and supports student growth through development of Communication internship opportunities throughout the Treasure Valley.

Bennett is engaged in the community and volunteers as the Chair of the Canyon County Friends of the Fair Foundation. Bennett loves gardening, cooking, and horseback riding with husband Mark. Michelle has three children Brandon (Kuna Fire Fighter), Alexis (Editor), and Mitchell (US Army).

Stephen Crumrine

Best Overall Faculty Member, English
Best Teacher, English
Most Helpful to Students, English
Best Researcher/Scholar, Liberal Arts, Social Sciences, and Humanities
Best Teacher, Liberal Arts, Social Sciences, and Humanities
Most Helpful to Students, Liberal Arts, Social Sciences, and Humanities
Best Overall Faculty Member, Liberal Arts, Social Sciences, and Humanities

Stephen Crumrine earned his B.A. in English from Linfield College and his M.F.A. in Creative Writing, Poetry from the University of Montana. Stephen serves as the Coordinator for the College Readiness Project, a program developed to help at-risk high school students earn college credit through CWI Dual Credit courses. He also teaches English composition and creative writing at CWI. Stephen is passionate about many things, including rowdiness, the South Fork of the Boise River, and shallots.

He lives on the outskirts of Boise, under a Sycamore tree, with his wife and boys.

Greg Heinzman

Best Researcher/Scholar, English

Johnny Rowing

Best Teacher, Communication

Johnny Rowing is a Communications instructor for CWI. He received both his Bachelor of Arts and Master of Arts in Communication from California State University Chico (2005, 2007). Johnny taught as a graduate assistant for CSUC and coached their nationally ranked Speech and Debate team. Johnny has taught for CWI since the fall of 2009 and has also served as a coach of CWI's two National Championship Speech and Debate teams.

Johnny served four years in the United States Air Force as an Avionics Technician on the F-16 Fighting Falcon. Although he enjoyed his time of service, it was always his dream to return to a Community College and teach. He is proud to be a part of this wonderful institution as it grows and meets the needs of the local community. Currently Johnny teaches fully online and face-to-face sections of Communication 101 (Fundamentals of Oral Communication), Communication 209 (Critical Thinking and Argumentation), and coaches Comm 105 (Intercollegiate Tour Speaking).

COLLIN COLLEGE

Linda Bennett

Best Overall Faculty Member, English

Betty Bettacchi

Most Helpful to Students, English

Debra Frazier

Best Teacher, English

Rozelle Gilman

Best Researcher/Scholar, Liberal Arts, Social Sciences, and Humanities

Diana Gingo

Best Teacher, English

Frozina Goussak

Best Teacher, Languages
Most Helpful to Students, Languages

I teach different levels of Spanish including Spanish Literature. I am originally from Ukraine. Spanish became my native language after Ukrainian and Russian. I lived in Havana, Cuba where I graduated from High School "Ciudad escolar Libertad" in Havana, Cuba. Upon returning to Ukraine, I studied in National University T.G. Shevchenko, Kyiv, Ukraine, department of Romance and Germanic languages were I earned a MA in Spanish language. I studied French, Italian and Portuguese. I taught in the National University T.G. Shevchenko and Kyiv and Pedagogical University of Foreign Languages, for twelve years. I am teaching in Collin College full time professor since 1994. My fascination with languages goes back to my childhood when I first listened to Spanish music. At that very moment, I decided to share my sentiments and my love for the Spanish language with as many people as I could. Teaching is my passion. All my life I have been dedicated to search for the best method by which to teach Spanish. Music enables me to awaken the students' interest, not only in language, but also in the culture of Spanish-speaking countries. My childhood dream has become a reality. Now I am here sharing my "Spanish fever" with my students.

Shannon Kearns-Simmons

Best Teacher, Liberal Arts, Social Sciences, and Humanities
Best Overall Faculty Member, Liberal Arts, Social Sciences, and Humanities

Melody Lucas

Most Helpful to Students, Liberal Arts, Social Sciences, and Humanities

Shirley McBride

Best Researcher/Scholar, English

COLORADO COLLEGE

Daniel Arroyo Rodriguez

Best Overall Faculty Member, Liberal Arts, Social Sciences, and Humanities

COLORADO SCHOOL OF MINES

Tonya Lefton

Most Helpful to Students, Liberal Arts

COLORADO STATE UNIVERSITY

Jane Kneller

Best Overall Faculty Member, Philosophy

Katharine E. Leigh

Best Researcher/Scholar, Fine Arts

Katharine E. Leigh, PhD, Associate AIA, LEED AP BD+C, NCIDQ is a Professor in the Interior Design program at Colorado State University, ranked #6 of 244 US programs in interior design offering a bachelor's degree. She holds a doctorate in Organizational Performance and Change focusing on creativity and performance in architectural practice. Her dissertation, entitled ORGANIZATIONAL CREATIVITY: The relationship between the creative work environment and performance in architectural practice, explores value propositions, the relationship of fiscal performance to rankings, and characteristics of creative innovation and collaboration in work, learning, and health environments. She has consulted as a strategic planner to organizations seeking change to maximize their creative innovation and performance.

Before joining the CSU faculty in 2001, she served as division director of interior design at the University of Oklahoma and prior to that appointment headed the Design Area at the University of Massachusetts/Amherst; she has also held appointments at Syracuse University, Rhode Island School of Design, the University of Tennessee/Knoxville and University of Maryland. She is past Editor in Chief of the Journal of Interior Design. Although recognized with advising, research, teaching and service awards in academic environments, she has been a design practitioner since 1971 in NYC, Washington, DC and Hartford, CT, holding major design leadership positions with the largest and top US and European architectural firms including Ellerbe, VVKR/Suter + Suter's Design Collaborative, and Greenwell Goetz with responsibilities for business development, administration, design, and finance for commercial, financial services, healthcare, R & D, higher education and institutional projects from 5,000 to more than 2 million square feet.

She has received awards from ASID, IIDA, AIA, AICAE, IDEC, NAIOP, recognizing design excellence and leadership in interior design practice and served as the Architectural lead for the Interiors team on the Tennessee Valley Authority's 1.4 million SF passive solar demonstration project in Chattanooga, TN. Her recent project, the d-lab reflects a research partnership with Herman Miller, and over $500,000 in funding she orchestrated through to completion of a 2,500 SF collaborative learning space serving as a campus classroom model for five learning modalities.

In 2010, Katharine was named by DesignIntelligence, as the only interior designer among the 25 Most Admired Design Educators exemplifying excellence in design education leadership. As Undergraduate Program Coordinator for the Interior Design area at CSU from 2001–2009, she continues to be responsible for program quality and accreditation, in addition to serving as thesis and dissertation advisor to graduate students in the areas of sustainability, creativity, and healthcare, teaching the senior capstone experience, materials, sustainable structures and systems, space planning and human factors, in addition to mentoring first year and honors students, and students planning to study abroad.

Dr. Leigh chairs CSU's Faculty Council Committee on Strategic and Financial Planning, the International Development Board, and co-chaired the Campus Leadership Committee focused on stadium strategies for the future

Her research publications are diverse, focusing on creativity in learning and work environments as well as research presentations (e.g., ARCHEX, AIA, Greenbuild Research Track, IDEC, Design Management Institute and EDRA). She and colleague, Dr. Laura Malinin, will receive the 2015 Interdisciplinary Scholarship Award for their emerging team international and national research activity with the Creativity Research Group, led by Leigh.

Jack Lovelace

Best Teacher, Journalism

Lori Peek

Best Teacher, Liberal Arts, Social Sciences, and Humanities

COLUMBIA COLLEGE CHICAGO

Nathan Bakkum

Most Helpful to Students, Music

He has presented papers at the national meeting of the Society for Ethnomusicology, the Experience Music Project Pop Conference, and other local and national conferences. His research has appeared in Black Music Research Journal, Jazz Perspectives, and Journal of Music History Pedagogy, and he has authored ancillary materials for the textbooks Music Then and Now and Jazz, both published by W.W. Norton and Company.

As a bassist, Bakkum has studied with jazz legend Richard Davis and Chicago jazz stalwart Larry Gray. He is active throughout the Chicago area as a performer and arranger.

Dale Chapman

Best Teacher, Fine Arts

Mariano Di Marco

Most Helpful to Students, Film

Ron Falzone

Best Overall Faculty Member, Fine Arts

Julian Grant

Best Researcher/Scholar, Film

"Born in London, England, I was an early cinema fan with "King Kong" and American comics being instrumental in my passion for telling stories with pictures. Emigrating to Canada in the 1970's, I worked diligently as a skateboard enthusiast, fantasy novel devotee (discovering Tolkein, Howard and the art work of Frazetta) and began creating worlds of pictures and stories. I graduated from York University in 1985 with a BFA in Film Production and immediately began working in 'Hollywood North'. Through the late 80's and early 90's I dedicated myself to independent production, learning the ropes of distribution (with both New World and Disney as employers!) and made my first feature film, "Bust A Move" before I was 30! This was the picture that started it all as I was introduced to Lionsgate and began making kickboxing and erotic thrillers for them. HBO came calling and I moved ever further up the production ladder creating three World Premiere cinema showcases for them and was able to begin the Millenium as the Producer/Director of the acclaimed mini-series, "Robocop: Prime Directives.

As reality television and production shortfalls continued in Canada, I turned to producing US Cable-driven feature films and beginning my academic career. Successful tenures at The Canadian Film Center, Sheridan College and a host of other educational facilities allowed me to transition to the USA and join the full-time faculty at Columbia College Chicago.

Today, I am an active working artist and professional educator passing on my 'no-holds-barred' production philosophies and continuing to live the dream. I make cinema for myself now and am an evangelist for DIY and self-distribution in this new digital age."

Award Winning Tenured Professor and published author. Recipient of Gemini Award for Outstanding TV Feature Film and evangelist for DIY microcinema filmmaking.

Del Harvey

Best Researcher/Scholar, Film
Best Teacher, Film

"Helpful, courteous, knowledgable, devoted to the student's learning."

Anna Kunz

Most Helpful to Students, Art & Design
Best Overall Faculty Member, Fine Arts

"Knowledge in Color Strategies class: positive, fun, personable; a successful, practicing fine artist."

Anna Kunz is a Chicago-based artist, exhibiting her interdisciplinary works in the US and abroad. Kunz received her BFA from The School of the Art Institute of Chicago and her MFA from Northwestern University. Kunz describes her works on paper, paintings, sculptures, and installations as "seeping out of the rectangle", and makes work often for other places and audiences outside of traditional gallery spaces. In addition to painting, the artist uses painted and dyed fabrics that function like nets to capture and manipulate light and color. These experiential works are often combined with objects or surfaces that add complexity and invite viewers to interact with the works in space. Kunz has earned honors and awards from the Union League Civic and Arts Organization, the School of the Art Institute of Chicago, Northwestern University, and the Department of Cultural Affairs in Chicago. Kunz has also served as an artistic collaborator for many organizations within the city and around the U.S. Most notably, the artist worked with dancers and musicians to design sets for various theatrical and dance productions, namely for the Chicago-based dance company SELDOMS in 2012 and the Merce Cunningham Dance Company in NYC. Anna has been nominated for awards from the Louis Comfort Tiffany Foundation, The Museum of Women in the arts, WA D.C, The Artadia Foundation, The Rema Hort Mann Fondation, NYC, The Rolex Foundation, an Excellence in Teaching Award from Columbia College Chicago, and others. She was recently awarded artist Residencies at The Marie Walsh Sharpe Fondation, NYC, The Skowhegan School of Painting and Sculpture, The Edward Albee Foundation. Kunz serves on exhibiton boards, professional organizations and as a curatorial consultant and juror for institutions across the Midwest.

Ilya Levinson

Best Teacher, Music

Jose Lopez

Best Teacher, Liberal Arts, Social Sciences, and Humanities

Carter Martin

Most Helpful to Students, Fine Arts

Mary Novak

Best Researcher/Scholar, Fine Arts

Natalie Rast

Best Researcher/Scholar, Dance
Best Teacher, Dance

Jim Rohn

Best Teacher, Film Most Helpful to Students, Film

Philip Seward

Best Overall Faculty Member, Music

Philip Seward has worked internationally as a composer, conductor, pianist, tenor and teacher. In 2013, How To Date A Coloratura, Philip's chamber opera was one of three finalists at the National Opera Association Chamber Opera Competition. In the same year, another opera, Eve, premiered as a part of Midwest New Musical's series of new works in concert. Highlights of Mr. Seward's compositions include productions of Les Dames á trios … et piano at the Kaye Playhouse in New York City and the Stages Festival in Chicago; High Fidelity at Merkin Hall in New York City and the Chicago Humanities Festival; and Jeff Award-winning Hans Brinker.

The world renowned Lyric Opera of Chicago commissioned numerous pieces that include, Stone Soup, A Noteworthy Tale, and African Stories. These works were produced in Chicago, Toledo Opera, Memphis Opera, and Pensacola Opera among others. The Lira Ensemble of Chicago also commissioned works by Philip Seward including Blessing premiering on WFMT radio. Another piece for chorus and orchestra, Sonnet, was performed at the Chicago's Symphony Center.

Upcoming recording releases include How To Date A Coloratura and The Rose Prologues both with soprano Patrice Boyd and conductor Gregory Buchalter. Both albums will be released in 2014 on iTunes and Amazon joining other albums of his including The Piano Album: Songs from Atonality, The Holiday Album: Songs from Atonality, Juan Peron's Hand, Home, Hans Brinker, and Les Dames à trios … et piano.

As a pianist Philip Seward has performed his compositions for piano as well as other contemporary composers all over the United States and in Europe including stops in Holland, Germany and Poland. As a singer, Seward sang the role of Eisenstein in Die Fledermaus for Intimate Opera; John Jasper in The Mystery of Edwin Drood; Lt. Cable in South Pacific; the prince in R & H Cinderella. Other productions featuring Philip Seward include The Beggar's Opera, A Chorus Of Disapproval, The American Clock, Happy End, and The Phantom of the Opera. Internationally, he performed Art Song recitals in Rzeszów and Wroclaw in Poland.

Philip Seward is the recipient of the Excellence in Teaching Award that he received during his teaching career at Columbia College Chicago. Aside from his composing, performing and teaching, Philip also conducts the choir at St. James Presbyterian Church, the Edgewater

Singers and is currently completing his DMA in Composition with Dr. Stephen Davismoon at the University of Salford.

Mr. Seward's publisher for his choral music – Porfiri & Horvath. Other music is available through his website: philipseward.com.

Jeffrey Jon Smith

Best Overall Faculty Member, Film Best Teacher, Film

"Jeffrey has been directing and acting in theater since 1969, and making films since 1999. he received an M.F.A. in film and video at Columbia College Chicago. He continues to be active in theater and film as a writer and director, and is currently developing a feature screenplay for his long time friend and collaborator Tekki Lomnicki. In addition to writing and directing more films, he is also interested in directing more live theater in Chicago.

"I have had the privilege of sitting in on a few classes that Jeffrey Jon Smith has taught over the past thirteen years at Columbia College. He brings a passion, absolute joy and inspiration to his students. His preparation for each class is thorough which allows for his lessons to be clear and yet when in front of his class, he is improvisational which creates which creates a powerful and exciting learning space.

Jeffrey Jon Smith is an exceptional instructor who deserves great recognition. Columbia College is so fortunate to have him teaching in the Cinema Science + Art Department.

Sincerely, Cheryl Graeff"

COLUMBIA STATE COMMUNITY COLLEGE

Jeff Hardin

Best Teacher, Liberal Arts, Social Sciences, and Humanities

devotion to his writing and his teaching — also he is humble and kind to all teachers"

Stuart Lenig

Best Overall Faculty Member, Humanities

"Enthusiasm for arts and learning as well as tireless in his outreach to adjunct as well as full time faculty, outstanding example of a life devoted to community college."

COMMUNITY COLLEGE OF PHILADELPHIA

Cynthia Giddle

Best Overall Faculty Member, English
Best Overall Faculty Member, Liberal Arts, Social Sciences, and Humanities

"She has an amazing amount of grace. She conveys respect for faculty and students. She is the kind of leader and colleague I want to make proud."

COMMUNITY COLLEGE OF RHODE ISLAND

Marilynne Plunkett-Soscia

Most Helpful to Students, English

"Marilyn has been extremely helpful to my students in the developmental English classes. She is patient with them and attentive to their individual needs. She also keeps close contact with me so that we can coordinate strategies for each student. She is very dedicated - even comes in on her day off!"

CONCORDIA COLLEGE

Elizabeth Arroyo

Best Teacher, Languages

Viann Pederson

Best Overall Faculty Member, Languages

Don Rice

Best Overall Faculty Member, Liberal Arts, Social Sciences, and Humanities

CONTRA COSTA COLLEGE - SAN PABLO

Jeff Michels

Best Overall Faculty Member, English

Elvia Ornelas

Best Teacher, English
Most Helpful to Students, English

Agustin Palacios

Best Teacher, Ethnic Studies

COUNTY COLLEGE OF MORRIS

Matthew Ayres

Best Teacher, English

Emily Birx

Best Overall Faculty Member, English

Philip Chase

Most Helpful to Students, English

Stephen Longo

Best Overall Faculty Member, Graphic Arts

Professor Longo has specialized in brand repositioning, new product/brand development, visual identity and packaging design since 1978. Overseeing every aspect of the design process, he has assisted clients in significantly increasing the value of their brand assets by developing original, beautiful and strategically sound design solutions.

Before opening his own design firm, he rose through the ranks from Junior Designer to Senior Design Director at several major international design groups such as CatoJohnson, Young & Rubicam, and Lippincott & Marguilies Inc.

His expertise includes branding identity programs for a wide range of clients. He has been responsible for the design of such blue chip accounts as Haagen-Dazs, Weight Watcher's Smart

Ones, Tang, Star Kist Tuna, Close-Up toothpaste, Ronzoni pasta, Borden's Cremora, Ore-Ida, 9-Lives cat food and numerous other national brand accounts.

His work has been featured in many trade and design publications and has garnered him many recognized awards from the American Institute of Graphic Arts, the Clios, Graphis, Package Design Council International, Graphic Design USA, Creativity, ADC Global, the Davey Awards, and the International Academy of the Arts to name a few.

CURRY COLLEGE

Marcy Holbrook

Most Helpful to Students, Communication

Sanford Kaye

Most Helpful to Students, Liberal Arts, Social Sciences, and Humanities

CUYAHOGA COMMUNITY COLLEGE

Susan Anderson

Best Teacher, Speech Communication

Ashley Brand

Best Overall Faculty Member, Liberal Arts, Social Sciences, and Humanities

Carol Ruby Demiray

Best Teacher, Fine Arts

George Kanieski

Best Overall Faculty Member, English
Best Teacher, English
Most Helpful to Students, English
Best Teacher, Liberal Arts, Social Sciences, and Humanities
Most Helpful to Students, Liberal Arts, Social Sciences, and Humanities

John Kerezy

Best Overall Faculty Member, Liberal Arts, Social Sciences, and Humanities

A native of Cleveland, John Kerezy has also worked in journalism and public relations in Indiana, Alaska, and Washington, DC, as well as in England. Kerezy holds a bachelor's degree from Wabash College, a master's degree in journalism from the Ohio State University, and a post-graduate certificate in college teaching from Kent State University. Kerezy has nearly 40 years of continuous writing and public relations experience, dating back to freelancing articles in a neighborhood newspaper in Cleveland in 1973.

Kerezy has worked as public relations officer at Wabash College, was a business press editor with Penton Media and served as an account executive with Alaska's largest advertising and public relations agency, now known as Bradley/Reid Communications, where he practiced health care and government affairs PR. He also worked as press secretary to the late Sen. Ted Stevens and was public relations and marketing director at Lakewood Hospital.

In 1988, John began Kerezy Communications and directed or coordinated public relations and marketing communications programs for private business, non-profit organization and the Cleveland Teachers Union and other clients.

Kerezy became an adjunct faculty member at Cuyahoga Community College in 2003 and joined the faculty full time in 2006. He serves on various college committees and holds positions on college journalism associations. Kerezy lives in Broadview Heights with his wife Kathy.

Samuel Lipuma

Best Researcher/Scholar, Philosophy

Thomas Lyon

Best Teacher, History

A. D. Mensah

Best Researcher/Scholar, Speech Communication

Justin Miller

Best Teacher, Philosophy

Henry Young

Best Overall Faculty Member, Speech Communication
Best Researcher/Scholar, Speech Communication

Henry Young started Tri-C as a student and believe it or not, actually failed his first semester of classes. Like many Community College students, he had the same false assumption that this experience was supposed to be easy. Years later after coming back as an adjunct, he applied for a full-time position and is now working on completing his Ph.D. Henry's journey is a inspiring example of Tri-C's slogan where futures begin and begin again.

DALLAS COUNTY COMMUNITY COLLEGE - TELECOLLEGE

Lois Parrott

Best Overall Faculty Member, Fine Arts
Best Overall Faculty Member, Fine Arts

"I have taught with Lois for 40 years. She is the best."

DALTON STATE COLLEGE

Judy Cornett

Most Helpful to Students, Liberal Arts, Social Sciences, and Humanities

Christian Griggs

Best Teacher, Liberal Arts, Social Sciences, and Humanities

"oustanding teacher and scholar who performed field research in England"

Professor Griggs grew up in North Carolina before heading to Utah for his undergraduate studies at BYU. He completed his BA in History teaching and spent a year teaching junior high. He then went to Purdue University in Indiana for his graduate work. At Purdue, he studied European history, specializing in Early Modern England. His research interests lie in the religious and political situation in late Stuart England.

Baogang Guo

Best Researcher/Scholar, Social Science

"Baogang Guo–outstanding scholar and teacher"

I was born and raised in a working class family in Zhengzhou, China. I came to the U.S. in 1986 and graduated with a PhD in Political Science from Brandeis University in 1994. I taught at Framingham State University, Bridgewater State University and Thomas University before joining faculty members at Dalton State College in August 2000. My teaching responsibilities include both political science and history courses. I served as the President of Association of Chinese Political Studies (2008–2010) and I am currently a Research Associate at the China Research Center in the Atlanta area. I take students and faculty members to my hometown in China every summer.

Kent Harrelson

Best Researcher/Scholar, English
Best Teacher, English
Most Helpful to Students, English

Kent Harrelson is a native of North Carolina. He received a B. A. in music from Gardner-Webb College (now Gardner-Webb University). He holds an M. A. in English from Appalachian State University and a Ph. D. in English from the University of Southwestern Louisiana (now the University of Louisiana at Lafayette). He has taught at DSC since 1994.

George Jones

Best Overall Faculty Member, Liberal Arts, Social Sciences, and Humanities

Marsha Matthews

Best Overall Faculty Member, English

Thomas Veve

Best Researcher/Scholar, Liberal Arts, Social Sciences, and Humanities
Best Teacher, Liberal Arts, Social Sciences, and Humanities

Dr. Thomas D. Veve, a native of the great state of New York, arrived at Dalton State College in the fall of 1991. He received his Bachelor's Degree from Hofstra University in 1974. He earned his Master's Degree in 1983 from Marquette University, and his Doctorate in Philosophy in 1990, also from Marquette. Dr. Veve, commissioned as an Infantry Officer after completing ROTC at Hofstra, retired from the United States Army Reserve as a Lieutenant Colonel in 2002, after 28 years of service. He has published one book, The Duke of Wellington and the British Army of Occupation in France, 1815–1818 (Greenwood Press, 1992). Dr. Veve has recently published numerous entries in a several encyclopedias, including Naval Warfare, Encyclopedia of the Second World War, French Revolutionary and Napoleonic Wars, The Encyclopedia of American Military History, and the Encyclopedia of the Age of Imperialism. He currently is active with the Conference on Baseball in Literature and Culture, and has had two articles published by that organization. Dr. Veve currently serves the college as an Associate Professor of History,

This home page opened in July, 2000, and remains under permanent construction. Feel free to roam around the course you are enrolled in to see if I have added any neat stuff designed to further your education. You'll find some additional reading material, some recommended readings, and some sample test questions along the way. Hope it helps!

DARTMOUTH COLLEGE

Aimee Bahng

Best Overall Faculty Member, English
Best Teacher, English

Michael Chaney

Most Helpful to Students, English

I specialize in 19th-century American and African American literature and culture, with a focus on race representation, mixed-race identity, and visual culture. Other research and teaching interests include comics and graphic novels, autobiography, and flash fiction.

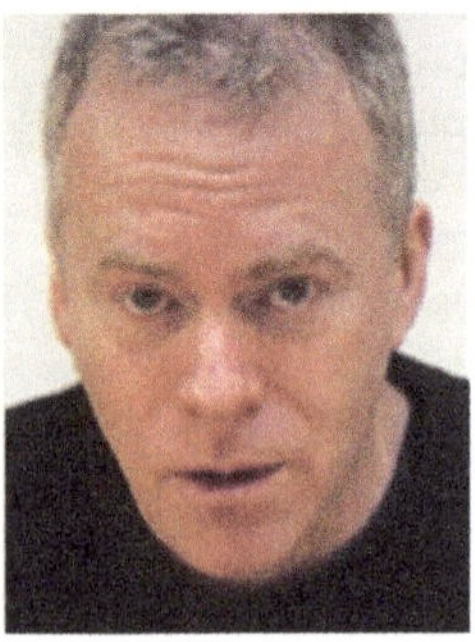

Andrew McCann

Best Researcher/Scholar, English

Studying literature as an undergraduate taught me to think outside of prevailing social and economic logic. It was a tremendously exciting undertaking. Whether I'm teaching Romanticism or contemporary experimental fiction, one of my goals is to reproduce the inherently critical impulse that I believe lies at the heart of literary expression.

DAYTONA STATE COLLEGE

Robert Fuller

Most Helpful to Students, Music

Robert Fuller has a Bachelor of Music degree from McGill University and Master of Arts degree from Columbia University. He has studied with Robert Nagel, James Ranti, and John Glasel. His professional career includes playing with numerous orchestras, Broadway shows, choral groups, Ice Capades, Ice Follies, Vienna Ice Revue, Shriners Circus, and Disney Productions. Mr. Fuller has backed entertainers such as Engelbert Humperdinck, Glen Campbell, Joey Heatherton, Jerry Vale, Peggy Lee, Connie Francis, Sonny and Cher, and Dean Martin. He has also performed with the Montreal Symphony, Grand Ballet of Canada, CBC and BBC Radio and T.V. and the National Film Board. Mr. Fuller is also the author of five music books. He was the former director of the Canadian Brass Quintet. He was the Coordinator of Music Studies at SUNY Ulster. He is now a resident of Palm Coast, Florida, freelancing, teaching privately and is a music instructor at Daytona State College.

Kathleen Lazarus

Best Teacher, English

Peter Waidelich

Best Researcher/Scholar, Liberal Arts, Social Sciences, and Humanities

DE ANZA COLLEGE

Matthew Abrahams

Best Researcher/Scholar, Communication

Elaine Lee

Best Teacher, Liberal Arts, Social Sciences, and Humanities

Kristin Skager

Best Teacher, English

Donna Stasio

Best Overall Faculty Member, Liberal Arts, Social Sciences, and Humanities

Donna Stasio is Facilitator Trainer, Speech Communication Instructor, and Speech Communication Department Chair at De Anza College.

She has experience in facilitator training, instructional skills development, curriculum development, assessment, and organizational leadership in both the public and private sectors.

Ken Weisner

Best Overall Faculty Member, English
Best Researcher/Scholar, English
Most Helpful to Students, English

DELTA COLLEGE

Dana Denney

Best Teacher, English

"Dedicated to student success and student learning, a team player."

DEVRY UNIVERSITY

Barbara Burke

Most Helpful to Students, Liberal Arts, Social Sciences, and Humanities

Robert Christie

Best Overall Faculty Member, Humanities
Best Researcher/Scholar, Humanities

Mark Geller

Best Overall Faculty Member, Liberal Arts, Social Sciences, and Humanities

Barbara Goldberg

Best Researcher/Scholar, Liberal Arts, Social Sciences, and Humanities

Rose Vydelingum

Best Teacher, Humanities
Most Helpful to Students, Humanities
Best Teacher, Liberal Arts, Social Sciences, and Humanities

DIABLO VALLEY COLLEGE

Jeannie Chiu

Best Researcher/Scholar, English

"She is always doing research and presenting her findings at conferences."

Laury Fischer

Best Overall Faculty Member, English

"Laury is a teacher's teacher. He leads us in pioneering new approaches to student success and understands and helps us when educational problems inevitably arise. His vast knowledge and total dedication to our field of work are a constant source of inspiration, and I know he thinks of us all as family."

Jackie Halm

Best Overall Faculty Member, Humanities

Tara Maginnis

Best Researcher/Scholar, Theater

DUQUESNE UNIVERSITY

Ronald Polansky

Best Overall Faculty Member, Philosophy
Best Researcher/Scholar, Philosophy
Best Teacher, Philosophy
Most Helpful to Students, Philosophy
Best Researcher/Scholar, Liberal Arts, Social Sciences, and Humanities
Best Teacher, Liberal Arts, Social Sciences, and Humanities
Most Helpful to Students, Liberal Arts, Social Sciences, and Humanities

James Swindal

Best Overall Faculty Member, Liberal Arts, Social Sciences, and Humanities

James Swindal is Professor of Philosophy and Dean of the McAnulty College of Liberal Arts. He specializes in Continental Philosophy, Action Theory, Ethics, and Catholic Philosophy.

Elochukwu Uzukwu

Best Researcher/Scholar, Theology

Fr. Uzukwu has authored three books, numerous articles and book sections that explore areas of the liturgy ecclesiology in continental Africa and Africa in the Diaspora. The Governing Council for the Spiritan International School of Theology in Enugu established the Uzukwu Award for Excellence in his honor to be granted to the best student of the year.

EAST TENNESSEE STATE UNIVERSITY

Alison Barton

Best Teacher, Human Development

Dr. Barton received her B.A. in Psychology from the University of Kentucky. She continued her academic career at Northern Illinois University with an M.A. in Clinical Psychology and a Ph.D. in School Psychology. She worked as a school psychologist in the Henderson County, KY school district for three years.

Dr. Barton is one of the first twelve faculty to have been awarded an INtopFORM fellowship, a program designed to promote excellence in seeking, using and communicating diverse and often complex information. Within the College of Education, she serves as the coordinator for the EDFN program and as e-Learning Liaison, advising other faculty on the design and development of online courses. Additionally, she supports faculty by giving seminars on best teaching practices with a focus on integrating technology, as well as co-teaching a course for faculty on pedagogical strategies for teaching college-level courses.

Current research interests include college student adjustment, primarily regarding factors related to college student persistence, retention, and mental health outcomes.

Dr. Barton has two sons and a daughter and is a volunteer for the Kingsport Hunger Task Force, for which she developed and manages the Feed Kingsport website: http://feedkingsport.org.

EASTERN CONNECTICUT STATE UNIVERSITY

Warren Osterndorf

Most Helpful to Students, Liberal Arts, Social Sciences, and Humanities

EASTERN KENTUCKY UNIVERSITY

Carroll Hale

Most Helpful to Students, Fine Arts

Jason Koontz

Best Researcher/Scholar, Music

"Extensive travels around the world as a percussionist, including South America, India, and Africa, makes him a standout in our department and university."

Dr. Jason Koontz has served as director of percussion studies at Eastern Kentucky University since 2000. He currently teaches applied percussion, percussion ensemble, percussion methods, music appreciation, world music, and directs both the EKU Percussion Ensemble and the EKU Steel Band.

He has traveled to perform and study world music/percussion in Europe (England, Holland, and Switzerland), Africa, India, Indonesia, and Brazil. Koontz has also performed with groups such as Global Rhythms, Blast, the Cincinnati Pops and Ballet Orchestras, Lexington Philharmonic, Richmond Symphony, Lexington Theater, the Art of Rice Traveling Theater (Indonesia), Tonus Percussion Trio, Resonance Percussion Sextet, Star of Indiana drum and bugle corps, Brass Theater, and the Southwind drum and bugle corps.

He is a member of the Percussive Arts Society International Convention world music planning committee, and is endorsed by Vic Firth Inc., Ludwig/Musser, Pearl, and Coyle Steel Drums. His compositions and arrangements for steel band and percussion ensemble are published by Drop 6 Media and Coyle Steel Drums, and Pan Ramajay productions.

EASTERN MICHIGAN UNIVERSITY

Jessica Kilbourn

Best Teacher, Women's Studies

Dyann Logwood

Best Researcher/Scholar, Women's Studies
Most Helpful to Students, Women's Studies

Pamela Speelman

Most Helpful to Students, Fine Arts

"Dr. Speelman is a visionary who seeks to the highest degree to make her students comfortable and skilled. She is a leader of the highest degree and not afraid to take on tough issues."

Peter Wood

Best Researcher/Scholar, Sociology

EASTERN WASHINGTON UNIVERSITY

Georgia Bazemore

Best Researcher/Scholar, Liberal Arts, Social Sciences, and Humanities

Matthew Binney

Best Researcher/Scholar, English

"Constantly researching, being published, editing for scholarly journals."

Auburn University, PhD, 2004

I am interested particularly in early modern notions of and responses to the foreign in British and European travel accounts and how these inform philosophical and cultural discourse. I am also a Contributing Editor for The Scriblerian: http://www.scriblerian.net/.

Michael Conlin

Best Researcher/Scholar, History
Best Overall Faculty Member, Liberal Arts, Social Sciences, and Humanities

"Serves in a variety of capacities."

Specializations: Early Republic, Civil War, History of Science
Specializations: Antebellum and U.S. Civil War, Slavery in the U.S., History and Philosophy of Science, Early American Republic, Atlantic History

Michael Conlin's research focuses on public memory, sectional identity, and political conflict during the Antebellum Era of U.S. History (1846–1861) as well as the difficulties in repeating experiments, the idea of an experimentum crucis, and the influence of nationalism on scientific practice.

Michael Conlin has published peer-review journal articles on Joseph Priestley's defense of Phlogiston Theory, Pierre-Auguste Adet's Revolutionary Diplomacy and Chemistry, the Reception of the Foucault pendulum, and the Smithsonian Abolition Lecture Controversy.

Michael Conlin's completed book manuscript "One Nation Divided by Slavery: Remembering the Founders while Marching toward the Civil War" examines how Americans in the two decades before the Civil War remembered their common past, in particular how they explained (or explained away) the presence of slavery during the American Revolution, in the lives of the Founders, and in the early republic. "One Nation Divided by Slavery" is currently under review by an academic press. Listen to podcasts {3.14.2010_mp3} {7.4.2011_mp3} from Anthony Flinn's radio show "Just a Theory" (KSFC FM 91.1) which discuss parts of "One Nation Divided by Slavery."

Michael Conlin has begun to write the book manuscript "South Carolina versus Massachusetts: Sectional Extremes in a Hegelian Regress to Civil War," which examines how caricatures and stereotypes of Northern Abolitionists and Southern Fire-Eaters - Boston Garrisonians and Palmetto Calhounites – exacerbated the sectional conflict, by crowding out the middle ground and reducing the other side to its most radical elements.

In addition to being named a Visiting Scholar by the Center for the Study of the American South at the University of North Carolina, Michael Conlin has won research fellowships from the Virginia Historical Society (Richmond, VA), the Maryland Historical Society (Baltimore, MD), the Filson Historical Society (Louisville, KY), the Gilder Lehrman Institute of American History (New York City, NY), the North Caroliniana Society (Chapel Hill, NC), and the Beckman Center for the History of Chemistry (Philadelphia, PA).

Tom Crofoot

Most Helpful to Students, Social Work

Kevin Decker

Best Overall Faculty Member, Philosophy
Best Researcher/Scholar, Philosophy
Most Helpful to Students, Philosophy
Best Researcher/Scholar, Liberal Arts, Social Sciences, and Humanities
Most Helpful to Students, Liberal Arts, Social Sciences, and Humanities
Best Overall Faculty Member, Liberal Arts, Social Sciences, and Humanities

Dana Elder

Most Helpful to Students, Liberal Arts, Social Sciences, and Humanities

Jane Ellsworth

Best Overall Faculty Member, Music

"Honest, a wonderful sense of collaboration, extremely brilliant and a lovely colleague!"

Jane Ellsworth is Associate Professor of Music at Eastern Washington University, where she teaches music history. Both a musicologist and a clarinetist, her scholarly and performing activities explore the intersection of various disciplines.

Dr. Ellsworth earned a PhD in musicology in 2004 from The Ohio State University, and also holds Bachelors and Masters degrees in clarinet performance from the Cleveland Institute of Music and a DMA from Ohio State.

Her recent research has been mainly in early American music; she has published on the history of the clarinet in early America, and is working on projects related to the status and livelihood of early American instrumentalists. Other areas of interest include English music of the nineteenth century, performance practice, the history of performance, and organology.

She is active with the International Clarinet Association, contributing regularly to its journal, and has also published in the Journal of the American Musical Instrument Society and elsewhere. She has performed and lectured on the clarinet throughout the United States and in Europe, South America and Asia, on both modern and historical instruments.

D. Ellsworth participates annually as a performer and faculty member at the Classical Music Festival in Eisenstadt, Austria. She also plays with the Spokane Symphony Orchestra. She currently serves as Treasurer of the Haydn Society of North America, and as Chair of the ICA Research Competition.

Scott Finnie

Best Teacher, Liberal Arts, Social Sciences, and Humanities

Dr. Scott Finnie, raised in the Bay Area, came to Spokane in 1975 on a basketball scholarship at Gonzaga University. He earned a B.A. in English and a Minor in Criminal Justice in 1979, a Master's degree in American History from Eastern Washington University in 1992, and a Ph.D. in Leadership Studies from Gonzaga University in 2000. His dissertation was entitled "The Impact of the Removal of Affirmative Action Upon the Freshmen Class at the University of Washington - Implications of a Ten Year Quantitative Study."

Dr. Finnie has been a faculty member of EWU's Africana Education Program and History Department since 1992. As a recipient of over 20 awards for excellence in teaching, leadership, civil rights and mentoring - Dr. Finnie has made over 40 presentations in the past ten years at Oxford-England, Mexico City, Atlanta, Baton Rouge, San Francisco, San Diego, Seattle as well as the Inland Northwest on such themes as civil rights, affirmative action, servant leadership, diversity and cross-cultural communication, the American criminal justice system and faculty hiring in higher education.

He has published a dozen articles in The National Social Science Journal, The Inland Northwest Health Services Training Manual, Investigating Diversity: Race, Ethnicity and Beyond, The Oxford Round Table Journal, The Journal of Intercultural Disciplines, Ethnic Studies Review, The Council on Undergraduate Research Quarterly and The International Journal of Servant Leadership.

Logan Greene

Best Overall Faculty Member, English

"Serves in a variety of capacities."

University of New Mexico, PhD in English: Rhetoric and Writing (2004) University of Oregon, MA English: Medieval Studies (1974) University of California, Berkeley, BA English Literature (1972)

I taught for many years at community colleges before entering the doctoral program at the University of New Mexico. After defending my dissertation, I joined the faculty at Eastern Washington University, where I work in the literature, composition and humanities programs. My dissertation analyzes the rhetorical strategies of five women from different historical periods, finding commonalities that constitute a discourse strategy characteristic of women's rhetoric. These women - Hildegard of Bingen, Margery Kempe, Aphra Behn, Sojourner Truth and Helene Cixous - created rhetorical strategies that made productive use of their positions at the margins of institutionalized power.

Gary Kenney

Best Researcher/Scholar, Liberal Arts, Social Sciences, and Humanities

Christopher Kirby

Best Teacher, Liberal Arts, Social Sciences, and Humanities

Dr. Kirby joined the Philosophy Program in the fall of 2008. The general focus of his teaching and scholarship is in the history of philosophy and comparative thought. Kirby (as his students call him) teaches upper division courses on ancient Greek and Roman philosophy, the philosophy of religion, Chinese Philosophy, and Environmental Ethics, as well as several introductory philosophy courses, including the recent offering... Zombies and Philosophy! In his research, he is most interested in thinkers who draw insight from exploring nature - such as Aristotle and the Daoist sage Zhuangzi - as well as more recent thinkers who combine those sorts of insights with scientific and/or aesthetic modes of thought, like Ralph Waldo Emerson, Henry David Thoreau, Aldo Leopold, Arne Naess, E.O. Wilson... and especially John Dewey.

Kirby also sponsors several clubs on campus: The Transcendental Apathetic (EWU's philosophy club), the Environmental Club, and the Mandarin Chinese Conversation Group. Contact him if you'd like more information on any of these organizations.

Current and Recent Activity: In 2014, Dr. Kirby published an edited volume of essays with Bloomsbury Press called Dewey and the Ancients: Hellenic and Hellenistic Themes in the Philosophy of John Dewey. He is currently developing an essay on service learning in environmental ethics courses and will be on leave during the fall quarter to draft the first few chapters of a book-length project that traces the philosophical history of the concept of impulse (hormé in Greek, conatus in Latin).

Ann Le Bar

Most Helpful to Students, Liberal Arts, Social Sciences, and Humanities

"Always available."

Judy Logan

Best Teacher, English
Most Helpful to Students, English
Best Teacher, Liberal Arts, Social Sciences, and Humanities

I have taught at EWU since 1996 and have directed the undergraduate and graduate programs in literature since 1998. Other research interests are Anne Tyler, Doug Marlette and Spenser.

Vernon Loke

Best Researcher/Scholar, Social Work
Best Researcher/Scholar, Liberal Arts, Social Sciences, and Humanities

M Soc Sci (Social Work) National University of Singapore, PhD George Warren Brown School of Social Work at Washington University. Scholarly interests include asset building and economic empowerment of disadvantaged populations, and program evaluation.

Romel Mackelprang

Best Overall Faculty Member, Social Work
Best Teacher, Social Work
Best Overall Faculty Member, Liberal Arts, Social Sciences, and Humanities

MSW and DSW University of Utah. Scholarly interests include disability studies and international social work.

Lavonna Reeves

Best Overall Faculty Member, English
Best Teacher, English
Most Helpful to Students, English

"Truly dedicated, generous with spirit and time.
Spends more time and dedicates more love than anyone else.
Committed equally to service, scholarship, and learning."

University of Nebraska, BA: French (History and English emphases) Columbia University, MA: Languages & Literature (TESOL) Indiana University of Pennsylvania, PhD: English/Linguistics and Rhetoric

MATESL founder, LaVona Reeves, has taught in New York City and Boise Public Schools, Boise State University, the University of Wyoming, Osaka University and Harvard University. As President of TESOL affiliate, Washington Association for the Education of Speakers of Other Languages, she initiated a teaching award to honor deceased co-founder of the undergraduate TESL Program, Sally Wellman. Dr. Reeves works closely with graduate students conducting original research in her writing classes and has published in composition, TESOL, literature and gender studies. Awarded EWU Professor of the Year in 2007, she has also been widely recognized for university and community service.

Peter Shields

Best Overall Faculty Member, Communication
Best Researcher/Scholar, Communication
Most Helpful to Students, Communication

*"Peter is a very engaged teacher scholar.
works hard to connect with students
contributes to faculty and students"*

Gregory Spatz

Best Researcher/Scholar, English

"Prolific and accomplished both as a musician and a writer-researcher."

Jeff Stafford

Most Helpful to Students, Communication

"Jeff is a wonderful advisor who gives tirelessly to his students. He seems to be always available and is very proactive about developing useful advising materials. His advising sessions integrate discussions of overall career goals and life plans with a discussion of appropriate coursework. This is why he is so popular with the student body."

Henry-York Steiner

Best Overall Faculty Member, English
Best Teacher, English
Most Helpful to Students, Liberal Arts, Social Sciences, and Humanities

Bob Sterling

Most Helpful to Students, Social Work

Lead Worker with Home Finders/Foster Parent support and retention unit. Placement coordination for children/teens going into emergency or long-term foster care. Conduct new foster parent orientation and training. Assists with foster home recruitment, retention and support activities. Case manage a guardianship caseload of ten families. Responsible for services, payments and court reviews as needed.

Steve Stewart

Best Teacher, Communication

"Steve works hard and is generally spoken of by students in high regard."

Rachel Toor

Best Researcher/Scholar, Liberal Arts, Social Sciences, and Humanities

"incredibly talented creative writing faculty"

Yale University, AB (1984) University of Montana, MFA (2006)
Formerly acquisitions editor at Oxford and Duke University Presses and admissions officer at Duke.

Beth Torgerson

Best Researcher/Scholar, English

Christina Valeo

Best Teacher, English
Most Helpful to Students, English

"Always willing to help! Motivates students"

Chris Valeo earned her BA from Brown University in 1992, and her MA in teaching English in 1992. She taught high school English in Montana for five years (Hays/LodgePole High School 1992–1994, Havre High School 1994–1997), leaving to pursue graduate-level work in literature at the University of Illinois at Urbana-Champaign (MA 1999, PhD 2003). She joined the faculty of Eastern in 2003. Her current research and teaching interests include British literature, English pedagogy, children's literature and popular romance.

David Weise

Best Teacher, Philosophy

Sue Wright

Most Helpful to Students, Liberal Arts, Social Sciences, and Humanities

"Sue works tirelessly for our students."

Sue Marie Wright, Ph.D., served as Director for the Children's Studies Program at Eastern Washington University from the program's inception in 2002 until the year 2013. She currently holds a position as Professor in the Department of Sociology and Justice Studies, teaching courses on children, gender, and family. Published articles include "Bridging Third-Wave Feminism and Family Pluralism" (2004).

J. William Youngs

Best Overall Faculty Member, Liberal Arts, Social Sciences, and Humanities

EASTFIELD COLLEGE

Courtney Carter

Most Helpful to Students, Liberal Arts, Social Sciences, and Humanities

Elizabeth Huston

Best Researcher/Scholar, English

Gloria Jackson

Best Teacher, Liberal Arts, Social Sciences, and Humanities

Phillip Ortiz

Best Overall Faculty Member, Liberal Arts, Social Sciences, and Humanities

Phillip Ortiz has held a full-time faculty position teaching Social Work courses at Eastfield College since January of 2000. He holds a bachelor's degree in psychology from the University of Dallas and a master's degree in social work from the University of Texas at Arlington. Mr. Ortiz is preparing to defend his dissertation for his Ed.D. degree in higher education from Texas A&M University-Commerce. He has been a licensed clinical social worker (LCSW) in Texas since 2000 and has extensive clinical experience in the areas of mental health, child welfare, family counseling, gang intervention, substance abuse, domestic violence and child custody disputes.

He is recognized by the Texas State Board of Social Work Examiners as a Continuing Education Provider and frequently conducts training seminars on cultural diversity, ethics, conflict resolution, clinical assessment and motivational interviewing in counseling-resistant clients. He periodically consults for Dallas Family and Protective Services, Dallas County Family Courts and local juvenile probation departments.

Bob Whisnant

Best Teacher, Arts & Humanities

EASTMAN SCHOOL OF MUSIC

James Vandemark

Best Teacher, Music

One of the most brilliant virtuosi ever to perform on the double bass, James VanDemark was hailed by the New York Times at his Lincoln Center recital debut as "an exceptionally gifted string player and a musician of taste, intelligence and the best spontaneous musical instincts, with an unerring sense for exact intonation." The San Francisco Chronicle praised his "wonderful facility for making really musical phrases, relaxing, building, shading with unlimited subtleties – and a capacity to dig into whole pages of rip-roaring coloratura and make every note count."

VanDemark began his musical studies at the age of 14 in his hometown of Owatonna, Minnesota, making such rapid progress that just 18 months later he made his solo debut with the Minnesota Orchestra. Subsequently, VanDemark has performed as soloist with the New York Philharmonic (Mehta), St. Paul Chamber Orchestra (Zukerman), Buffalo Philharmonic (Yoel Levi), Grant Park Symphony (James Paul), Rochester Philharmonic Orchestra (Peter Bay), the San Antonio Symphony (Barrios), the Mostly Mozart Festival Orchestra (Rampal), the Chautauqua Festival Orchestra (Hugh Wolff), the New Mexico Symphony (Lockington), the Quebec Symphony, the National Symphony of Mexico, the Netherlands Radio Symphony, and in numerous other concerto appearances.

VanDemark's duo recitals with André Watts, including one on Lincoln Center's Great Performer Series, and also with Samuel Sanders, Anthony Newman, Barry Snyder, and Robert Spillman have won him great acclaim. Chamber music collaborations with the Guarneri, Cleveland, Colorado, Muir, Ying, and Audubon Quartets, the Los Angeles Piano Quartet, Vienna Schubert Trio, Kandinsky Trio, Gryphon Trio and pianists Gary Graffman, Alfred Brendel, Anton Nel, Anton Kuerti, and Jeffrey Kahane highlight VanDemark's versatility.

The recipient of numerous commissioned works, including those by three Pulitzer Prize winners – Gian-Carlo Menotti, Joseph Schwantner, and Christopher Rouse – VanDemark also performed the American premieres of Nino Rota's Divertimento Concertante (Charlotte Symphony) and Edvard Tubin's Double Bass Concerto (Queens Symphony). VanDemark's most recent commission is a solo double bass work by the noted composer/violist Adrienne Elisha.

VanDemark recently premiered and recorded the recital work Dana la Colora by composer/cellist Emilio Colon. He also recently premiered the concerto Shiva Shakti by composer Todd Coleman, winner of the Scorch Music Competition, and performed and recorded Jerod Sheffer Tate's Iyaaknasha' (The Medicine Man and His Helper) with the Columbus (OH) Pro Musica Chamber Orchestra.

Also in considerable demand as a narrator with orchestra, VanDemark has appeared on numerous occasions with the Rochester Philharmonic, the Texas Festival Orchestra, the Eastman Philharmonia, the NEO Ensemble, and many others. In June, 2011, VanDemark will premiere a new narrated work with orchestra, A Young Rodent's Guide to the Orchestra (for which VanDemark wrote the text and Emilio Colon the music), with the Texas Festival Orchestra.

As a sought-after guest artist at summer festivals, VanDemark performs at the Mostly Mozart, Spoleto, Seattle Chamber Music, Montreal Chamber Music, Round Top, Maverick, Norfolk, South Bank (London), and Newport festivals.

An important direction in VanDemark's career has been his involvement with Native American performers in Circle of Faith, composed by Alton Clingan. VanDemark commissioned this unique musical and cultural collaborative work, developing it in conjunction with respected Native artists and elders. He also produces the work, which has had more than two dozen performances since its 1992 premiere with the Muir Quartet.

Appointed Professor of Double Bass at the Eastman School in 1976, at age 23, VanDemark became the youngest person ever to hold such a position at a major music school. VanDemark is recognized as a renowned teacher; his students hold positions with many of the world's major orchestras – Cleveland, Pittsburgh, St. Louis, Minnesota, San Francisco, Singapore, Taiwan, and the Vienna Chamber Orchestra, as well as the Rochester Philharmonic, New World Symphony, Oregon Symphony, and the orchestras of Syracuse, Buffalo and Wichita. Currently, VanDemark serves as Co-Chair of the String Department at the Eastman School, and Chair of the Musical Arts Major, Eastman's interdisciplinary academic honors program.

As a recording artist, VanDemark can be heard on d'Note Records, Philips, Telarc, Vox, Pantheon, and NEXUS.

VanDemark has been profiled in such diverse media as Connoisseur magazine, the New York Times, on PBS's MacNeil/Lehrer NewsHour, National Public Radio's Morning Edition and All Things Considered, and in the Lakota Times, the largest Native American newspaper in the United States.

VanDemark graduated in 1976 from SUNY Buffalo (BFA, Magna cum Laude). His principal teachers include bassist James Clute and cellist Paul Katz, with additional study with bassist Gary Karr and cellists Gabor Rejto and Leonard Rose.

EDINBORO UNIVERSITY OF PENNSYLVANIA

Jean Snyder

Best Teacher, Music

EL CAMINO COLLEGE

William Hoanzl

Best Overall Faculty Member, Humanities
Best Teacher, Humanities
Most Helpful to Students, Humanities

EL CENTRO COLLEGE

Arumugam Devarani

Best Researcher/Scholar, Liberal Arts, Social Sciences, and Humanities
Best Overall Faculty Member, Liberal Arts, Social Sciences, and Humanities

Janet Dodd

Most Helpful to Students, Liberal Arts, Social Sciences, and Humanities

During my Graduate studies, I passed a few summers in Paris and later on spent a school year in Lorraine (North Eastern France) teaching English to school children. Throughout the year, so many people invited me to dinners, took me on outings, and let me become a part of their lives and families. Leaving them was one of the hardest things I've ever had to do.

I teach French now because it keeps me connected to my French friends and fond memories. However, I have made some wonderful French-speaking friends here in Texas! As you know, the Dallas Fort Worth Metroplex boasts a very culturally diverse bunch of Texans from all over the world. So consider this: Nearly half of the countries in Africa use French as one of their main languages. There is a whole group of islands in the Pacific, as well as in the Caribbean, where French is an official language. Add Canada, French Guyana, and a few more countries in Europe, Asia, and the Middle East – where French is one of their staple languages – and you will understand why French is the most universally spoken language after English and why, believe it or not, there are a bunch of French speaking people living right here. When you meet them, be sure to tell them, "Bonjour!"

I encourage everyone to learn another language – French or otherwise. You cannot imagine the doors it will open up to you – the experiences you can have, and the friends you can make! Learning another language opens up your mind – not just to the cultures of the peoples who share that same language, but to all people across the globe. Learning to speak another's language is a step towards understanding that person. Also, you will learn a remarkable amount about yourself. I hope you will take that step, start on a marvelous adventure, and fall in love with Language. Bon Voyage!

(Note: Professor Dodd is also an instructor in French at the University of Texas in Arlington.)

Angela Taylor

Best Teacher, Liberal Arts, Social Sciences, and Humanities

EL PASO COMMUNITY COLLEGE

Rebekah Bell

Best Overall Faculty Member, Liberal Arts, Social Sciences, and Humanities

"Rebekah provides numerous outside resources to her students to keep them abreast of how communication impacts them globally. She supplements her lectures with collaborative learning techniques obtained during her attendance at the El Paso Community College Teachership Academy. She effectively uses technology in the classroom and empowers students to take responsibility for their own learning. As District Wide Coordinator for the Speech Discipline, Rebekah's leadership is unparalleled. She confidently accepted the position during SACS Accreditation and was instrumental in developing, implementing, and aligning core competencies with Student Learning Objectives."

ELIZABETH CITY STATE UNIVERSITY

John Luton

Best Researcher/Scholar, Liberal Arts, Social Sciences, and Humanities
Most Helpful to Students, Liberal Arts, Social Sciences, and Humanities
Best Overall Faculty Member, Liberal Arts, Social Sciences, and Humanities

A native of Elizabeth City, Dr. Luton began teaching as an adjunct at ECSU in 1997. He joined the full-time faculty in January 2001. An Endowed Professor of Communication Studies, Luton also serves as Assessment Coordinator for the B.A. in Communication Studies program.

Dr. Luton's research interests include the development of orthographies (alphabets) for languages which do not presently have writing systems. He has also contributed to Bible translations in more than a dozen languages, and has taught English as a Foreign Language (EFL) in several countries in Africa and Asia. A recent paper entitled "Language and the Preservation of Culture," co-authored with Dr. Jose S. Gil, has been published in the International Journal of Interdisciplinary Social Sciences. Luton has also served as associate editor of several other international, peer-reviewed journals.

Christopher Palestrant

Best Teacher, Liberal Arts, Social Sciences, and Humanities

An accomplished composer, Christopher Palestrant's works and arrangements have been performed from New York City to the Kennedy Center in Washington, D.C and throughout the United States. Prizes and awards include the Third Biennial Michael Hennagin Prize, the Turner Classic Movies Young Film Composers' Competition and the Randolph S. Rothschild prize in composition. His music and writings have also been selected for presentation for the College Music Society national conferences, the Society for Ethnomusicology and the Intercollegiate Music Association. Chris' compositions include the orchestral works The Wind in His Mane, Caligula and Festival Hebridean as well as numerous chamber pieces, choral music and one-act operas. Dr. Palestrant also performs on solo keyboard regularly throughout the area and with the award-winning blues band, Uphill.

His primary teachers in composition include Nicholas Maw, Morris Cotel, Tom Benjamin, Jack Gallagher and Adolphus Hailstork, and master classes with John Corigliano, George Rochberg, Samuel Adler and John Maxwell Geddes. He earned degrees from The College of Wooster, Wooster OH (B.A.); New York University (M.A.) and the Peabody Conservatory of Johns Hopkins University (M.M. and D.M.A.)

EMORY UNIVERSITY

Niall Slater

Best Researcher/Scholar, Classics

Determining that acting was too hard a way to make a living, Niall Slater opted for the leisure of the theoried class. After receiving a B.A. from The College of Wooster and a Ph.D. from Princeton, he started out as a SLACer, at Concordia College (Moorhead, Minnesota). Despite the warmth of his colleagues there, one Fargo/Moorhead winter made southern California look really good, and he moved on to USC. The earthquakes and mudslides weren't so bad, but he did decamp just ahead of the riots for Emory. At Emory he served as chairman of the department until 1994, then later as director of Emory's Center for Language, Literature, and Culture. He received the Emory Williams Award for Distinguished Teaching at Emory in 1999.

His research interests include the ancient theater and its archaeology, the ancient novel, gender studies, and recently warfare and its cultural impacts. His books include Euripides: Alcestis (in the Bloombury Companions to Greek and Roman Tragedy series); Spectator Politics: Metatheatre and Performance in Aristophanes (University of Pennsylvania Press, 2002); Reading Petronius (The Johns Hopkins University Press, 1990); and Plautus in Performance: The Theatre of the Mind (Princeton University Press, 1985). His translations of various Middle and New Comedy poets are included in The Birth of Comedy: Texts, Documents, and Art from Athenian Comic Competitions, 486-280, edited by Jeffrey Rusten (The Johns Hopkins University Press, 2011). He is a past president of the Classical Association of the Middle West and South as well as past president of the national Phi Beta Kappa Society.

EVERGREEN VALLEY COLLEGE

Keith Aytch

Best Overall Faculty Member, Liberal Arts, Social Sciences, and Humanities

"Keith Aytch is a great teacher, and a caring administrator."

Katharine Giles

Most Helpful to Students, English As A Second Language

"Katharine works in the lab and is willing to help students with more than the subject matter. She will research career, training and financial aid options with them. She offers counseling on whatever is going on in their lives."

Alexandra White

Best Teacher, English

"A fabulous teacher, always growing and expanding her teaching skills. She is very caring, and puts her students first."

FAIRFIELD UNIVERSITY

Mary Ann Carolan

Best Overall Faculty Member, Languages

"Wide mastery of Italian and terrific teaching, from the most elementary college-level course to advanced undergraduate and graduate-level courses on Dante, American Studies, Italian-American cinema and more, buttressed by a well-received book on this cinema topic just published in 2014, 'The Transatlantic Gaze' (State University of New York Press)."

Mary Ann McDonald Carolan is a graduate of Dartmouth College and Yale University Graduate School. She received her Ph.D. in Italian Language and Literature with a dissertation entitled "Power and Language in Alessandro Manzoni's I promessi sposi." After completing her doctoral work, Prof. Carolan taught at Wesleyan University and Yale University before joining the Fairfield faculty in 1997.

At Fairfield, Prof. Carolan directs the Italian Studies program, established in 1998, and is Associate Professor of Modern Languages & Literature. She teaches courses on all aspects of Italian culture, including:

Literature courses on Dante and the Italian novella Film courses entitled Italian Cinema, Contemporary Italian Culture, and Visions of Italy and America in Film Interdisciplinary courses such as Rome in the Cultural Imagination and The Italian American Experience

She has taught in the Honors Program, the Ignatian Residential College Program, the Catholic Studies program as well as in the graduate program in American Studies.

Prof. Carolan has published articles on Italian authors Dante Alighieri, Elsa Morante, Grazia DeLedda, and Alessandro Manzoni as well as on film directors Michelangelo Antonioni, Sergio Leone, Ettore Scola and Ferzan Ozpetek. She has also written about the role of women in Italian American comedies such as Moonstruck, My Cousin Vinny, and Married to the Mob.

In addition, she has published on the teaching of language and literature in Italian Studies curricula. Her work has appeared in Modern Language Notes, ADFL Bulletin, Literature Interpretation Theory, Rivista di studi italiani, Romance Languages Annual and Quaderni d'italianistica.

Prof. Carolan's book, The Transatlantic Gaze: Italian Cinema, American Film (State University Press of New York, 2014), documents the sustained and profound artistic impact of Italian cinema upon filmmakers in the United States from the postwar period to the new millenium. This analysis acknowledges the influence of Italian directors Michelangelo Antonioni, Vittorio DeSica, Federico Fellini, Sergio Leone, and Roberto Rossellini on the transformations and innovations of genres such as neo-realism, comedy, western, and art film in the works of American directors Woody Allen, Lee Daniels, Brian DePalma, Neil LaBute, Spike Lee, and Quentin Tarantino. Her current book-length project, Orienting Italy: China through the Lens of Italian Filmmakers, examines the ways in which Italian directors have employed documentary, historical fiction, and fictional narratives to represent China and its people both at home and abroad in Italy.

A member of the Modern Languages Association, the American Association of Teachers of Italian and the American Association of Italian Studies, Prof. Carolan regularly presents papers at universities and conferences in the United States and in Europe.

Robert Epstein

Most Helpful to Students, English

"With patience and expertise, Robert Epstein has illuminating dialogues with his students in his office."

Liz Hohl

Best Teacher, History

"Outstanding care and attention to students while remaining current on the latest research in her field. Work above and beyond the job description continuously for almost three decades!"

Laura Nash

Best Researcher/Scholar, Music
Best Teacher, Music

Emily Orlando

Best Teacher, English

"Emily Orlando imparts her knowledge in a gentle, pleasing manner, and her students feel at ease to participate in class."

Emily J. Orlando is Associate Professor of English at Fairfield University. She is the author of Edith Wharton and the Visual Arts (Choice Outstanding Academic Title) and has published essays in American Literary Realism, Women's Studies: An Interdisciplinary Journal, New Voices on the Harlem Renaissance, Memorial Boxes and Guarded Interiors: Edith Wharton and Material Culture, Edith Wharton's The Custom of the Country: A Reassessment, Edith Wharton in Context, and Twentieth-Century Literature (forthcoming).

She co-directed the 2012 "Edith Wharton in Florence" conference and currently serves as Vice President of the Edith Wharton Society and Book Review Editor for the Edith Wharton Review. She earned her B.A. at Saint Anselm College and her Ph.D. at the University of Maryland. From 2002 to 2007 she taught as an Assistant Professor of English at Tennessee State University.

Gavriel Rosenfeld

Best Researcher/Scholar, History

Brian Torff

Best Overall Faculty Member, Music
Most Helpful to Students, Music

Brian Q. Torff is a renowned bassist and composer, and is currently Music Program Director at Fairfield University. He is a featured bass soloist, leading his own trio, and is the musical director for the Django Reinhardt New York Festival, which has appeared at Lincoln Center, the Kennedy Center and Birdland in New York City. Torff has performed at Carnegie Hall for Fiddle Fest where he appeared with Mark O,Connor, Dave Grusin, Regina Carter, Yo Yo Ma, Itzhak Perlman and Pinchus Zukerman.

He performs in jazz festivals around the U.S., has founded the Fairfield University Summer Jazz Workshop and has served as co-chair person for the music advisory board for the National Endowment for the Arts.

Brian Torff's professional career began in 1974 when bassist Milt Hinton offered him the opportunity of touring with Cleo Laine. During the late 70's, Torff recorded and performed with pianists Mary Lou Williams and Marian McPartland, and toured Australia, New Zealand, and Hong Kong with the jazz violin virtuoso, Stephane Grappelli. He played in pianist Erroll Garner's last group and worked in the big bands of Oliver Nelson, and the Thad Jones-Mel Lewis Orchestra. In 1979, Brian Torff joined in a duo with pianist George Shearing. In the course of their 3 1/2 year collaboration, they toured throughout the U.S., Europe, Brazil, and South Africa and were featured on the NBC Tonight Show, Merv Griffin, and their own PBS special from the Café Carlyle in New York City. They received worldwide acclaim, and were invited to perform at the White House in 1982 for President Reagan. Their third album won a Grammy for vocalist Mel Tormé.

Brian Torff is a noted composer who has contributed works for records with George Shearing, Larry Coryell, Union trio, and his own Recordings - Life in East Bumblepuck, Hitchhiker of Karoo, Manhattan Hoe-Down, and Workin' On a Bassline. He has written scores that have been performed by the Boston Pops, Los Angeles Philharmonic, and the Pittsburgh Symphony, and has appeared as conductor, composer, and clinician for numerous high school and college jazz festivals. He is a published writer on music and musicians, composes for big band, and directs the Summer Jazz Workshop at Fairfield University. Torff is listed in the Groves Dictionary of Jazz and has been featured in the New York Times, Chicago Tribune, Downbeat Magazine, Jazz Times, and many other music publications.

FARMINGDALE STATE COLLEGE

Barry Capella

Best Overall Faculty Member, Liberal Arts, Social Sciences, and Humanities

"He is a great professor with outstanding personality and well liked and respected by his peer and students"

FASHION INSTITUTE OF TECHNOLOGY

Richard Balestrino

Best Teacher, Communication
Most Helpful to Students, Communication

FERRIS STATE UNIVERSITY

Tracy Busch

Most Helpful to Students, History

Dr. Tracy Nichols Busch is currently a Visiting Assistant Professor of History in the Humanities Department. She received her BA from Vassar College in 1991 and her PhD from Georgetown University in 2004. Her PhD dissertation focused on a society of volunteers that promoted automobile production and road construction in the early Soviet period. This semester Dr. Busch is teaching World History and American History.

Tracy Busch brings a great deal of practical experience to her academic work. Most recently, she managed a Russia program for the Federal Highway Administration that brought U.S. and Russian states together to conduct technology transfer. She has also managed a U.S. Agency for International Development grant that promoted civil society in Russia after the fall of Communism.

Although her academic interests are many and varied, Dr. Busch is currently most interested in the connection between technology and culture. Her article on Russia s eighteenth century roads is about to be published and several others are under development. Her next large project is the expansion of her dissertation into a book titled Stalin's Roads: Highways as Workshops of Soviet Culture in the Interwar Period.

Kimn Carlton-Smith

Best Overall Faculty Member, History

I have been teaching at Ferris since August of 1990. My undergraduate degree in Television Production is from UNC-Chapel Hill and my Masters & Doctorate Degrees in American History are from Rutgers University. While at Rutgers I had the pleasure to study under two noted African American scholars, David Levering Lewis and Deborah Gray White. During my time here at Ferris I have been actively involved in teaching and developing courses addressing the African American experience. My teaching interests compliment my scholarly activities of the last decade. My research has focused on African American visual artists of the 1920s and 1930s. At present, I am writing a book on Harlem-based artists of this period.

Gary Huey

Best Overall Faculty Member, History
Best Teacher, History
Most Helpful to Students, History

Education:
PhD, Washington State University
MS, Illinois State University
BS, MA, Bradley University
Employed at Ferris since: Fall 1986

Christian Peterson

Best Researcher/Scholar, History

Dr. Christian Peterson is currently a Visiting Assistant Professor of History in the Humanities Department. He received his BA from Yale in 1999 and MA from Utah State University in 2001. In 2009, he earned his PhD in history from Ohio University. His dissertation focused on how private citizens and government officials used the issue of human rights violations as a Cold War weapon from 1968–1989. This semester Dr. Peterson is teaching courses on U.S. foreign policy, historical methods, and American history. He is also looking forward to teaching a class on the Vietnam War during the upcoming Spring Semester.

Christian Peterson brings a wide variety of experience to his academic work. He has authored Ronald Reagan and Antinuclear Movements in the United States and Western Europe, 1981–1987 (2003) and Globalizing Human Rights: Private Citizens, the Soviet Union, and the West, which Routledge will release on 9 September 2011. He has also submitted a wide variety of articles to scholarly journals for publication. In the past, he has taught courses in areas as diverse as World History, Western Civilization, and the Humanities. At Utah State, he worked as an Editorial Fellow for the Western Historical Quarterly. He also played Division I varsity hockey at Yale during his undergraduate studies.

Although Christian has a wide variety of academic interests, he is currently researching how the U.S. government worked with labor organizations to promote human rights and democracy across the globe during the 1970s and 1980s. He is also examining how dissenters in the Soviet Union and Eastern Europe shaped the policies of peace activists in the United States and Western Europe during the Cold War.

FITCHBURG STATE UNIVERSITY

Kelly Morgan

Best Teacher, Theater

FLORENCE-DARLINGTON TECHNICAL COLLEGE

Mark Rooze

Best Teacher, English

English 101, English 102, English 160, English 205, English 260, Religion 103
Education:
B.A. in English, Angelo State University M.A. in English, Angelo State University M.Div., Greenville Presbyterian Theological Seminary Interests: Reformed theology Playing cello, guitar, and bass Contemporary hymnody and its composition Mathematics (I was actually a double major until the last semester of my BA) Computing (my first computer was a TRS-80 Model III) Golf

Angela Summerlin

Best Teacher, Liberal Arts, Social Sciences, and Humanities

Ms. Edwards is a native of Florence, South Carolina. Since 1998, she has lived and taught in South Carolina. She has been a full-time instructor at FDTC since 2001, teaching both day and evening classes at FDTC, in addition to distance learning and online college courses. She also serves as the Humanities Curriculum Coordinator.

In her time away from FDTC, she is a full-time wife and mother. While travel, hiking and reading were once favorite pastimes, now coloring books and puzzles, silly songs, the Sprout channel and sunny afternoons at the playground fill non-work hours as she strives to keep up with her daughter. (She is still waiting to introduce her daughter to Tutenstein, which is Ms. Edwards' favorite silly "history" cartoon.)

FLORIDA AGRICULTURAL AND MECHANICAL UNIVERSITY

Yussif Dokurugu

Best Overall Faculty Member, Liberal Arts, Social Sciences, and Humanities

Yussif Mijirah Dokurugu, Dr.PH, MPH, M.A Dr. Yussif Dokurugu is currently a faculty member in the Department of Sociology and Criminal Justice, an Affiliated Faculty Member and Program Coordinator of the Center for Health Equity, Institute of Public Health, International Education and Development and Center for Global Security and International Affairs. He also serves as an Adjunct Assistant Professor in the School of Allied Health and Barry University Tallahassee campus. Dr. Dokurugu began his career in the Sociology and Criminal Justice Department as an Adjunct Instructor in 2005 and a Visiting Assistant Professor in 2008 before obtaining a regular faculty status. Dr. Dokurugu worked as a Prostate Cancer Research Coordinator from 2004 to 2007 in the Prostate Cancer Center which funded by the Department of Defense (DOD) and a Research Coordinator from 2007 to 2009 in the Minority Health and Health Disparities International Research Training (MHIRT) Program funded by National Institutes of Health (NIH) in the College of Pharmacy and Pharmaceutical Sciences (COPPS) at FAMU. With the Department of Sociology and Criminal Justice, Dr. Dokurugu special interests are in Global Health, Medical Sociology, and Aging in America, Social Epidemiology, The Sociology of Globalization and Research Statistics. Dr. Dokurugu earned both his Masters of Public Health (MPH) and a Doctor of Public Health (Dr.PH) in Epidemiology and Bio-statistics from FAMU College of Pharmacy & Pharmaceutical Sciences. Dr. Dokurugu also earned a Master of Arts (MA) and a Bachelor of Arts (BA) in Sociology from the University of Ghana, Legon. Dr. Dokurugu is an alumna of the Minority Training Program in Cancer Control (MTPCCR) in UCLA and the Emerging Leaders in Public Health in University of North Carolina, Chapel Hill. He has successfully led 155 FAMU students, faculty, staff and alumina to Ghana, Nigeria and Kenya, in Africa for international education, research and cultural immersion since 2007. Other training certificates received are listed below: GIS for Epidemiology – University of Michigan School of Public Health Cancer Epidemiology – University of Michigan School of Public Health Emerging Leaders in Public Health – University of North Carolina, Chapel Hill School of Public Health Cultural Competency – FAMU Human Resources Organizational Development Cancer Culture and Literacy – University of South Florida College of Medicine & Moffitt Cancer C. Health Disparities in America – Famu College of Pharmacy & Pharmaceutical Sciences Minority Training in Cancer Control – University of California, Los Angeles (UCLA) School of Public H. Skills-building on community planning – Jackson State University, Mississippi Assessments Training and Research Institute – FAMU

Edward T. White

Best Teacher, Architecture

FLORIDA ATLANTIC UNIVERSITY

Jane E. Caputi

Best Overall Faculty Member, Women's Studies

Kyle Prescott

Best Researcher/Scholar, Music

"World-class artistry, intellect, sensitivity, leadership, mentorship and impact"

Kyle Prescott is the Director of Bands at Florida Atlantic University. His responsibilities include conducting the University Wind Ensemble and Chamber Winds, teaching the graduate instrumental conducting sequence and coordinating all aspects of FAU's comprehensive university bands program. He holds his undergraduate degree from the University of Idaho, and the Master of Music and the Doctor of Musical Arts degrees in Wind Conducting from The University of Texas at Austin, where his principal conducting teacher was Jerry F. Junkin. Additionally, he has studied with exemplary conductors through fourteen international conducting symposia.

In the summer of 2013, Dr Prescott was appointed Coordinator of Community Engagement in the Performing Arts for the Dorothy F. Schmidt College of Arts and Letters. In this administrative capacity, he works to foster artistic relationships between the university and various agencies, including non-profit, public, and private.

Prior to his work at The University of Texas, Dr. Prescott conducted bands and orchestras in the Pacific Northwest. Ensembles under his direction have been invited to perform at the Northwest Regional Conference of the National Association for Music Education in 1997 and 1999, the Southern Conference of the College Band Directors National Association in

2008 and the Florida Music Educators Association Conference in 2009. A strong proponent of new music for wind ensemble, Dr. Prescott has conducted nine world premieres in recent years, including works by Arthur Weisberg and Libby Larsen. In 2016, Dr Prescott will serve as the conductor of the Florida All State 9–10 Concert Orchestra, in Tampa. His research in score analysis and rehearsal has been published in five separate volumes of the Teaching Music Through Performance in Band series. He is the current Florida State Chair of CBDNA, Conductor of the Philharmonic Orchestra of the Palm Beach Youth Orchestras, Vice President for Strategic Planning for The Symphonia: Boca Raton, and Conductor and Music Director of South Florida's professional wind band, the Florida Wind Symphony.

Mark Rose

Best Researcher/Scholar, History

Mark Rose studies and teaches the social and policy bases of American life in the 20th Century. He is the author of more than forty articles in such journals as the Journal of Urban History, Technology and Culture, The Journal of Policy History, and the Journal of Planning History. Rose's books include Interstate: Highway Politics and Policy Since 1939, 3d.ed.rev (with Raymond Mohl, and Cities of Light and Heat: Domesticating Gas and Electricity in Urban America. For another book, The Best Transportation System in the World: Railroads, Trucks, Airlines and American Public Policy in the Twentieth Century, Rose served as Senior Author (with Paul Barrett and Bruce E. Seely). Rose's research has been supported by grant agencies including the National Endowment for the Humanities, the National Science Foundation, and the Woodrow Wilson International Center for Scholars.

Rose's current projects include a book-length study of US bank politics since the 1970s (under contract with the University of Pennsylvania Press); and the American city since 1945 (with Roger Biles). Rose co-edits, "American Business, Politics, and Society," a book series published by the University of Pennsylvania Press. Up to February 2015, the press has published eight books in the series, two are in production, and additional manuscripts are under contract. In 2013, Rose prepared "The President and Transportation Policy: FDR to Obama" for the University of Virginia's Miller Center, following its biannual Goode transportation conference (http://web1.millercenter.org/conferences/report/conf_2013_transportation.pdf).

Mark is active in several professional associations. Since 2000, he has held elected and appointed positions in the Business History Conference, the Organization of American Historians, the Society for American City and Regional Planning History, and the Urban History Association. Rose edited or co-edited issues of Enterprise & Society, the Journal of Planning History, and the Journal of Urban History, where he also served as a member of the Editorial Board. Mark is past president of the Business History Conference.

FLORIDA GULF COAST UNIVERSITY

Jon Braddy

Best Researcher/Scholar, Communication

Kim Huff

Best Overall Faculty Member, Communication
Best Researcher/Scholar, Communication
Best Teacher, Communication
Most Helpful to Students, Communication
Best Researcher/Scholar, Liberal Arts, Social Sciences, and Humanities
Best Teacher, Liberal Arts, Social Sciences, and Humanities
Most Helpful to Students, Liberal Arts, Social Sciences, and Humanities
Best Overall Faculty Member, Liberal Arts, Social Sciences, and Humanities

"Dr. Huff's research and area of scholarship has broadened my understanding of the discipline. She offers unique approaches and perspectives with rigor and the highest standards of our field. Dr. Huff is an excellent researcher in methodology and application; she deserves, without question, the Research Award.

Since Dr. Huff has joined the faculty, she has inspired both students and her collegues to be better; and we have."

Kimberly Jackson

Best Researcher/Scholar, Liberal Arts, Social Sciences, and Humanities

Professor Jackson began as Assistant Professor of English at FGCU in the fall of 2005. In addition to Composition I and II and Introduction to Literature, she has taught upper level undergraduate and graduate courses on nineteenth-century British literature (focusing on mad scientist texts and the gothic), 20th century horror, and literary theory. Professor Jackson's first monograph, titled Technology, Monstrosity, and Reproduction in Twenty-First-Century Horror, was published with Palgrave in 2013. She is now working on her second monograph, "Gender and the Nuclear Family in Twenty-First Century Horror," due to come out in 2015.

Miles Mancini

Most Helpful to Students, Communication

Amanda Parke

Best Teacher, Communication

"Professor Parke has made vast strides in pedagogical approaches in the classroom. Her scholarship has undergone radical growth from a positivist perspective to a critical approach. When discussing assignments, she always demonstrates with examples. For instance, in Public Speaking, she doesn't just assign a topic, she performs a speech. Ms. Parke has excellent interaction with her class and she gets to know her students personally; the mark of a great teacher."

FLORIDA INTERNATIONAL UNIVERSITY

Maria Marino

Best Overall Faculty Member, Liberal Arts, Social Sciences, and Humanities
Best Researcher/Scholar, Liberal Arts, Social Sciences, and Humanities

Asher Milbauer

Best Overall Faculty Member, English
Best Teacher, English

FLORIDA MEMORIAL UNIVERSITY

Martha Dawson

Best Overall Faculty Member, Liberal Arts, Social Sciences, and Humanities

Dr. Martha Dawson joined the faculty at Florida Memorial University in 1995. She serves as advisor to Iota Phi Theta Fraternity, Inc and Sigma Tau Delta, an international honor society for students of English. Her teaching interests include College Writing, Advanced Grammar and Composition, World Literature, Literary Criticism, and British Literature. She is an editor of the Florida Memorial University literary magazine Imagine and was recently published in Teaching English in the Two Year College.

FLORIDA STATE UNIVERSITY

Laura Arpan

Best Overall Faculty Member, Communication

Attitudes and persuasion; message processing and effects, especially environmental and health campaign messages; perceptions of bias Teaching Interests

Persuasion, communication campaigns, research methods, and media & society

Janet Berry

Best Overall Faculty Member, Social Work

Jim Dawkins

Best Overall Faculty Member, Design

"Brings professional expertise to education"

Jim Dawkins is an assistant professor of Interior Design at The Florida State University where he serves as an instructor in both undergraduate and graduate design studios and courses in graphic techniques, specifically hand drawing and sketching, used for design ideation and visual communication. He is a registered architect in several states, having earned his BA in Design and Master of Architecture degrees from Clemson University prior to practicing for twenty years as an architect, designer and corporate officer with design firms in Atlanta, GA and Vail, CO. Dawkins' primary research interest focuses on the notion of thinking through drawing – graphic facilitation, communication, and mediation of ideas through hand-drawing techniques and their realization in traditional hand and digital hybrid forms of computer aided design. Secondary research currently involves the analysis of and instruction in hand sketching skills through the cognitive framework of expertise theory.

Adam Gaiser

Best Teacher, Religion

Matthew Goff

Best Researcher/Scholar, Religion
Most Helpful to Students, Religion

Ray Hattaway

Most Helpful to Students, Languages

"There is nobody in the German division of MLL who does more for or spends more time with the students. Coordinating graduate teaching assistants in German by itself is a big enough task, but in addition to this, Ray is a dedicated teacher, who dedicates his energy not only on his classes, but also participates actively in extracurricular activities that help the German division."

Ray Hattaway is an Adjunct Instructor of German, and the Coordinator of the German Basic Language Program. He holds the BA in German language and Literature, Valdosta State College and Masters Degrees in Foreign Language Education and German Language and Literature from Florida State University. Ray specializes in German Language acquisition, Foreign Language instruction techniques, as well as German Literature: Middle-High German Epic, 20th Century (Brecht, Grass, Mann, Hesse), German Poetry, Drama, Tragedy, and Theater.

John Kelsay

Best Overall Faculty Member, Religion

John Kelsay focuses on religious ethics, particularly in relation to the Islamic and Christian traditions. His current work deals with religion and politics. Professor Kelsay serves as editor of Soundings: An Interdisciplinary Journal, as well as Director of FSU's Center for Humanities and Society.

Mark Pietralunga

Best Overall Faculty Member, Languages

"Mark has been and still is the one person most responsible for the overall success and standing of the department. As chairperson of MLL for over 10 years he provided leadership and enhanced recognition of MLL within the FSU community. As colleague and teacher he is respected and liked. His scholarly work speaks for itself, and when the department recently asked him once more to assume the position of chair he agreed to take on this task."

His research interests include 20th century Italian literature, translation studies, post war Italian narrative, Italian American studies, and book publishing. He has served as resident associate director and as summer director of Florida State University's study abroad program in Florence, Italy.

Jennifer Proffitt

Best Teacher, Communication

I am an Associate Professor in the School of Communication, College of Communication & Information, Florida State University.

My research interests include political economy of media and broadcast regulation and history. I teach courses in Political Economy of Media, Media Ethics, and Communication Regulation & Policy.

My research has appeared in the Journal of Broadcasting & Electronic Media, Journal of Radio & Audio Media, Journal of Communication Inquiry, Journal of Mass Media Ethics and others.

Arthur Raney

Best Researcher/Scholar, Communication

Art Raney is the James E. Kirk Professor of Communication and Director of Doctoral Studies in the School of Communication. His research primarily examines how and why we enjoy various media entertainment content, with specific attention to the role morality plays in those processes. His writings on these issues have been published in various anthologies, as well as Journal of Communication, Media Psychology, Communication Theory, and Mass Communication and Society. His edited books include Media and Social Life (with Mary Beth Oliver; Routledge, 2014) and Handbook of Sports and Media (with Jennings Bryant; Routledge, 2006). He currently serves as Associate Editor for Journal of Media Psychology.

Julie Stirwalt

Best Overall Faculty Member, Liberal Arts, Social Sciences, and Humanities

Mark Zeigler

Most Helpful to Students, Communication

FORT HAYS STATE UNIVERSITY

Brenda Craven

Best Overall Faculty Member, English
Best Teacher, English

Brenda received her M.A. in English from Fort Hays State University. Her interests range from travel and wilderness camping to politics and current events. While much of her reading focuses on works by 20th century Midwestern and Southern writers, Brenda relaxes by reading espionage novels set during the Cold War. She joined the KHC TALK program as a discussion leader in 2008.

Eric Leuschner

Best Researcher/Scholar, English

Sharon Wilson

Most Helpful to Students, English

Ethnic Literature, especially Native-American and Hispanic; Young-Adult Literature

FRANCISCAN UNIVERSITY OF STEUBENVILLE

John Bergsma

Best Researcher/Scholar, Theology
Best Researcher/Scholar, Liberal Arts, Social Sciences, and Humanities

Dr. John Bergsma is Associate Professor of Theology at the Franciscan University of Steubenville, in Steubenville, Ohio. He holds the M.Div. and Th.M. degrees from Calvin Seminary, Grand Rapids, Michigan, and served as a Protestant pastor for four years before entering the Catholic Church in 2001 while pursuing a Ph.D. in Theology from the University of Notre Dame. He specialized in the Old Testament and the Dead Sea Scrolls, graduating with high honors in 2004. His major study of the interpretation of the Year of Jubilee in ancient times is published as The Jubilee from Leviticus to Qumran (Brill Academic, 2007).

Dr. Bergsma's articles, some co-authored with Dr. Scott Hahn, have appeared in several academic journals as well as Lay Witness, the magazine of Catholics United for the Faith, and he has published a book, Bible Basics for Catholics: A New Picture of Salvation History (Ave Maria Press, 2012). He has appeared as a guest on EWTN's The Journey Home, Franciscan University Presents, and Relevant Radio's Drew Mariani Show. Twice voted Faculty of the Year by graduating classes of the Franciscan University of Steubenville, Dr. Bergsma is a popular teacher who inspires his students with a love of Scripture. He and his wife Dawn reside with their eight children in Steubenville, Ohio. - See more at: http://johnbergsma.com/biography.html#sthash.YzhM6cHB.dpuf

Steven Hildebrand

Best Overall Faculty Member, Theology
Best Overall Faculty Member, Liberal Arts, Social Sciences, and Humanities

Mark Miravalle

Best Teacher, Theology
Best Teacher, Liberal Arts, Social Sciences, and Humanities

Michael Sirilla

Most Helpful to Students, Theology
Most Helpful to Students, Liberal Arts, Social Sciences, and Humanities

FULLERTON COLLEGE

Tony Mazzaferro

Best Overall Faculty Member, Music

"Head of the Music Department, Fullerton College"

Dr. Anthony Mazzaferro was born and raised in San Francisco. He received his Bachelor of Music degree from San Francisco State University. During his undergraduate years, he served as the Principal Tubist in the Symphonic Band and Symphony Orchestra. In addition to this, he was the Assistant Conductor for both the Symphonic and Concert Bands. After graduating with honors in 1978, Dr. Mazzaferro attended Northwestern

University. He pursued a Master of Music degree in Instrumental Conducting where he studied with John P. Paynter. In 1986 he received his Doctor of Musical Arts degree from Arizona State University. As a Graduate Assistant, Dr. Mazzaferro worked with the Band and Music Education departments. While at Arizona State, Dr. Mazzaferro studied with Richard Strange, Robert Fleming, Daniel Perantoni, Ronald LoPresti and Barrie Wells.

Previous to his appointment at Fullerton College, Dr. Mazzaferro taught for the Mount Diablo and Palo Alto Unified School Districts. His assignments included both elementary and secondary instrumental ensembles. His guest conducting and adjudication assignments have included Honor Bands and Orchestras throughout the Western United States, Canada, and New Zealand. In 1994 and 2001 Dr. Mazzaferro was selected to conduct the California All-State Honor Band. In 2000 Dr Mazzaferro was selected to conduct the Alberta, Canada All-Provincial Honor Band. In 2007 Dr. Mazzaferro will conduct the California All-State Junior High Symphonic Band and the Northern California High School Honor Band. As a clinician, Dr. Mazzaferro has worked with Concert Bands and Orchestras from The United States, Japan, Singapore, New Zealand, Canada, and China.

On his principal instrument, tuba, Dr. Mazzaferro has performed with numerous bands and orchestras throughout California. This includes the Diablo Symphony Peninsula Symphony, Symphony of the Mountain, Berkeley Symphony, Santa Cruz Symphony, Santa Ana Symphony, and the Crystal Cathedral "Hour of Power" Orchestra.

Dr. Mazzaferro is active in several other musical affiliations outside of Fullerton College. His most recent activities include serving on the Board Of Directors for the Cazadero Performing Arts Camp in Northern California. Dr. Mazzaferro is a member of, CBDA, CBDNA, MENC SCSBOA, TUBA and WASBE.

Dr. Mazzaferro is currently the Director of Bands at Fullerton College. In addition to his academic assignments, Dr. Mazzaferro conducts the Symphonic Wind Ensemble, Community Band, and instructs Applied Euphonium and Tuba. He is also completing his eleventh year as Principal Conductor of the Claremont Symphonic Winds.

Bruce Miller

Best Researcher/Scholar, Music

Jaime Perez

Best Researcher/Scholar, Fine Arts

Jamie Shew

Most Helpful to Students, Music

Whether on stage or in the classroom, these are words that vocalist Jamie Shew seeks to achieve. Isn't that really what all vocalists are trying to accomplish? To be able to project our personal emotions and experiences through song?

Jamie grew up in the Tri-Cities, a group of towns located in south central Washington State. She began her musical training at the age of four, when she was first introduced to the piano. As a young child, she remembers having a special interest in the music of Frank Sinatra, Nat King Cole, and Dean Martin. She often played radio games with her parents in the car, trying to guess which one was singing while her father was impersonating along.

The desire to sing was very strong for Jamie throughout her childhood, however, she was always needed as an accompanist. It was during her senior year in high school that her jazz choir director gave her the opportunity to sing a Karen Carpenter solo. This was a pivotal moment - she was hooked.

Jamie earned a Bachelor's Degree in Music Education from Washington State University and a Master's Degree in Vocal Performance/Jazz Studies at Western Michigan University. As a graduate assistant at WMU, she had the opportunity to study privately with one of her favorite jazz vocalists, Sunny Wilkinson, and perform in Gold Company, the internationally acclaimed vocal jazz ensemble.

As a jazz vocalist, Jamie continues to enjoy a dynamic musical career among many internationally recognized jazz musicians such as Darmon Meador of the New York Voices, Sunny Wilkinson, Nancy King, Steve Christofferson, Royce Campbell, Bruce Babad, Joe Jewell, Tom Knific, Steve Zegree, Shawn "Thunder" Wallace, Doug Bickel, Evan Stone, and Josh Nelson.

Recently, Jamie has finished intensive work recording her debut album, The Answers Are You, featuring her husband and bassist, Roger Shew, guitarist Royce Campbell, percussionist Chris Earley, and Rick Holland on the flugelhorn. The album highlights arrangements by Jamie, along with her original composition "My Only Obsession is You."

Currently, Jamie is Director of Vocal Jazz Studies at Fullerton College in Fullerton, CA and performs regularly in the greater Los Angeles area. Before coming to LA, she was an Assistant Professor of Jazz Voice at Virginia Tech University in Blacksburg, VA.

John Tebay

Best Teacher, Music

Mr. Tebay received his Bachelor of Music Education Degree from Biola University and is pursuing a Master's Degree from California State University, Long Beach. Prior to joining the Music Department Faculty at Fullerton College, he held the position of Choral Director at El Dorado High School in Placentia from 1981 to 1986. He was named Teacher of the Year in 1985. Mr. Tebay's responsibilities include directing the 70 voice Concert Choir, Bravo, a song and dance ensemble, as well as teaching voice, and History of Rock Music. John has been a guest conductor throughout California and his choirs have performed in ten states, Canada, Mainland China and at the American Choral Directors Association National Conference. John is also busy as an actor and with his Dickens Carolers.

FURMAN UNIVERSITY

Lane Harris

Best Overall Faculty Member, History

After growing up in the Minnesota, Lane J. Harris attended Drake University in Des Moines, Iowa, where he received a B.A. in History. Upon graduation, Harris was selected by the United Board for Higher Education in China to teach English at Northeast Normal University in Changchun, Jilin, China for a year. Returning to the United States, he attended Washington University in Saint Louis where he earned a Master's degree in East Asian Studies with a concentration in Republican (1911–1949) Chinese history. For his Ph.D., Harris attended the University of Illinois, Urbana-Champaign. His research focused on late imperial and Republican Chinese postal history. After spending a year in Nanjing researching postal history at the Second Historical Archives of China, Harris returned to the United States and moved to Ellensburg, Washington where he taught World History. He joined the faculty at Furman University in 2009.

Marian Strobel

Most Helpful to Students, History

"Professor Strobel willingly commits untold hours advising students, talking with them about their problems and keeping up with years after they have graduated. She is an excellent advisor and helps students more than any other colleague I know. Her office is always filled with students. I can think of no one in our department more deserving of this award."

A native of Rochester, New York, Professor Strobel attended Mount Holyoke in Massachusetts and graduate school at Duke University. She taught at the University of North Carolina at Charlotte, the University of Montana, the University of Virginia, and Phillips Exeter Academy before coming to Furman. She has served on a myriad of committees at Furman dealing with the status of Furman's students and faculty, as well as the institution as a whole. From 1999–2010, she served as Chair of the History Department. Professor Strobel has been the recipient of the Meritorious Teaching Award and the Maiden Invitation Award for excellence in the classroom. She has also been an active participant in the First Year Seminar program and was a member of the original task force that implemented that project. Currently, she is a Shi Sustainability Fellow. Professionally, Professor Strobel studies the history of women's higher education and American politics after World War II, as well as African-American history. She has presented her research in sessions at such prestigious venues as the annual conferences of the Organization of American Historians, the Southern Historical Association, and the American Historical Association. She has also been a member of special teaching based seminars such as ones sponsored by the Mellon

Foundation and the Gilder-Lehrman Institute and also been part of Furman faculty foreign study trips to Canada, Jamaica, Cuba, and Mexico. During May Term 2014, Professor Strobel co-directed a study away class on "War and Remembrance" that commemorated the 100th anniversary of the outbreak of World War I and traveled to England, France, and Belgium.

GAINESVILLE STATE COLLEGE - OAKWOOD

Ric Kabat

Best Overall Faculty Member, Liberal Arts, Social Sciences, and Humanities

Jeffrey Pardue

Best Overall Faculty Member, History

Ben Wynne

Best Researcher/Scholar, History

A native of Florence, Mississippi, Dr. Wynne specializes in antebellum American History, History of the U.S. South, and Civil War history.

GALLAUDET UNIVERSITY

Jeffrey Brune

Best Researcher/Scholar, History

I have been at Gallaudet since 2005 and am passionate about U.S. history because everyone's identity and politics, whether or not they realize it, is based on their interpretation of the past. Because our perception of history is so important, it is vital that we think critically and deliberately about the past. It is because of this that I enjoy teaching a wide variety of history courses, including disability history, urban history, immigration history, colonial history, African American history, research methods and historiography, and both halves of the U.S. history survey.

Disability history has become my passion. Since I began to focus on disability in 2008, one year after completing my dissertation, I have gained a new sense of purpose for my scholarship. Disability studies, which approaches the topic as a social and cultural phenomenon rather than as a medical condition, has tremendous potential to improve our society, politics, culture, and scholarship. While the field has already made a significant impact, there is still much more to be done. I have never felt as passionate about research and scholarship as I have since joining this field.

Robert Sanchez

Best Overall Faculty Member, History
Best Teacher, History
Most Helpful to Students, History

GANNON UNIVERSITY

Jeff Bloodworth

Best Overall Faculty Member, History
Best Researcher/Scholar, History
Best Teacher, History
Most Helpful to Students, History

Hi. I'm Jeff Bloodworth. I am an associate professor of history at Gannon University. Gannon is located right on Lake Erie (well, about 1/2 mile from it) in Erie, Pennsylvania. One of the reasons I love Erie is the snow . . . we get a lot of it. For its size, Erie has quite a bit of diversity. From Iraqi and Bosnian refugees to Sudanese and Burundians, not to mention the Poles and Slovaks, Erie boasts thousands of citizens from all corners of the globe.

Currently, I am the head of the department of history and archaeology. I teach modern US history and advise the History Club & Campus Democrats. As an adviser, I have led student trips to Washington, D.C., Philadelphia, New York City, the FDR Library, London, Prague, Berlin, Copenhagen and the Middle East. History majors have fun and do lots of great stuff!! I really love teaching here at Gannon.

My research specialty centers on contemporary United States Political history. My book, The Wilderness Years: A History of American Liberalism, 1968–1992 is coming out in the fall of 2013. In the long term, I am very excited about my book project, a biography on the Speaker of the House, Carl Albert. This will be years in the making–but at this time my favorite thing about Carl Albert is his hometown, Bug Tussle, Oklahoma.

In my spare time, aside from watching television, I follow my beloved St. Louis Cardinals and anxiously await the publication of any book by Paul Berman.

I was born in Memphis, Tennessee and lived throughout the South as a kid. From Atlanta and Birmingham to Knoxville to eventually Springfield, Missouri, I lived in a lot of places and have continued those nomadic ways until I settled down in Erie.

MC Gensheimer

Best Overall Faculty Member, Liberal Arts, Social Sciences, and Humanities

MC Gensheimer started her career, in 9th grade, with a soldering gun, a can of orange spray paint, and a camera.

Her photographic and graphic work has been accepted regionally for publication and gallery exhibits. Most recently her work was accepted by the Erie Philharmonic for the celebration of their 100th year, the work "The Chair-ished Violin will be exhibited summer 2013. Recently, she returned from a workshop at the Denver Art Museum which explored the geometric work of

Native American artists and craftsmen. On Friday July 5, she finished her first MOOC, on "The History of Rock and Roll." She finished the course- "with Distinction."

She currently serves as a member of the new GannonKnight sculpture group. And she is working on compiling the history of the Gannon Knight tradition. She lives in Erie, with her husband, two cats- Digit and Gracie and a dog named Cali.

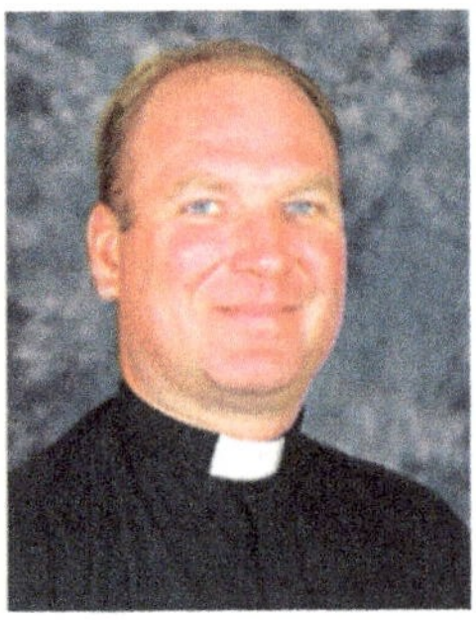

Jason Glover

Most Helpful to Students, Liberal Arts, Social Sciences, and Humanities

I was ordained a Roman Catholic priest for the Diocese of Erie at Saint Peter's Cathedral by Bishop Donald Trautman on June 7, 2002. After three years of wide and varied ministry at Notre Dame Church in Hermitage, Pennsylvania, I was assigned to continue my advanced theological studies at Saint Mary's Seminary & University in Baltimore, Maryland. After completing my Licence in Sacred Theology (STL) in 2007, the bishop assigned me to teach and reside at Gannon University.

As an instructor in the Department of Theology, I have the privilege of interacting with our students in the formal classroom setting. As a resident-priest living on campus, I also have the unique opportunity to be immersed in the daily life of our student community in a far more casual way. Outside of the classroom, I can often be spotted almost everywhere on campus, including our sporting events, fraternity/sorority functions, and faith-sharing gatherings organized by students.

I am an avid reader and traveler. I enjoy any type of athletic competition and am a diehard fan of the New York Yankees, Pittsburgh Steelers and Penguins, and The Ohio State Buckeyes. I am frequently seen around campus with Drake, my 3-year old black lab (and "unofficial" mascot of Gannon), and I also a keeper of saltwater fish aquariums. I am proud of my German heritage and celebrate it, in part, through my hobbies of cooking and homebrewing. I enjoy most all music (except country and rap), with the Dave Matthews Band topping my charts. I was addicted to House, and now find The Big Bang Theory one of the most entertaining commedy series on television.

Of all of these joys and pleasures of life, my first and true love is my priesthood. I love being a priest and strive everyday to bring others to Christ.

Doug King

Best Teacher, English
Best Teacher, Liberal Arts, Social Sciences, and Humanities

I have been teaching at Gannon since 2001 and absolutely love it. Aside from the wide range of courses I teach, I also lead student trips overseas every year. In the past three years, we have gone to: Greece (2010); southern Italy; Nice, France; Monaco; and Barcelona (2011); and London, Paris, Florence, and Rome

(2012). Next year I have planned trips to England and to Japan. Students and family members love these tours; one can also get credit for going.

I love helping out with my son's boy scout troop, which gets us outdoors and away from it all. My other interests include mountain biking, acting, and playing music. I write songs, sing, play guitar, and have a band in which we play original, indie-type music. On occasion, I do break out the guitar in class. Courses Taught

Drama of Shakespeare
The Art of Film
The Spiritual Struggle in Literature
Modern Drama
Comedic Literature
European Literature
Prose Fiction
College Composition
and many more!

Graduate:

Shakespeare: Tragedies Shakespeare: Comedies & Histories Tudor and Stuart Drama The American Novel Post-World War II Literature and Film (online)

Educational History

Ph. D. English, Duquesne University (2002). Dissertation: Shakespeare as Adapted for Children.

M.A. Education, The George Washington University (1994).

B.A. English, State University of New York at Buffalo (1982).

Professional Societies

The Shakespeare Association of America

Scholarship/Research

I have done articles, book chapters, reviews, and conference presentations on a variety of topics, including:

Shakespearean Adaptations for Children; Shakespeare in Performance; Film Acting; European Literature and Film; and Literature Adapted into Song

Janet Laboda

Most Helpful to Students, English
Most Helpful to Students, Liberal Arts, Social Sciences, and Humanities

Sally Levan

Best Overall Faculty Member, English
Best Overall Faculty Member, Liberal Arts, Social Sciences, and Humanities

Laura E. Rutland

Best Researcher/Scholar, English
Best Researcher/Scholar, Liberal Arts, Social Sciences, and Humanities

Anjali Sahay

Best Researcher/Scholar, Liberal Arts, Social Sciences, and Humanities

Brent Sleasman

Best Teacher, Liberal Arts, Social Sciences, and Humanities

I am originally from Fayette County in southwestern Pennsylvania and my family currently lives in Millcreek Township. Prior to coming to Gannon University in 2008, I taught at West Chester University of Pennsylvania. In addition to my teaching experience, I have served as a youth pastor, church planter, and pastor of a small congregation in Beaver County, PA.

GARDNER-WEBB UNIVERSITY

Joseph Webb

Best Teacher, Communication Studies

GEORGE MASON UNIVERSITY

James R. Carroll

Best Teacher, Music

Olga Arans

Most Helpful to Students, Classics

M.A. in Classics from the Saint Petersburg University, Russia. Ph. D. in Classical Philology from the University of Illinois at Urbana-Champaign. Adjunct professor in Classics at GMU; UMW; NOVA Community College.

Lorrie Berkshire-Brown

Best Researcher/Scholar, Music

Amy Best

Best Overall Faculty Member, Sociology

Amy L. Best is Professor of Sociology and Chair in the Department of Sociology and Anthropology at George Mason University. Her research focuses on the study of youth identity formation, youth well-being, and social inequalities, with a particular interest in how gender, ethnicity, sexuality, race and class differently shape the social experiences of contemporary American youth. She is interested in qualitative and feminist approaches to social research. Best is author of Prom Night: Youth, Schools and Popular Culture (2000 Routledge), which was selected for the 2002 American Educational Studies Association Critics' Choice Award and Fast Cars: Cool Rides: The Accelerating World of Youth and Their Cars (NYU Press 2006), and editor of Representing Youth: Methodological Issues in Critical Youth Studies. (NYU Press, 2007). Her forthcoming book is Fast Food Kids: Youth and Changing Food Landscape of School and Home (NYU Press, Forthcoming 2015).

William Carey

Best Teacher, Classics

Stanley Engebretson

Best Overall Faculty Member, Music

Dan Joyce

Best Researcher/Scholar, Fine Arts

"Dan Joyce is an accomplished dance scholar and historian. He is accurate in his research and relates the contact to his audience (students)."

Dan Joyce is an associate professor in the School of Dance at George Mason University in Fairfax, VA. A graduate of the North Carolina School of the Arts, he began his professional career with the Maryland Dance Theater and Concert Dance Company of Boston. In 1988, he was selected to join the Mark Morris Dance Group with whom he performed for ten years. Dan appeared in over thirty works choreographed by Mark Morris, including several film and opera projects. Since joining the GMU faculty in 1999, where he received his MFA in 2003, Dan has choreographed for the GMU Dance Company and the Theater of the First Amendment. A popular teacher for numerous regional and national festivals, he has also served on the faculties of the Virginia Governor's School, the School of the Washington Ballet, and the Citydance Center at Strathmore. He has participated on funding panels and advisory committees for the Arts Council of Fairfax County, the Arts and Humanities Council of Montgomery County, the Alexandria Council for the Arts, and Dancemetro/DC. In 2006–07, Dan returned to the stage in an international collaborative project involving American and Russian dancers, performed in Washington, DC, Moscow, Russia, as well as in Siberia. He makes his home in Arlington, VA.

Brian Platt

Best Overall Faculty Member, History
Best Teacher, History

Brian Platt is Associate Professor of History and is currently serving as Department Chair. He is a specialist in Japanese history, with a research focus on the 18th and 19th centuries. He received his Ph.D. from the University of Illinois at Urbana-Champaign in 1998, and is the author of Burning and Building: Schooling and State Formation in Japan, 1750–1890 (Harvard, 2004). He is the recipient of grants from the Fulbright Foundation, the National Endowment for the Humanities, the Spencer Foundation, the National Academy of Education, and the Association for Asian Studies. His current research project deals with historical commemoration and autobiography in 18th and 19th century Japan. He has taught various courses on Japanese and Asian history, as well as comparative courses on such issues as modernization, memory, gender, and national identity.

Elaine Rendler

Most Helpful to Students, Music

Elaine Rendler-McQueeney, received a Doctor of Musical Arts in organ with minors in theory and liturgy from The Catholic University of America under the tutelage of Daniel Roth (St. Sulpice, Paris) and Conrad Bernier. Before coming to CUA she studied organ with Alexander McCurdy while serving as principal organist at the Cathedral Basilica of SS. Peter and Paul, Philadelphia.

Dr. Rendler – teacher, author, choir director, composer is known principally for her work as a national and international clinician in church music and pastoral liturgy. She has served in music ministry at Georgetown University, the US Naval Academy, Annapolis, and Bellarmine Chapel at GMU. In 2008, she received a National Award for Outstanding Contributions to the Liturgical Life of the Church from the Georgetown University Center for Liturgy. She is the editor of Today's Liturgy, a quarterly publication for church musicians and author of several books, including In the Midst of the Assembly, and This is the Day. Her Keyboard Praise collections are published by OCP Publications. She is also the founding director of the Georgetown Chorale, whose mission is to provide for the musical excellence of members and community service through music and whose appearances include the Kennedy Center with the Washington Chamber Symphony. She has taught in the music department at Mason since 2000.

Martin Winkler

Best Researcher/Scholar, Classics

GEORGE WASHINGTON UNIVERSITY

Allyson Currin-Stokes

Best Researcher/Scholar, Theater

Allyson Currin Stokes is a two-time Helen Hayes-nominated playwright and D.C. area actress. Her plays have been produced locally at Theatre Alliance, Charter Theatre, Source Theatre Company, Washington Shakespeare Company and Church Street Theatre. As an actress she has appeared at Catalyst Theatre, Everyman, Rep Stage, Signature, Round House, Washington

Stage Guild, Washington Shakespeare Company, Studio, Olney, and more, in addition to her work in television, commercials and film. She is a member of Actors Equity, the Screen Actors Guild and The Dramatists Guild.

Maria De La Fuente

Best Overall Faculty Member, Spanish

Alexander Dumbadze

Best Researcher/Scholar, Art History

Alexander Dumbadze teaches courses on contemporary art, theory, and historiography. He is the author of Bas Jan Ader: Death Is Elsewhere (University of Chicago Press, 2013) and co-editor and co-author, with Suzanne Hudson, of Contemporary Art: 1989 to the Present (Wiley-Blackwell, 2013). He is currently at work on a book manuscript titled Jack Goldstein: All Day Night Sky. In 2011, he received a Creative Capital Warhol Foundation Arts Writers Grant. A founder of both the Society of Contemporary Art Historians and the Contemporary Art Think Tank, Prof. Dumbadze's recent essays include "Of Passivity and Agency: Jack Goldstein" in Jack Goldstein x 10,000 (Orange County Museum of Art, 2012), "Spectacle and Death" in September 11 (MoMA PS1, 2011), and "Can You Hear the Lights?" in Art, History and the Senses: 1830 to the Present (Ashgate, 2010). In spring 2012, Prof. Dumbadze was a Visiting Professor of Art History at the Université Paris 8, Vincennes-Saint Denis.

Dolores Perillan

Best Teacher, Spanish
Most Helpful to Students, Spanish

Erin Speck

Best Researcher/Scholar, Interior Design

"Always available and supportive even semesters after they have taken her classes."

Nashwa Taher

Best Teacher, Arabic

Alan Wade

Best Overall Faculty Member, Theater
Best Teacher, Theater
Most Helpful to Students, Theater

"Professor Wade is unequivocally the best teacher. His experience as a director, his critical thinking skills and his ability to articulate salient information to students, production teams, colleagues and other stakeholders make him an excellent candidate as the best teacher. He is always teaching even when he isn't in front of a class. He was my professor 25 years ago and I still learn from him all the time.

Professor Wade is an excellent candidate. He offers specific advice and well thought out information to students and his colleagues.

Professor Wade is a consistent presence in the department. He is an outstanding teacher. He practices his craft in theatre and films. His ongoing experiences give him a special skill set. He is very smart."

A professional actor and director, Mr. Wade most recently directed for The Washington Stage Guild Shaw's Overruled! and for The Bay Theatre, Annapolis, A.R. Gurney's Love Letters. For The Olney Theatre Center he directed Steinbeck's Of Mice and Men, as well as nine productions for the Center's National Players Touring Company. Over the last 20 years he has directed five critically-acclaimed productions for The Washington Stage Guild including Beckett's Happy Days with noted Washington actress June Hansen. Mr. Wade directed a staged reading of Anastasia Traina's In Audela with TV's Scott Cohen (Law and Order) for The Royal Academy of Dramatic Art in London. At GW his directing credits include, among many others, a workshop premiere of Rick Chafe's Shakespeare's Dog produced in association with The Manitoba Theatre of Canada.

As an actor Mr. Wade has appeared off-Broadway at The Atlantic Theatre and regionally at Baltimore's Center Stage and several Washington, DC, theatres including Arena Stage, The Studio Theatre, Signature Theatre, The Potomac Theatre Project (for which he received a Helen Hayes Award nomination), The Washington Stage Guild and The Olney Theatre. On network television he has been seen on Homicide, A Man Called Hawk and John F. Kennedy and His Times. Mr. Wade's movie credits include The Pelican Brief and Major League II. He has local and national voice-over credits.

He has lectured at The Royal Academy of Dramatic Art, at Richmond University, London, and The University of Sussex in Brighton, England. He is a former editor of Literature in Performance and a former associate editor of Text and Performance Quarterly. Mr. Wade is a member of the Board of Directors of The Olney Theatre Center for the Arts and serves on the advisory boards of The Washington Stage Guild and The Potomac Theatre Project.

At GW Mr. Wade teaches "Introduction to Acting" and several scene study courses including contemporary and classical text. He is the immediate past director of Columbian College's signature program The Dean's Scholars in Shakespeare, a former department chair, and a former co-director of The University Teaching Center.

GEORGIA COLLEGE

Aurora Castillo

Best Researcher/Scholar, Languages
Best Teacher, Languages
Most Helpful to Students, Languages
Best Overall Faculty Member, Languages
Best Teacher, Liberal Arts, Social Sciences, and Humanities
Most Helpful to Students, Liberal Arts, Social Sciences, and Humanities
Best Overall Faculty Member, Liberal Arts, Social Sciences, and Humanities

"Very professional. For her, students come first."

Dr. Castillo, a Colombian native, joined the GC faculty in August 2007. She came to the United States to study English at West Virginia University. She moved to Milledgeville after living in West Virginia for eight years. She holds an M.A. in Spanish, an M.A. in Secondary Education, and an Ed.D. in Technology Education from West Virginia University. Her research interest focuses on Instructional Technologies in Foreign Languages, Technology Education, and Foreign Language Methodology. She has been working on the application of authentic activities through Instructional Technologies in Foreign Language teaching and learning. Most recently, Dr. Castillo has used new teaching approaches such as implementing virtual conferencing with students in her native country, Colombia, and virtual worlds in her Spanish classes. At the International Conference on College Teaching and Learning, she was distinguished with the Award for Innovative Excellence in Teaching, Learning and Technology for her work with virtual worlds.

Hedwig Fraunhofer

Best Overall Faculty Member, Languages

Richard Greene

Best Overall Faculty Member, Music
Best Researcher/Scholar, Music
Best Teacher, Music
Most Helpful to Students, Music
Best Researcher/Scholar, Liberal Arts, Social Sciences, and Humanities
Best Teacher, Liberal Arts, Social Sciences, and Humanities

Most Helpful to Students, Liberal Arts, Social Sciences, and Humanities
Best Overall Faculty Member, Liberal Arts, Social Sciences, and Humanities

Professor of Music (1997), is a classical guitarist, composer and analytical musicologist. Dr. Greene received a Bachelor of Music in Guitar Performance degree from Loyola University in 1970, a Master of Fine Arts in Guitar Performance degree from Tulane University in 1976 and a Ph. D. in Analytical Musicology from the University of Leeds in 1992. He is one of the world's leading authorities on the music of the English composer Gustav Holst, whose composition, "Jupiter" (from "The Planets") is a favorite of most bands and orchestras. Dr. Greene's books on Holst and music analysis are published worldwide. As a composer, Dr. Greene has been commissioned to write for many different media and styles, from opera to musical comedy, and from solo guitar to orchestral works. As a performer, he has been involved in solo and chamber recitals from coast to coast, and he has recorded a CD of his own compositions. Dr. Greene teaches courses in music history and literature, music theory, classical guitar and interdisciplinary studies.

Eustace Palmer

Best Researcher/Scholar, Liberal Arts, Social Sciences, and Humanities

Eustace Palmer, Professor, (1993) College of Arts and Sciences, English and Rhetoric PhD, EDINBURGH UNIVERSITY; MA, EDINBURGH UNIVERSITY

GEORGIA GWINNETT COLLEGE

Susan Bussey

Best Overall Faculty Member, English
Most Helpful to Students, English
Most Helpful to Students, Liberal Arts, Social Sciences, and Humanities
Best Overall Faculty Member, Liberal Arts, Social Sciences, and Humanities

Michael Fournier

Best Researcher/Scholar, English

Dr. Fournier was born and raised in Maine. He received a B.A. and an M.A. in English from the University of Maine and a Ph.D. in English from Georgia State University. He has been proud to teach at the University of Maine, Kennesaw State University, Georgia State University and the Georgia Institute of Technology, but is proudest of being able to teach at Georgia Gwinnett College. His poetry and articles have appeared in a number of journals and his plays have been produced in Ann Arbor, Michigan, Kyoto, Japan, and Aberdeen, Scotland.

Fournier was a founding editor of the now-defunct poetry journal Chants and has edited The Collected Poems of Louis Ginsberg, a critical edition of Parker Tyler's long poem, The Granite Butterfly and Take This Waltz: A Celebration of Leonard Cohen. He has delivered conference presentations on the Beat Generation, Edmund Spenser, Kenneth Fearing, Herbert Read and Louis Zukofsky. His academic interests include Renaissance poetry, 20th century poetry in English and technical communication. In addition to his academic work, he has also worked in publishing and as a technical writer and instructional designer for IBM, Convergent Technologies and BMW motorcycles.

Mazin Tadros

Best Researcher/Scholar, Liberal Arts, Social Sciences, and Humanities

Daniel Vollaro

Best Teacher, English
Best Teacher, Liberal Arts, Social Sciences, and Humanities

Dr. Daniel Vollaro is an assistant professor of English, scholar and communication consultant who faithfully practices experiential learning pedagogy and service learning in his classes at GGC. Believing that students learn best by doing, his classes often revolve around semester-long projects that require students to interact with people and organizations outside of the college. He wants his students to experience something real in his classes, whether it be successfully building and launching a blog or completing a community service project off-campus. He believes that education works best when students are given more choice and autonomy in the classroom.

Vollaro is interested in the relationships between power and literature. His dissertation explored how anthologies of American literature construct historical origin stories about the United States. He has published articles on Harriet Beecher Stowe's meeting with Abraham Lincoln, Briton Hammon's Narrative, the Transcendentalist publication The Dial, and Utopian Communities in Nineteenth Century America. He is currently working on a series of articles about the confluence between anarchism and American literature. He serves on the boards of the digital journal Notes on Teaching English and the literacy organization Read Aloud Chattanooga.

With 15 years of teaching experience in the humanities and extensive professional experience as a journalist and an editor/writer for the publishing industry, Vollaro brings a praxis-oriented approach to his teaching of professional writing courses. His writing classes are often built around service learning projects wherein students work with nonprofits to improve their internal and external communication. His journalism courses train students to be citizen journalists who publish their stories online.

He holds a master's degree in Jewish-Christian studies from Seton Hall University and has taught a wide variety of humanities courses, including American and world literature, religious studies, ethics and philosophy.

GEORGIA INSTITUTE OF TECHNOLOGY

Kelly Comfort

Best Overall Faculty Member, Modern Languages

"She is wonderful! She is so dedicated to her job, students and research. I wish I could be more like her!"

Dr. Kelly Comfort received her PhD in Comparative Literature with a designated emphasis in Critical Theory from the University of California, Davis. She joined the Georgia Tech faculty as an Assistant Professor of Spanish in the School of Modern Languages in 2005 and became Associate Professor in 2012.

A specialist in Latin American literature and transatlantic modernisms, Dr. Comfort's research agenda focuses primarily on the intersections between Latin American modernismo and contemporaneous turn-of-the-century literary movements in Europe such as aestheticism and decadence.

Her book, European Aestheticism and Spanish American Modernismo: Artists Protagonists and the Philosophy of Art for Art's Sake (Palgrave MacMillan, June 2011, 192 pages) examines the changing role of art and the artist during the turn-of-the-century period and considers the multiple dichotomies of art and life, aesthetics and economics, production and consumption, and center and periphery.

Her edited volume, Art and Life in Aestheticism: De-Humanizing and Re-Humanizing Art, the Artist, and the Artistic Receptor, (Palgrave MacMillan, 2008) rethinks the relationship in aestheticism between the aesthetic and the human realms over the past two hundred years.

She published an article entitled "Art for the Artist's Sake or Artist for Sale: Lulu's and Else's Failed Attempts at Aesthetic Self-Fashioning" in the 2006 issue of Women in German Yearbook: Feminist Studies in German Literature and Culture. Another article, "The Critic as Artist and Liar: The Reuse and Abuse of Plato and Aristotle by Wilde"—which appeared in the January 2008 edition of The Wildean—considers how Wilde's "The Decay of Lying" and "The Critic as Artist" engage in a curious fin-de-siècle rereading of Plato's Republic and Aristotle's Poetics. Kelly has also published three articles related to Latin American literature: "Masculinidad rechazada: El artista recluído y no productivo en Rubén Darío y José Asunción Silva" (Latin American Literary Review. 37.73 (2009) 26–46), "The Clash of the Foreign and the Local in Martí and Carpentier: From 'Misplaced Ideas' to 'Trasculturation'" (Hipertexto. 11 (2010) 51–62), and "Colonial Others as Cuba's Protonational Subjects: The Privileged Space of Women, Slaves and Natives in Gómez de Avellaneda's Sab" (MESTER: Journal of Spanish and Portuguese Studies.32 (2003) 179–194). Another article, "The Artist as Impressionistic Critic in José Asunción Silva's De sobremesa: Transatlantic Borrowings from Walter Pater, Oscar Wilde, and British Aestheticism," is forthcoming with Revista de Estudios Colombianos (37 (2010)).

Dr. Comfort frequently teaches courses on Spanish conversation and Latin American literature as well as her "Spanish Service-Learning in the Hispanic Community" and "Hispanic Community Internship" classes. She is also faculty director of Georgia Tech's International House (a Think Big Living and Learning Community) as well as faculty advisor of the student organization "Gringos y Latinos: Atlanta's Spanish Service Society" (GLASSS). Dr. Comfort received the 2010 CETL/BP Junior Faculty Teaching Excellence Award.

GEORGIA PERIMETER COLLEGE - DECATUR

Martha Bowles

Best Teacher, English
Best Teacher, Liberal Arts, Social Sciences, and Humanities

Otto Burianek

Best Researcher/Scholar, Liberal Arts, Social Sciences, and Humanities
Best Overall Faculty Member, Liberal Arts, Social Sciences, and Humanities

Valerie Crawford

Best Researcher/Scholar, Liberal Arts, Social Sciences, and Humanities

Evelyn Hughes

Best Teacher, Music

"Ethics, professionalism, loyalty, focus, doing the right thing, follow through,"

Evelyn Hughes received her Master of Music degree in Vocal Performance from West Virginia University, with further graduate study at the University of Virginia, Westminster Choir College, and the American Institute of Musical Studies in Graz, Austria. She has performed early music throughout Europe and has appeared locally with the DeKalb Baroque Ensemble, Heritage Singers, Atlanta-Emory Orchestra, DeKalb Symphony Orchestra, DeKalb Music Theatre, Atlanta Opera, Baroqueworks, and DaSalo Solisti Chamber Orchestra.

Mrs. Hughes is a member of the Georgia Music Teachers Association, holds National College Faculty Certification through the Music Teachers National Association, and has been recognized as an outstanding teacher by Who's Who Among America's Teachers. She is also an active member of the National Association of Teachers of Singing. Mrs. Hughes has adjudicated a variety of voice competitions locally and throughout the southeast, and has presented workshops in vocal technique and diction. She is an Associate Professor of Music at Georgia Perimeter College where she teaches Voice, Diction, Opera, and Survey of Music Literature, and also serves as the Music Honors Coordinator.

Karen McKinney

Best Overall Faculty Member, English
Most Helpful to Students, English
Most Helpful to Students, Liberal Arts, Social Sciences, and Humanities
Best Overall Faculty Member, Liberal Arts, Social Sciences, and Humanities

Jack Riggs

Best Researcher/Scholar, English

Jack C. Riggs is the author of two prize-winning novels, "When the Finch Rises" and "The Fireman's Wife," and holds the rare position of a writer-in-residence at a two-year institution, Georgia Perimeter College near Decatur. He is also associate professor of English at the college. His two books, both set in the South of the 1960s and '70s, are centered on character development and relationships, one on the ties between two young boys, and the other on the strained bonds of marriage. "When the Finch Rises" was chosen for inclusion on the Georgia Center for the Book's 2010 list of "25 Books All Georgians Should Read." - See more at: http://www.georgiacenterforthebook.org/Georgia-Literary-Map/Georgia-Author-Detail.php?record_id=136#sthash.htU6AJ00.dpuf

Shawn Sublett

Best Overall Faculty Member, Sociology
Best Researcher/Scholar, Sociology
Best Teacher, Sociology
Most Helpful to Students, Sociology
Best Teacher, Liberal Arts, Social Sciences, and Humanities
Most Helpful to Students, Liberal Arts, Social Sciences, and Humanities

GEORGIA STATE UNIVERSITY

Andrew Altman

Best Researcher/Scholar, Philosophy

Dr. Altman's scholarly activity is mainly in the areas of legal and political philosophy, with a focus on contemporary topics such as human rights, discrimination, genocide, and freedom of expression. His current projects include co-editing and writing the introduction for the book, Targeted Killings: Law and Morality in an Asymmetrical World (Oxford University Press) and coauthoring the book, Pornography: For and Against (Oxford University Press). His articles currently in press include, "Genocide and Crimes Against Humanity: Dispelling the Conceptual Fog," Social Philosophy and Policy and "Freedom of Expression and Human Rights Law: The Case of Holocaust Denial," in Maitra and McGowan, Speech and Harm (Oxford University Press). His recent publications include, "Discrimination," Stanford Encyclopedia of Philosophy, and A Liberal Theory of International Justice, C.H. Wellman, coauthor (Oxford University Press).

Stephen Jacobson

Best Teacher, Philosophy

Steve Jacobson has wide-ranging interests in philosophy. He has taught in the areas of epistemology, metaphysics, Indian Philosophy, philosophy of language, metaethics, among others. His research and publications have concerned a variety of topics in epistemology, including contextualism, epistemic norms, the internalist/externalist debate, the value of truth, and Cartesian epistemology. His articles have appeared in journals and anthologies, such as Synthese, Principia, Pacific Philosophical Quarterly, and Australasian Journal of Philosophy. He has also published on aspects of Yoga philosophy. His recent research concerns the epistemic status of science for non-experts.

Soon Ho Kim

Best Overall Faculty Member, Hospitality
Best Researcher/Scholar, Hospitality
Best Teacher, Hospitality

Soon-Ho Kim specializes in leisure participant behaviors and formulating effective marketing strategies. More specifically, his research spans three areas: 1) understanding leisure participant behaviors for designing and implementing activities; 2) identification of motives and their interrelationships with media consumption patterns (televised sport watching, uses of internet for information gathering, reading about leisure in magazines, etc.); and 3) development of scales to measure service quality, perceived value, and destination loyalty.

Kim's research has been published in the top hospitality and tourism journals. He has been engaged in funded research projects, such as assessing the recreational facilities and programs of the Suwannee River Water Management District.

Daniel Robin

Best Teacher, Film

"Has consistently received outstanding students evaluations and raving comments about his pedagogy, advising, and all-around passionate and encouraging relationship to the students. Consistently praised for allowing students to develop a personal artistic voice."

Robin's recent film, "my olympic summer", won the 2008 Sundance Film Festival Jury Prize for Short Filmmaking. The film also won the Grand Jury Award for Best Documentary

Short at the Florida Film Festival, Best Documentary Short at the Nashville Film Festival, and best documentary short at several other festivals. The film was also selected to screen at the prestigious New Directors/New Films Festival in New York. His other films have twice screened at Sundance, at the SXSW Film Festival and festivals in Bogota, Cartagena, Abu Dhabi, Ireland, Israel, Italy, Spain, Philadelphia, London, Atlanta, Baltimore, Humboldt, Ann Arbor, and elsewhere.

Filmmaker Magazine recently named Robin one of the "25 New Faces of Independent Film," a major honor. Robin is an innovative online artist, having pioneered web video in 2001, and with 6 episodic web series to date available at his website www.neighborhoodfilms.com. Robin recently wrote an essay about his experiences with web video for Movement, the NYU Tisch School journal. Daniel has been invited to present web video work at the Amsterdam Worldwide Video Festival and elsewhere. His projects have received support from the San Francisco Arts Commission, Apple Computers, the Fleishbacker Foundation, and the Film Arts Foundation.

My film/video work is primarily made for two distinct exhibition venues: film festivals and the Internet (web series). My short films are personal and, for the most part, interrogate relationships I have with women, my family, and Judaism. While the majority of my filmmaking can be read within a documentary framework, I'm more inclined not to use labels to categorize my work. I have a deep love and appreciation for narrative cinema and have incorporated some fictional devices in my more recent films. I use narrative strategies both as a means to critique how we consume documentaries and, in some sense, as a way to liberate my work from the constraints of conventional non-fiction storytelling. The most important aspect of filmmaking, for me, is to arrive at an emotional truth that resonates with the viewer. I began making web series at my site http://www.neighborhoodfilms.com in 2000. My first series, the valet chronicles, set the thematic standard for each new series, my relationship with a specific neighborhood, and or community. My work here maintains an aesthetic and tonal consistency that aligns itself with familiar tropes of documentary practice, yet I consider these web series an extension of my personal filmmaking.

Michael Wien

Most Helpful to Students, Hospitality

Mike Wien is a part-time instructor in the School of Hospitality Administration. He was among the Atlanta runners who participated in a run to raise funds for victims of the Boston Marathon bombing. Wein discussed his experience at the marathon and gave his medal to a fellow runner who was unable to complete the race when it was halted due to the explosions.

When it comes to defining success in business and in life, Mike Wien has mastered the powers of endurance, discipline and concentration to make the shift from ordinary to the extraordinary and has been helping entrepreneurs and business leaders do the same for decades. As a motivational speaker with over 35+ years of marketing and sales experience, confidence and charisma behind a podium and an extraordinary Ironman story. Mike brings a unique and entertaining perspective to any corporate event or motivational speech, From a young age Mike learned that success requires more than talent, connections or sheer luck - and that the only way to win is through a sustained effort.

Mike began a career in brand management where he quickly discovered two things: first, the most successful businesses differentiate themselves and are not afraid to be specific; and second, the path to success is defined by a consistent and committed effort.

Today, as a world-class athlete, motivational speaker and consultant. Mike has the opportunity to combine his corporate background and Ironman experiences to share his truly unique personal philosophy of "Sustained Effort Wins". He shares his best thinking with entrepreneurs and business leaders, helping them apply the same principles to focus on solutions for improving growth and increasing profits.

Mike earned an MBA in Marketing from the Kellogg School at Northwestern University and a B.S. in business from Colorado State University. He currently serves on the USA Triathlon board (under the US Olympic Committee), the Atlanta Track Club and is an Adjunct Professor of Marketing at Georgia State University where he shares his wisdom on marketing and business development strategies with the next generation.

Mike Wien is the founder and senior consultant at Wien & Associates, a hospitality marketing company. As a senior marketing executive, he has developed marketing programs that increase sales and market share for such major organizations as Deloitte & Touche, Cole Taylor Bank, Citibank, Omni Hotels and PepsiCo (Frito-Lay and Pepsi-Cola Bottling Group). Mike has a bachelor's degree in marketing from Colorado State University and an MBA from Northwestern University.

GLENDALE COMMUNITY COLLEGE

Roger Bowerman

Best Researcher/Scholar, History

Mary Jane Onnen

Best Teacher, English

GLOUCESTER COUNTY COLLEGE

Erica Mercado Moore

Most Helpful to Students, Liberal Arts, Social Sciences, and Humanities
Best Teacher, Liberal Arts, Social Sciences, and Humanities

I currently reside in NJ with my husband, two children, and two furbabies. I teach at Rowan College at Gloucester County and teach at two different online universities. I have a BA in History, and MA in Ancient and Classical history. I am currently working on my PhD in Higher Education Leadership specializing in inclusion. I am also a member of the Academic Standing Committee, Golden Key International Honor Society, Association of Ancient Historians, and Kappa Delta Pi honor society.

GONZAGA UNIVERSITY

Jane Rinehart

Best Teacher, Liberal Arts, Social Sciences, and Humanities

"Jane is a master teacher who leads classes through a process of shared inquiry. She insists that students read and be prepared for classes, but she also values and supports every voice in the room. She challenges, engages, and promotes a desire for further inquiry and analysis. She has been a teacher who changes students lives as well as a mentor who supports students as they follow their passions."

GOUCHER COLLEGE

Gail Husch

Best Researcher/Scholar, Fine Arts

Zahi Khamis

Most Helpful to Students, Liberal Arts, Social Sciences, and Humanities

Kathryn StOurs

Best Overall Faculty Member, Languages

"Kathryn is a top notch teacher in the classroom, an empathetic adviser, a team-worker in the department and a brilliant scholar."

GOVERNORS STATE UNIVERSITY

Dave O'Donnell

Best Teacher
Most Helpful to Students, Liberal Arts, Social Sciences, and Humanities

For more than 22 years, I have been a full-time Senior Lecturer in the Addictions Studies program in the College of Health and Human Services. As a GSU graduate (Bachelor's Degree in Liberal Arts and Master's Degree in Health Science), I really enjoy teaching at GSU. Having come through the GSU culture as a student myself, I have developed a strong commitment to the students' professional and academic development. It gives me a tremendous sense of pride and accomplishment when I attend conferences and workshops and see so many of our graduates in leadership roles in the addiction field presenting evidence-based topics or sharing opinions in panel discussions regarding the direction of our field.

Moreover, I have had the pleasure of working in an environment that encourages creativity in the classroom arena, as well as flexibility with regard to teaching styles and evaluation. I have had the good fortune to have been with the program long enough to be a part of its dynamic growth — the addition of an advanced clinical track in 2012 and the program going fully online, as projected for Spring, 2015.

My background and experience comes in part from life experience dating back to the 70's, when I was a young adult actively participating in a cultural revolution made famous by Dr. Timothy Leary ("Turn On - Tune In - Drop Out") — better known as the "Hippie Generation." After surviving that period, I returned home from the West Coast and came to GSU to complete my education. My personal experience helped in the development of counseling skills and awareness to become a therapist in the addiction/mental health field. My education in Addictions Studies helped me develop expertise to become the best therapist and lecturer I could. As an instructor, I will frequently draw from my clinical experience to present casework to our students to formulate and practice clinical interventions in the classroom. Students always enjoy having input on strategy with real clients.

My most valuable community experience is working with fellow department faculty members, the Department of Alcohol and Substance Abuse (D.A.S.A.) and the Substance Abuse Mental Health Service Administration (S.A.M.H.S.A.) as a training instructor in the Recovery Coaching Model. Recovery coaching is a community-based training model of paraprofessionals who are successful in overcoming problems with addiction and related issues and who want to help others with similar problems. The model was originally funded for three years by D.A.S.A., with GSU as the lead training provider. Results have been very positive and we are now beginning our fifth year of training Recovery Coaches to work in the field.

Of personal interest, I am an avid gardener, specializing in roses, and have enjoyed spending winter holidays in Bucerias, Mexico for the past 15 years.

GRAND VALLEY STATE UNIVERSITY

Deanna Morse

Best Teacher, Communication

Jim Persoon

Best Teacher, English

Peter Zhang

Best Teacher, Communication

Dr. Zhang is interested in the intersection of rhetoric, media ecology, cultural studies, and the Chinese culture. He has written on Gilles Deleuze, Marshall McLuhan, minor rhetoric, communication and ethics, the English language, and the Chinese culture. Currently, he serves on the editorial board of the Journal of Contemporary Rhetoric. He also serves as a manuscript reviewer for Explorations in Media Ecology, and as a Board member of the Grand Rapids Chinese Language School (GRCLS).

GREENVILLE TECHNICAL COLLEGE

Sybil Davis

Most Helpful to Students, Hospitality

HARDING UNIVERSITY

Britt Lynn

Best Overall Faculty Member, Theater

"Overall good relation with students, faculty and administration. Multi-talented. Balances work, family and social/faith a activities well."

HARPER COLLEGE

Barbara Bowker

Best Teacher, Music

Barb began teaching music at Harper part time in 1979, and was appointed Assistant Professor in 2002. She directs the theory, aural skills, and literature courses. Barb initiated and runs the Composer's Forum, a once-per-semester meeting at which composers of all styles can perform and get feedback on their work.

Barb is an active member of several music professional organizations, including The College Music Society, The Society for Music Theory, and Music Theory Midwest. She has presented papers and appeared on panels on music theory, theory pedagogy, and community college music education across the country, and performs as a pianist in concerts in the Chicago area, primarily featuring compositions by Harper College alumni.

Barb received the Bachelor of Arts (with Academic Honors) and Master of Arts in Teaching degrees from Beloit College, the Master of Music from the University of Wisconsin-Madison, and the Ph.D. in music theory from Northwestern University, which also nominated her for membership in Pi Kappa Lambda, the National Music Honor Society.

Patricia Ferguson

Most Helpful to Students, Communication

Cathy Hanahan

Best Overall Faculty Member
Most Helpful to Students
Most Helpful to Students, Liberal Arts, Social Sciences, and Humanities
Best Overall Faculty Member, Liberal Arts, Social Sciences, and Humanities

Perry Pollock

Best Teacher, Fine Arts

"Perry is a knowledgeable instructor and is helpful with his students in achieving their best."

I make small wall-mounted objects that emphasize interrelationships between form and surface. My forms are rectangular with subtle deviations such as rounded corners or simple appendages. Their surfaces are painted geometrically, some areas treated to look gently weathered or residual. I am interested in making objects that cause viewers to look into, across and around their surfaces and lead them to experience a collection of subtle harmonious relationships. Recently my work has taken on an interactive element requiring participants to open doors that pivot or slide to reveal interior surfaces and spaces.

L'Tonya Shumpert

Best Teacher, Design

HARTNELL COLLEGE

María Teutsch (Tabor)

Best Overall Faculty Member, English

HARVARD DIVINITY SCHOOL

David Carrasco

Best Overall Faculty Member, Religion

HASTINGS COLLEGE

Brett Erickson

Best Teacher, Liberal Arts, Social Sciences, and Humanities

HAWAII PACIFIC UNIVERSITY

John Hart

Best Teacher, Communication

Chair, Department of Communication Professor Teaching and research fields: media criticism, persuasion, rhetoric, argumentation, and debate.

Russell Hart

Best Overall Faculty Member, History

"Russell Hart goes out of his way for his students and the faculty in his department. He is a tireless advocate for reaching our military students, and he does an excellent job listening to and representing faculty."

Dr. Russell Hart earned his PhD in History from Ohio State University. He is the author of Clash of Arms: How the Allies Won in Normandy, 1944 (Boulder, Colorado, 2001) and Guderian: Panzer Pioneer or Mythmaker? (Dulles, Virginia, 2006) and is the co-author of four additional works. He has published articles on topics ranging from armored warfare, coalition command, combat effectiveness, and reconstruction of wartime memory. Dr. Hart's research interests include modern history, military history, diplomatic history, and genocide studies. Dr. Hart teaches U.S. Military History, History of Military Thought, the Second World War, History of Genocide, Modern European Diplomatic History and Research Methods.

Paul Tran

Best Overall Faculty Member, Social Work

"Engages students at all levels in and out of the classroom."

HENDRIX COLLEGE

Rebecca Resinski

Best Researcher/Scholar, Languages

"Excellent quality of research work"

HENRY FORD COMMUNITY COLLEGE

Kalvin Harvell

Best Overall Faculty Member, Sociology
Best Researcher/Scholar, Sociology
Best Teacher, Sociology
Most Helpful to Students, Sociology

Bill Norris

Best Researcher/Scholar, Liberal Arts, Social Sciences, and Humanities
Best Teacher, Liberal Arts, Social Sciences, and Humanities
Most Helpful to Students, Liberal Arts, Social Sciences, and Humanities
Best Overall Faculty Member, Liberal Arts, Social Sciences, and Humanities

HIGH POINT UNIVERSITY

Scott Raynor

Best Teacher, Fine Arts

HORRY-GEORGETOWN TECHNICAL COLLEGE

Alyssa Wood

Best Teacher, English
Most Helpful to Students, English

HOSTOS COMMUNITY COLLEGE

Lewis Levine

Best Overall Faculty Member, English
Best Researcher/Scholar, English
Best Teacher, English

Elyse Zucker

Most Helpful to Students, English

HOUSATONIC COMMUNITY COLLEGE

Drew Denbaum

Best Teacher, English
Most Helpful to Students, English

Drew began teaching Literature and Composition at Housatonic Community College in 2010, and was elected to the College Senate in 2014. Drew's goal as a teacher is to build confidence and motivate students to improve their writing skills and achieve higher levels of critical thinking, analysis, and self-expression.

In 2007, Drew earned a Master of Teaching degree and Connecticut certification (English, Grades 7–12) from Sacred Heart University and began teaching at-risk high school students in English, Film Studies, and Dramatic Literature and Performance, heavily influenced by the progressive education theories of Ken Robinson and Native American Circle of Courage principles.

Since graduating cum laude with honors from The Lawrenceville School and Yale University (where he was awarded the Saybrook Fellows Prize), Drew has also been an award-winning professional writer with many credits as a playwright, screenwriter, and published author.

HOUSTON BAPTIST UNIVERSITY

Russell Hemati

Most Helpful to Students, Philosophy

His writing interests include: Metaphysics of Free Will, Philosophy of Time, Medieval Philosophy. While at HBU, Dr. Hemati has served on the Honors College Committee, and is the faculty sponsor for the BYX fraternity. Dr. Hemati has been with Houston Baptist University for 1 year. He is a member of the Society of Christian Philosophers. He enjoys leading film groups and spending time with his family.

Jerry Walls

Best Researcher/Scholar, Philosophy

"High productivity with world-class publishers."

HOUSTON COMMUNITY COLLEGE (ALL CAMPUSES)

Corey Ackelmire

Best Researcher/Scholar, Fine Arts
Best Teacher, Fine Arts

Griffin Anita

Most Helpful to Students, English

Glenna Bell

Most Helpful to Students, English

Sandra Gin

Best Teacher, Communications

"Has 30-plus years of real world, hands-on experience in the field and brings this knowledge to the classroom; winner of 17 Emmy awards and nominations in television news, respected for mentoring students entering broadcast journalism, excellent performance ratings as an educator and role model."

Gloria Greene

Best Researcher/Scholar, Cosmetology

Linda Griffin

Best Teacher, Liberal Arts, Social Sciences, and Humanities
Most Helpful to Students, Liberal Arts, Social Sciences, and Humanities
Best Overall Faculty Member, Liberal Arts, Social Sciences, and Humanities

John Harvey

Best Researcher/Scholar, Liberal Arts, Social Sciences, and Humanities

Stanley Kaminski

Best Teacher, Fine Arts

Jason Kishell

Best Teacher, Fine Arts

Robert Lunday

Best Teacher, English

Betty Martinez

Best Teacher, Cosmetology

Karl Mayes

Most Helpful to Students, Sociology

"Very charasmatic, charming and intelligent"

Karl Mayes, M.A. is a passionate Sociology professor at the Houston Community College (Central Campus). He graduated from Prairie View A&M University with an undergraduate major in Communications (RA/TV) and minor in Marketing. He then went right into completing a Master of Arts degree in Sociology. After a few years working in the mental health industry, he decided to move into the industry of higher education. Karl has worked as a Director of Education as well as undergraduate and graduate advising. Karl's interests are in people, hence why he teaches Sociology so passionately. Within the domain of Sociology, he has been led to motivating others to live to their greatest potential. Karl is a national motivational speaker where he goes by "Karl Mayes the King Motivator". He has orated on a number of stages across the country and continues to expand his brand while continuing to get students to embrace the discipline of Sociology. One of Karl's goals is to become one of the world's greatest orators. Another goal is to complete the doctoral degree that he is currently working on. Embrace the journey of Karl Mayes by visiting his website at www.KingMotivator.com

Joe McDade

Best Researcher/Scholar, English

Jaime Olivares

Best Overall Faculty Member, Liberal Arts, Social Sciences, and Humanities

Melinda Payne

Best Overall Faculty Member, English

Patricia Porcynaulk

Best Overall Faculty Member, Fine Arts

Michele Snelson

Best Overall Faculty Member, Cosmetology
Most Helpful to Students, Cosmetology

James Sparks

Best Teacher, Liberal Arts, Social Sciences, and Humanities

Portia Tinnermon

Best Teacher, Sociology

David Tsai

Most Helpful to Students, Fine Arts

Ashley Vining

Best Teacher, English

"Very thorough. An excellent teacher."

June Woest

Best Researcher/Scholar, Fine Arts

Maria Zambrano

Best Overall Faculty Member, Cosmetology

AAS San Jacinto College, Houston, Texas
Instructor, 1995–Present
Houston Community College Salon
Experience: • Stylist/Manager 1991 to 1999
License: • Cosmetology Operator Instructor, Texas Department of Licensing & Regulation, Austin, Texas
Awards and Recognition: • 2010 NISOD Excellence Award Recipient • 2011 Certificate of Achievement Award College Leadership Program

HOWARD UNIVERSITY

Algris Bolton

Most Helpful to Students, Languages

Yasmin Degout

Best Overall Faculty Member, English

Dr. Yasmin Y. DeGout specializes in African American and Caribbean literatures. She earned the Ph.D. from Howard University in 1998 with a dissertation entitled "The Fiction of James Baldwin: Revisioning the Autobiographical, Gendered and Christian Selves." She remains ABD at Yale University and continues work on her dissertation project entitled "(Post) Colonialism and Cultural Production in the Danish West Indies/United States Virgin Islands."

She has published on the slave narratives of Frederick Douglass and Harriet Jacobs, the short fiction of James Baldwin and his novel Giovanni's Room, the poetry of Maya Angelou, and the Black Arts Movement drama of Ed Bullins. She also participated in a roundtable discussion on Whiteness Studies published by MLS, and she has provided encyclopedia articles on topics ranging from The Negritude Movement and Leopold Sedar Senghor to St.-John Perse for a forthcoming companion to 20th-century world poetry by Facts on File. Her academic interests include postcolonial studies, gay and lesbian studies/queer theory, and gender studies, and her presentations have covered a range of topics, including Toni Morrison's Tar Baby, photographic representation of the Danish West Indies, Merle Hodge's Crick Crack, Monkey, African American homo-eroticism, as well as identity and aesthetics in Sugar Cane Alley.

She serves as a member of the Caribbean Studies Program of the College of Arts and Sciences, and has served the Department of English as faculty advisor to the Sterling Allen Brown English Society and as member of a variety of departmental committees.

Cerue Diggs

Best Teacher, Languages

"Very dedicated"

Annette Dunzo

Best Overall Faculty Member, Languages

"Very dedicated"

Andre Esters

Most Helpful to Students, English

"He is there every morning before 7 a.m. and has a constant stream of students even at that early hour!"

Curdella Forbes

Most Helpful to Students, English

Dr. Forbes taught at the University of the West Indies, Mona, before coming to Howard in 2004. Her research interests include postcolonial theory; Caribbean poetics; Shakespeare and post-colonialism; Caribbean literature and culture with particular focus on issues of gender, performance, and orality; and the (postcolonial) issues of diaspora and globalization. She earned her Ph.D. in 2000 from the University of the West Indies. Her doctoral dissertation is titled "Through the Lens of Gender: A Revisionary Reading of the Novels of Samuel Selvon and George Lamming."

Dr. Forbes is the author of From Nation to Diaspora: Samuel Selvon, George Lamming and the Cultural Performance of Gender (2005) three works of fiction: A Permanent Freedom (2008), Flying with Icarus and Other Stories (2003) and Songs of Silence (2002). Her numerous scholarly articles appear in Literature and Psychology: A Journal of Psychoanalytic and Cultural Criticism,Journal of West Indian Literature, Small Axe, Aspekt (a Slovak feminist periodical), Anthurium, Postcolonial Text, and Caribbean Quarterly. Her short stories have appeared in journals and edited collections. She has presented scholarly papers at conferences in the Caribbean, the United Kingdom, Austria, Canada and here in the USA. She was an editor for the online peer reviewed journal Postcolonial Text, is book review editor for The Journal of West Indian Literature, and a member of the advisory board for Anthurium, an online peer-reviewed journal of original Caribbean works and critical studies of Caribbean literature, film, art, and culture. She serves as peer reviewer for a number of refereed journals, including Social and Economic Studies and Caribbean Quarterly. Dr. Forbes is also a curriculum developer. She has served and continues to serve on panels and examining committees for the Caribbean Examinations Council, the regional examining body in the Caribbean. Dr. Forbes has been involved in planning various symposia and conferences, including the conference on George Lamming (The Sovereignty of the Imagination), which took place at Mona, Jamaica, 5–7 June 2003 and the Walter Rodney 25th Anniversary Conference which took place here at Howard in September 2005.

Altaf Husain

Best Teacher, Social Work

Elisa Oh

Best Teacher, Liberal Arts, Social Sciences, and Humanities

Elisa Oh received her B.A. from Smith College; her M.A. in English Language and Literature from the University of Virginia; and her Ph.D. in Early Modern English Language and Literature from Boston University. Her dissertation, "Defining Absence: Reading Female Silence in Early Modern English Literature, 1580–1640," sought to broaden the definition of literary silences and to theorize them as active modes of gendered signification in early modern drama and prose romance. She teaches Shakespeare, a British Renaissance literature survey, a British Literature survey, British Drama, and introductions to the English major and literary theory. She has reviewed a number of books for Sixteenth Century Studies and Seventeenth-Century News, and a recent, prize-winning article in Explorations in Renaissance Culture compares the resistant intentional silences of Cordelia in King Lear and Mariam in The Tragedy of Mariam. Informed by feminist, new historicist, and cultural materialist critics, her research interests focus on sixteenth- and seventeenth-century public drama, closet drama, women writers, representations of the racial Other in Jacobean court masques, the Sidney circle, concepts of subjectivity, letter-writing, and social dance practices. In terms of other professional activities, she has participated in two research seminars at the Folger Shakespeare Library and presented a dozen papers at conferences such as the Sixteenth Century Studies Conference and the Shakespeare Association of America. Before joining the Howard University faculty, she taught at Montgomery College, James Madison University, and the University of Maryland.

Norman Sandridge

Best Overall Faculty Member, Classics
Best Researcher/Scholar, Classics
Best Teacher, Classics
Most Helpful to Students, Classics
Best Researcher/Scholar, Liberal Arts, Social Sciences, and Humanities
Best Teacher, Liberal Arts, Social Sciences, and Humanities
Most Helpful to Students, Liberal Arts, Social Sciences, and Humanities
Best Overall Faculty Member, Liberal Arts, Social Sciences, and Humanities

My current book, Loving Humanity, Learning, and Being Honored: The Foundations of Leadership in Xenophon's Education of Cyrus (HUP 2012), explores the conceptual and cultural foundations of leadership. I have published on the emotion of pity in three tragedies of Sophocles. At the 2012 American Philological Association's annual meeting, I presented on the emotion of envy as it pertained to leadership. I have a forthcoming chapter in the Cambridge Companion to Apollonius of Rhodes on Jason's "democratic" leadership of the Argonauts. For AY 2010–2011 I was on a fellowship at Harvard's Center for Hellenic Studies in Washington, DC.

Since coming to Howard in 2005, I have taught courses on Leadership in the Ancient World (also listed as Ideas in Antiquity), Greek Literature, Greek Civilization, Greek Drama, Ancient Comedy and Satire, an Independent Study on the Athenian Statesman Pericles, Elementary Greek, Xenophon's Cyropaedia, Euripides' Medea, selected speeches of Lysias, Catullus' Carmina, and selections from Cicero's Pro Caelio. In the summer of 2007 I took nine Howard students on a two-week tour of mainland Greece and the Greek islands.

Outside the field of Classics, I enjoy reading books about neuroscience, playing the guitar (poorly!), running marathons, and gardening.

Jean-Jacques Taty

Best Researcher/Scholar, Languages

"thorough researcher"

Postcolonial Studies, Theoretical and Aesthetic Canon of Cinema as a Visual Medium, Arts and Literature as Vehicle and Tool for New Cultural Identity in Postcolonial Africa (especially in francophone countries), Arts, Culture and Social Change.

Douglas Taylor

Best Teacher, English

"Prison Slang and the Poetics of Imprisonment." Prose and Cons: Essays on Prison Literature in the United States. Ed. Quentin Miller. Jefferson, North Carolina: McFarland, 2005.

Richard Wright's Black Boy (American Hunger): A Casebook Eds. William L. Andrews and Douglas Taylor. New York: Oxford University, 2003.

"From Slavery to Prison: Benjamin Rush, Harriet Jacobs, and the Ideology of Reformative Incarceration." Genre, Prisoners Writing Special Issue, 35.3/4 (2002).

"Interview with John Edgar Wideman," The Carolina Quarterly 53.3 (2001).

J Toungara

Best Overall Faculty Member, Liberal Arts, Social Sciences, and Humanities

Dr. Toungara is an Africanist scholar who offers courses on West Africa, the African Diaspora and Women in Africa. Her research interests include West African intellectual history, indigenous thought, precolonial state formation, political culture and democratization within francophone African nations, with particular emphasis on Côte d'Ivoire. She has carried out archival research in several countries, including France (Colonial Archives at Aix-en-Provence), Senegal (National Archives and French West Africa Archives), Mali (National Archives) and Côte d'Ivoire (National Archives of Côte d'Ivoire, and holdings at the Ministry of Women's Condition, the National Assembly and Odienne Prefecture). In pursuing her interest in indigenous ideologies and the cultural foundations of contemporary African politics, Professor Toungara has collected hundreds of hours of oral testimonies in Senegal and Côte d'Ivoire. She is the recipient of study and research awards from the Ford Foundation Doctoral Fellowship, the National Endowment for the Humanities, Fulbright-Hayes Program, Ministry of National Education of Côte d'Ivoire, Carter G. Woodson Institute, University of Virginia, the West African Research Association, and Howard University.

Dr. Toungara is the author of over a dozen articles on the history of women, family, education, economy and politics of various West African communities. Her work has appeared in the Journal of Modern African Studies, Journal of African Studies, Journal of Social History, Annales de l'Université d'Abidjian, African Studies Review, Les cahiers de la paix, West African Research Association Newsletter, and others. She has contributed essays and chapters in African Women: States of Crisis (ed., Gwendolyn Mikell, 1996), The Younger Brother in Mande (eds. Jansen and Zobel, 1996), and The Oxford Companion to the Politics of the World (1993) and is working on a book-length manuscript on political culture in the northwestern region of Côte d'Ivoire. She has presented numerous papers at the annual conferences of the African Studies Association, conferences at the University of Virginia, Howard University, the Smithsonian Institution, and others in Senegal, Côte d'Ivoire, Niger, Chad, France and the Netherlands. Professor Toungara actively supports peace and development efforts in sub-Saharan Africa, and in June 2000 she was awarded a grant by the U.S. Institute for Peace for the training of 15 African women in the gender studies conflict management and peace building.

Clint Wilson

Best Researcher/Scholar, Journalism

HUDSON COUNTY COMMUNITY COLLEGE

Dorothy Anderson

Best Overall Faculty Member, Liberal Arts, Social Sciences, and Humanities

Joseph Colicchio

Best Overall Faculty Member, English

Javedd Khan

Best Teacher, English

Barry Tomkins

Most Helpful to Students, Liberal Arts, Social Sciences, and Humanities

Barry Tomkins grew up in England and later moved to the United States. Since 1980 he has been teaching writing and literature – including Science Fiction – at Hudson County Community College in Jersey City, New Jersey.

HUNTER COLLEGE

Morgan Horowitz

Best Teacher, Philosophy
Best Teacher, Liberal Arts, Social Sciences, and Humanities

Mike Kilivris

Most Helpful to Students, Philosophy

Moti Mizrahi

Best Researcher/Scholar, Philosophy
Best Researcher/Scholar, Liberal Arts, Social Sciences, and Humanities
Most Helpful to Students, Liberal Arts, Social Sciences, and Humanities

Has been a member of the Philosophy Department since 2012.

Monique Whitaker

Best Overall Faculty Member, Philosophy
Best Overall Faculty Member, Liberal Arts, Social Sciences, and Humanities

Monique Whitaker is a PhD candidate in the home Philosophy Program at the City University of New York Graduate Center, writing on the semantics and pragmatics of the conditional. The focus of her dissertation is on a providing a comprehensive account of the logical structure underlying the varied everyday usage of the ordinary-language conditional, using a novel possible worlds semantics. She also has research interests in the metaphysics of modality, on questions of the metaphysical reality of possibility; ethical issues in critical animal studies, on the reproduction of discriminatory argument tropes in justifications of speciesism and problematising rights-based approaches to animal ethics; and on the applicability of feminist epistemology to aspects of formal logic. Monique has a keen interest in the pedeagogy of philosophy, and has worked as a Writing Fellow to help academic staff to develop and improve their course materials and lecturing, and presented on strategies to help retain typically marginalised students in philosophy.

INDIANA STATE UNIVERSITY

Matthew Brennan

Best Overall Faculty Member, Liberal Arts, Social Sciences, and Humanities
Best Researcher/Scholar, Liberal Arts, Social Sciences, and Humanities

Terry Dean

Best Overall Faculty Member, Music
Best Researcher/Scholar, Music
Best Teacher, Music
Most Helpful to Students, Music

Dr. Terry Dean joined the faculty of Indiana State University in 2008 where he currently serves as Assistant Professor of Musicology. He holds the M.A. and Ph.D. in historical musicology from The University of Georgia. Entitled "The Nineteenth-Century Russian Operatic Roots of Sergei Prokofyev's War and Peace," his dissertation assesses Prokofyev's opera in the context of its relationship with the nineteenth-century Russian opera tradition.

Other areas of interest include Socialist realist aesthetics, primitivist aesthetics, issues of gender in Russian and Soviet opera, contemporary American musical theater, and music history pedagogy. In addition to his work in musicology, Dr. Dean completed graduate course work in horn performance at West Virginia University with Virginia Thompson and remains an active performer of chamber and orchestral music.

Prior to his appointment, Dr. Dean served on the faculties of Gainesville State and Piedmont Colleges in Georgia. He also taught courses through the University System of Georgia's Independent and Distance Learning Program and worked as part of The University of Georgia Writing Intensive and Honors Programs. At ISU, Dr. Dean teaches courses in music history, popular music, world music, and research methodologies/music bibliography.

INDIANA UNIVERSITY - PURDUE UNIVERSITY

Lewis Dibble

Best Researcher/Scholar, English
Best Teacher, English
Most Helpful to Students, English
Best Researcher/Scholar, Liberal Arts, Social Sciences, and Humanities
Best Teacher, Liberal Arts, Social Sciences, and Humanities
Most Helpful to Students, Liberal Arts, Social Sciences, and Humanities
Best Overall Faculty Member, Liberal Arts, Social Sciences, and Humanities

My undergraduate studies at Yale and the University of Massachusetts were in English, Philosophy, and Symbol Sciences/Linguistics, with ephemeral competence gained in German, French, Latin, Japanese, and Chinese. I received my MA and my PhD (1997) in Comparative Literature from the Graduate Program in Comparative Literature at Indiana University with minors in Semiotics and Victorian Studies, qualifying exams in Nineteenth Century, Epic and Romance, and Semiotics and Literary Theory (Memory); my dissertation was on "Symmetry and Memory, From Proust to Stevenson.

INDIANA UNIVERSITY BLOOMINGTON

Troy Byler

Best Teacher, Languages

"Troy is an outstanding pedagogue at every level."

INDIANA UNIVERSITY-PURDUE UNIVERSITY INDIANAPOLIS

Anita Ashendel-Morgan

Most Helpful to Students, History

"Excellent director of undergraduate studies: knows the students and challenges them to excel and become recognized for good work."

David Beck

Best Teacher, English

Jon Eller

Best Researcher/Scholar, English

Didier Gondola

Best Overall Faculty Member, History

"In his tenure as chair, he has exhibited the best qualities of a faculty member–primus inter pares."

Daniella Kostroun

Best Researcher/Scholar, History

"commitment to excellence in research (methods and theory) and scholarship (dissemination of findings to audiences that are generally interested and that can build on it)"

Krista Longtin

Best Researcher/Scholar, Communication Studies
Best Researcher/Scholar, Liberal Arts, Social Sciences, and Humanities

Krista Hoffmann-Longtin is an Assistant Professor of Communication Studies, and an Assistant Dean in the IU School of Medicine Office of Faculty Affairs and Professional Development. Her work focuses on faculty development, organizational/professional identity, and strategic communication in academic health settings. Before returning to university life, Krista worked as an advertising account executive. She earned a B.A. in Telecommunications from Ball State University in Muncie, IN and an M.A. in Communication and Creative Arts from Purdue University, and a PhD in Education Leadership and Policy Studies at Indiana University.

Jennifer Medcalfe

Best Overall Faculty Member, Communication Studies

J.J. Stenzoski

Most Helpful to Students, English

Kate Thedwall

Best Teacher, Communication Studies
Most Helpful to Students, Communication Studies
Best Teacher, Liberal Arts, Social Sciences, and Humanities
Most Helpful to Students, Liberal Arts, Social Sciences, and Humanities
Best Overall Faculty Member, Liberal Arts, Social Sciences, and Humanities

INDIANA WESLEYAN UNIVERSITY

Robert Curfman

Best Overall Faculty Member, Fine Arts

IOWA STATE UNIVERSITY

Sherry Berghefer

Best Teacher, Liberal Arts, Social Sciences, and Humanities

"Innovative, creative and challenges students"

Education Ph.D., Human Computer Interaction, Iowa State University, Anticipated Spring 2016 M.S., Journalism and Mass Communication, Iowa State University, 2006. B.A., Journalism and Mass Communication, English minor, Iowa State University, 1999.

Michael Bugeja

Best Researcher/Scholar, Liberal Arts, Social Sciences, and Humanities

"He teaches and publishes relevant and cutting edge research for media and education. His research is internationally and nationally recognized. If one teaches and publishes valuable research in one's discipline, they should not be excluded from this list."

Barbara Caldwell

Most Helpful to Students, Art

Dennis Chamberlin

Best Overall Faculty Member, Journalism

Michael Dahlstrom

Best Teacher, Journalism

Daniela Dimitrova

Most Helpful to Students, Journalism

Joel Geske

Best Researcher/Scholar, Journalism

Gang Han

Best Overall Faculty Member, Liberal Arts, Social Sciences, and Humanities

Brenda Jones

Best Teacher, Art

Amy Slagell

Most Helpful to Students, Liberal Arts, Social Sciences, and Humanities

IVY TECH COMMUNITY COLLEGE: BLOOMINGTON

Donn Hall

Best Teacher, History

IVY TECH COMMUNITY COLLEGE: MUNCIE

Jeff Meyers

Best Overall Faculty Member, English
Best Teacher, English
Most Helpful to Students, English

IVY TECH COMMUNITY COLLEGE: NORTHWEST

Heather Clark

Best Teacher, English
Most Helpful to Students, English

"Heather provides digital access to course materials for both online and in class for student success. Student feedback is given a top priority. Heather is constantly revising her methods to reflect the current trends in language education. Additionally, she consistently exceeds expectations in mentoring and sharing her methods.

Heather provides digital access to course materials for both online and in class for student success. Student feedback is given a top priority. Heather is constantly revising her methods to reflect the current trends in language education. Additionally, she consistently exceeds expectations in mentoring and sharing her methods."

Scott Guffey

Best Overall Faculty Member, English
Best Researcher/Scholar, English
Best Teacher, English
Most Helpful to Students, English
Best Researcher/Scholar, Liberal Arts, Social Sciences, and Humanities
Best Teacher, Liberal Arts, Social Sciences, and Humanities
Most Helpful to Students, Liberal Arts, Social Sciences, and Humanities
Best Overall Faculty Member, Liberal Arts, Social Sciences, and Humanities

JACKSON COMMUNITY COLLEGE

Loren Wingblade

Best Teacher, Social Science

"has 60 reports on rate you professor.com"

JACKSON STATE COMMUNITY COLLEGE

Andrew Kelley

Best Overall Faculty Member, English

Education: B.A., Rhodes College; M.A., Memphis State University; D.A., University of Michigan at Ann Arbor.

JACKSONVILLE STATE UNIVERSITY

Legare McIntosh

Most Helpful to Students, Music

Dr. Legare McIntosh received his undergraduate degree from the University of South Carolina where his teachers were Hugh Williamson, piano, and Robert Van Doren, organ. His graduate degrees were earned at Columbia University where he studied with Martin Canin, piano, and Alec Wyton, organ. He was also a pupil of Edwin Hughes. He appeared as soloist in the southeast and with the Salisbury (NC) Symphony Orchestra.

Dr. McIntosh has been the recipient of grants from the National Endowment of the Humanities, and Music Department Chair since 2001.

JAMES MADISON UNIVERSITY

L. Stephen Chappell

Most Helpful to Students, History

Dr Chappell was born and bred in Great Britain and is a devout Yorkshireman. He is the Director of the North of England Programme (NOEP) based in York, Chester and Ripon. The first trip was Summer 2014. The dates for Summer 2015 are 11th July–9th August.

Timothy Fitzgerald

Best Teacher, History

H. Michael Gelfand

Best Overall Faculty Member, History

"Dr. Gelfand devotes many hours to working with students, both as an academic and as a student organization adviser. He is also a very dedicated and original, thought-provoking and challenging teacher."

Michael Gubser

Best Researcher/Scholar, History

Michael Gubser's second book, The Far Reaches: Phenomenology, Ethics, and Social Renewal in Central Europe, was published in 2014 by Stanford University Press. His current research focuses on the history of ahistoricism in international development theory and practice.

Andrew Witmer

Best Overall Faculty Member, History

JEFFERSON COLLEGE

Scott Holzer

Best Overall Faculty Member, History
Best Researcher/Scholar, History
Best Teacher, History
Most Helpful to Students, History
Best Researcher/Scholar, Liberal Arts, Social Sciences, and Humanities
Best Teacher, Liberal Arts, Social Sciences, and Humanities
Most Helpful to Students, Liberal Arts, Social Sciences, and Humanities
Best Overall Faculty Member, Liberal Arts, Social Sciences, and Humanities

Scott Holzer has taught at Jefferson College since August of 1997. His previous teaching experience includes two years as an assistant professor at Alcorn State University and two years as an instructor at the University of Mississippi. He holds a B.A. from Humboldt State University, and an M.A. and a Ph.D. from the University of Mississippi. Dr. Holzer is a member of the Organization of American Historians, the Society for History Education, the Missouri Historical Society, the State Historical Society of Missouri, and the Southern Historical Association. He is the Dual Enrollment Coordinator for history, serves on the Tenure Appeals Committee, served on the Academic Affairs Committee for six years (two as chairman), and served as club advisor to Phi Theta Kappa and later Habitat for Humanity. In 2003, Dr. Holzer received the Emerson Excellence in Teaching Award.

JEFFERSON COMMUNITY AND TECHNICAL COLLEGE

Morgan Broadhead

Best Teacher, History

"Very creative teacher who makes European history live."

David Cooper

Best Teacher, Liberal Arts, Social Sciences, and Humanities

"David is a wonderful teacher and has received a number of statewide awards."

Marie Good

Best Teacher, Languages

"Marie teaches French and focuses on teaching the language practically allowing students to learn language more naturally."

Stacy Taylor

Best Teacher, Liberal Arts, Social Sciences, and Humanities

JEFFERSON COMMUNITY COLLEGE

Lisa Scrivanni-Tidd

Best Researcher/Scholar, Music

"Lisa has contributed greatly to the historical documentation of music through her publications. She is an outstanding musician and teacher. She is responsible for the development of the music department and it's course offerings at JCC. She is highly respected and sought for advice from her students."

JOHN JAY COLLEGE OF CRIMINAL JUSTICE

David Brotherton

Best Researcher/Scholar, Sociology

"the depth of knowledge, the diversity of interests, the many publications and studies conducted."

Dr. Brotherton grew up in the East End of London, England where he worked in various blue-collar jobs while organizing labor and youth. He came to the United States in the 1980's and later worked toward his Ph.D. degree at the University of California, Santa Barbara while teaching public high school in San Francisco. Dr. Brotherton gained his doctorate in Sociology in 1992 and began work on street gang subcultures at U.C. Berkeley in the same year. In 1994, Dr. Brotherton came to John Jay College of Criminal Justice where he continued his research on youth resistance, marginalization, and deportation co-founding the Street Organization Project in 1997. He has received numerous research grants from both private and public agencies and has published widely in journals, books, newspapers and magazines. In 2003 and 2004, Dr. Brotherton co-organized the first academic conferences on deportation in the Caribbean and the United States respectively. He was named Critical Criminologist of the Year in 2011 and has been nominated for the 2011 George Orwell Prize in England and the C. Wright Mills Award in the United States. Among his recent books, published by Columbia University Press, are: Banished to the Homeland: Dominican Deportees and Their Stories of Exile, with Luis Barrios (2011); Keeping Out The Other: A Critical Introduction to Immigration Control, edited with P. Kretsedemas (2009); and The Almighty Latin King and Queen Nation: Street Politics and the Transformation of a New York City Gang, with Luis Barrios (2004). Dr. Brotherton's current projects include a comparative study of the occupied movements in Europe, the development of holistic policing strategies vis a vis gangs and organized crime in Sweden, and community responses toward the social problems of gangs and deportation in the Caribbean.

David Brotherton

Best Researcher/Scholar, Liberal Arts, Social Sciences, and Humanities

"Two critical ethnographic research studies with street Ganga in New York and with Dominicans deportees."

Lesley Hansen

Most Helpful to Students, English

"Painstakingly shows students how and why the mechanics and grammar of the English language are utterly crucial to mastering if they are ever going to get by in this life."

After getting three bachelor's (English, philosophy, and chemistry) and two master's degrees, Dr. Hansen received the Ph.D. in English from Columbia University. He taught at Queens College (CUNY), the Fashion Institute of Technology (SUNY), Pace University, The City College of New York, and John Jay College, the last three for 26 years each, devoting his time to students and their writing. He is working on a philosophical motorcycle memoir and a novel. For the last 22 years, he has been practicing the soft martial arts and has traveled around America on his Harley-Davidson, putting in—so far—a little over one-hundred-and-forty-thousand miles.

Liliana Soto-Fernandez

Best Teacher, Languages
Best Overall Faculty Member, Languages
Best Researcher/Scholar, Languages
Most Helpful to Students, Languages
Best Researcher/Scholar, Liberal Arts, Social Sciences, and Humanities
Best Teacher, Liberal Arts, Social Sciences, and Humanities
Most Helpful to Students, Liberal Arts, Social Sciences, and Humanities
Best Overall Faculty Member, Liberal Arts, Social Sciences, and Humanities

"She is a caring, talented teacher who is devoted to her academic work as well as her teaching."

Barry Spunt

Best Teacher, Liberal Arts, Social Sciences, and Humanities

JUNIATA COLLEGE

Alison Fletcher

Most Helpful to Students, History
Best Teacher, Liberal Arts, Social Sciences, and Humanities

Alison Fletcher joined the Juniata College faculty in 2007 as an assistant professor of history. She previously worked at as assistant professor of history at Kent State University from 2003 to 2007.

She earned a bachelor's degree in history in 1992 from Bryn Mawr College in Bryn Mawr, Pa. She went on to earn a master's degree in history in 1995 and a doctorate in history in 2003, both from The Johns Hopkins University, in Baltimore, Md.

Fletcher is currently working on a book, Faith in Empire: The London Missionary Society and the Building of British Colonial Modernity, which is based on her doctoral research centering on how British evangelical missionaries functioned as part of the British empire in southern Africa and Madagascar and how returning missionaries and their converts came to England and became independent influences on colonial policy.

At Juniata, she teaches introductory and upper-level courses on European history and has previously taught courses on British history, the British Empire, Women in Britain and The Historian's Craft. She has published several articles in professional journals, including Minerva: Journal on Women and War and History of Religions.

She started her teaching career as a graduate student at Johns Hopkins, working as a teaching assistant from 1994 to 1995. She held a series of jobs as an instructor in history from 1997 to 2003 at the University of Pennsylvania (1997, 2001–2003), St, Joseph's University (1997–1998), Cedar Crest College (1998), and Bryn Mawr College (2000).

She has presented her research at various conferences, including the International Congress of Historical Sciences at the University of New South Wales in Sydney, Australia, and the Conference on Imperial and Post-Colonial Historical Studies at Michigan State University in Lansing, Mich.

Fletcher received the 2006 Graduate Applause Teaching Award from Kent State and was named a Teaching Scholar in 2004. In 1992 she received the Helen Taft Manning History Prize.

David Hsiung

Best Teacher, History
Best Researcher/Scholar, Liberal Arts, Social Sciences, and Humanities
Best Overall Faculty Member, Liberal Arts, Social Sciences, and Humanities

David C. Hsiung earned a B.A. from Yale University (1983) and his Ph.D. in History from the University of Michigan (1991). He has taught at Juniata College ever since, and now is the Charles and Shirley Knox Professor of History.

David Hsiung teaches the introductory "U.S. History to 1877" survey and the first-year writing course "College Writing Seminar," as well as advanced courses that examine early North America, the American Revolution, North American environmental history, and public history. He also teaches with biologist Dr. Jay Hosler an interdisciplinary colloquium called "Comics and Culture."

David Hsiung received Juniata College's 1995 Junior Faculty Award for Excellence in Teaching and the 2007 Beachley Award for Distinguished Teaching. The Carnegie Foundation for the Advancement of Teaching and the Council for Advancement and Support of Education named David Hsiung the 2000 Pennsylvania Professor of the Year. The Appalachian Studies Association and the University Press of Kentucky jointly conferred the 1996 Appalachian Studies Award on his book, Two Worlds in the Tennessee Mountains: Exploring the Origins of Appalachian Stereotypes (1997). His article in the New England Quarterly, "Food, Fuel, and the New England Environment in the War for Independence, 1775–1776" (December 2007), won the Forest History Society's 2008 Theodore C. Blegen Award. Most recently, A Mountaineer in Motion: The Memoir of Doctor Abraham Jobe, 1817–1906 - a work David Hsiung edited and annotated - was published by the University of Tennessee Press in 2009. The journal Pennsylvania History published a special issue (Fall 2012) on the environmental history of the Mid-Atlantic region, an issue he co-edited with Allen Dieterich-Ward of Shippensburg University.

When not grading essays, advising students, going to meetings, presenting at conferences, or trying to figure out the computer, David Hsiung tries to stay in shape by hiking, but instead often winds up gorging himself on his wife Rachel's desserts, learning about new music from his son Benjamin, and baking cupcakes with his daughter Rebecca.

David Sowell

Best Researcher/Scholar, History

Professor Sowell's scholarship focuses upon the social history of Latin America. His initial research into labor history resulted in The Early Colombian Labor Movement: Artisans and Politics in Bogota, 1832–1919 (Temple University Press, 1992). An ongoing project on 'A History of Social Violence in Latin America' has been supported by an American Council of Learned Societies Fellowship and by Juniata's William P. and Frances

Clemens Nyce Endowment for Faculty Development. His latest book, The Tale of Miguel Perdomo Neira: Medicine, Ideologies, and Power in the Nineteenth Century Andes, explores the history of medicine in Colombia and Ecuador during the 19th century. Sowell's 1992 lecture series, 'Contact, Conquest and Consequences' commemorating the Quincentennial of 1492, was published in Juniata Voices.

From 1996 until 1999 Dr. Sowell served as Assistant Academic Dean and Director of International Programs. Since coming to Juniata in 1989, he has been active in the 'internationalization' of the campus, having served on the 1993 Task Force on Internationalizing Juniata, the International Education Committee, and the Middle States Task Force on Internationalization. After service in the International Programs Office, he has been part of the International Studies Committee.

Professor Sowell lectures on 'American history,' offering courses on both Latin and Anglo-America. In 1994, he received the Junior Faculty Award for Distinguished Teaching. He graduated with an A.B. from Western Kentucky University (1975), a B.A. from Grand Valley State College (1976) and an M.A. and Ph.D. from the University of Florida (1980, 1986). His research and studies have been supported by the National Endowment for the Humanities, the Fulbright and Doherty Foundations, the Organization of American States, and the Tinker Foundation.

Dr. Sowell is a member of the American Historical Association, the Latin American Studies Association, and the Conference on Latin American History (having been secretary and president of the Gran Colombian Committee).

David and his wife Chris are the proud parents of a daughter, Emily.

Belle Tuten

Best Overall Faculty Member, History
Most Helpful to Students, Liberal Arts, Social Sciences, and Humanities

It's safe to say that Juniata historian Belle Tuten isn't a big fan of "the great man" theory of history. For her, that methodology, with an emphasis on dates and focus on leaders of nations, is as boring as this sentence. For Tuten, history is all about stories—the more unlikely and humorous the better. "History is the visceral connection of putting yourself in the story and saying, 'What if that happened to me?' " Tuten says. "My students often think I go off on strange tangents but often that's the best part of the story." Often, Tuten's scholarly interest is piqued by events that otherwise might have appeared in a Monty Python movie. "I love the moments in history where people do things that are completely hilarious," she says, smiling. "I love the names too. I've come across people named Odo Donkeyneck and Guido the Serf Twister." Tuten's appreciation of both the sublime and the ridiculous inspired her to investigate a fistfight between an order of monks and an order of nuns that occurred in the aptly named Angers, France between 1050 and 1150. Her research revealed a window into monastic disputes that has implications on how medieval laws were formed. By following a single story, Tuten was able to find an interesting dissertation topic (she earned a master's degree and doctorate in history from Emory University). She is planning to further illuminate the fallout from this tale of religious fisticuffs in a book or series of articles on monastic disputes. Historical stories, however sublime or ridiculous, have held fascination for Tuten since her childhood. Growing up in Birmingham, Ala., she read book after book on all topics, but she found a particular affinity for medieval stories, such as Robin Hood and the King Arthur legends. "I liked stories where the girls got to do all the neat stuff," she says. "Those stories always seemed to be set in the distant past." After starting as a theater major at the College of Charleston in Charleston, S.C., her love of reading and stories eventually led her to the history department, where she found herself doing some pretty neat stuff. She also found time to meet the man who is now her husband, James Tuten, a historian and assistant provost at Juniata. They both came to Juniata in 1997. "She creates a sense of bringing history alive," says Andy Murray, professor of peace and conflict studies, who team-teaches "Napoleon to Monnet" with Tuten and physicist Jim Borgardt. "She is able to pull the students into the moment." Judging from her course titles ("Sword and Scimitar," "Wives, Nuns and Witches," "Filming the Dark Ages"), Tuten's storytelling gifts make for compelling history. She also has kept up her reading habit, preferring British mysteries these days. She's also found a mystery of her own. It seems a medieval bishop, whose poems are renowned in literary circles, wrote most of his poems to cloistered nuns. Tuten suspects that some of his correspondents were nuns in the same Angers convent she has studied. She's thinking of writing a book or article on that—a medieval "You've Got Mail." Still, what really energizes her is standing in front of a class and teaching–telling stories. "I like that moment in class when I see on someone's face that they've grasped the concept for the first time," she says.

KANKAKEE COMMUNITY COLLEGE

Chelise Slowik

Most Helpful to Students, Art History

KEENE STATE COLLEGE

Joseph Darby

Best Overall Faculty Member, Music
Best Researcher/Scholar, Music

Heather Gilligan

Most Helpful to Students, Music

An active composer and performer, Dr. Heather Gilligan has recently written works for the Keene State College Chamber Singers, the Keene State College Concert Band, The Chamber Singers of Keene, The Rivers School's "Seminar On Contemporary Music For the Young," Juventas! New Music Ensemble, and Lorelei Ensemble. She serves as accompanist and Composer-in-Residence for the Chamber Singers of Keene and was appointed Composer of the Year in 2010 by the New Hampshire Music Teachers Association. Before coming to Keene State College, she taught music theory and aural skills at Boston University and keyboard harmony at the Massachusetts Institute of Technology. She also taught piano and music theory classes at the Brookline Music School, where she served as coordinator of both the piano and theory departments. At Keene State College she coordinates theory and aural skills. She also teaches theory and composition.

Dr. Gilligan received her D.M.A. in Composition from Boston University in 2008, where she won the Malloy Miller Prize in Composition, a Dean Scholarship, and a Department Honor Award. At BU she studied composition with Richard Cornell, Martin Amlin, and Ketty Nez. She earned her M.M. in Composition from the Longy School of Music, where she studied composition and piano. In summer of 2002 she completed, with distinction, a course at La Schola Cantorum in Paris, under the auspices of the European American Musical Alliance. She also holds a bachelors degree in Chemistry from Lehigh University.

Maura Glennon

Best Teacher, Music

KENNESAW STATE UNIVERSITY

Stephanie Foote

Best Overall Faculty Member, Liberal Arts, Social Sciences, and Humanities

Stephanie M. Foote, Ph.D. is the Director of the Master of Science in First-Year Studies graduate program and an Associate Professor of Education in the Department of First-Year and Transition Studies at Kennesaw State University. Dr. Foote teaches various first-year seminars, sophomore career exploration seminar, and transfer student seminars at Kennesaw. She also currently teaches graduate courses in student development theory, advising and counseling in higher education, and the contemporary college student in the Higher Administration Program at Stony Brook University. She is a past recipient of the NODA Outstanding Research Award for her dissertation study of the perceived effects of first-year seminar participation on the experience of students in their first semester of college. Dr. Foote is the lead author of College Students in Transition: An Annotated Bibliography, and her current research focuses on self-authorship development in transfer students. In addition to publishing her research on college students in transition, Dr. Foote has presented nationally and internationally on her work; she has also made other invited presentations on her research. Additionally, Dr. Foote has developed and taught the online course, Fostering First-Year Student Success for the National Resource Center for The First-Year Experience and Students in Transition® annually since 2010. Dr. Foote is currently a guest co-editor of a special "Fostering Success for Students in Transition" issue of the Journal of College and Student University Housing, she currently serves on the editorial board for the Journal of Student Affairs Research and Practice (JSARP), and is the editor for the Journal of College Orientation and Transition (JCOT).

Donna Freeman

Most Helpful to Students, Liberal Arts, Social Sciences, and Humanities

Dee Grindell

Best Teacher, Liberal Arts, Social Sciences, and Humanities

Diane "Dee" Grindell is an Instructor in the KSU First-Year and Transitions Department. She also serves as Graduation Coach for the nationally recognized, award-winning Thrive Program. She holds a Master's of Science Degree in Education/Human Resources Development and has been with Kennesaw State University for six years.

Ken Hill

Best Researcher/Scholar, Liberal Arts, Social Sciences, and Humanities

Ken Hill earned his M.A. in Education and Human Development from the George Washington University in 1993. In 2009 he completed his TESOL certification through Kennesaw State University and has been teaching in the Intensive English Program since its inception. While he enjoys teaching grammar and listening/speaking, reading is probably the subject that he is most drawn to. Beyond his academic interests, his hobbies include sailing and trail running.

Ayokunle Odeleye

Best Teacher, Fine Arts

Ayokunle Odeleye is a professional sculptor who has specialized in the creation of three-dimensional art for thirty years. He was trained at Virginia Commonwealth University in Richmond, Virginia and at Howard University in Washington, D.C. He received a BFA Degree from Howard University in 1973 and the MFA Degree in 1975. Mr. Odeleye has consistently been teaching art since 1973. He has held teaching positions at Dunbar High School, The Duke Ellington School for the Arts, Howard University, Spelman College, Woodland Middle School, Georgia State University and Kennesaw State University. His personal work involves the creation of large-scale sculpture for public environments. He works both abstractly and figuratively. He maintains two separate state-of-the-art studio facilities: one for woodworking, and another for metal fabrication. He is represented nationally by several galleries.

Robin Saunders

Best Overall Faculty Member, University Studies
Best Researcher/Scholar, University Studies
Best Teacher, University Studies

Brian Wooten

Most Helpful to Students, University Studies

KENT STATE UNIVERSITY

Susanna Fein

Best Researcher/Scholar, English

"Her complicated research engages scholars and students alike. Her productivity in research is outstanding."

Susanna Fein is Professor of English and Coordinator of the A&S Minor in Ancient, Medieval, and Renaissance Studies (AMRS). She has served terms as English Department Chair, Undergraduate Studies Coordinator, and Chair of the Ph.D. Program in Literature. She was 2010 Bloomfield Fellow in Medieval English Studies at Harvard University. In Fall 2014, she will be Visiting Professor of English at the University of Notre Dame.

Professor Fein's research focuses on the literature, languages, and manuscripts of medieval England, ca. 1100–1500. She edits the learned quarterly journal The Chaucer Review and is the recipient of numerous grants from the NEH and elsewhere. Professor Fein serves on the advisory/editorial boards of Oxford Bibliographies in Medieval Studies, the Archive of Early Middle English, the TEAMS Middle English Texts Series, and The Journal of the Early Book Society. She has also served as a Trustee of the New Chaucer Society and held terms on the KSU Press Editorial Board and the KSU Foundation Board of Trustees. A 2001 KSU Retention Grant allowed the English Undergraduate Studies Center to be established.

Professor Fein frequently undertakes archival research in England on medieval manuscripts. What she discovers there renews her teaching of the works of Geoffrey Chaucer, the legends of King Arthur, and manuscript studies. She looks for students who enjoy literature, history, and languages and who are curious about human differences across time.

In summer 2014, Professor Fein will direct an NEH seminar on Chaucer's Canterbury Tales. Designed for American high school teachers, the seminar will be held in London with excursions to Canterbury and Oxford.

Robin Vande Zande

Best Researcher/Scholar, Fine Arts

KENYON COLLEGE

Jianhua Bai

Best Overall Faculty Member, Languages
Best Researcher/Scholar, Languages
Best Teacher, Languages
Most Helpful to Students, Languages

Jianhua Bai, professor of Chinese, teaches linguistics and Chinese at all levels as well as training and supervising teaching assistants in the Department of Modern Languages and Literatures. He also served as Chair of the department and as Chair of the Asian studies Program. Beyond Kenyon, he has served on the executive board of the Chinese Language Teachers Association (CLTA) for two terms (1995–98) (2001–2004) and President of the CLTA. He also served chair of the Committee of Curriculum and Articulation of the Ohio Foreign Language Association. He directed two study-abroad programs in China–the Associated Colleges in China (ACC) Program in Beijing and the Hopkins/Nanjing/CET Summer Program at Nanjing University. During the summer, he directs the Chinese School at Middlebury College. He is also actively engaged in other professional activities such as organizing international conferences, evaluating conference papers and journal articles in his field, serving as a reviewer for program assessment and faculty reappointment and promotion at other institutions, serving as Senior Advisor (flagship curriculum) for American Council of International Education, chairing (2004–2010) the AP Chinese Language and Culture Exam Development Committee, and maintaining a discussion list that promotes communication among teachers, researchers, and students of Chinese. His recent research projects include developing proficiency-based materials for advanced Chinese; schema theory and reading Instruction; how web-based and distance learning technology can be used to enhance Chinese language pedagogy; the implications of pragmatics to Chinese pedagogical practice; and action research, documenting excellence in teaching Chinese as a foreign language.

KILGORE COLLEGE

Terry Booker

Best Teacher, Liberal Arts, Social Sciences, and Humanities

David Stroud

Best Researcher/Scholar, Liberal Arts, Social Sciences, and Humanities

Jeanne Tagg

Best Overall Faculty Member, Communication
Best Teacher, Communication
Most Helpful to Students, Communication

LAFAYETTE COLLEGE

George Rosa

Most Helpful to Students, French

Deborah Rosen

Best Overall Faculty Member, Liberal Arts, Social Sciences, and Humanities

"Most serious adviser of students and model for them; outstanding research scholar and author; superb and dedicated teacher"

LAKE MICHIGAN COLLEGE

David Keith

Best Teacher, Liberal Arts, Social Sciences, and Humanities
Best Overall Faculty Member, Liberal Arts, Social Sciences, and Humanities

LAKELAND COMMUNITY COLLEGE

Kelly Madsen

Most Helpful to Students, Languages

LAMAR STATE COLLEGE AT ORANGE

Arlene Turkel

Most Helpful to Students, English

Professional Experience Past president: Texas College Reading and Learning Association. Faculty consultant: Educational Testing Service (English Language AP scoring for 7 years). Manuscript reviewer: Longman and McGraw Hill. Research Interests: Peer tutoring, developmental education.

LANE COMMUNITY COLLEGE

Cynthia Scweigert

Best Teacher, Writing

"Cindi Sweigert goes above and beyond being helpful to her students. She really wants to make sure they understand the material and are ready for the next writing class!"

LANSING COMMUNITY COLLEGE

Sharon Park

Best Teacher, Art History

LAS POSITAS COLLEGE

Cindy Ahre

Most Helpful to Students, English

LAWRENCE TECHNOLOGICAL UNIVERSITY

Michelle Belt

Best Researcher/Scholar, Interior Design

LEE UNIVERSITY

Megan Moe

Best Researcher/Scholar, Communication

Highly skilled College Professor with twenty-one years of experience teaching a broad range of communications courses in a number of college levels and business settings. Demonstrates strong capabilities in teaching effectiveness as well as versatility in course design and implementation.

Specialties: academic, advertising, analytical skills, basic, business presentations, communication skills, concept development, customer relations, meeting facilitation, Microsoft Powerpoint, modems, organizational skills, persuasion, programming, public speaking, research, speech writing, strategic, teaching, theater, writing.

LEHMAN COLLEGE

Sharon Jordan

Best Teacher, Fine Arts

"Her courses are not generic. She carefully researches her topics allowing for an original approach to teaching each subject."

B.A., Boston Univ.; Ph.D. Art History, Graduate Center of CUNY

LENOIR-RHYNE UNIVERSITY

Jeff Delbert

Best Researcher/Scholar, Communication

"One of the hardest working researchers I have run across in my 30 years of higher education and teaching."

LETOURNEAU UNIVERSITY

Fredna Stuckey

Most Helpful to Students, English

"Fredna makes every effort to be available to students. She spends countless hours one-on-one with students, helping them with their writing."

Fredna Stuckey brings 21 years of experience teaching middle school and high school English, as well as supervisory and administrative experience, to her work at LeTourneau University.

"Although I had taught English for more than 20 years, teaching at LeTourneau is definitely the high water mark of my professional career. The sharing, caring attitude of the faculty and staff makes planning and preparing for instruction an exciting, pleasant task. The actual teaching brings unspeakable joy as I watch students' faith and learning grow simultaneously. Student/teacher ratios are low enough to ensure accessibility to students; yet LeTourneau offers numerous advantages and opportunities often found only at much larger universities. Opportunities to prepare for your chosen vocation and to reach the world for Christ go hand-in-hand at LeTourneau. I love working together with students as we seek to fulfill the Great Commission."

Besides teaching, Mrs. Stuckey enjoys sewing, camping, swimming, and water skiing. She and her husband David are also passionate about missions. One of their favorite ministry sites is an orphanage in Mexico. They have also participated in mission trips to Barbados, the Dominican Republic, and many destinations within the US.

Jim Watson

Best Researcher/Scholar, English

"Jim's multidisciplinary scholarship has resulted in creative learning opportunities for students. The most recent of these is his work on the World War II lecture series in which he integrates literature with history. Jim's research and work in these kinds of projects contributes to integrated University learning."

LEWIS UNIVERSITY

Therese Jones

Best Overall Faculty Member, English
Best Teacher, English
Most Helpful to Students, English

Therese Jones, Assistant Professor with a Masters in Literature and Composition Studies from St. Xavier University, had been a part-time member of the Department since 1995 and was promoted to full time in 2011. She served as the editor of Lewis' Arts and Ideas Magazine (Fall-Spring of 2002–2003), and is the current Editor, Designer, and Coordinator of Lewis' annual journal, Windows Fine Arts Magazine (2003 to present). She was the Writing Assistant for the Dean of the College of Arts and Sciences (2003–2004) and collaborated with Dr. Nancy Workman on co-editing Dimensions of Curiosity: Liberal Learning in the 21st Century (2004). Therese petitioned and was awarded Title III funding for a 2004 summer committee on the viability of directed self-placement for writing placement. She presented "Plagiarism" at the 2005 ACCA Scholarship of Pedagogy and continually researches data on innovative assessment procedures for writing placement as the Director of Writing Placement. Therese collaborated with Dr. Sheila Kennedy and co-authored the Writing Manual for Lewis' Bridge Program (2008). She presented "A Change in Pedagogy for Lewis SUCCESS Students Studying Writing" at Lewis' ACCA Scholarship of Pedagogy in 2009. Therese also has worked on various subcommittees, such as the College Writing II Rubric, Program III, WAC, and Bridge. While the primary area of specialization that Therese contributes to the English Department is her skill with assessment, particularly as the Director of Writing Placement since 2002, and her expertise in Composition, she also regularly teaches literature courses in the general education curriculum, most notably "Experience of Literature," and she also serves as an advisor for 4–5 English majors each year and through her leadership of Windows, she offers English majors encouragement in and opportunities for publication of their creative and scholarly writing. In 2015, she received the faculty of the year award.

LEWIS-CLARK STATE COLLEGE

Marlowe Daly-Galeano

Best Teacher, English

"Marlowe is a dedicated teacher. Her door is always open for her students. She is an inspiration for me to improve my own teaching."

Marlowe Daly-Galeano teaches American literature, writing, and humanities classes in addition to the Hells Canyon Institute class. Her interests include nineteenth-century American literature, women authors, the Alcott family, and regional literature. She is the "Year in Conferences" Director for ESQ: A Journal of the American Renaissance. She also serves as the faculty advisor of the Ambassador Honor Society and is one of the organizers of Women's History Month.

LIBERTY UNIVERSITY

Stephanie Blankenship

Most Helpful to Students, Languages

Mrs. Blankenship began her Spanish studies at Marshall University. During her time at Marshall, she studied abroad in Madrid, Spain. Upon completion of both her B.A. and M.A., she moved to Lynchburg, VA to begin her teaching career. Before coming to Liberty in 2006, Mrs. Blankenship taught Spanish at various levels from middle school to college. Since beginning at Liberty, she has taught both lower and upper division Spanish courses and now serves as an Assistant Professor of Spanish.

Chris Foley

Best Teacher, Languages

Dr. Chris Foley has taught at Liberty University since 1995. Before coming to Liberty, Dr. Foley taught French Immersion and Spanish at every level, from elementary school through high school to university, in Canada. In the U.S., he has taught French and Spanish at Lynchburg University, George Mason University and Central Virginia Community College. He has also worked as a technical translator of French and Spanish for more than 14 years.

Degrees

Ph.D. in French, University of Manitoba

M.A. in French, University of Manitoba

M.A. in Spanish, George Mason

B.Ed. in Fench Immersion and Spanish, Collège universitaire de Saint-Boniface

B.A. in French and English, University of Saskatchewan

Publications and Presentations

Dissertations and Publications

"Don Quijote: un análisis de sus síntomas esquizofrénicos." MA Thesis George Mason U, 2005. Print.

Baudelaire, traducteur de Poe: précurseur de la traduction moderne. Diss. U of Manitoba, 1998. Winnipeg: UM, 1998. Print.

Molière, juge du système judiciaire de son temps. MA Thesis U of Manitoba, 1985. Winnipeg: UM, 1985. Print.

"La casada infiel de Garcia Lorca: una explicación del texto." Hispanic Culture Review. 10(2004) 43–52. Print.

"The Struggle between heaven and hell for the soul of Arthur Rimbaud in Une Saison en enfer." Journal of Christianity and Foreign Languages. 9(2008) 49–60. Print.

Presentations "García Lorca's Yerma: The Quest for Love is Empty, the Verdict is Guilty." The 2009 National Conference of the National Association of Hispanic and Latino Studies, Baton Rouge, LA. February 12, 2009.

"Cervantes' errant knight: his schizophrenic behavior." The 2009 National Conference of the National Association of Hispanic and Latino Studies, Baton Roug, LA. February 13, 2009.

"Baudelaire: traducteur du vernaculaire noir du Sud américain." The Mountain Interstate Foreign Language Conference, Furman University, Greenville, SC. October 8, 2009.

"Don Quijote: un análisis de sus sintomas esquizofrénicos." The Mountain Interstate Foreign Language Conference, Furman University, Greenville, SC. October 8, 2009.

Teaching and Research Interests

Translation

Seventeenth and Nineteenth-Century French Literature

Dramatic Literature of El Siglo de Oro

Molière, Voltaire, Maupassant and Zola

Cervantes and Lorca

S. Hanlahn

Best Overall Faculty Member, Languages

Taeseong Kim

Best Overall Faculty Member, Liberal Arts, Social Sciences, and Humanities

Associate Professor of Piano in the School of Music, Taeseong Kim received his Bachelor of Music degree in piano from Liberty University, his Master of Music degree in piano pedagogy from the University of Tennessee, Knoxville, and his Doctor of Musical Arts degree in Piano Pedagogy from the University of South Carolina, Columbia. He studied piano with Charles Fugo, David Northington, David Ehrman, and Beverly Henkel, and piano pedagogy with Scott Price, Stephen Taylor, and Fay Adams. He joined the Liberty faculty in 2006.

He strongly believes in the value of music as an effective tool for glorifying and serving God. He is currently serving as the music director of Korean Baptist Church of Lynchburg.

Bruce Kirk

Best Teacher, Communication

"A born and experienced leader; courage and integrity; works hard to always make things better for the students. He cares."

Mr. Bruce Kirk spent 33 years in the broadcasting field, working in 8 markets and with 12 stations in both radio and television. His goal at Liberty is to be able to transfer some of his experiences into meaningful opportunities for his students, and to prepare them for "how the business works."

Mr. Kirk began his studies by pursuing a career in Pharmacy, but earned his bachelor's degree in Liberal Arts. After completing his undergraduate work, Mr. Kirk discovered his passion for broadcast, and began to pursue this path. He later earned his master's degree from Spring Arbor University, and is currently working on his doctorate.

Mr. Kirk is the winner of five television Emmys, two of which were won for Investigative Journalism. He has held most every position in the field, including Assignment Editor, Producer, Executive Producer, Editor, Photographer, Reporter, Field Anchor, Studio Anchor, Field Producer, Special Projects producer, Executive Producer, Managing Editor, Assistant News Director and News Director. Mr. Kirk's work has allowed him to travel internationally to cover news in Australia, Brazil, Canada and Mexico, as well as across the United States. He has covered Super Bowls, and NBA and NHL finals.

Mr. Kirk's political work is extensive. He has interviewed U.S. Presidents Jimmy Carter, George H.W. Bush, Bill Clinton, and Vice President Walter Mondale. Mr. Kirk has done features on political figures and events like President Ronald Reagan; presidential candidates Sen. John McCain, Sen. Ted Kennedy and Sen. John Glenn; and numerous national party conventions.

Mr. Kirk's vast industry experience has enabled him to mentor many that have gone on to have successful careers with Fox, NBC, CBS and CNN; and in markets such as Dallas, Phoenix, Washington, D.C., New York, Detroit and San Diego. Mr. Kirk taught broadcast media and journalism at the Walter Cronkite School of Journalism at Arizona State University. He also taught as an adjunct professor at Liberty University for about 5 years, before returning in 2009 as a full-time professor.

Mr. Kirk is married to his wife June, and has four kids, two boys and two girls.

Mr. Kirk enjoys his work at Liberty, where he works to convey the importance of media to all of his students:

"The media is so critical in all of our lives, that we must know more about it to be successful advocates for Christ."

Mr. Kirk also oversees all internships within the School of Communication & Creative Arts.

LINCOLN COLLEGE

Michelle Baldwin

Best Overall Faculty Member, English
Best Researcher/Scholar, English
Best Teacher, English

Michelle Baldwin is an alum of Lincoln College, earning her A.A. degree in 1989. Her educational background also includes a B. S. degree in English and a M. S. degree in Education from Illinois State University. She has taught English and Literature at Lincoln College since 1996 and was the recipient of the Dr. Jack D. Nutt Educator of the Year award in 2008. Most recently, she co-presented on the potential use of Twitter as a teaching tool to increase student responsibility, engagement, and critical thinking for use with course management, collaboration, and class discussion. Her research interests include: authentic writing strategies, writing across the curriculum, and the use of technology to engage students and critical thinking.

Steven Schreiber

Best Researcher/Scholar, Liberal Arts, Social Sciences, and Humanities

Mark Tippett

Most Helpful to Students, Liberal Arts, Social Sciences, and Humanities

Coach Tippett has been head coach of the Lincoln College Volleyball program since 2001. During his tenure, Mark Tippett has brought a high level of excellence to his program, both in the classroom and on the court.

In addition to coaching the volleyball team, Tippett is the Social Sciences Division Chair and teaches Introduction to Education.

Before joining Lincoln College, Tippett was the Assistant Volleyball Coach at Lincoln Christian College. While at LCC, Tippett contributed to the success of the Angels appearing in the 1998 NCAA Division 2 tournament, in which they achieved their goal by becoming the National Champions.

Tippett holds a B.A. degree in English from Millikin University and a Masters in Education from the American College of Education. He and his wife, along with their two daughters, live in Lincoln.

LINCOLN COLLEGE: NORMAL

Kratz Jeffrey

Best Researcher/Scholar, Liberal Arts, Social Sciences, and Humanities

"Thorough and Reasoned in approach. Concerned about the students experience and education."

LINDENWOOD UNIVERSITY

Kay Hammond

Best Teacher, Liberal Arts, Social Sciences, and Humanities

"She is always available to her students. She gives clear instructions. She is strict but fair."

Kay Hammond earned her MBA from Lindenwood in 1996 Magna Cum Laude. She took early retirement from Boeing and started teaching the very next day. She taught Information Systems, Microcomputer Applications, HTML, and now teaches Oral Communications. She is full of energy and thoroughly appreciates her second career. She models parts of her class after Toastmasters, incorporating table topics and current events. She is adamant about properly formatted paperwork and does not tolerate late assignments. She encourages the students to make the class interesting by their candid interaction, interesting speeches, and by their personal evaluations of fellow student speeches. It is a student- driven class. Taboo words are prohibited and the clicker is a constant reminder of this. She does value laughing and the associated camaraderie that comes with atmosphere of a speech class, but classroom etiquette is a must! She is a social butterfly and rarely turns down an invitation.

Ben Scholle

Best Overall Faculty Member, Communication
Most Helpful to Students, Communication
Best Overall Faculty Member, Liberal Arts, Social Sciences, and Humanities

Ben Scholle, Professor of Communication, holds a BA from Washington University and received his MFA in Film and Electronic Media from American University. He has worked on a variety of independent films and television programs, and his short films and documentaries have played in festivals nationwide. His documentaries have also aired on The Documentary Channel, the BET Network, and Free Speech TV. In 2008, Ben's documentary, HairKuTT, was selected to screen at the Museum of Modern Art. Ben Scholle serves as Department Chair for Cinema and Television.

Andrew Smith

Best Researcher/Scholar, Communication
Best Teacher, Communication
Best Researcher/Scholar, Liberal Arts, Social Sciences, and Humanities
Best Teacher, Liberal Arts, Social Sciences, and Humanities

Edward Voss

Most Helpful to Students, Liberal Arts, Social Sciences, and Humanities

Edward Voss earned his MA in Communications Management from Webster University as well as his BA in Radio, TV, and Film Production from the University of Missouri – Columbia. He served in the U.S. Navy were he was graduated from the Defense Information School, earning diplomas in their Broadcast and Journalism programs. Upon an Honorable Discharge he worked for Group W Cable, directing and producing a number of sports and community programs. Later, he worked as a freelance production technician for Maritz Communications, Cardinals Baseball, KMOV, KDNL, Civic Center Corporation, The Federal Reserve Bank-St. Louis, Ameren, Video Nine, Rosenthal Productions and KETC Ch. 9 were he was Emmy nominated for his work on post production audio. Afterwards he was employed by KTVI FOX2 as an operations technician performing a number of tasks including live coverage of Pope John Paul II's St. Louis visit, Rams Super Bowl victory parade, several presidential elections and coverage of Hurricane Katrina. Voss has taught video production at St. Louis Community College at Forrest Park. He works in the School of Communications as an Assistant Professor and Director of Operations of LUTV.

LINN-BENTON COMMUNITY COLLEGE

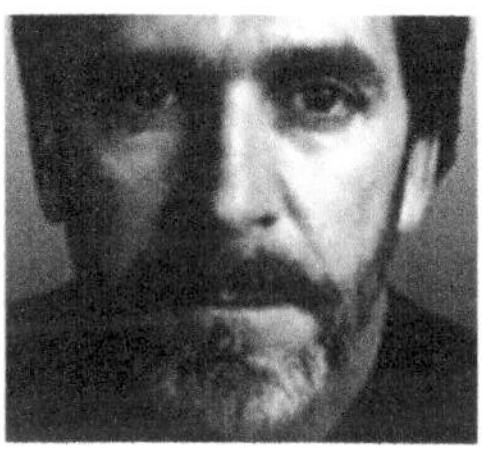

Terrence Millet

Best Overall Faculty Member, Writing

"Terrence is a brilliant individual with a flair for teaching creative writing, poetry, fiction, and anything regarding literature."

I have an M.F.A. in Creative Writing: Fiction, from Oregon State University, where I studied under Tracy Daugherty, Marjorie Sandor, and Ehud Havazelit, all award-winning writers. My B.A. in English Literature is from Huron University in Ontario, Canada, where I put in another post-grad year of honors studies in English Literature, and my M.A. is from the University of Western Ontario.

I teach creative writing workshops here at LBCC for short fiction, script writing, and poetry writing. Because the workshop model works well, I use it for my literature courses as well, so the classes are discussion driven and very collaborative.

My professional specialties are fiction and poetry writing, and my work has been published in various journals and small presses. Most recently, my stories have appeared in the issues of The Hopkins Review, published out of the Johns Hopkins University.

I've just completed the first in a series of three novels, Becoming Finn, and plan to have the set completed in 2016.

Karelia Stetz-Waters

Best Teacher, Writing

"Karelia is kind, patient and intelligent. She has a calmness and grace about her that make her easy to approach and she has a presence in the classroom that comes with experience, understanding of others and knowledge of her subject matter."

I am the English department chair. My professional specialties are English composition and technical writing, but I am also an avid creative writer. My work includes The Admirer, The Purveyor, Forgive Me If I've Told You This Before,and Something True. Over the years, I have worked with the ROV club, the Open Source Club, Lutheran Campus Ministry at OSU, and the Gay Straight Alliance. I grew up in Corvallis, and I love the Willamette Valley. In my free time I like to write, garden, travel with friends and spend time with my wife and my parents.

Jane Walker

Most Helpful to Students, Liberal Arts, Social Sciences, and Humanities

"Jane is one of the kindest and most considerate individuals I have met. She always has time for students and shows patience and concern for them at all times."

LONE STAR COLLEGE (ALL)

Jared Cootz

Best Overall Faculty Member, Sociology
Best Researcher/Scholar, Sociology
Best Teacher, Sociology
Most Helpful to Students, Sociology

Education Ph.D. Educational Administration Human Resource Development (focus on Organizational Development). Currently ABD. Texas A&M University. 2000–Present. Level II Gerontology Certificate. Post-graduate work. Our Lady of the Lake University. 2002. M.A. Human Sciences (Psychology and Sociology). Our Lady of the Lake University. 1997–1999. B.A. Psychology, Minor in English. Our Lady of the Lake University. 1995–1997. A.A. Associate of Arts. North Harris College (Lone Star College-North Harris). 1993–1995.

Experience

- Professor of Sociology. Lone Star College-Montgomery. 08/2001–Present
- Department Chair, Behavioral Sciences. Lone Star College-Montgomery. 01/2008–08/2012 Lecturer of Sociology and Psychology. Our Lady of the Lake University. 08/2005–05/2011 Lecturer of Sociology and Social Sciences. The University of Houston Downtown. 2001–2002
- Case Manager/Counselor. Star of Hope. Houston, Texas. 04/1998–08/2001
- Teacher. Psychology, Sociology, Biology. South West High School.
- American Home Services. Home remodeling. 1993–1998
- U.S. Army. 101st Airborne, Fort Campbell, Kentucky. MI (LI) BN, Soto Cano Airbase, Honduras. Goodfellow AFB, San Angelo, Texas. Fort Dix, New Jersey. 1989–1992

Leyel Hudson

Most Helpful to Students, Sign Language

Paul Nimon

Best Teacher, English

Alex Qian

Best Overall Faculty Member, Music

"Alex Qian is a servant leader for his students and the faculty. He handles every issue with profound professionalism. He respects all and deserves recognition. Sincerely, John Wolfskill"

Alex Qian is the founder of the LSC-CyFair Music Department, where he has produced and conducted musicals, operas and major choral works. Qian has appeared as a festival guest singer, conductor and workshop presenter in Japan, Korea, Singapore, Malaysia, Hong Kong, Europe, China and the United States. A doctoral student of University of Houston and a graduate of both Shanghai Conservatory and Sam Houston State University, he has studied musical theatre, opera performance and choral and orchestral conducting with Jingzu Bian, Shaozeng Xie, Geshuen Ma, Walter Foster, Kent Hatteberg, Steve Smith, Charles Haussmann and Katherine Ciesinski. He also studied in summer workshops with Robert Shaw, Richard Miller and Ingo Titze

Riffat Saeed

Most Helpful to Students, English/Developmental English

"Respects/assists the students."

LONG BEACH CITY COLLEGE

Anthony Carreiro

Best Researcher/Scholar, Theater
Most Helpful to Students, Theater

A professional actor since that time, Mr. Carreiro has appeared in almost one hundred television shows including recurring roles on Frasier, Sister Sister and Deadly Games. Tony was a series regular on the award winning CBS show: Doctor, Doctor. He has done scores of television commercials and over a half dozen major motion pictures, including: Lethal Weapon II and Liar Liar. Tony has worked in regional theatres and Shakespeare festivals as an actor, director and fight choreographer up and down the West Coast and in New York. He also teaches workshops on "Theatrical Combat," "The Business of Acting" and "Acting for the Camera."

Cathy Crane

Best Overall Faculty Member, Theater
Best Teacher, Theater
Most Helpful to Students, Theater

Cathy Crane has been a faculty member at LBCC since 1984 and has been designing costumes for Theatre, Dance, Film and Opera for over 30 years. She was nominated for an Ovation Award in 1994 for her designs of Opera Comique at International City Theatre and a LA Weekly Theater Award in 2001 for The Three Sisters at the El Portal Center Theatre in North Hollywood. Other professional costume designs were seen at Pasadena Shakespeare Company's Tartuffe, El Portal's Shooting Craps(with Harold Gould) and Out of Order(directed by Ray Cooney), Shakespeare Orange County's Twelfth Night, Richard III and A Midsummer Night's Dream, Court Theatre's Spaghetti & Apple Pie, Back Alley Theatre's Amazing Grace, ICT's Autumn Elegy, Vanishing Points, Quiet End, Eden Court, Song of Singapore, Romance, Romance and Tennessee in the Summer. Her dance design works include Long Beach City College, Goldenwest College and C.S.U. Los Angeles She has been guest costume designer at Chapman University for Lysistrata, Los Angeles City College for Charlie's Aunt and C.S.U. Bakersfield for What the Butler Saw. In addition, she designed the opera Summer & Smoke at C.S.U. Long Beach with Michael Carson. Her film credit includes costume design and wardrobe for the short Trade Day by Dawson Williams that won best short film at the 2001 Long Island International Film Festival.

Peter Knapp

Best Researcher/Scholar, Liberal Arts, Social Sciences, and Humanities

Greg Mortensen

Best Researcher/Scholar, Theater

Mr. Mortensen is a graduate of the Advanced Drama program of The Juilliard School for the Performing Arts in New York City. He holds a B.A. and M.F.A. in both Acting & Directing, and is a Member of Actor's Equity Association, Screen Actors Guild, A.F.T.R.A. and The Dramatists Guild. As an actor, his Broadway credits include "MacBeth" at The Circle in the Square and "How It All Began" (which he co-authored) for Joseph Papp at the New York Shakespeare Festival, at The Lincoln Center for Electra & Orestes, The Ghost Sonata, The Alchemist, The Wood Demon and The Comedy of Errors. Regional Theatre credits include seasons at The Denver Center for the Performing Arts where he portrayed, Jack in The Impotanance of Being Earnest, Julian in Spokesong, Captain Bluntschli-Arms and the Man and Wilson- The Front Page. Other seasons include Stage West Terra Nova, Berkeley Shakespeare Festival; Hamlet, As you Like It, Henry IV, California Actors Theatre's Henry V, Scapino, Wild Oats. Colorado Shakespeare Festival; A Winters Tale, California Repertory, Tom Jones, Taming of the Shrew and Dylan to name a few. Other Seasons include The Barn Musical Theatre, San Jose Music Theatre, The Grove Theatre, Sand Harbor and The New York Public Theatre. His film, television, commercial & voice over credits include recurring and guest starring roles on Knots Landing, Remington Steele, Dragnet, The New Adam-12, The Deliberate Stranger, Safe at Home, The Munsters Today, Hollywood Chaos, Godspeed-The Gail Deavers Story, Dick Tracy, Punks, S.O.L., Murder She Wrote, A.T.T, Ford Motor Co., Disney & Miller Brewing Co.

Daytime audiences may remember him as Paul Stewart on Days of Our Lives and Martin Tyler on General Hospital. His directing credits include, A Cry of Players, Indians, The Tempest, Trinity, Unexpected Company and three Columbia Artists national musical tours which he also authored The Americus Civil War Brass Band- A History with Music, The Dodge City Cowboy Band and The Riverboat Revue.

Mr. Mortensen is also a produced playwright and screen writer; some credits include, How It All Began, The Scoundrel, Hold-Outs, A Point of Honor and The Man in Blue.

His other related professional experience includes, consultant for Walt Disney's Beauty and the Beast at The Shubert Theatre in Los Angeles, Co-Producer B&B on the Arts for P.B.S., Producer for HCTV 26 Campus Collage, Producing Director for Shakespeare L.A.-98. Related professional teaching credits are: Lecturer and Adjudicator for USF/SUU Competition at The

Utah Shakespeare Festival, Lecturer for Lincoln Center Outreach Program, Member of Tadashi Suzuki Classical Theatre Exchange, Guest Lecturer for Denver Center for the Performing Arts, Speaker- The Shanghai Drama Academy Cultural Exchange and Director/Producer for Kirti Management Young Professionals for seven years. Academic teaching credits include, California State University Long Beach, Long Beach City College, Cerritos College, Los Angeles City College and OCHSA. Mr. Mortensen is a four time Drama-Logue Award winner and Outstanding Graduate winning teacher for Theatre Arts & Communications from the International Communications Association.

Stephanie Powell

Most Helpful to Students, Fine Arts

Professor Stephanie Marie Powell is a California native and a tenured full Professor in Dance at Long Beach City College. Throughout her academic teaching career she has taught courses in Dance Appreciation, Choreography, Composition, Classical Ballet, Contemporary Ballet, Pointe, Modern Dance, Jazz, Turns, Zumba, Dance Conditioning, Stretch and Relaxation, Dance History, Hip Hop Dance History, Performance Workshop, Yoga and Pilates. Professor Powell is a certified instructor of the New York City Ballet Workout, Zumba, Qiginition and Body Arts and Science International Pilates with a Dance Specialization. Ms. Powell has taught as a part-time instructor at the University of California, Irvine in Ballet, Modern Dance and the Etude Ensemble. She is the principal interpreteur of the choreographic works for the legendary choreographer Donald McKayle and stages his work around the world.

Ms. Powell's dance training began with Cindy Trueblood at the Civic Dance Center in Bakersfield, California and her studies continued on scholarship with Pacific Northwest Ballet in Seattle, Washington and the School of American Ballet at Julliard in New York City. She received her B.A. from the University of California at Berkeley in Sociology and Education and a Master of Fine Arts in Dance with distinction from the University of California at Irvine as a Chancellor's Fellow, where she was honored with the Celebration of Teaching Award and the Distinguished Alumnus Award from the Claire Trevor School of the Arts. She danced professionally and toured internationally with major dance companies such as the Oakland Ballet Company, the San Francisco Opera, The Dance Theatre of Harlem, Donald Byrd/The Group, Las Vegas Contemporary Dance Theater and Alvin Ailey American Dance Theater. Ms. Powell was a performer in the Los Angeles Pantages Theater production of the Broadway show "Lion King" where she danced featured principal roles as well as performed as a principal singer and actress. Additionally, Powell has modeled internationally for Redken and Trevor Sorbie of the UK. She has appeared on television in the Grammy's, on the Jay Leno Show and has danced commercially for artists such as MC Hammer, Kenny G., Janet Jackson, and Beyonce Knowles.

For the LBCC Dance Program, Powell has fundraised over $40,000 by organizing student involvement in car washes, obtaining hundreds of opportunity drawing prize donations, corporate monetary gifts, costume donations, organizing TV tapings, t-shirt, roses and concession sales. These funds have afforded students and faculty the opportunity to travel out of state to the American College Dance Festivals and purchased Pilates Reformer equipment for students.

She was instrumental in attaining a $50,000 ASB grant to install a state of the art sprung dance floor in the Q-building Dance Studios. As a scholarly researcher, she is responsible for creating two new online courses for Dance (Dance History and Hip Hop Dance History) to give more students accessible opportunities to attain an A.A. degree from LBCC through the Distance Learning Program.

Powell is the only full-time African-American faculty member in the Theatre, Dance and Film Department, and was selected by Human Resources to inspire diversity among new faculty at LBCC in the advertisement brochure. In addition to her course load, she created the "Dance Showcase" which features all of the course offerings in Dance at LBCC. She incorporates the work of other colleagues in disciplines such as Science, Photography, Graphic Design, Music, Radio, Television, Fashion, Film and Theater to name a few. She is responsible for establishing the first cable television premiere of the "Dance Ensemble in Concert," photographic marketing of Dance Students for publicity and has consistently been recognized by the Viking Newspaper for encouraging student success through Dance. She has directed two "High School Dance Invitational" events where she has recruited the students of twelve (12) local high school Dance programs to perform on the LBCC stage. Powell serves as the Dance Club Advisor and works closely with the campus Outreach and Recruitment Coordinators to support President Eloy's wish for all 4th graders to experience a campus tour. She is also on several College Service committees such as the Fine Arts Association, the 85th Anniversary committee, the Distance Learning Plan Implementation Oversight Task Force and she is a Student Learning Outcome Committee Officer for her department.

As a testament to her reputation and contributions, Powell was a recipient of the Long Beach Post Award for Best Teacher in the City of Long Beach and in March of 2010 she was a nominee for the 2010 "Woman of the Year" Award by the 54th Assembly District's Assembly member Bonnie Lowenthal and most recently the Black Leadership Advancement Coalition Alumni Award for 2012.

Professor Powell continues to dance as a performing artist today and has taught and choreographed her own work in contemporary dance and ballet throughout the US, Australia, Europe and Asia. She serves as a volunteer for the NAACP and is on the committee that will organize the International Association of Blacks in Dance Festival in 2014. Powell attends conferences regularly for continued education in her art form to bring up to date information, dance techniques and digital technology to her students in the LBCC classroom. She has taught classical ballet at the world renowned "Debbie Allen Dance Academy" in Los Angeles and performed in her production of the "Hot Chocolate Nutcracker" at Royce Hall-UCLA. Powell was also be featured in an upcoming article as a leading figure in the community in the Cultural Arts Guide for the Long Beach Business Journal. Professor Powell is currently creating a new lively and high energy choreographic work for eight men as a faculty choreographer and is the Assistant Artistic Director of the Fall Dance Ensemble in Concert in the LAC campus Auditorium.

LONG ISLAND UNIVERSITY

Melisa Antinori

Most Helpful to Students, English

Michael Bokor

Best Researcher/Scholar, Liberal Arts, Social Sciences, and Humanities
Best Teacher, Liberal Arts, Social Sciences, and Humanities

I am from Ghana in West Africa. Some of my hobbies include: sports (soccer, volleyball and ping-pong); music (especially gospel songs and reggae); gardening; and bird watching.

LONG ISLAND UNIVERSITY - C.W. POST

Dave Hugo

Most Helpful to Students, Theater
Best Overall Faculty Member, Liberal Arts, Social Sciences, and Humanities
Best Teacher, Liberal Arts, Social Sciences, and Humanities

David Hugo received a BFA in acting from Syracuse University and moved directly to NYC where he worked as an actor in regional theater, national tours, and on Broadway. He was last seen on Broadway playing the role of Grantaire in Les Miserables where he also had the honor of understudying Terrance Mann in the role of Javert. After twelve years working professionally, he returned to school to receive a master's degree in theater in order to teach aspiring young actors. He received his master's degree from LIU Post in 2006, where he had the distinct pleasure of studying under Maria Porter, Benjamin Moore, and Cara Gargano. He is thrilled to be able to start his teaching career among such inspiring professors and artists.

LOS ANGELES CITY COLLEGE

Irene Kim

Most Helpful to Students, Music

Dr. Kim joined the LACC music faculty in Fall 2000 as the Director of Choral Activities. Over the years, she has taught introductory theory courses, piano courses levels 1–4, and all choral groups. Prior to her position at LACC, she served as choral director at various schools, community groups, and churches including San Gabriel High School and the University Campus Choir at UCLA.

Dr. Kim is a graduate of University of Arizona where she earned a Ph.D. in music education. She received her M. M. and B.A. from UCLA in Choral Conducting, Music Education, and Piano Performance.

LOS ANGELES VALLEY COLLEGE

Jim Marteney

Best Overall Faculty Member, Speech Communication
Best Researcher/Scholar, Speech Communication
Best Teacher, Speech Communication
Most Helpful to Students, Speech Communication
Best Researcher/Scholar, Liberal Arts, Social Sciences, and Humanities
Best Teacher, Liberal Arts, Social Sciences, and Humanities
Most Helpful to Students, Liberal Arts, Social Sciences, and Humanities
Best Overall Faculty Member, Liberal Arts, Social Sciences, and Humanities

Phil Poulos

Most Helpful to Students, English

William Wallis

Best Overall Faculty Member, English

Dan S. Watanabe

Best Overall Faculty Member, Media Arts
Best Teacher, Media Arts
Most Helpful to Students, Media Arts
Most Helpful to Students, Fine Arts

Brian Yablon

Best Teacher, English

LOUISIANA STATE UNIVERSITY AT SHREVEPORT

Kenna Franklin

Most Helpful to Students, Liberal Arts, Social Sciences, and Humanities

Alex Mikaberidze

Best Overall Faculty Member, History
Best Researcher/Scholar, Liberal Arts, Social Sciences, and Humanities

Born in the Soviet Kazakhstan, Dr. Mikaberidze grew up in the newly independent Republic of Georgia. He holds an advanced degree in international law from Tbilisi State University (Republic of Georgia, 1999) and worked as an international law expert at the Ministry of Foreign Affairs of Georgia (1996–2000), where he handled the human rights issues and relations with the Council of Europe. In 2000, Dr. Mikaberidze moved to the US to pursue his dream of studying the Napoleonic Era (1799–1815). He joined the prestigious Institute on Napoleon and the French Revolution at Florida State University where he finished his Ph.D. in history in 2003. He has taught European, World and Middle Eastern history at Florida State University and Mississippi State University and lectured on strategy and policy for the U.S. Naval War College. He is the general editor of the Selected Papers of the Consortium on Revolutionary Era, as well as editor-in-chief of the periodical The Napoleonic Scholarship. He has been elected Fellow of Royal Historical Society (United Kingdom) and awarded the International Napoleonic Society's Legion of Merit Medal and La Renaissance Française's Médaille d'or du Rayonnement Culturel for his contributions to the Napoleonic studies. He joined LSUS in 2007 and was chosen by students as one of Top 20 Professors in 2009.

LOUISIANA TECH UNIVERSITY

Brenda Heiman

Best Overall Faculty Member, Speech

Cheryl Leachman

Best Researcher/Scholar, Speech

Chad Lyles

Best Teacher, English

Education: MA Louisiana Tech

Robert Rudnicki

Best Researcher/Scholar, English

C. Strebeck

Most Helpful to Students, English

LOYOLA MARYMOUNT UNIVERSITY

Linda Bannister

Most Helpful to Students, English
Most Helpful to Students, Liberal Arts, Social Sciences, and Humanities

Linda Bannister, professor of English at Loyola Marymount University, has a Ph.D. in Rhetoric, Linguistics, and Literature from the University of Southern California. She was the inaugural holder of an endowed chair, The Daum Professorship of the Bellarmine College of Liberal Arts, LMU, 2008–09. She works primarily in these areas: Rhetoric, Composition and Stylistics; contemporary literature by underrepresented groups; Journalism and professional writing; and playwriting. She has published over 100 works in the areas of rhetoric, composition, contemporary writers, and journalism, and has most recently focused her attention on playwriting. She is also the Director of the LMU Journalism Certificate Program and The Secondary Teacher Preparation Program in English. She and her writing partner, James E. Hurd, Jr., had their one-act play "One Sunday in Mississippi" selected for presentation at the 2003 National Black Theatre Festival in Winston-Salem, NC. Their full-length drama, Turpentine Jake, was selected for reading at the 2007 National Black Theatre Festival, also in Winston-Salem. She produced the world premiere run of Turpentine Jake at the Del Rey Theater in Los Angeles, August 1–24, 2008, which was nominated for two 2009 NAACP Theatre Awards, and won the award for Best Actress. Turpentine Jake was a featured performance at the Great Plains Theatre Conference 2010. Bannister and Hurd's work has also been seen at Theatre Unlimited in North Hollywood, First Stage in Hollywood, Stage 52, The Greenway Court Theatre, The ACME Theatre and The Actor's Gang, all in Los Angeles. Their work has also been seen on several stages in Florida, Georgia, New York and New Jersey. Bannister and Hurd have co-authored an award-winning short film, Poet of the Swingin' Blade, which was selected as a finalist at the 2007 San Diego Black Film Festival, the 2007 Arizona Black Film Showcase, the 2007 Big Island Film Festival in Kona, HA, the 2007 Hollywood Black Film Festival, Reel Black Men 2007, 14th Annual The African American Film Marketplace & S.E. Manly Short Film Showcase 2007, and The Fort Omaha Film Festival and Conference 2009. Poet of the Swingin' Blade won Best Message Film at the San Diego Festival 2007 and The Fort Omaha Film Festival and Conference 2009. Bannister and Hurd are the Producing and Artistic Directors, respectively, of Los Angeles based Theater Company, The Kohl Players: www.kohlplayers.com.

Constance Chen

Best Teacher, History

Dr. Constance Chen received her Ph.D. in History from the University of California, Los Angeles. Exploring the development of Asian Pacific histories and cultures within a transnational context, her research interests include comparative racial and gender discourses, museums and art collecting, and the politics of visual culture. Dr. Chen's work has appeared in the Journal of Asian American Studies, Amerasia Journal, the Journal of American Studies as well as other journals and anthologies. One of her current projects examines the ways in which cross-cultural encounters shaped the development of ethnic-racial identities in nineteenth- and twentieth-century China, Japan, and the United States.

James Faught

Best Overall Faculty Member, Liberal Arts, Social Sciences, and Humanities

Rick Hadley

Best Overall Faculty Member, Film & Television

If there's a question on American film Dr. Hadley can't answer, the students and faculty have yet to find it. Whether walking the halls of the School of Film and Television building, or strolling across Alumni Mall, Dr. Hadley is always willing to engage you in a conversation that can range from the latest Hollywood blockbuster to the relative merits of Billy Wilder versus Sam Peckinpah. After earning his BA from the University of California, Berkeley, and his MA from the University of Wisconsin, Madison, Dr. Hadley went on to complete a both an MFA and a PhD at the University of Southern California. In addition to his teaching duties, Dr. Hadley is also the Director of the Graduate Programs in Film, Television, and Screenwriting. But, his true calling is in the classroom, where he mixes his knowledge of cinema with equal parts careful scholarship and wry humor. Over the years, his classes in film genres and film authors have delighted, and educated, students and colleagues alike. As much as he enjoys watching films, Dr. Hadley also enjoys writing them. His film, Quotations From Chairman Stu, was screened at Edinburgh and many other film festivals. He has written documentaries on Woody Guthrie, Andrew Carnegie, Charles Lindbergh, Paul Robeson, Mao-Tse-Tung and Samuel Adams for Nickelodeon's Award-winning show, Against The Odds and the pilot, Pop Quiz, for the Disney Channel.

Paul Harris

Best Overall Faculty Member, English
Best Researcher/Scholar, English
Best Overall Faculty Member, Liberal Arts, Social Sciences, and Humanities

PH's preferred creed: the extended letter-centered text effect perfected. He, the well-versed geek, the respected exegete, esteems letter-centered texts, the genre where letters per se get stressed. He prefers perverse Escher verses, he references Tweedledee's speech, Beckett's empty Zen lessness. He eschews severe, well-set symmetry; he never renders texts level; he reverses terms, skews sense, bends the text's edges, then extends them.

Terry Lenihan

Most Helpful to Students, Fine Arts

Terry Lenihan holds an MFA from Claremont Graduate University. Her sculpture and installation artwork has been featured in numerous gallery and museum venues both locally and internationally, including the Irvine Fine Art Center, Sherry Frumkin Gallery, Brand Art Gallery, and the Museum of Modern Fine Art, Minsk, Belarus.

Professor Terry Lenihan is an arts education specialist and Director of the LMU ARTsmart Service Learning Program. With over twenty years of experience in art education, Lenihan has directed site-specific art projects with young artists that have been sponsored by the Museum of Contemporary Art, Los Angeles (MOCA), and the Metropolitan Transit Authority (MTA).

Juan Mah Y Busch

Best Researcher/Scholar, English

Through the Departments of English and Chicana/o Studies, Professor Mah Y Busch offers courses on topics such as American autobiography and fiction, postcolonial theory, moral philosophy, and U.S. Latina/o cultural studies. He publishes in the fields of Chicana/o literature and ethics. He currently is researching Latina American philosophy, narrative ethics and humanist geography in order to complete his book, Conciencia: Materializing a Chicana/o Spirit, in which he examines agency through the lens of value in Chicana/o aesthetics.

Ron Marasco

Best Teacher, Liberal Arts, Social Sciences, and Humanities

Ron Marasco studied acting at the Moscow Art Theatre and has a B.A from Fordham University and an M.A and Ph.D. in Theatre from UCLA. He is a Professor of Theatre at Loyola Marymount University in Los Angeles. As an actor, on television, Marasco has appeared on Entourage, Medium, The West Wing, The X-Files, Family Law, Judging Amy, Murphy Brown, Good-Vs-Evil, Charmed, Angel, Martial Law, Enterprise, Nip/Tuck, Lost, Bones, and in the movie The 40-Year Old Virgin. He played the role of Mr. Caspar on the critically acclaimed Freaks and Geeks. He played opposite screen legend Kirk Douglas in the movie Illusion, for which he also wrote the award-winning screenplay.

Onstage, in New York and Los Angeles, Marasco has appeared in such plays as Sophistry, A Midsummer Night's Dream, American Iliad, School For Wives, Modigliani, Ghetto at the Mark Taper Forum, Casual Sex at the Groundlings and in the Los Angeles premiere of the Stephen Flaherty-Lynn Ahrens musical Lucky Stiff. Among his directing credits are Julius Caesar, Fifth of July, The Man Who Came to Dinner, Twelfth Night, The Black Pope, Noises Off, Hamlet, Rumors, and most recently The Elephant Man. He won the 1997 LA Weekly Award as BEST DIRECTOR for his production of Modigliani. The play also won for BEST REVIVAL. For the same production he won a DRAMA-LOGUE AWARD for directing and one for acting in the production.

His screen play for the movie Illusion won the Best Screenplay Award at the East Hampton Film Festival. His book Notes to an Actor was named by the American Library Association as one of the "Outstanding Academic Books of 2008." He is the co-author of "About Grief" (2010) about teaching the subject of grief to students.

Aimee Ross

Most Helpful to Students, English

Aimee Ross-Kilroy teaches courses in British literature, Renaissance literature including Shakespeare, composition, fiction and children's literature. She also serves as the Associate Director of the Freshman English Program. Her research interests include early modern revenge tragedies, and she has an article forthcoming in the journal Renaissance and Reformation/Renaissance et Reforme entitled "'The Very Ragged Bone': Dismantling Masculinity in Thomas Middleton's The Revenger's Tragedy," and is at work researching purgatory and its absence in the English Renaissance. In her spare time, she finds herself watching a great deal of youth soccer and entertaining the notion of someday writing children's literature of her own.

Stephen Shepherd

Best Researcher/Scholar, English

Damon Willick

Best Overall Faculty Member, Art History
Best Researcher/Scholar, Art History
Best Teacher, Art History
Most Helpful to Students, Art History

Damon Willick is Associate Professor of Modern and Contemporary Art History. He has written for journals including X-TRA, Art Lies, Spiritus, East of Borneo, and NKA Journal of Contemporary African Art and contributed to exhibition catalogues like the Norton Simon Museum's Proof: The Rise of Printmaking in Southern California. He served on the editorial board of X-TRA Contemporary Art Quarterly from 2006–2013 and is currently a contributing editor of the journal. His book and exhibition on the San Fernando Valley's contribution to Los Angeles art is forthcoming from Angel City Press and California State University, Northridge in 2014.

Kelly Younger

Best Teacher, English

KELLY YOUNGER is an award-winning playwright with work staged off-Broadway, regionally, and internationally. Select works include: Banished Children of Eve (Off-Broadway, Irish Repertory Theatre); This World We Know (Winner, Firehouse Theatre Festival of New American Plays; Naked Angels/Furious Theatre); In the Fold (Finalist for the Joanne Woodward/Paul Newman Drama Award, Last Frontier Theatre Conference); Tender (Finalist for the Laurents/Hatcher Award, part of Manhattan Theatre Club's 2011 "7@7" reading series, New Repertory Theatre, Gloucester Stage, IRNE nomination Best Play 2010); Once a Marine (PlayFest at Orlando Shakespeare Theater, The Blank Theatre); I Think You Think I Love You (Playscripts; Smith and Kraus anthology "Best Plays of 2005"); Forgive me, Father (JAC publishing); Lady Gregory's Ingredients (JAC publishing), winner of the Ireland National Lady Gregory Playwriting Award; Off Compass (Ensemble Studio Theatre/LA) winner of the John Gassner New Play Award; Epiphany Cake (Ensemble Studio

Theatre/LA, Moving Arts); Why Wyoming (Three Graces NY), Critics' Choice Samuel French off-off-Broadway Festival.

Younger is also working in the television and feature worlds, developing projects with DreamWorks Animation, Mandeville Films, Junction Films, as well as a television comedy for the producer of Clueless. He placed in the top 10% in the 2013 Academy Nicholl Fellowships in Screenwriting, and in the top 5% in the 2012 Warner Brothers Television Writers Workshop. Professionally, he is repped by The Gersh Agency in Los Angeles and New York, and Washington Square Arts and Films in New York.

At LMU, Younger conducts workshops in play writing, teaches courses in the survey of drama, and leads seminars in modern and contemporary drama, Irish drama, classical drama in translation, performance studies, drama theory, as well as a seminar in Fairy Tales. Critical book publications include Dionysus in Ireland: Irish Adaptations of Greek Tragedies (Mellen) and Beth Henley in an Hour (Smith and Kraus). He has also published numerous interviews, book reviews, pedagogy essays, and multiple articles on plays and playwrights. For the university, Younger serves as faculty moderator of Alpha Sigma Nu Honors Society. For the department, Younger serves as the faculty advisor for the Screenwriting Minor. Prof. Younger earned an MA in Classics at Loyola University Chicago and PhD in Drama Studies from University College Dublin in Ireland. He is a member of the Pacific Residents Theatre Writers Group, The Dramatists Guild of America, and serves on the board of the international Comparative Drama Conference.

Katerina Zacharia

Best Researcher/Scholar, Liberal Arts, Social Sciences, and Humanities

LOYOLA UNIVERSITY CHICAGO

Maria Nanos

Most Helpful to Students, Social Work

Adriaan Peperzak

Best Researcher/Scholar, Philosophy

Adriaan Theodoor Peperzak was born in Java (Indonesia) as a Dutch citizen. He studied philosophy in Venraai and theology in Alverna and Weert (The Netherlands). He obtained a licentiate in philosophy at the Higher Institute of Philosophy of Louvain University (Belgium) and a Ph.D. in the Humanities at the University of Paris (Sorbonne). His doctoral dissertation, Le jeune Hegel et la vision morale du monde (director: Paul Ricoeur), was published in 1960 and republished in 1969.

Peperzak taught at various universities of The Netherlands (including those of Amsterdam, Utrecht, and Nijmegen) and, as visiting professor, at the universities of Bandung (Indonesia), Mallorca (Spain), the Scuola Normale of Pisa (Italia), the Istituto degli Studi Filosofici of Naples (Italia), the University of Nice (France), Duquesne University (Pittsburgh), Pennsylvania State University (State College), Boston College, Loyola University Chicago, Villanova University, and Stanford University, and he lectured at many others. From 1991 to the present, he has been the Arthur J. Schmitt Professor of Philosophy at Loyola University Chicago.

His research in the history of philosophy has focused on Hegel (six books and numerous articles) and Emmanuel Levinas (two books and three others edited). He also published on Plato, Aristotle, Bonaventura, Descartes, Heidegger, and Ricoeur, and on thematic questions in ethics, social and political philosophy, meta-philosophy, and philosophy of religion.

On Hegel he wrote Philosophy and Politics; Selbsterkenntnis des Absoluten: Grund-linien der Hegelschen Philosophie des Geistes; Hegels praktische Philosophie; and Modern Freedom, Hegel's Legal, Moral, and Political Philosophy. On Levinas he published To the Other: An Introduction to the Philosophy of Emmanuel Levinas; and Beyond: The Philosophy of Emmanuel Levinas. Some of his more personal books are Before Ethics; Faith in Reason; Platonic Transformations; The Quest for Meaning; Elements of Ethics; and Philosophy between Faith and Theology.

Besides Dutch, he speaks and writes English, French, German, Spanish, and Italian; he also reads ancient Greek, Latin and Por tuguese

Dr. Peperzak is general editor of the Purdue University Series in the History of Philosophy and member of the board of editors of the book series Perspectives in Continental Philosophy, published by Fordham University Press. His current works accepted for publication include: "Sincerely yours. Toward a Phenomenology of Me" in a volume on Levinas, in 2008. "Gods, 'God,' God," in Festschrift for Edith Wyschogrod, New York, Fordham University Press, in 2008. Sei Studi Hegeliani(manuscript of ± 200 pp.), Napoli, Istituto di Filosofia, in 2009. "Sacrifice or willing the Will," in Archivio di Filosofia (Roma), in 2008. "Paul Ricoeur: Reader, Responder, Writer," in a book on the philosophy of Paul Ricoeur, Continuum in 2009." Ontology and Ethics from the Perspective of Religion," in Archivio di Filosofia, in 2008. "Desire" for Festschrift for Remo Bodei. Other books include: Philosophy between Faith and Theology. Addresses to Catholic Intellectuals. Notre Dame: Notre Dame University Press, 2005, 216 pp., Thinking. From Solitude to Dialogue and Contemplation, New York: Fordham University Press, 2006, 178 pp., Aanspraak en Bezinning (Speaking and Recollection), Kampen, Klement Publishing, 2007, 280 pp.

LOYOLA UNIVERSITY MARYLAND

Margaret Haggstrom

Best Teacher, Languages
Most Helpful to Students, Languages

"She has received the best student evaluations over 25 years, and works well with colleagues to produce an excellent, attractive and appropriate French program.

Spends long hours working with students, developing courses."

Thomas Ward

Best Overall Faculty Member, Languages
Best Researcher/Scholar, Languages

"He has received excellent student evaluations and consistently publishes well and frequently.

As chair, got new lines for the department; as colleague encourages all, and as teacher, excellent!"

LUBBOCK CHRISTIAN UNIVERSITY

Jana Anderson

Most Helpful to Students, English

Kenneth Hawley

Best Teacher, English
Best Teacher, Liberal Arts, Social Sciences, and Humanities
Best Overall Faculty Member, Liberal Arts, Social Sciences, and Humanities

Dr. Hawley has spent much of his academic time with medieval and renaissance literature, especially the epic poem Beowulf, vernacular translations of the Bible, Chaucer's tales and poetry,

Milton's major works, and various other religious poets. His dissertation on The Boethian Vision of Eternity in Old, Middle, and Early Modern English Translations of De Consolatione Philosophiæ includes analysis of works by King Alfred the Great, Chaucer, and Queen Elizabeth I (see link below). In addition to his courses in Composition, Literature, and Technical Writing, he also enjoys studying and teaching film, particularly westerns, adaptations of novels, historical epics, and biopics–his M.A. thesis was on Scottish nationalism and Braveheart (see link below).

While at the University of Kentucky, he earned a graduate certificate in Electronic Texts, with an emphasis on Computing for Humanities, a discipline interested in addressing teaching, learning, and research in the Humanities with the tools, methods, and technologies of Computer Science. He was associated at that time with organizations and projects such as the Collaboratory for Research in Computing for Humanities, The Digital Atheneum, ARCHway, The Electronic Boethius Project, and The Student Edition of Electronic Beowulf (see links below).

Here at LCU, he teaches Composition (ENG 1301), Writing About Literature (ENG 1302), Masterpieces of Literature (ENG 2301), Making Connections: Literature and Life (HON 2301), Classical and Contemporary Rhetoric (ENG 3307), Technical Writing (ENG 3308), Christianity and Literature (ENG 4306), Comparative Literature (4314), Major British Writers to 1800 (ENG 4315), Literature and Film (ENG 4326), and Senior Seminar (HUM 4380).

He and his wife Deborah have three children: William (Will), Bennett (Ben), and Coleman (Cole).

Staci Patty

Most Helpful to Students, Liberal Arts, Social Sciences, and Humanities

Dr. Patty came to LCU in 1992. With degrees from Harding Graduate School, Union Theological Seminary, and Baylor University, he has taught courses in areas from Philosophy to Comparative Religions, from Ethics to Theology. He travels frequently to research religious culture, especially in the regions of Turkey and India. As Director of The University Honors Program, Dr. Patty oversees the interdisciplinary program for LCU's most outstanding students.

Ronna Privett

Best Overall Faculty Member, English
Best Researcher/Scholar, English

Dr. Privett graduated from Lubbock Christian University 1983, earning her B.A. in Art and secondary teaching certificate in Art and English. In 1985, she married Bryan Privett and they have two children, Austin and Emily. Dr. Privett returned to school when her children were small, earning her Master's in English (British and American Literature) in 1995 and her Ph.D. in English (American Literature) in 1999, both degrees from Texas Tech University. Dr. Privett has had the joy of teaching English and Humanities in the Humanities department at LCU since 1999, and currently serves as chair of the department.

Co-chair of the Task Force for Academic Reorganization Presented paper entitled "Reading Moby-Dick in Context: The Final Scene" at the Conference for College Teachers of English in Beaumont (March 5–7, 2010) Will be presenting a paper entitled "Finding Thea Kronborg's Voice: "One Passion and Four Walls" at the national College English Association conference in San Antonio (March 24–27, 2010) Plans to present at the LCU Scholars Colloquium in April 2010 Current Classes American Literature African-American Literature Multi-cultural Literature Comparative Literature Children's Literature Composition

Mark Sneed

Best Researcher/Scholar, Liberal Arts, Social Sciences, and Humanities

He transferred with FedEx to here. It is here that he met Dr. Jesse Long, who was finishing his dissertation. This developed into a lasting friendship. His sons Brian and Luke were born during this time. He returned to Tennessee and taught as a part-time instructor for David Lipscomb University for a year and a half. He also served as pulpit minister for a small congregation

MACALESTER COLLEGE

Linda Schulte-Sasse

Best Overall Faculty Member, German

Linda Schulte-Sasse, Department Chair, teaches in Cultural Studies as well as German Studies. Her areas of specialization include German and American film, German literature, and the analysis of American political discourse. She teaches in the cinema studies (most recently Nazi Cinema and Film and the Fantastic), eighteenth- and twentieth-century German literature, and the German language. In 1996 Duke University Press published her book Entertaining the Third Reich: Illusions of Wholeness in Nazi Cinema. The book, which appeared in the series "Post-Contemporary Interventions" edited by Stanley Fish and Fredric Jameson, was named one of the outstanding scholarly works of 1997 by CHOICE. Her recent work has been on love letters written to Adolf Hitler; moreover, she is completing a monograph on Italian director Dario Argento. She also likes journalistic writing, and publishes occasionally for Scripps-Howard News Service or for the Star-Tribune.

MANHATTANVILLE COLLEGE

Michael Posnick

Best Overall Faculty Member, Fine Arts
Best Researcher/Scholar, Fine Arts
Best Teacher, Fine Arts
Most Helpful to Students, Fine Arts

Michael Posnick served as Director of the Department of Dance and Theatre at Manhattanville College from 1994 to 2012. He has taught and directed both graduate and undergraduate levels at Yale and Hunter College, worked with the National Theatre of the Deaf for more than 25 years and served as Artistic Director of Mosaic Theatre at the 92nd Street Y. He taught at the National Theatre Institute at the Eugene O'Neill Theatre Center and was a visiting artist with The Lincoln Center Institute. Mr. Posnick has directed nearly one hundred theatrical and musical productions at venues which include: The Manhattan Theatre Club, Yale Rep, the 92nd Street Y, and the New York Philharmonic Orchestra at Lincoln Center. He recently served as dramaturge with Pilobolus Dance Theatre for Davenen, which premiered at the Kennedy Center in Washington D.C. and the Joyce Theatre. He is co-editor of Nine Contemporary Jewish Plays. He holds an MFA from Yale Drama School and an MS in Education from Yeshiva University.

Randy Williams

Best Teacher, Fine Arts

Randolph Williams maintains two careers: One as an active visual artist and another as an educator developing new ways to present the visual arts to children. A professor of art at Manhattanville College, Prof. Williams has worked as an educational consultant to the Metropolitan Museum of Art, the Museum of Modern Art, The Jewish Museum, Storm King Sculpture Park, and the Neuburger Museum. He is also the program director of the New York State Summer School of the Visual Arts. Prof. Williams holds a B.S. in art education from New York University and a master's degree in art education from Sir George Williams University in Montreal.

MANSFIELD UNIVERSITY

Bradley Holtman

Best Teacher, Liberal Arts, Social Sciences, and Humanities

Jean Anne Teal

Most Helpful to Students, Music

MARANATHA BAPTIST BIBLE COLLEGE

Michael Zwolanek

Best Overall Faculty Member, History
Best Researcher/Scholar, History
Best Teacher, History
Most Helpful to Students, History

Mr. Michael Zwolanek and his wife, Lyssa, have two children, Daniel and Natalie. Mr. Zwolanek enjoys traveling with his family—especially to historic sites and museums. One of his favorite verses for the classroom is Philippians 1:6. He believes that "we all need to remember that God is continuing to work with us, through us. He also knows that flexibility is the teacher's friend." Mr. Zwolanek and his family attend Calvary Baptist Church in Watertown, where they are active in many ministries.

MARIST COLLEGE

John Knight

Best Researcher/Scholar, Philosophy

James Snyder

Best Overall Faculty Member, Philosophy
Best Teacher, Philosophy
Most Helpful to Students, Philosophy
Best Researcher/Scholar, Philosophy

I grew up on Mount Ida in Troy, NY. After a decade in New York City, I now live in Millbrook, NY with my wife, son, and daughter.

My research focuses mainly on Renaissance and early modern philosophies. At present I am studying the natural philosophy of the Renaissance Platonist Marsilio Ficino (1433–1499). I am especially interested in his theories of prime and corporeal matter, his view of natural change, and the purgative function that he considered natural philosophy to have on those who studied it. In the past few years, I have been examining the role that the Epicurean poet Lucretius plays in Renaissance and early modern philosophies.

MARQUETTE UNIVERSITY

Stephen Beall

Most Helpful to Students, Languages

Associate Professor Stephen Beall was born in Detroit and graduated from the University of Detroit Jesuit High School. He earned his B.A. in Latin and Medieval Studies at the University of Michigan, and advanced degrees in Classics at the University of California, Berkeley. Under the direction of Professor Danuta Shanzer, he completed a doctoral dissertation entitled Civilis eruditio: Style and Content in the Attic Nights of Aulus Gellius (1988).

Dr. Beall joined the Marquette faculty in 1993, and received tenure in 2000. He is a member of Phi Beta Kappa, the Fellowship of Catholic Scholars, and the Society for Catholic Liturgy. He is also a Fourth-Degree Knight of Columbus.

Teaching Fields: Latin (all levels) Greek (all levels) Classical Civilization and Literature

Terrence Crowe

Best Teacher, Theology

"Terrence Crowe as an Adjunct faculty member has had a major connection and influence in undergraduate theology over the years."

Sarah Gendron

Best Researcher/Scholar, Languages

Ulrich Lehner

Best Researcher/Scholar, Theology

"Original work in historical theology that affects the field of history as well as theology."

Ulrich L. Lehner (Dr. Theol., University of Regensburg 2005), [Historical], specializes in the study of Religious History and Historical Theology from the late 15th to the early 19th century, especially the so called Catholic Reformation/Catholic Reform. Dr. Lehner is an internationally recognized expert of Early Modern Religious History and Theology. He was Member/Herodotus Fellow of the Princeton Institute of Advanced Studies, School of Historical Studies (2009), inaugural fellow of the Notre Dame Institute of Advanced Study/University (2010), fellow of the International Research Center for Comparative History of Religious Orders, Germany (2008), and research fellow of the Earhart Foundation (2013). He was also a member of the Andrew Mellon Research Group (History) "Religion Across the Disciplines" of the University of Notre Dame (2011–2012). In 2013 he was elected a member of the European Academy of Arts and Sciences, category "World Religions."

Dr. Lehner serves on the advisory board of Eighteenth Century Thought, the Journal of Baroque Studies, the Oxford Handbooks of Religion-Series, and the Institute for Baroque Studies in Malta.

D. Stephen Long

Best Overall Faculty Member, Theology

"Excellent publications, outstanding teaching, consistently among best mentors"

D. Stephen Long (Ph.D., Duke University, 1991), [Theology/ Ethics], is professor of Systematic Theology at Marquette University. Previously, he worked at Garrett-Evangelical Theological Seminary, St. Joseph's University and Duke Divinity School. He is an ordained United Methodist and served churches in Honduras and North Carolina. He has published eleven books, Living the Discipline: United Methodist Theological Reflections on War, Civilization, and Holiness (Grand Rapids, Michigan: William B. Eerdmanns, 1992), Tragedy, Tradition, Transformism: The Ethics of Paul Ramsey (Boulder, Oxford: Westview Press, 1993), Divine Economy: Theology and the Market (London and New York, Routledge, 2000), The Goodness of God: Theology, Church and Social Order (Brazos Press, 2001), John Wesley's Moral Theology: The Quest for God and Goodness (Kingswood, 2005), Calculated Futures (Baylor, 2007),

Theology and Culture (Cascade, 2007), Keeping Faith: An Ecumenical Commentary on the Articles of Religion and Confession of Faith in the Wesleyan Traditions (Cascade, 2012), Hebrews: Belief, A Theological Commentary on the Bible (WJKP, 2011), Christian Ethics: A Very Short Introduction (Oxford, 2010), and Speaking of God;Theology, Language and Truth (Eerdmans, 2009). An eighth work is in process entitled Speaking of God: Theology, Truth and Language (Eerdmann, 2008).

Steve was baptized by the Anabaptists, educated by the evangelicals, ordained and pastorally formed by the Methodists and given his first position as professor of theology by the Jesuits, which makes him either ecumenically inclined or theologically confused. He is married to Ricka and they have three children, Lindsey, Rebecca and Jonathan.

Timothy McMahon

Best Teacher, Liberal Arts, Social Sciences, and Humanities

Jason Meyler

Best Teacher, Languages

Phillip Rossi

Most Helpful to Students, Theology

"Dr. Rossi's influence has been indirect yet massive as department Chair and interim Dean of Arts & Sciences."

Philip J. Rossi, S.J. (Ph.D., University of Texas [1975]), specializes in the philosophy of religion and Christian ethics; he has published extensively on the theological import of the work of Immanuel Kant. He has been visiting professor at Sogang University, Seoul, Korea (1985), the Ateneo de Manila University, Philippines (1998), and Arrupe College, Harare Zimbabwe (2014), a research fellow at the Institute for Advanced Studies in the Humanities, University of Edinburgh (1992, 1999), visiting scholar (1979–80) and visiting fellow (2004–05) at the Woodstock Theological Center, Washington, DC, and Visiting Fellow, Faculty of Theology and Religious Studies, KU Leuven, Belgium (2013). He served as Interim Dean of the Helen Way Klingler College of Arts and Sciences from 2010–2013, Associate Dean for Graduate Affairs from 2005–2008, and ten years as Chair of the Theology Department. He is author of The Social Authority of Reason: Kant's Critique, Radical Evil and the Destiny of Humankind, (State University of New York Press, 2005), Together Toward Hope: A Journey to Moral Theology (University of Notre Dame Press, 1983), editor of God, Grace, and Creation (Orbis, 2010), co-editor (with Michael J. Wreen) of Kant's Philosophy of Religion Reconsidered (Indiana University Press, 1992) and co-editor (with Paul Soukup, S. J.) of Mass Media and the Moral Imagination (Sheed and Ward, 1994). He has presented papers at meetings of the American Philosophical Association, the American Academy of Religion, the Catholic Theological Society of America, the College Theology Society, the Society of Christian Ethics, the Russian Kant Society, six International Kant Congresses, and The Parliament of the Worlds' Religions in Barcelona (2004) and Melbourne (2009). He has published more than fifty articles in books and professional journals, was editor of Philosophy & Theology (1993–2000) and served on the board of editorial consultants for Theological Studies (1991–98). He was a member of the Board of Trustees of Creighton University (Omaha) from 1985–2013. He was the Executive Director of National Conventions for the College Theology Society (2005–09) and served two terms as a member of its Board of Directors. His current research focuses on Kant's anthropology as a resource for a post-modern theology of grace, the theological appropriation of the work of the Canadian philosopher Charles Taylor, and the theological warrants for human rights discourse and for just war theory in a globalized, post-modern culture.

Jennifer Vanderheyden

Best Overall Faculty Member, Languages

My research and teaching interests include Eighteenth and Nineteenth Century French Literature and esthetics, especially theories of theatrical and artistic representation (mimesis). My research focuses on the writings of Denis Diderot in the Eighteenth Century, and the Nineteenth Century novel. I have a doctorate in French Literature from the University of Washington, and a Master's Degree in French Literature and Civilization from the University of Cincinnati. I also hold an undergraduate degree in Education from Otterbein College, as well as a Supervisor's Certificate in secondary Foreign Language Education.

My approach to literary studies is multidisciplinary: philosophical, anthropological, artistic and psychoanalytical. In my book, The Function of the Dream and the Body in Diderot's Works (June, 2004, Peter Lang), I argue that the body's visual transmission of meaning, such as gestures and other nonverbal expressions, communicates more efficiently and spontaneously than the mediated spoken word. More specifically, I explore the capabilities of the body's "painted" nonverbal communication to convey thought, especially the body as portrayed artistically in literature through the concept of the tableau vivant: in effect, a literary painting in which the narrator portrays the characters as if suspended in a state of oscillation between paralysis and movement. Because the subject of the tableau vivant resembles the subject of the dream in many aspects, I begin my book with an investigation of the possibility of analyzing the body's representation based on Freud's theories of dream interpretation. My second book project involves a detailed study of Diderot's novel Les Bijoux indiscrets, a text often diminished by scholars as simply a frivolous satire of the court of Louis XV.

MASSACHUSETTS COLLEGE OF LIBERAL ARTS

Mark Miller

Best Overall Faculty Member, English

"Mark has a wonderful ability to work with both students and faculty. His calm and thoughtful patience makes him easily accessible,and his ability to share his prodigious sense of humor makes him easy to talk to. Aside from his capacious knowledge about the subjects he teaches, his extensive experience, personal interest and involvement in what happens in the English Department as well as in the college as a whole, makes him a very special human being as well as a teacher."

MASSASOIT COMMUNITY COLLEGE

Kevin MacWade

Best Teacher, History

"The best, straight forward history teacher at the college"

Tracey Schaub

Best Researcher/Scholar, English

MCDANIEL COLLEGE

Ira Domser

Most Helpful to Students, Theater

M.F.A., Boston University), a specialist in design and technical production who coaches theatre technology students as they explore puppetry, animatronics, special effects and flying often at U.S. Institute of Theatre Technology conferences, teaches courses in his areas of expertise: design, technical production, theatre appreciation and theatre management as well as a Jan Term theatre study trip to England.

MCHENRY COUNTY COLLEGE

Lindsay Carson

Best Overall Faculty Member, Liberal Arts, Social Sciences, and Humanities

Justin Hoy

Best Teacher, Liberal Arts, Social Sciences, and Humanities

Anne Humphrey

Best Researcher/Scholar, Liberal Arts, Social Sciences, and Humanities

Kathrine Poe

Best Teacher, English
Best Overall Faculty Member, Liberal Arts, Social Sciences, and Humanities

Tara Singer

Most Helpful to Students, Liberal Arts, Social Sciences, and Humanities

Tara has appeared as a solo artist and bandleader, as well as with a wide variety of ensembles, from small combos to large swing bands, at countless venues in the greater Chicago area, southern Wisconsin and throughout the Midwest. Her performance credits as a bandleader include engagements at the Palmer House, the Chicago Marriot, the Continental Plaza, the Ambassador East, the Drake Hotel, the Lincolnshire Marriot and the Hyatt Regency O'Hare, as well as established Chicago jazz venues such as Redford's and the Jazz Bulls. Ms. Singer performed as pianist and vocalist for the Chicago Grandstand Big Band from 1979–1981, with the Jazz Wave Big Band from 1994–1996 and was the resident solo pianist at Lake Lawn Resort in Delavan, Wisconsin from 1990–2005. She was a featured vocal soloist for the Chicago White Sox for three seasons and a headliner for the Chicago Tribune Concert Series. In addition to her commercial music experience, Ms. Singer has accompanied church choirs and performed for a wide variety of church sponsored events and services.

MCNALLY SMITH COLLEGE OF MUSIC

Randy Sabien

Most Helpful to Students, Music

"A phenomenal teacher."

Randy Sabien founded and chaired the jazz string department at the Berklee College of Music, where he also performed with, and produced albums for, folk musician Jim Post (of the 1960s group Friend & Lover). Randy Sabien is one of those rare string musicians to be hailed as both a violinist and a fiddler. Sabien has been a frequent guest on "A Prairie Home Companion" and "Austin City Limits" and has toured the world playing in different styles and with many of music's great stars. Randy is also a highly-regarded educator and author of instructional texts.

Cliff Wittstruck

Best Researcher/Scholar, Liberal Arts, Social Sciences, and Humanities

"Although let go for financial reasons, his dedication to the Institution, the teachers and the students is legendary."

Cliff has played with the Minnesota Orchestra, Art Garfunkel, Bernadette Peters, Jennifer Holliday, Chet Atkins, The Guy's All-Star Shoe Band (the house band on A Prairie Home Companion) The Rupert's Orchestra, The R-Factor, Mick Sterling and the Stud Brothers, and the Wolverines Classic Jazz Orchestra. He also performed throughout the U.S., Far East, Mexico, and Caribbean with the group Synergy. Additionally, Cliff performed with the R&B group, Inside Straight, opening for such nationally-renowned performers as James Brown, Roy Buchanan, Junior Walker and the All Stars, Sam and Dave, and Sun Ra.

MCNEESE STATE UNIVERSITY

Kim Bowman

Best Researcher/Scholar, Languages

"Dr. Kim Bowman is always ready to help our Spanish students by giving them the best cultural experience in her courses. She obtains our grants, organizes our lab and generally is the heart of our program. We do so appreciate your hard work, Kim!"

Rita Costello

Best Overall Faculty Member, Liberal Arts, Social Sciences, and Humanities

Michael Crawford

Best Teacher, History

Tiffany Dvorske

Best Teacher, Liberal Arts, Social Sciences,
and Humanities

Phillipe Girard

Best Researcher/Scholar, History
Most Helpful to Students, History
Best Researcher/Scholar, Liberal Arts, Social Sciences,
and Humanities

Scott Goins

Best Overall Faculty Member, Languages

Molly Martin

Most Helpful to Students, Liberal Arts, Social Sciences,
and Humanities

"Malory's Launcelot and Gwenyver in the Twenty-First Century Classroom," forthcoming in Studies in Medieval and Renaissance Teaching. "Vision and Sex in the Iconography of the Old English Genesis Manuscript," forthcoming in "Goode myrth ye shall here": Telling Tales and Crafting Books, Essays in Honor of

Thomas H. Ohlgren. Eds. Dorsey Armstrong, Shaun F. D. Hughes, and Alexander L. Kaufman. Kalamazoo: Medieval Institute Publications. Review of Fiona Tolhurst, Geoffrey of Monmouth and the Feminist Origins of the Arthurian Legend (New York: Palgrave Macmillan, 2012). Forthcoming in The Medieval Review (online). Review of Norris J. Lacy and James J. Wilhelm, eds. The Romance of Arthur: An Anthology of Medieval Texts in Translation, 3rd ed. (London and New York: Routledge, 2012). Forthcoming in Arthuriana. "Castles and the Architecture of Gender in Malory's 'The Knight of the Cart.'" Arthuriana 22.2 (2012): 37–51. Vision and Gender in Malory's Morte Darthur (Boydell & Brewer, 2010). "Troilus's Gaze and the Collapse of Masculinity in Romance." Men and Masculinities in Chaucer's Troilus and Criseyde. Eds. Marcia Marzec and Tison Pugh (Boydell & Brewer, 2008).

Michael Smith

Best Overall Faculty Member, History

MEMPHIS COLLEGE OF ART

Fred Burton

Best Teacher, Fine Arts

"Fred Burton is admire by faculty and students alike for his knowledge in his area of drawing and great inspirational skills."

Professor Fred Burton taught at the Memphis College of Art between 1987 and 2014. His paintings, drawings and woodcuts have been exhibited in Paris and Vence, France, London, New York City, Chicago, Houston, Kansas City, Lincoln and Orlando. Burton has held residencies at the Edward Albee Foundation, Montauk, Long Island; the Michael Karolyi Foundation, Vence, France; Altos de Chavon in the Dominican Republic; the Tyron Guthrie Center in County Monaghan, Ireland; the Millay Colony for the Arts, Steepletop, Austerliz,

New York; the Ragdale Foundation, Lake Forest, Illinois; the Woodstock School of Art in Woodstock, New York, the Morris Graves Foundation in Goleta, California and the Vermont Studio Center in Johnson, Vermont. He has written art criticism for Art Papers, The Commercial Appeal, and Number. Burton is listed in Who's Who in the Humanities, Who's Who Among America's Teachers, Who's Who in Tennessee Education, and in the 15th edition of Men of Achievement.

Tom Lee

Best Researcher/Scholar, Fine Arts

"Tom Lee is very knowable in his area of sculpture and a magnificent teacher and artist and greatly admired by students and faculty alike."

Tom Lee has been teaching Sculpture and Foundations at Memphis College of Art since 1989. He has received two individual state arts grants, in Mississippi and Tennessee, and has exhibited work locally, regionally, and nationally. Tom's community work includes set and prop design for Theater Memphis, Opera Memphis, Ballet Memphis and docent training lectures at Memphis Brooks Museum of Art. His home and studio are in Coldwater, Mississippi where he pursues the elusive happiness of creative living.

Remy Miller

Most Helpful to Students, Fine Arts

"During a very important time for our Memphis College of Art he stepped up to the responsibilities as the Dean of our college being instrumental in guiding the faculty to a successful re-accreditation. Remi Miller is a great supportive member of our faculty, a great teacher and an inspiration to faculty and students alike."

MERCER COUNTY COMMUNITY COLLEGE

Sharmilla Sen

Best Overall Faculty Member, English
Best Researcher/Scholar, English
Best Teacher, English
Most Helpful to Students, English

Best Researcher/Scholar, Liberal Arts, Social Sciences, and Humanities
Best Teacher, Liberal Arts, Social Sciences, and Humanities
Most Helpful to Students, Liberal Arts, Social Sciences, and Humanities
Best Overall Faculty Member, Liberal Arts, Social Sciences, and Humanities

MERCER UNIVERSITY

Fernando Palacios

Most Helpful to Students, Languages

Deneen Senasi

Best Researcher/Scholar, English
Most Helpful to Students, English

Trained as a classical ballerina before attending graduate school, Senasi earned B.A. degrees in English and Dance Performance at Birmingham-Southern College (1993), an M.A. at the University of Alabama (1996) and a Ph.D. from the Hudson Strode Program in Renaissance Studies at the University of Alabama (2003). She came to Mercer in 2007. Dr. Senasi's scholarship has appeared in The Journal for Early Modern Cultural Studies, Religion and Literature, and the book collection, Desire of the Analysts: Psychoanalysis and Cultural Criticism in the New Millennium (SUNY 2008). Some of her recent and upcoming conference presentations include a study of gendered silences and the law in Tudor England at the 2008 Modern Language Association and an exploration of the visual rhetoric of the self from medieval heraldry to polymer preserved bodies for the 2009 South Atlantic Modern Language Association meeting. She is currently completing a book on gendered performance, material culture, and the name in early modern England. Her research and teaching interests include sixteenth and seventeenth century British literature and culture, in particular the works of Shakespeare, Middleton, Donne, and Milton. She also works in the areas of Humanist pedagogy, Renaissance law, the lives of early modern women, and literary theory, in particular gender and performance theory and psychoanalysis. Other interests include Victorian literature, especially the works of Dickens and Rossetti. Dr. Senasi teaches the following courses: ENG 332 (Shakespeare I: Comedies and Histories); ENG 333 (Shakespeare II: Tragedies and

Romances); ENG 335 (Milton); ENG 342 (Seventeenth-Century Literature); ENG 301 (Introduction to Literary Studies); ENG 263 (Survey of British Literature: Beginnings through the Eighteenth Century); ENG 235 (Study of Poetry); FYS 101 (First Year Seminar: Composing the Self); FYS 102 (First Year Seminar: Engaging the World); ENG 108 (Composition I); and ENG 109 (Composition II).

METROPOLITAN COMMUNITY COLLEGE

Kris Fulkerson

Best Overall Faculty Member, Liberal Arts, Social Sciences, and Humanities

METROPOLITAN STATE COLLEGE OF DENVER

Katia Campbell

Best Teacher, Speech Communication

David Kottenstette

Best Researcher/Scholar, Speech Communication

Professor Kottenstette, M.A. teaches in the area of communication theory. You know, all those courses that have a foundation in interpersonal communication theory—Conflict, Culture, Gender, Family, Elderly Nonverbal. He has also taught classes in Rhetoric, Persuasion, and Research Methods.

David enjoys teaching and has a passion for student centered learning, creative and innovative projects that integrate theory and application. He considers himself a mastery learning teacher, one who is willing to work with tenacious students to help them set learning goals. Being a mastery learning teacher also means that he is willing to review and guide a student through the on going process of an assignment so that the final project meets with success.

David's research and professional development interests include, interpersonal conflict process, teaching practice, auto-ethnography, death and grief process, and performance studies. Those intellectual forays have lead to some fascinating professional conferences, as a panelist, workshop presenter and scholarly writer. While he is active in local, regional and national conferences, David will tell you that intellectual and frivolous fun can be had at international conferences in places like Denmark, Germany, Scotland and Russia.

Before teaching in the department of Communication, David taught for 10 years in theatre, while Theatre was still part of CAS. He was the scenic designer and technical director during that time. He has also directed numerous theatrical productions on the Auraria Campus, from "Hot L Baltimore" in 1985 to "A Man for All Seasons" in 2002. David received the Outstanding Higher Education Theatre Teacher of the Year in 1998 from the Alliance of Colorado Theatre.

METROPOLITAN STATE UNIVERSITY

Rachael Kulick

Best Researcher/Scholar, Liberal Arts, Social Sciences, and Humanities
Best Overall Faculty Member, Liberal Arts, Social Sciences, and Humanities

Jenni Runte

Most Helpful to Students, Liberal Arts, Social Sciences, and Humanities

B.A. in literature and humanities, Metropolitan State University; M.A. in English, University of St. Thomas; Ph.D. student, University of Wisconsin, Milwaukee; Metropolitan State University Center for Academic Excellence director.

Margaret Vaughan

Best Teacher, Liberal Arts, Social Sciences, and Humanities

MIAMI DADE COLLEGE (ALL)

Preston Allen

Best Researcher/Scholar, Liberal Arts, Social Sciences, and Humanities

Jennifer Basile

Best Overall Faculty Member, Fine Arts
Best Teacher, Fine Arts
Best Overall Faculty Member, Fine Arts

Wayne Bumpers

Best Overall Faculty Member, Music
Best Researcher/Scholar, Music
Best Teacher, Music
Most Helpful to Students, Music
Best Teacher, Liberal Arts, Social Sciences, and Humanities
Most Helpful to Students, Liberal Arts, Social Sciences, and Humanities
Best Overall Faculty Member, Liberal Arts, Social Sciences, and Humanities

"He is always positive and always takes the high road."

Ed Calle

Best Overall Faculty Member, Liberal Arts, Social Sciences, and Humanities

Miami's Ed Calle has been dazzling audiences with his bountiful musical gifts for about a quarter of a century. From the start, it was obvious that he was a natural player and now, Calle has been heard on thousands of recordings both as a sideman and a soloist, and his resume includes tour dates with some of the biggest names in the business. Even if Calle's name is unfamiliar to you, it is guaranteed that you have heard his music. His fiery tenor has graced the work of Gloria Estefan from the earliest days of the Miami Sound Machine, and he's heard on Grammy-award-winning recordings by Arturo Sandoval, Vicky Carr and pop singer Jon Secada. You have also heard him with Julio Iglesias, Vanessa Williams, Bob James, Frank Sinatra and many others, as well as on television and in motion picture soundtracks. The man is not only gifted, but versatile whether it be rock, jazz or pop. Calle has done it all and done it well.

His passion, however, is jazz, an affection he discovered upon hearing a Michael Brecker recording back in his teen years as a student at the University of Miami. Since then he has never looked back. Born in Caracas of Spanish parents, the Latin connection is also a part of his sound, an ingredient that adds just the right touch of spice for his solo recordings.

The essence of music, he feels, is melody; a lesson he learned from his late father, who played Mantovani records in the family home. It's a lesson Calle passes on to his students today when he teaches in public schools. "Melody is first and foremost" my father said, "and then there has to be enough harmonic interest to create a solid foundation for inventive soloing."

Besides instilling a love for melody, Calle also credits his father for launching his career in music. "We came to America in 1966," he recalled, "and I remember my dad suggesting that I take a class in music. I decided that his advice was good and that I wanted to play the sax. It was all my dad's idea for me to play—it was something that I didn't choose for myself."

Calle took to the sax with incredible ease: "It didn't take me long to figure it out. I had a sound that was different from everyone else, and I seemed to learn faster than most of the other students from the first day I began playing." The realization that he wanted to make music for a living did not come until his days at the University of Miami. Calle recalls spending every hour he could spare inside a practice room. It was not rare for him to practice eight, 12, even 14 hours in a day. "I decided I wanted to be a musician, to play the sax, and to play it until it smokes."

His early inspiration was John Coltrane, but he absorbed Charlie Parker, Dizzy Gillespie, Joe Henderson, Sonny Rollins- the list of players he studied, in his own words, "goes on forever." The great players became a motherlode of ideas for him, but his desire was not to play like them, but rather to get inspiration from them. "I was after what it was that made them sound that way, but I didn't want to copy them. I already had an idea of how I wanted to play."

Even before he left the University of Miami, Calle had already hooked up with Gloria Estefan and the Miami Sound Machine, and also with Julio Iglesias. "After college, I went on tour with Bob James, Julio, and Gloria all at the same time." Calle was with Estefan before she hit the big time and he learned first hand what it was like to be a part of a group that took off like a rocket. "I see a little bit of that same good fortune starting to happen with me," he noted. "We perform around the world with my own band, and I can see a similar positive reaction from the crowd happening with us."

Besides playing, writing is vital to Calle's creative instincts. "Writing is a big part of your own voice as a musician," he said, "and it's the vehicle you use to state your case musically. If you have a song that people can remember, you'll get a lot more mileage out of it than just standing up and playing 32 bars of something with no melody."

Second to a passion for music is a passion for mathematics. Besides his master's degree in jazz performance, Calle also has a bachelor's degree in math from Florida International University. "I've studied jazz and harmony from a mathematical perspective," he said. "And there is a strong connection between music and math. I'm convinced that most of the great figures in math history were also musicians, because the same guys who gave us geometry also gave us the modes. But even though the same part of the brain is dedicated to both math and music, I still know a lot of guys who are lousy in math but really great musicians. They probably just had bad teachers."

Ed Calle strives to be both an artist and an entertainer. "As a person, I'm very light-hearted. I'm not the musician who wears the artsy shirt or whose house is painted black. I'm pretty average—except when there's a saxophone in my face. Then I become something else. When I'm eight or twelve bars into what I'm doing, the artist in me takes over. When I'm in that zone, you could slap me in the face and I wouldn't even realize it. I'm only thinking about the music." www.professorcalle.com

Paul George

Best Overall Faculty Member, History

Magdalena Lamarre

Most Helpful to Students, Social Science

Amy Lund

Most Helpful to Students, Philosophy

Amy is a native of the Midwest, born in Chicago and raised in Minnesota. She attended college at Trinity University in San Antonio, TX, where she studied Philosophy, English, and Cognitive Science. She went on to complete her MA and PhD at the University of New Mexico. Her areas of specialization include Philosophy of Art, Kant, and the nature of human judgment. Areas of competency include Continental philosophy, 19th Century Philosophy, and Feminist Philosophy. Amy moved to Florida two years ago, and has now been through 6 hurricanes.

Maritza Marrero

Most Helpful to Students, Hospitality

Maritza started her career in the Hospitality industry in 1981 as a travel agent. Through out the years she worked for Club Med Resorts, Continental Airlines, Amadeus and she owned and operated a travel agency. Her interest in the industry started after she went on a student exchange trip to Europe right after college. Maritza managed an agency and was a partner in a travel agency. In that capacity, she managed a staff, coordinated marketing events, obtained ARC, IATA and CLIA certifications, and negotiated contracts with suppliers. She also worked in Amadeus (a global information system) in their sales and training departments. As a trainer, she traveled throughout the United States and Central America training travel agents. As a sales manager, she met with owners of agencies and negotiated new contracts. In 1998 Maritza came to Miami Dade College to start a new phase of her career in education. In addition to teaching, Maritza is the Hospitality Student Association advisor. Maritza has a B.S. and an M.S. degree in Hospitality Management from Florida International University.

Marcia Mello

Best Overall Faculty Member, Architecture

Amnel Molina

Best Teacher, Philosophy

Mark Neunder

Best Overall Faculty Member, Philosophy

Dr. Mark Neunder received his PhD in Philosophy from the University of Miami in 2001, for a dissertation on the ethics of genetic engineering. Material from this study was published as "Can Genetic Enhancement Be Obligatory? Four Arguments" in the Journal of Philosophical Research. His interests include applied ethics, ancient philosophy, skepticism, argumentation theory, rhetoric, and the philosophy of religion.

David Schroeder

Best Overall Faculty Member, English
Best Researcher/Scholar, English
Best Teacher, English
Most Helpful to Students, English

"creatively engages students in the writing process"

Matthew Schuh

Best Researcher/Scholar, Philosophy

Timothy Starkey

Best Teacher, Social Sciences

Barbara Sussman

Best Teacher, Liberal Arts, Social Sciences, and Humanities

Theodore Syder

Best Researcher/Scholar, History
Best Teacher, History
Most Helpful to Students, History

Ana Szolodko

Best Overall Faculty Member, Liberal Arts, Social Sciences, and Humanities

Marisol Varela

Best Teacher, English

Hector Vazquez

Best Overall Faculty Member, Humanities

Victor Vazquez-Hernandez

Best Teacher, Social Science

Kimberly Yantis

Best Teacher, Art

Kim Yantis, an MFA from Florida International University, began her career in Philadelphia, Pennsylvania where she received a BFA from Moore College of Art and Design in Painting and Printmaking. Her body of work draws inspiration from hypertext theory and incorporates non-linear narrative . This physical search for information is represented in her mixed media works which combine digital and hand drawn imagery, collage, stitching, in various formats. She recently was awarded a one year Artist Residency position at the Deering Estate at Cutler in Miami, Florida where she became the Artist in Residency Liaison and initiated several fine arts programs for the public and the home-school community. Her other accomplishments include a curatorial debut, Sharing the Press: Miami, an international show in the Dominican Republic, a juried exhibition Phantasm through the University of Miami, and several two person shows and collaborative performances in Austin, Texas and Miami, Florida. Her work was recently acquired by the Frost Museum of Art and The Girls Club of Fort Lauderdale Florida Art Collection/Francie Bishop Good and David Horvitz.

MICHIGAN STATE UNIVERSITY

Alan Rudy

Best Researcher/Scholar, Liberal Arts, Social Sciences, and Humanities

MIDDLE TENNESSEE STATE UNIVERSITY

Don Aliquo

Best Teacher, Music
Most Helpful to Students, Music

Don Aliquo is Professor of Saxophone at Middle Tennessee State University.

Originally from Pittsburgh, a city with a long-standing jazz tradition, Aliquo grew up listening and learning from his father, a well-respected jazz artist and educator. Since his appointment at MTSU, he has had extensive experience as an educator and performer. Recent concert appearances include the International Association of Jazz Educators Conference in New York City, the University of Northern Colorado Jazz Festival and the North American Saxophone Alliance International Conference. He has also performed on concerts and recordings with some of the nation's leading jazz artists including Rufus Reid, Gary Burton, Clay Jenkins and Marvin Stamm.

An active recording artist, Aliquo's 4 th solo CD was released in October 2006 and features Rufus Reid, Clay Jenkins, Dana Landry and MTSU alum Jim White. Aliquo's previous release Another Reply (CAP 964) broke the top 20 for airplay on the Jazz Week charts and showcases his compositional style and modern tenor approach within the genre.

Aliquo also performs and co-leads the Middle Tennessee Jazz Orchestra. Aliquo holds the M.M. and the B. MusEd. from Duquesne University and has studied at the Berklee College of Music. He has studied saxophone with Joseph Viola, Marino Galluzzo and Eric Kloss among others.

Ron Bombardi

Best Teacher, Philosophy

"A long-time dedicated teacher, devoted to student learning and success, passionate about his subject and vocation."

Ron Bombardi earned his Ph.D. in philosophy from Marquette University in Milwaukee, Wisconsin in 1984. With his dissertation on the empirical foundations of clinical psychology, he initiated a series of interdisciplinary studies in the philosophy of modern science that still characterizes the tenor of his contemporary work. Capitalizing on extensive scholarship in the history and philosophy of Western science and mathematics, Bombardi has advanced curriculum development in the discipline while teaching at MTSU with the creation of state-of-the-art courses in philosophical psychology and the philosophy of music. Bombardi is an active web-designer who routinely utilizes Internet resources in his teaching and scholarship; his hypertext edition of Spinoza's Ethics has received international acclaim. His current projects are focused on topics in neurophilosophy.

Mark Doyle

Best Teacher, History

"Has the best rapport with students and is the most innovative in the classroom."

Mark Doyle was born and raised in Oklahoma City, OK, where he developed a lifelong allergy to the state of Texas. He attended college at Tulane University in New Orleans and spent his junior year in Ireland at Trinity College, Dublin. It was here that the history bug bit him: a research project on agrarian violence in nineteenth-century Ireland took him to Dublin's National Archives, where he discovered the thrill of holding (and deciphering) old police reports and crumbling letters. Since then he has lived in Boston (where he got his PhD), Belfast, Dublin (again), Philadelphia, and Northampton, MA. He currently lives in Nashville with his wife Kate, daughter Rose, and two ill-trained cats named Worthington and Chesterfield.

Philippa Holloway

Best Researcher/Scholar, History

"Most disciplined researcher in the department, and most productive."

Pippa Holloway is a professor of history and the Director of Graduate Studies for the history department.

Robert Hunt

Best Overall Faculty Member, History

I began my career as a Southern historian, examining the relationship between plantation slavery and antebellum social reform (the public school movement). Over the years, I have shifted to become an historian of modern war. By this I do not mean just describing and recounting battles, or examining the latest weapons' system. I explore how different societies, through time, have understood war, defined it, prosecuted it, and remembered it. I study the history of war-making, not simply battles and leaders.

Kristine McCusker

Most Helpful to Students, History

"Goes beyond the merely necessary to help students excel and succeed."

Ph.D. Indiana University, History, Ethnomusicology and Folklore. MTSU 2000–present.

Janice Porth

Best Teacher, English

John Sanborn

Most Helpful to Students, Social Work

Stephen Smith

Best Overall Faculty Member, Music

After the completion of a Bachelor of Arts Degree in Political Science at Davidson College, H. Stephen Smith studied voice at the Eastman School of Music in Rochester, New York, completing a Bachelor of Music Degree in Vocal Performance and Literature before continuing on at Eastman to receive a Master of Music Degree and a Performer's Certificate in Opera.

Smith has performed leading roles on numerous opera stages with artists such as Renée Fleming, June Anderson, Elisabeth Söderström, and Jerome Hines, to name a few, and while residing in Stockholm, Sweden from 1988–98, he studied privately with Nicolai Gedda. Smith made his tenor debut with the Pittsburgh Opera Theater, singing the role of Pinkerton in Puccini's MADAMA BUTTERFLY. He has performed numerous leading tenor roles with opera houses in Scandinavia, Switzerland, Poland, Britain, Latvia, and Portugal.

His operatic work in America includes performances with opera companies in New Orleans, San Francisco, Houston, Mobile, Tulsa, Sacramento, Austin, Augusta, Jacksonville, Sarasota, Chattanooga, Norfolk, Shreveport, Columbus, Boston, Nashville, and New York.

Smith has recorded with Musica Sveciae and Caprice Records, including the roles of Salvatorre in Franz Berwald's ESTRELLA DE SORIA, Atis in Joseph Martin Kraus' PROSERPIN, and Cardinal Rohan in Daniel Börtz's MARIE ANTOINETTE. He also appeared as the male lead, Mats, in a full-length, feature film adaptation of Ture Rängström's opera, KRONBRUDEN, which has been viewed by television audiences throughout Scandinavia, Australia, and Germany. More recently, he served as the featured vocalist on Rocky Streams, a compact disc compilation of the works of Dr. Paul Osterfield, released in 2011 on Albany Records.

Smith made his Carnegie Hall (Weill Recital Hall) debut in March of 2014, performing the world premiere of Michael Linton's Franchetti Songs, a recording of which is scheduled for release later in 2014.

He joined the music faculty at Middle Tennessee State University in August of 1998, where he serves as Professor of Voice and Assistant Director of the School of Music. In addition to his teaching duties at MTSU, Smith remains active as a recitalist, and has appeared as the tenor soloist in numerous oratorio performances.

Paolo Volpe-Rinonapoli

Best Overall Faculty Member, Spanish
Best Teacher, Spanish

Paolo Volpe-Rinonapoli

Best Overall Faculty Member, Foreign Languages & Literature
Best Teacher, Foreign Languages & Literature

MIDDLESEX COUNTY COLLEGE

Ralph Tellone

Most Helpful to Students, Hospitality

MIDLANDS TECHNICAL COLLEGE

Helen Baldwin-Kingkade

Best Researcher/Scholar, Liberal Arts, Social Sciences, and Humanities

Janice Jake

Most Helpful to Students, Liberal Arts, Social Sciences, and Humanities

MILLERSVILLE UNIVERSITY

Jill Craven

Best Overall Faculty Member, Liberal Arts, Social Sciences, and Humanities

Chair of English, 2014–present, and Professor of Film Studies.

David Johnson

Best Overall Faculty Member, Social Work
Best Teacher, Social Work

Karen Rice

Best Researcher/Scholar, Social Work

Dr. Karen M. Rice. (At MU since 2006). B.A., Millersville University, 1991; M.S.W., Temple University, 2000; Ph.D., University of Maryland, 2011. Assistant Professor of Social Work/Department Chair.

Interest areas: child welfare, research, statistics, diversity

Major teaching areas: Social Work & Child Welfare, Social Work Research, Advanced Research Methods, Social Work Statistics, Encounters in Human Diversity.

Kristin Sims

Best Teacher, Music

"Outstanding teacher and performer."

Kristin Sims has appeared as a soloist with orchestras and opera companies throughout the United States. Ms. Sims received her B.M. and M.M. in vocal performance from the University of Cincinnati College-Conservatory of Music. Ms. Sims has sung leading operatic roles with Opera Lancaster, the Harrisburg Symphony, Bel Canto Lyric Opera, and Trinity Opera Theatre. Ms. Sims appeared with Allegro Chamber Orchestra as the soprano soloist in Beethoven's Ninth Symphony, along with the Lancaster Symphony Chorus in Haydn's Lord Nelson Mass and the Mozart Requiem. Kristin appeared as Mimi in Opera Lancaster's performance of La Bohème and Nedda in Pagliacci, Rosina in The Barber of Seville in Philadelphia and as Sister Constance in The Dialogues of the Carmelites in Cincinnati.

MILLIKIN UNIVERSITY

Eduardo Cabrera

Best Teacher, Languages

"Dr. Cabrera is always open and very attentive with students and academic standards. He makes sure everyone gets a fair chance to do their best."

MINNESOTA STATE UNIVERSITY

Alisa Eimen

Best Teacher, Art History

Hahn Huy Phan

Most Helpful to Students, Liberal Arts, Social Sciences, and Humanities

"Professor Phan is very helpful to students"

Mavis Richardson

Most Helpful to Students, Mass Communications

"Extremely helpful to students. She is a popular teacher and students feel comfortable asking her any questions. She makes herself available to students."

Mavis Richardson earned a bachelor's and a master's degree from North Dakota State University and a Ph.D. in mass communication from the University of Minnesota. She has taught at MSU for two years, and holds the rank of assistant professor. Richardson's scholarly research includes coverage of American Indian protest events by various print media. She has public relations and journalism professional experience and teaches introductory mass communication, public relations, ethics and writing courses.

Marshel Rossow

Best Overall Faculty Member, Mass Communications
Best Researcher/Scholar, Mass Communications

"Very well educated, asks wonderful questions that lead to his research. His writing is excellent so his research is compelling and clear.

He asks researcher questions that are valuable and fascinating. His research is solid and thorough. Marshel does longitudinal research, so it's particularly useful to the field. His writing is clear and concise so his work is highly readable and enjoyable."

Marshel Rossow has taught at MSU since 1984 and has earned tenure and the rank of professor. He received an undergraduate degree in technical journalism, a master's in journalism and mass communication from Iowa State University, and a Ph.D. in mass communication from the University of Wisconsin - Madison. Before joining the MSU faculty, he spent three years as a reporter, photographer and editor at the Fort Dodge (Iowa) Messenger and 12 years as an editor at the Waterloo (Iowa) Courier. At MSU, he teaches introductory mass comm courses as well as basic and advanced reporting, editing, and layout and design. He served as department chair for more than 10 years.

MISSOURI STATE UNIVERSITY

Diana Botsford

Best Teacher, Fine Arts

Edwin Carawan

Best Researcher/Scholar, Modern Languages

Steve Fielding

Most Helpful to Students, Fine Arts

Carol Maples

Best Overall Faculty Member, Theater

"Outstanding and trusted educator. Does everything in her power to get the students to succeed."

James Moyer

Best Overall Faculty Member, Religion

Eric Nelson

Best Researcher/Scholar, History

Timothy White

Best Researcher/Scholar, Fine Arts
Best Overall Faculty Member, Fine Arts

MOLLOY COLLEGE

Robert Kinpoitner

Best Overall Faculty Member, English

MONMOUTH UNIVERSITY

Rosemary Barbera

Best Overall Faculty Member, Social Work

"very well rounded, serious but can have fun and works well with everyone. A wealth of information"

Kim Hanbury

Most Helpful to Students, Social Work

"goes out of her way to help them but holds them accountable."

B.A., Loyola College; M.S.W., Monmouth University.

Elena Mazza

Best Teacher, Social Work

"Ellie's relationships with students is one to emulate"

Elena Mazza, joined the faculty in the fall of 2004. She has worked extensively in the area of mental health as a clinician and agency administrator.

Michelle Scott

Best Researcher/Scholar, Social Work

"Michelle is completing cutting edge research about suicie prevention on campus."

Michelle Scott is an Instructor in the School of Social Work. She holds a Ph.D. and MSW from the University of California–Berkeley. Her dissertation examined the effects of behavioral health managed care on youths' involvement with the juvenile justice system. She earned her BA in Psychology from Clark University in Worcester, Massachusetts. Dr. Scott has extensive research experience in the areas of adolescent mental health, adolescent alcohol use, mental health services, and suicide prevention.

MONROE COMMUNITY COLLEGE

Ilene Benz

Best Teacher, Fine Arts

Associate Professor, Speech/Media Communication NISOD Award, 2004 Inspiring Woman Award, 2013 Toastmasters International, ATM-S, AL

Marisol Galarza-Ruiz

Most Helpful to Students, Languages

Bethany Gizzi

Best Overall Faculty Member, Liberal Arts, Social Sciences, and Humanities

Maria Kendig

Best Overall Faculty Member, English

Tony Leuzzi

Best Researcher/Scholar, Liberal Arts, Social Sciences, and Humanities

"Tony Leuzzi is a great intellectual."

Njeru Murage

Best Researcher/Scholar, History

Theresa Murano

Best Teacher, Languages

Cristina Rowley

Best Overall Faculty Member, Languages

Bridgette Yaxley

Best Overall Faculty Member, English
Best Researcher/Scholar, English
Best Researcher/Scholar, Liberal Arts, Social Sciences, and Humanities

MONTCLAIR STATE UNIVERSITY

Fawzia Afzal-Kahn

Most Helpful to Students, Liberal Arts, Social Sciences, and Humanities

Fawzia Afzal-Khan is Professor of English, University Distinguished Scholar, and Director of Women and Gender Studies at Montclair State University. Afzal-Khan received her BA from Kinnaird College for Women, Lahore, Pakistan, and her MA and PHd in English Literature from Tufts University, MA. She is a cultural materialist who works at the intersection of Feminist Theory, Cultural and Performance Studies, and Postcolonial Studies. She created and teaches the first course on writings by and about Muslim Women at MSU, and has published extensively on this topic. She is author of 5 books, her latest a controversial memoir entitled Lahore With Love; Growing Up With Girlfriends Pakistani-Style (Syracuse University Press 2010; rep Insanity Ink 2010; can be ordered at amazon). She is a frequent contributor to Counterpunch and Express Tribune, and a published poet and playwright. She is a Contributing Editor to TDR

(The Drama Review), and serves on the Advisory Board of SAR (the South Asian Review. She is also a trained vocalist in the North Indian Classical tradition. Her music videos exploring themes of gender, religion and class, set in Pakistan, can be viewed on youtube (FAK Lahore, FAK Payal, FAK Smokescreen, FAK Sacrifice). She was a founding member of the experimental theatre collective Compagnie Faim de Siecle, with whom she toured and performed in Europe and North America. Her current research work is focused on Pakistani Popular Culture, with a focus on Pakistani Female Singers. She won an National Endowment of the Humanities grant in 2011 to make a trailer for a documentary on this subject, called "From the Melody Queen to the Muslim Madonna: Pakistani Female Singers 1947–Present"

Gina Balestracci

Most Helpful to Students, Music

Bob Browning

Best Overall Faculty Member, Art & Design

Heather Buchanan

Best Teacher, Music

Alexandra M. Camacho-Luna

Most Helpful to Students, Sign Language

Dr. Alexandra Camacho-Luna, a native Puerto Rican, received her Bachelor of Arts Degree in Special Education from the University of Puerto Rico. She received her Master of Arts in Audiology from the University of Indiana. Dr. Camacho-Luna completed her doctoral degree, also in Audiology from Montclair State University in New Jersey. Dr. Camacho-Luna has more than ten years of experience in Audiology. She has been working with both the pediatric and adult population in this field. Some of the testing includes auditory evaluation, pediatric audiological evaluation, auditory evoked potential evaluation, middle ear analysis, vestibular evaluation, amplification evaluation, cochlear implant evaluation, as well as cochlear implant mapping. Dr. Camacho-Luna has been working at Hackensack University Medical Center since June 2003 and at Hackensack Audiology and Hearing Aid Associates since January 2009. She is currently an Adjunct Professor for the Linguistic Department at Montclair State University for the American Sign Language courses. Dr. Camacho-Luna is fluent in Spanish and American Sign Language (ASL). When not at work, Dr. Camacho-Luna enjoys the time with her husband and her son, Dylan. She also loves cooking, traveling and exploring other cultures.

Richard Conway

Best Overall Faculty Member, History

"Dr. Conway has an exceptional command over his area of expertise, a fantastic control over the classroom, and yet proves to be among the most approachable of instructors. His humility, despite his vast knowledge makes him very well liked by both students and faculty alike."

David Del Principe

Best Researcher/Scholar, Italian

Associate Professor of Italian at Montclair State University. He is the author of the book Rebellion, Death, and Aesthetics in Italy: The Demons of Scapigliatura (1996) and various articles on 19th- and 20th-century Italian and European, Gothic literature and culture. His current research interests lie in a new field of study, the EcoGothic, that embraces Ecofeminist and Vegan Theory and Praxis, Critical Animal Studies, and Post-Humanism. He recently served as Guest Editor of a special issue of the British journal Gothic Studies

Laura Dolp

Best Researcher/Scholar, Music

Mary English

Best Researcher/Scholar, Liberal Arts, Social Sciences, and Humanities
Best Teacher, Liberal Arts, Social Sciences, and Humanities
Best Overall Faculty Member, Liberal Arts, Social Sciences, and Humanities

Eileen Fitzpatrick

Best Researcher/Scholar, Liberal Arts, Social Sciences, and Humanities
Best Overall Faculty Member, Liberal Arts, Social Sciences, and Humanities

Kathleen Hajdamacha

Best Teacher, Sign Language

Marta López-Luaces

Best Researcher/Scholar, Spanish

Susanna Pastorino

Best Teacher, Italian

Jennifer Perlis

Best Overall Faculty Member, Sign Language
Best Researcher/Scholar, Sign Language

Ruth Rendleman

Best Overall Faculty Member, Music

Harriet Saxon

Most Helpful to Students, Liberal Arts, Social Sciences, and Humanities
Best Teacher, Liberal Arts, Social Sciences, and Humanities

Marisa Trubiano

Best Overall Faculty Member, Italian

Bob Whitney

Best Researcher/Scholar, English

MONTGOMERY COUNTY COMMUNITY COLLEGE (ALL)

Neil Goldstein

Best Overall Faculty Member, Communication
Best Researcher/Scholar, Communication
Best Teacher, Communication
Most Helpful to Students, Communication

During over twenty-five years in the communication industry, Mr. Goldstein has produced and directed close to one hundred television, radio and theatrical dramas, documentaries, training films and entertainment specials. Mr. Goldstein is a member the National and Hollywood academies of Television Arts and Sciences, the Society of Motion Picture and Television Engineers and the Director's Guild of America and is currently on the faculty and serves as Coordinator for the Communication Program at Montgomery County Community College in Blue Bell, Pennsylvania. He holds a Master's degree in a communication related field, along with a Bachelor's in English from Washington University in St. Louis, Missouri, and has taken additional courses at The American Film Institute, University of Southern California and UCLA.

Mr. Goldstein was creator and executive producer of Superfest, a television special produced for public television with KCET-TV Los Angeles. Superfest was honored with a Los Angeles Area Emmy Award for Best Host. Breaking Ground, and Amnesty: The Dream Fulfilled?, are information specials made for KTLA, Tribune Television in Los Angeles, and produced and directed by Mr. Goldstein. Breaking Ground received two Los Angeles Area EMMY nominations and was honored with the Media Access Award for Best Documentary along with a Presidential Commendation.

With New Jersey Network (NJN), Mr. Goldstein produced, directed and wrote the documentary Lyme Disease: In Our Own Backyards. He also developed and managed a research project at the University of Pennsylvania, Annenberg School of Communications concerned with the image of persons with disabilities in the media.

For Thomas Jefferson University, Mr. Goldstein developed a school age child care project to integrate children with disabilities into the mainstream of existing day care facilities and completed a series of five training tapes and a documentary about traumatic brain injury. He recently finished A Chance To Hear, about children learning to hear for the first time after receiving a cochlear implant for Children's Hospital in Oakland California and is currently working on a piece about the integration of persons with disabilities into elementary school for Thomas Jefferson University College of Health Professions.

From 1982 until 1985 Mr. Goldstein developed and then sold a television station on California's Central Coast and from 1975 until 1982 was Director of Instructional Technology at Children's Hospital in Los Angeles. Mr. Goldstein is an active member of his Pennsylvania community and has served on county-wide and township committees. In addition to his current position at Montgomery County Community College, Mr. Goldstein also consults and teaches at various local educational institutions including Drexel University, Thomas Jefferson University and Chestnut Hill College.

MOREHOUSE COLLEGE

Linda Zatlin

Best Researcher/Scholar, Liberal Arts, Social Sciences, and Humanities

"Linda Zatlin has been a faculty member in the Department of English for 47 years. She has published four books, two with Oxford and Cambridge University Presses and has curated an exhibition in Japan on Aubrey Beardsley, about whom she writes. Yale University Press (London) will release her next book, in 2-volumes, in June 2015. In addition to bringing her knowledge into the classroom, Zatlin teaches students and student assistants ways to do primary research and therefore be successful in graduate school."

Linda Zatlin teaches Literary Form, British Literature, the Victorian Novel, Theory of the Novel, and World Literature, of which the recent subject has been War in Literature and Film. Professor Zatlin's research centers on the art and literature of the 1890s, notably the work of Aubrey Beardsley and Oscar Wilde, and also includes the Victorian novel, Pre-Raphaelite art, and Japanese art. The co-founder of the Nineteenth-Century Studies Association, an interdisciplinary scholarly organization which she currently serves as a member of the Senior Advisory Council, she is on the editorial board of Nineteenth-Century Studies and has been Sotheby's Beardsley Consultant since 1993. Twice the winner of the United Negro College Fund-Henry McBay Research Fellowship, the National Endowment for the Humanities Fellowship, and the Southern Fellowships Funds Fellowship, as well as numerous other research grants, Professor Zatlin was awarded the Historians of British Art Best Book Award for 1997, and in 2004, the Nineteenth-Century Studies Association's inaugural President's Award for Professional Achievement. A member of Phi Beta Kappa, faculty mentor for the Ford Foundation and the Dana Foundation Scholars Programs, Dr. Zatlin is keenly interested in training students to do primary research.

MORGAN STATE UNIVERSITY

Cynthia Bragg

Best Teacher, Sociology

Shirley Basfield Dunlap

Most Helpful to Students, Liberal Arts, Social Sciences, and Humanities

SHIRLEY BASFIELD DUNLAP, Associate Professor and Coordinator of Theatre Arts returns to Morgan State University after being an Associate Professor of Theatre at Iowa State University and Towson University. She has professionally directed August Wilson's Pulitzer Prize and Tony Award winning play FENCES. At Madison Repertory Theatre, featuring in the role of Troy Maxson, Roscoe Orman (who plays Gordon on the long running children's television show Sesame Street). Mr. Orman played the role of Gabe in FENCES on Broadway. She has directed the story of singer Alberta Hunter, COOKIN' AT THE COOKERY, at the Herberger Theatre in Phoenix, Arizona and Pearl Cleage's FLYIN' WEST. Tony Award winner Melba Moore starred in two of the six productions of Dr. Endesha Ida Mae Holland's FROM THE MISSISSIPPI DELTA which Ms. Dunlap has directed at regional theatres throughout the country including Milwaukee and Madison Repertory Theatre. Her production of HAVING OUR SAY: THE DELANY SISTERS' FIRST 100 YEARS originated at the Milwaukee Repertory Theatre before moving to Studio Arena in Buffalo, New York, The Dallas Center Theater and Delaware Theatre Company in Wilmington, DE. Also at Buffalo's Studio Arena, Ms. Dunlap directed John Henry Redwood's THE OLD SETTLER. Ms. Dunlap directed the late Ossie Davis in SPEAK TO US OF CHILDREN in Baltimore, MD. Her work has also been seen at The Human Race Theatre Company in Dayton, Ohio, The Hippodrome State Theatre in Gainesville, Florida, and The Tampa Bay Performing Arts Center to name but a few. Directing credits also include OAK AND IVY (the life of Paul Laurence Dunbar), MA RAINEY'S BLACK BOTTOM, FORTINBRAS, THE MEDEA MYTH, and SHIRLEY VALENTINE. She directed A RAISIN IN THE SUN with Daughters of the Dust (film) Cheryl Lynn Bruce at the University of Illinois-Urbana/Champaign in Spring, 2004 in commemoration of the 50th anniversary of the Brown v. Bd of Education.

Ms. Dunlap's leadership skills included developing and coordinating several workshops and symposia from the first Iowa International Storytelling Conference, the first African American Mimetics Conference in Maryland to a Maryland Hate Crime Symposia. Ms. Dunlap is a member of The Stage Directors and Choreographers Society (SDC). Her favorite productions? ... Stacie and Wesley Dunlap!

Stella Hargett

Best Overall Faculty Member, Sociology

Education:
B.A. Sociology, Minor Social Work, Fayetteville State University, 1973 M.A. Sociology, University of North Carolina-Chapel Hill, 1974 Ph.D. Sociology/Social Psychology, University of North Carolina-Chapel Hill, 1976 Postdoctorate, Educational Research, Michigan State University, 1981

Teaching Interests: Social Psychology, Social Theory, Research Methods, Applied Sociology, Mass Media, Social Problems, Program Evaluation

Research Interests: Transportation Patterns of African-American Elderly, Type II Diabetes in African-Americans, Intimate Partner Violence, Minority Access to Post- Graduate STEM Program, Improving Internal Program Evaluation Processes

Recent Publications/Reports/Presentations:

Smith, Robert & Stella Hargett. 2011. "Second Parenthoods: The Influence of Custodial Care of Children Among African-American Elderly on Their Travel Behaviors and Transportation Needs." DOT National Transportation Center, Morgan State University.

Hargett, Stella. 2009. "All Things to All People: Experiences and Challenges of an Internal Evaluator at a Historically Black College/University (HBCU)." Annual Conference of the American Evaluation Association, Orlando, Florida.

Hargett, Stella & Robert Smith. 2003. "Factors Influencing the Transportation Patterns of African-American Elderly: The Effects of Age, Gender, Household Structure and Residential Location." DOT National Transportation Center, Morgan State University.

Lois McMillan

Best Teacher, Liberal Arts, Social Sciences, and Humanities

Melvin Miles

Best Overall Faculty Member, Liberal Arts, Social Sciences, and Humanities

Melvin N. Miles, Jr., Director of University Bands and Instructor of Music at Morgan State University, is the conductor of the Morgan State University Symphonic Winds, Concert Band and the MSU Jazz Ensemble. In addition, his responsibilities include directing the Morgan State University Marching Band "The Magnificent Marching Machine", the "Bear Band"- MSU's Pep Band, and the MSU Jazz Combo. As an instructor in the Fine Arts Department, Mr. Miles teaches Instrumental Music Methods and coordinates the wind, percussion and string faculty. Melvin Miles has served on the Morgan State University faculty since 1973.

A native of Baltimore, Maryland, Melvin attended Frederick Douglass High School and received his Bachelor of Science and Master of Arts degrees in Music Education from Morgan State University. He attended The Catholic University of America for doctoral studies in Music Education. Mr. Miles is an active arranger and maintains an energetic performance schedule guest conducting Symphonic and Jazz Honor Bands through out the nation, as well as participating in adjudicating activities and presenting workshops and seminars on all styles and types of music and music education. He has written numerous arrangements performed by the Morgan State University Marching and Pep Bands and heard throughout the nation during the football and basketball seasons. For many years, "Miles" composed, adapted and arranged musical scores and served as the Musical Director for the "Encore Dinner Theater". In addition, he arranged and performed several selections for as well as appeared in Barry Levinson's movies "Diner" and "Avalon".

Melvin Miles is a member of Kappa Kappa Psi National Honorary Band Fraternity (Past National President and Chairman Board of Trustees), and Tau Beta Sigma National Honorary Band Sorority. In March 2003, he received Kappa Kappa Psi's highest award the "Distinguished

Service to Music Medal". In addition, he holds active membership in the Council of Higher Education in Music (Past President), Music Educators National Conference, Maryland Music Educators Association, College Band Directors Association (State Chairman), Maryland Band Directors Association, HBCU-National Band Directors Consortium, International Association of Jazz Educators, Promethean Kappa Tau, Alpha Psi Omega National Honorary Society and Alpha Phi Alpha Fraternity.

Melvin Miles resides in Baltimore, Maryland with his wife Cathy.

Anita Pandey

Best Researcher/Scholar, Liberal Arts, Social Sciences, and Humanities

Professional Communication at Morgan State University in Baltimore, USA. She was born and raised in a bilingual home in Africa. She picked up Hindi, English, Yoruba, Hausa, and Nigerian Pidgin in her childhood, and learned French and Spanish as a teenager—primarily from children. She earned her doctorate from the University of Illinois at Urbana-Champaign. She holds two M.A. degrees (the first in Teaching English to Speakers of Other Languages), and is Associate Editor of the Journal of English as an International Language, and guest editor of two forthcoming issues of the International Journal of Communication.

Natasha Pratt-Harris

Best Researcher/Scholar, Sociology

Pratt-Harris, Natasha (2011, February). Graduation and Release: The Intersection of Race, Gender, Socioeconomic Risk, & Culture. Paper presented at the annual meeting of the Eastern Sociological Society, Philadelphia, PA.

Pratt-Harris, Natasha (2012, February 25). Pardon Me: An Introduction to a Case Study Series of the 'Falsely' Accused Black Male Juvenile, Charged as an Adult. Paper presented at the annual meeting of the Eastern Sociological Society, New York, NY.

Pratt-Harris, Natasha (2012, March 15). Access to Justice and Disproportionate Minority Contact, A Meta-analysis 1961–2011. Paper presented at the annual meeting of the Academy of Criminal Justice Sciences, New York, NY.

Dr. Pratt-Harris works with the Department of Juvenile Services where she helps with the coordination of DMC efforts in Maryland, namely the reporting of efforts to increase awareness of DMC and to reduce and eliminate DMC in Maryland. She also serves on the Board of Directors for the Baltimore International Academy.

Carlene Turner

Most Helpful to Students, Sociology

MOTT COMMUNITY COLLEGE

Michele Dunnum

Best Overall Faculty Member, English
Best Researcher/Scholar, English
Best Teacher, English
Most Helpful to Students, English
Best Overall Faculty Member, Liberal Arts, Social Sciences, and Humanities
Best Researcher/Scholar, Liberal Arts, Social Sciences, and Humanities
Best Teacher, Liberal Arts, Social Sciences, and Humanities
Most Helpful to Students, Liberal Arts, Social Sciences, and Humanities

Sharon Eisen

Most Helpful to Students, Social Science

Elizabeth Lafray

Best Overall Faculty Member, Social Science

Mary Nieuwenhuis

Best Overall Faculty Member, Music
Best Teacher, Music
Most Helpful to Students, Music

Paul Rozycki

Best Researcher/Scholar, Social Science

Charles Tommasulo

Best Teacher, Social Science

Charles P. Tommasulo has been a lifelong advocate and educator for the needs and concerns of children and older adults. Since 1979, he has been employed by Family Service Agency of Mid Michigan (FSA).

There he has moved from Clinical Social Worker to Director of Professional Services to hold the position of Executive Director, which he assumed in 1990 and continues in today. He holds Masters Degrees in Social Work and Public Administration from the University of Michigan. During his time with FSA, he has focused on seniors, families and children with a special emphasis on intergenerational programming. Under his leadership, FSA's in-home eldercare services, guardian/conservatorship, representative payee, visually impaired center, senior companion and foster grandparent programs have all flourished.

Mr. Tommasulo helped bring county prosecutors, hospital officials and social service providers in mid-Michigan together to create the Adult Abuse Protocol, a comprehensive tool for dealing with elder abuse. He also encourages his staff to join local, regional and national groups working for seniors and he shares success stories about how to better serve seniors and children. In addition to these local programs, Mr. Tommasulo works at the state and national level with organizations such as the Corporation for National Service to bring other senior program directors together, and to keep a close eye on the political activities at all levels of government.

In addition, Mr. Tommasulo is an Adjunct Faculty member at Mott Community College. Since 1986, he has taught classes focused on social welfare and community development.

MOUNT SAINT MARY'S UNIVERSITY

Trudy Conway

Best Teacher, Liberal Arts, Social Sciences, and Humanities

David McCarthy

Best Researcher/Scholar, Theology

Dr. McCarthy is the Fr. James M. Forker Professor of Catholic Social Teaching. He started teaching at the Mount in 1998, and he has taught a variety of courses in the theology department and the University's Veritas Program (the core curriculum).

MOUNTAIN VIEW COLLEGE

James Behan

Best Overall Faculty Member, Fine Arts

MT. SAN ANTONIO COLLEGE

Lida Anderson

Best Researcher/Scholar, Child Development

"Just completed a relevant research in the discipline for a doctoral degree. Men in Early Childhood education."

Teacher for over 35 years. Graduated from Mt SAC in 1998. One of the first inductees into the honor society of Phi Delta Kappa. Graduated from Pacific Oaks College with a BA and MA in human development. Working with Mt San Antonio College for about 11 years as an Adjunct faculty in child development. Graduated from Walden University with a PhD in K–12 leadership with a 4.0 grade average. Enjoys doing research and writing articles.

Brooke Bain

Best Teacher, Sociology

John Brantingham

Best Overall Faculty Member, English

"John constantly goes above and beyond to help his students and to foster a creative community both on campus and in the surrounding cities. His tireless enthusiasm is an inspiration."

Andrea Diem

Best Overall Faculty Member, Philosophy
Best Teacher, Philosophy
Most Helpful to Students, Philosophy

Allison Frickert

Best Researcher/Scholar, Liberal Arts, Social Sciences, and Humanities

Marlene Gallarde

Most Helpful to Students, Sociology

Randy Ramal

Best Researcher/Scholar, Philosophy

Linda Rillorta

Best Overall Faculty Member, Sociology

MUHLENBERG COLLEGE

Joan Marx

Best Overall Faculty Member, Languages

Dr. Joan F. Marx is Professor of Spanish and Director of the Spanish Program. She specializes in contemporary Latin American literature, specifically, the Mexican and Mexican-American narrative. Her professional work includes scholarly publications in national and international literary journals as well as presentations of her work at national and international literary meetings. In addition, she has presented scholarly papers on the use of foreign language technology in the classroom as part of her work on the Mellon Project at national and international meetings. A graduate of the class of 1977, Dr. Marx has been a member of the faculty at Muhlenberg College since 1984. She received the Christian R. and Mary F. Lindback Foundation Award for Distinguished Teaching at Muhlenberg College (1991), the Louis Bevier Graduate Fellowship at Rutgers University (1983–84), and the Phi Sigma Iota Romance Language Award when she was a student at Muhlenberg College (1977). Dr. Marx teaches courses from the beginning to the advanced levels, which include Intermediate Spanish I and II, The Literature of Conquest and Colonization in Spanish America, Postcolonial Realities in Spanish American Literature, Border Literature, Human Rights Literature in the Americas, and the Capstone Seminar, Writing the Jewish Diaspora from Spain to the Americas. Over the years, she has brought to campus a number of speakers for public talks and participation in departmental Student Symposiums, such as Salvadoran writer and human rights activist Mario Bencastro, documentary filmmaker Peter Sanders (The Disappeared), Chilean poet and human rights activist, Marjorie Agosín, and Chicana writer and activist, Helena María Viramontes. From 1991–93, Dr. Marx served as Assistant Dean of the College and, from 1993–97, as Head of the department of Foreign Languages and Literatures. During her first term as Head of the department, she designed and implemented the $300,000 Mellon Grant for the Development of Foreign Language Multimedia Computer Programs, in which all full-time department members of every language program participated in the design and development of multimedia software. The Mellon Project Software, a series of grammar-based and cultural programs, is housed in the Language Learning Center and is incorporated into the department's language-level classes along with other technology-based applications. She has served two terms as Chair of the Department of Languages, Literatures and Cultures, the second from 2005 to 2014. She has been a member of the Modern Language Association (MLA), the American Association of Teachers of Spanish and Portuguese, the American Council on the Teaching of Foreign Languages (ACTFL), and of the Middle Atlantic Council for Latin American Studies (MACLAS). She currently serves on the Scholarship Committee of Phi Sigma Iota, the International Foreign Language Honor Society, and she represents Muhlenberg College on the Board of Advisors of CC-CS the Center for Cross-Cultural Studies.

MUSKEGON COMMUNITY COLLEGE

Nicholas Budimir

Best Overall Faculty Member, Social Science
Best Researcher/Scholar, Social Science
Best Overall Faculty Member, Liberal Arts, Social Sciences, and Humanities
Best Teacher, Liberal Arts, Social Sciences, and Humanities

In addition to teaching at Muskegon Community College, professor Budimir has taught sociology at Ferris State University and Western Michigan University while working on his graduate degree. As an active leader in the Michigan Sociological Association since 2006, Budimir organizes conferences, speakers and panels. Professor Budimir's specialties in sociology include economic sociology, the intersection of class, race, and gender, the sociology of education, and the sociology of labor.

NAPA VALLEY COLLEGE

A Gross

Best Overall Faculty Member, Communication

I received both my B.A. in Interpersonal Communication and my M.A. in Rhetoric from UC Davis. After teaching in a variety of college settings, I began teaching full time at NVC in 1994.

On the home front, I am married, with two children. We live in the country in Lake County with our dogs and cat. In my spare time I enjoy Bikram yoga, playing the piano, hiking, cooking and watching movies.

I love everything about communication and hope to share some of that enthusiasm with you! I encourage you to meet with me throughout the semester. My job is to help you be successful in the class and in college.

NATIONAL UNIVERSITY

Carl Boggs

Best Researcher/Scholar, Social Science

"Carl Boggs remains NU's most prominent and productive scholar. He has published thirty books, two in the last year alone from prominent publishers."

Carl Boggs is professor of social sciences at National University in Los Angeles, where he has taught for 26 years. With a Ph.D. at the University of California, Berkeley (political science), he is the author of 22 books and more than 300 articles and reviews. He has taught at Washington University in St. Louis, UCLA, USC, and several other universities. He is on the editorial board of several journals, including THEORY AND SOCIETY, where he is book-review editor. He is recipient of the Charles McCoy Career Achievement Award by the American Political Science Association (2007).

Ramie Tateishi

Best Researcher/Scholar, English

NAZARETH COLLEGE

Monica Weis

Best Overall Faculty Member, Liberal Arts, Social Sciences, and Humanities

NEBRASKA WESLEYAN UNIVERSITY

David Iaquinta

Best Overall Faculty Member, Sociology
Best Researcher/Scholar, Sociology
Best Teacher, Sociology
Most Helpful to Students, Sociology

NEUMANN UNIVERSITY

Terence Gleeson

Best Overall Faculty Member, Communication
Best Teacher, Communication

NEVADA STATE COLLEGE

Teri Romeo

Best Teacher, Speech

"Outstanding teacher."

Teri Romeo has been working as a speech-language pathologist since 1985 within school districts in Nevada. She holds the Certificate of Clinical Competence from the American Speech/Language Hearing Association (ASHA). Her specialty areas include Childhood Apraxia of Speech (CAS); phonemic speech sound disorders and teaching spoken language/listening skills to individuals with hearing loss.

NEW JERSEY CITY UNIVERSITY

Jaqueline Ellis

Best Overall Faculty Member, Women's Studies
Most Helpful to Students, Women's Studies

Dr. Jacqueline Ellis joined the faculty at NJCU in 2001. She received her Ph.D. in American Studies from the University of Hull in England, and is the author of Silent Witnesses: Representations of Working Class Women in the United States (1998). Her writing on constructions of working-class and gender identity in visual and popular culture has been published in the Feminist Review, and the History of Photography. Her latest article "Working Class Women Theorize Globalization" is forthcoming in the International Journal of Feminist Politics. She is also the co-editor of Transformations: The Journal of Inclusive Scholarship and Pedagogy and co-ordinates "Speaking Our Stories: The NJCU Oral History Project."

Antoinette Ellis-Williams

Most Helpful to Students, Liberal Arts, Social Sciences, and Humanities

Liza Fiol-Matta

Best Teacher, Women's Studies

Liza Fiol-Matta joined the NJCU faculty in 1999. She holds a Ph.D. in Critical Literary and Cultural Studies from The Union Institute and University (1996), an M.A. degree in English Literature from the University of Puerto Rico (1981), and a B.A. in English from Sacred Heart University in Santurce, Puerto Rico (1974). Her book, Women of Color and the Multicultural Curriculum: Transforming the College Classroom (1994), coauthored with Miriam K. Chamberlain, received the Gustavus Myers Center award as Outstanding Book on the Subject of Human Rights in North America. A second book, Telling to Live: Latina Feminist Testimonios (2001), coauthored with the members of the Latina Feminist Group, received a Gustavus Myers Center Human Rights and Bigotry Award in 2002.

Her essay, "Beyond Survival: A Politics/Poetics of Puerto Rican Consciousness," was reprinted in Word: On Being a [Woman] Writer (2004). Dr. Fiol-Matta's research interests include curriculum transformation in higher education; Latina/o identity and politics; critical cultural studies with a focus on race, class, ethnicity, and gender; and the literature of Latinas/os in the United States. Dr. Fiol-Matta served as Vice President of the National Women's Studies Association in 2005–2006. From 2002–2008, Dr. Fiol-Matta was Dean of the William J. Maxwell College of Arts and Sciences.

Demond Hargrove

Best Researcher/Scholar, Humanities

Irma Maini

Best Overall Faculty Member, Liberal Arts, Social Sciences, and Humanities
Best Teacher, Liberal Arts, Social Sciences, and Humanities

Catherine Raissiguier

Best Researcher/Scholar, Women's Studies
Best Researcher/Scholar, Liberal Arts, Social Sciences, and Humanities

Catherine Raissiguier joined the NJCU faculty in 2003. She completed her undergraduate education in France and holds a M.A. in Women's Studies/American Studies and a Ph.D. in Comparative and International Education from SUNY/Buffalo. Dr. Raissiguier's first book, Becoming Women/Becoming Workers: Identity Formation in a French High School, was published by SUNY Press in 1994. She is currently at work on a book manuscript that explores how concepts of citizenship and nationality intersect with gender and immigration in twentieth-century France. Re-Inventing the Republic: Gender, Migration, and Citizenship in France, is under contract with Stanford University Press. She has also published in such journals as Signs, the International Feminist Journal of Politics, the Radical History Review, and Educational Foundations.

NEW YORK UNIVERSITY

Peter Rea

Best Researcher/Scholar, Film & Television

NEW YORK UNIVERSITY - SCHOOL OF SOCIAL WORK

S. Lala Straussner

Best Teacher, Liberal Arts, Social Sciences, and Humanities

Dr. S. Lala Straussner is a professor who teaches in both the practice and HBSE areas and the doctoral dissertation seminar in the PhD program. She has taught an international study abroad course on trauma for NYU Silver School of Social Work students in Jerusalem, and will be teaching a substance abuse course for Silver School students in Italy in summer 2013. She also directs the School's Post-Master's Certificate Program in the Clinical Approaches to the Addictions. Currently Dr. Straussner serves as the interim chair of the practice area, and previously served three years as chair of that area. Dr. Straussner has been on the faculty at the annual Summer Institute on Addictions Studies, International School for the Humanities and Social Sciences, University of Amsterdam, The Netherlands.

Dr. Straussner is the founding editor of the Journal of Social Work Practice in the Addictions. During the 2006–07 academic year, Dr. Straussner was a Fulbright Senior Scholar at The Academy of Labour & Social Relations in Kiev, Ukraine, a visiting professor at Warsaw University in Poland, and recipient of the Lady Davis Fellowship to Hebrew University in Jerusalem. Previously, she was a Fulbright Senior Scholar to Israel (2003) and a visiting professor at Omsk State University, Omsk, Siberia, Russia (2000). She was the 2003 recipient of the Individual Distinction in Addictions Education and Training Award given by the Institute of Professional Development in Addictions, and was selected as Social Worker of the Year by the NASW ATOD Section in 2000. She received the Award for Lifetime Leadership To Addictions Community, from the New York City Chapter of NASW in 2008, and in 2007 Dr. Straussner was inducted as a Distinguished Scholar into The Social Work Academy of The National Academies Of Practice. Dr. Straussner was also nominated as the 2008 Outstanding Teacher from the NYU Silver School of Social Work, was listed in the 2011 edition of Who's Who in America, and has been twice nominated for the CSWE Significant Lifetime Achievement In Social Work Education Award.

Dr. Straussner is a current member of the New York State Office Of Alcoholism And Substance Abuse Services (OASAS) Talent Management Committee. From 2008–2010 she was an invited reviewer, Alcohol, Drug Abuse, and Social Work Peer Review Medical Research Program Panel, US Army Medical Research and Material Command, American Institute of Biological Sciences, and was a consultant/board member on the Expand & Enhance Asian American Substance Abuse Treatment Services (EASTS) Project, Hamilton-Madison House, SAMHSA grant from 2009–2011. She also served on the National Center on Substance Abuse Treatment panel on workforce issues, and was the Northeast regional director for the multi-million dollar federally funded national Project Mainstream, an Inter-disciplinary Project to Improve Health Professional Education on Substance Abuse sponsored by AMERSA/HRSA/CSAT. She is a past chair and a founding member of the NASW Section on Alcohol, Tobacco, and Other Drugs and has chaired the New York City Women's Issues Committee of the Employee Assistance Professionals Association and been a 2nd vice president of the NYC Chapter of NASW. During 2002, she was a keynote speaker for the Betty Ford/Hazelden Women's Healing conferences throughout the US, and during 2006, she chaired the National Association for Children of Alcoholics' (NACoA) Social Work Initiative to develop social work core competencies for working with children of alcohol and drug abusing parents.

NORFOLK STATE UNIVERSITY

Wanda Brockington

Most Helpful to Students, Communication

"Goes out of her way to assist students."

Michael Cotter

Best Overall Faculty Member, Liberal Arts, Social Sciences, and Humanities

"Exemplifies the qualities of the consummate professional in and out of the classroom. Innovative teaching approach."

Michael J. Cotter is a native of Chicago and currently resides in Virginia Beach. He earned a B.A. degree in American literature at Chicago State University and an M.S. degree in print and broadcast journalism at Northwestern University's Medill School of Journalism. He has taught courses ranging from African-American, American, British, and world literature to technical writing and cable, radio and television production. His international teaching experience includes courses in composition and the novel at Mohamed V University in Rabat, Morocco. He is also a former instructor in the Program for Afloat College Education (PACE), having taught aboard the U.S.S. Wasp in the Persian Gulf. Currently in his 14th year at NSU, Mr. Cotter is the interim director of the writing program in the Department of English and Foreign Languages. Mr. Cotter's favorite authors are Nathaniel Hawthorne and Sir Arthur Conan Doyle.

Ira Falls

Best Researcher/Scholar, Liberal Arts, Social Sciences, and Humanities

Master of Arts degree program in Urban Affairs is a multidisciplinary program, utilizing the resources of the Social Science disciplines. Applicants may be admitted for enrollment in the fall and spring terms.

Cathy Jackson

Best Researcher/Scholar, Communication

"Nationally recognized."

Page Laws

Best Researcher/Scholar, Liberal Arts, Social Sciences, and Humanities

Daniel Pearlman

Best Teacher, Liberal Arts, Social Sciences, and Humanities

"Students love him. They say he's the best!"

Mr. Daniel Pearlman was born in Norfolk and reared in Virginia Beach. He returned to Norfolk to teach at NSU in 1996, after having earned his MFA from George Mason University. His poetry has appeared in many journals, including most recently Dos Passos Review, Beltway Poet's Domain, and Washington Review. His band, The Dunes, released a CD, Spirits of the Revolution, in 2006 and has played in New York; Washington, D.C.; Norfolk; Virginia Beach; and other locations. His favorite books are Leaves of Grass by Walt Whitman and 1984 by George Orwell.

Dottie Perry

Best Teacher, Liberal Arts, Social Sciences, and Humanities
Most Helpful to Students, Liberal Arts, Social Sciences, and Humanities
Best Overall Faculty Member, Liberal Arts, Social Sciences, and Humanities

Ms. Dottie Demory Perry, a graduate of Norfolk State University, has taught English at Norfolk State University for twenty-five years. She holds a bachelor of arts degree in English and a master of arts degree in communication. She has done further study at The College of William and Mary. In addition to teaching writing, literature, and methods courses, Ms. Perry serves as the Department's writing specialist for the writing center. In that capacity, she provides supplementary instruction to many students, including those preparing for the Examination of Writing Competency, those taking writing courses, prospective graduates preparing for licensure examinations, and graduate students preparing for discipline-specific writing tasks.

NORTH CAROLINA CENTRAL UNIVERSITY

Lolethia Underdue

Best Overall Faculty Member, Communication
Best Teacher, Communication
Most Helpful to Students, Communication

Lolethia Underdue casually chose radio as an extracurricular activity in college, got bitten by the broadcasting bug, and switched her major from biology to English in order to get more involved in radio. She worked at various stations in North Carolina before earning her Master's in broadcast management at Norfolk State University. She is now the general manager of AudioNet, the campus wide cable TV student- staffed radio station that is also broadcast via the Internet; additionally, she is a visiting lecturer in the Department of Mass Communication at NCCU.

NORTH CAROLINA STATE UNIVERSITY

Joseph Gomez

Best Teacher, English

"Genuinely inspiring teacher!"

NORTH CENTRAL TEXAS COLLEGE

Charles Williams

Best Overall Faculty Member, History
Best Teacher, History
Most Helpful to Students, History
Best Teacher, Liberal Arts, Social Sciences, and Humanities

NORTH DAKOTA STATE UNIVERSITY

Gina Aalgaard Kelly

Most Helpful to Students, Sociology

Patty Corwin

Best Teacher, Sociology
Best Teacher, Liberal Arts, Social Sciences, and Humanities

Mrs. Patty Corwin has been a member of the North Dakota State University faculty since 1979. Her primary teaching responsibilities are introduction to sociology, minorities and race relations, sociology of gender and sociology through literature.

Mrs. Corwin's research interests include minorities and race relations, gender, and service learning.

Ms. Corwin does a variety of seminars and workshops for businesses.

Gary Goreham

Best Overall Faculty Member, Sociology
Most Helpful to Students, Liberal Arts, Social Sciences, and Humanities
Best Overall Faculty Member, Liberal Arts, Social Sciences, and Humanities

Dr. Goreham's research interests include rural sociology, rural communities and churches, social and ethical impacts of agro-biotechnology, farm financial stress, sustainable agriculture, agricultural cooperatives, and rural poverty.

Richard Rathge

Best Researcher/Scholar, Sociology
Best Researcher/Scholar, Liberal Arts, Social Sciences, and Humanities

Dr. Rathge is a demographer, research methodologist, and rural sociologist. He has a joint teaching/research appointment at NDSU in the departments of Sociology/Anthropology and Agribusiness and Applied Economics. He also serves as the State Demographer for North Dakota as he directs the North Dakota State Data Center and he is the State Coordinator for the Federal-State Cooperative Program for Population Estimates (FSCPE) and Projections (FSCPPE). In addition, he is a policy analyst for the North Dakota Kids Count program which is a research center that monitors the health and well-being of children. Dr. Rathge joined the faculty at NDSU in 1981.

NORTH IDAHO COLLEGE

Dave Hoskins

Best Overall Faculty Member, Communication
Best Researcher/Scholar, Communication
Best Teacher, Communication
Most Helpful to Students, Communication

NORTHAMPTON COMMUNITY COLLEGE

Randy Boone

Best Teacher, English

Susan Petrole

Best Teacher, Liberal Arts, Social Sciences, and Humanities

Susan Petrole is an adjunct instructor of English at Northampton Community College, DeSales University, and Lehigh Carbon Community College. Having received adjunct faculty of the year at Lehigh Carbon Community College and nominations at Northampton Community College, pedagogy at the two-year college is her primary focus; however, she is also pursuing a PhD in Literature and Criticism at Indiana University of Pennsylvania. In addition to literary criticism, Susan also writes fiction and has published numerous short stories in literary magazines, including Epiphany and New England Writer's Network. She resides in Bethlehem, Pennsylvania.

NORTHAMPTON COMMUNITY COLLEGE - MONROE

Barbara Benner

Best Teacher, Humanities

Mike Catrillo

Most Helpful to Students, Liberal Arts, Social Sciences, and Humanities

"The students reports are rather excellent."

Heather Jones

Best Researcher/Scholar, Liberal Arts, Social Sciences, and Humanities

"Seems to connect very well with the students"

NORTHEAST IOWA COMMUNITY COLLEGE

Jim Maddox

Best Overall Faculty Member, Communication
Best Researcher/Scholar, Liberal Arts, Social Sciences, and Humanities

"excellent instructor, caring, dedicated, accessible to students, life long learner and educator"

NORTHEAST MISSISSIPPI COMMUNITY COLLEGE

Carla Falkner

Best Researcher/Scholar, Liberal Arts, Social Sciences, and Humanities
Best Teacher, Liberal Arts, Social Sciences, and Humanities
Most Helpful to Students, Liberal Arts, Social Sciences, and Humanities
Best Overall Faculty Member, Liberal Arts, Social Sciences, and Humanities

Originally from Texas, Falkner moved to Mississippi in 1977 to pursue a graduate degree at the University of Mississippi. Today, she has over thirty years of experience as a Mississippi educator. She joined the Northeast faculty in 1989 after teaching high school social studies in Batesville and New Site. In addition to her teaching responsibilities, Falkner serves as the Assistant Division Chair and is active on the NE Cultural Arts Committee.

Harris Ray

Best Overall Faculty Member, Music
Best Researcher/Scholar, Music
Best Teacher, Music
Most Helpful to Students, Music

NORTHEASTERN ILLINOIS UNIVERSITY

Peter Chang

Best Researcher/Scholar, Liberal Arts, Social Sciences, and Humanities

Sarah Hoagland

Best Overall Faculty Member, Philosophy

"Dr. Sarah Hoagland is not only a very knowledgeable professor she is dedicated to her students professional well being and just an all around great person,"

Luis Ortiz

Best Teacher, Liberal Arts, Social Sciences, and Humanities

Jade Stanley

Best Overall Faculty Member, Social Work
Best Researcher/Scholar, Social Work
Best Teacher, Social Work
Most Helpful to Students, Social Work
Most Helpful to Students, Liberal Arts, Social Sciences, and Humanities
Best Overall Faculty Member, Liberal Arts, Social Sciences, and Humanities

NORTHERN ARIZONA UNIVERSITY

Julene Boger

Most Helpful to Students, Hospitality

Rita Borden

Most Helpful to Students, Music

Rita Borden joined the Northern Arizona University School of Performing Arts faculty in 1995 where she teaches piano, accompanying and chamber music. Ms. Borden is a pianist and coordinator with the NAU Faculty Chamber Players, and is an active collaborator on faculty recitals and with guest artists on the NAU Performing Arts Series. She has extensive experience as a collaborative pianist, including performances with: San Diego Symphony Pacific Symphony Chamber Music Sedona Colorado String Quartet San Diego Master Chorale Pacific Chorale Flagstaff Symphony Flagstaff Festival of the Arts. Ms. Borden coordinates the piano accompanying/chamber music emphasis at NAU and supervises the piano scholarship accompanying students. She has over twenty years of experience in all areas of the accompanying field including a staff accompanist position at California State University-Fullerton (1980–1989) and has accompanied many workshops and auditions such as the USC Piatigorsky Seminar, National Cello Institute, Metropolitan Auditions and Richard Miller Seminar. She has been on the faculty of the NAU Music Camp, Idyllwild Arts and USC-ISOMATA.

Julie Hedges Brown

Best Teacher, Music

Julie Hedges Brown graduated from Yale University in 2000 with the Ph.D. in musicology. Her research emphasizes nineteenth-century music, especially the history, biography, reception, and analysis of Robert Schumann and his music. Her publications have appeared in the Journal of the American Musicological Society, 19th Century Music, Journal of Musicology, and Rethinking Schumann (Oxford University Press). She has presented papers at numerous regional, national, and international conferences, including ones sponsored by the American Musicological Society, the International Musicological Society, the College Music Society, the North American Conference on 19th-Century Music, and the International Conference on Nineteenth-Century Music. A recipient of the Robert M. Leylan Dissertation Fellowship, Dr. Brown has also received two grants from the Mellon Foundation, along with three faculty research grants and a College of Arts and Letters research grant from Northern Arizona University

Galen Collins

Best Overall Faculty Member, Hospitality
Best Researcher/Scholar, Hospitality
Best Teacher, Hospitality

Galen Collins, Ph.D., has been with HRM since its beginnings as a free-standing school in 1987. Collins is noted for his breadth of knowledge on hospitality information technology and his ability to convey complex technological concepts in an easily understood manner. He was an early adaptor and advocate of distance learning as a radical tool for bringing HRM instruction to far-flung locales and diverse students. He continues to be a visionary in this area of emerging technologies. He is the founding publisher of the Information Technology in Hospitality journal and co-founder of HITA, the Hospitality Information Technology Association. He received his B.B.A. and MS degrees from Florida International University and his Ph.D. from the Graduate School of Computer and Information Sciences at Nova Southeastern University. Collins has co-authored the following texts: Hospitality Information Technology: Learning How to Use It (5th edition) and Overcoming the Customer Service Syndrome: A Practical Guide to Managing Customer Service. He also has published articles in trade and academic journals, including the Cornell Quarterly, Hospitality Education and Research Journal, Hotel and Motel Management, FIU Hospitality Review, Journal of Teaching in Travel and Tourism, Journal of Distance Learning Administration, and Hospitality and Tourism Educator.

Leilah Danielson

Best Researcher/Scholar, History

"New book coming out–Leiiah is a careful researcher and an excellent scholar, perhaps the department's best. She also excels in the classroom both as an undergraduate instructor and talented mentor for graduate students."

My teaching reflects my interests in cultural and intellectual history; politics and social movements; American empire and race; labor and working-class history; and history/social studies education. Courses taught The U.S. in the World (graduate seminar) U.S. Politics and Culture in the 20th Century (graduate seminar) Labor and the Left in American History (graduate seminar) American Cultural History (graduate seminar) The United States and the Cold War U.S. Culture and Thought since 1865 American Intellectual History Labor and Working-Class History of North America Women, Work, and Culture The History of American Radicalism U.S. History since 1865 History/Social Studies Teaching Methods Teaching and Learning History and Geography Practicum for History/Social Studies Teacher Candidates My teaching extends to the mentorship of graduate students and undergraduate honor's students who share similar interests. Over the years, my advisees have conducted research on topics such as: Hitch-hiking and post-war American politics and culture Gender, sexuality, and the history of the American pacifist movement The history of the new left and Chicano movement at ASU (subsequently published in the Journal of Arizona History). The history of Mormonism and the modern conservative movement. The relationship between the white new left and the black power movement. The cultural politics of the anarchist movement at the turn of the century Masculinity and whiteness in postwar popular culture

Rodrigo Detoledo

Best Overall Faculty Member, Liberal Arts, Social Sciences, and Humanities

Rodrigo is a visual artist and designer teaching and working on the moving edge of digital media, with an appreciation for the traditional foundation of the visual arts and graphic design. As a designer, Rodrigo worked with various clients, artists, advertising agencies, design studios, production houses, scientists and marketing consulting firms, including Barclay's Bank, NASA, MTV, NEC, Palm, Apple and GM. The term User Experience Architecture describes his 360 view of the current state of integration in the profession. His projects include motion graphics, experience design, TV commercials, printed publications, web sites and interactive multimedia. His work was recipient of research grants, the NASA Software of the Year and the Forbes best of the Web awards, and it was featured in printed, televised and online media, as well as exhibited in museums, galleries and festivals around the globe. Current creative interests include interactive illustrated fiction for mobile devices and gouache illustrations.

Martin Kalb

Best Teacher, History

"His syllabi, assignments and self-evaluation show that he is a thoughtful and rigorous instructor and committed to the education of his students. He introduces innovative assignments, seeks out new source material that will enlighten and challenge those who take his courses. His student evaluations support what we hear in the hallways: he is engaging, energetic, challenging, fair and interesting, and wildly popular."

I am a historian of Modern European history somehow 'stranded' in the American Southwest. Born and raised near Bayreuth, Germany, I originally came to Arizona on a scholarship from the Federation of German-American Clubs. I studied at Northern Arizona University (NAU) for two semesters (2004/05), and then returned to Germany to finish my M.A. in history, political science, and English (American cultural studies) at Friedrich-Alexander University Erlangen-Nuremberg (2007). I attended NAU again for my Ph.D. program, completed my dissertation in 2011, and have been a full-time lecturer here in Flagstaff ever since.

The history of everyday life (Alltagsgeschichte) guides my research and teaching. My first book focuses on supposed young delinquents in Munich (forthcoming in 2016). I also completed research on environmental justice and the persecution of homosexuals during the Holocaust. My current research project engages with daily life in German Southwest Africa (modern day Namibia), a former German colony. Everyday experiences are also in the center of my courses on European and World history here at NAU. I emphasize the voices of the silenced and disenfranchised, and highlight that history is, above all, an argument about the past. I also create hands-on learning experiences as ways to involve students beyond the classroom.

Anne Scott

Best Overall Faculty Member, English
Best Teacher, English

Anne Scott has been affiliated with NAU's Honors Program for 15(!) years (as teacher, associate director, interim director, freshmen writing coordinator, and coordinator for academic enhancement programs). Her specialties include medieval literature, multi-ethnic literature, and film. She has co-edited 3 volumes of interdisciplinary essays on various topics in the Middle Ages and Early Modern Period. Currently Honors Professor, Dr. Scott very much enjoys teaching and helping to administer the Honors Program, spending time with her son, Gavin (16 years old), riding her many bicycles (but not all at once), drinking good coffee and watching Alfred Hitchcock thrillers, and collaborating with her husband, Everett Akam (also a teacher in Honors), on projects and classes.

NORTHERN ILLINOIS UNIVERSITY

Alexandra Bennett

Best Teacher, English
Most Helpful to Students, English
Best Teacher, Liberal Arts, Social Sciences, and Humanities
Most Helpful to Students, Liberal Arts, Social Sciences, and Humanities

NORTHERN KENTUCKY UNIVERSITY

David Dunevant

Best Overall Faculty Member, Music

"David Dunevant has a long history of outstanding service to the Department of Music and as an outstanding teacher. He has been recognized for his excellence in advising and has been a leader in the department in that area. David has successfully taught classes and conducted ensembles in a broad cross section of departmental offerings. He has served as assistant department chair and as department chair, as well as assistant dean of the College of Arts and Sciences and President of Kentucky Music Educators Association. He has distinguished himself as a performer in faculty recitals, the Faculty Brass Quintet, and as principal trombone in the Kentucky Symphony Orchestra. He as served the students as adviser to Collegiate MENC and Phi Mu Alpha Sinfonia Fraternity. He has done all of this with integrity and fairness. There is no individual I have worked with who could equal David's excellent record."

David Dunevant has been an active member of Northern Kentucky University's faculty since 1984. As a performer, he has served as principal trombonist of the Cincinnati Chamber Orchestra, the Middletown Symphony, the Hamilton Symphony, and the Kentucky Symphony Orchestra. He is also active in brass chamber music as a long-time member of the NKU Faculty Brass Quintet.

In 1990, Professor Dunevant received the Outstanding Junior Faculty Award in the College of Arts and Sciences. He was named College Teacher of the Year by the Kentucky Music Educators Association and in 1995 was selected by the Who's Who Among America's Teachers. He is a member of the International Trombone Association, the Kentucky Music Educators Association, the Music Educators National Conference, and Phi Mu Alpha Sinfonia.

NORTHERN MICHIGAN UNIVERSITY

Steve Leuthold

Most Helpful to Students, Art & Design

NORTHERN VIRGINIA COMMUNITY COLLEGE

Laura Casal

Best Overall Faculty Member, English
Best Teacher, English

Jean Goodine

Most Helpful to Students, English

Zac Jackson

Best Researcher/Scholar, Fine Arts
Best Teacher, Fine Arts
Best Overall Faculty Member, Fine Arts

"He is selfless in his dedication to his students and building the sculpture program."

Ann Kuhta

Best Researcher/Scholar, English

Barbara Lash

Best Teacher, Fine Arts

Elizabeth Lynch

Best Overall Faculty Member, Art
Best Researcher/Scholar, Art
Best Teacher, Art
Most Helpful to Students, Art

Elizabeth B. Lynch is an assistant professor of art history. She brings 10 years of teaching experience, including three years at NOVA. She earned a doctorate from George Washington University, a master's degree from Pennsylvania State University and a bachelor's degree from the University of Delaware. She has taken several foreign language courses at NOVA and "had great instructors for all of them." She lives in Manassas with her husband, John. They have three children, Nicholas, Annie and Ryan.

Molly Lynch

Best Researcher/Scholar, Liberal Arts, Social Sciences, and Humanities

I began teaching at NOVA in 2003. I was an adjunct professor at the Loudoun and Manassas Campuses before I started teaching full time. I am now the head of the Psychology department at the Manassas Campus.

I taught at a much larger institution before I came to NOVA; I really enjoy the smaller environment NOVA has to offer. I went from teaching classes with 180 students, to teaching classes with 30 students. It has been a delight getting to know each of my students on a much closer level.

I love what I do. I work hard to try to relay that enthusiasm to my students. I like to engage them in various activities so that it's a lot easier for them to remember different theories and apply them to life. It's really fulfilling to see students begin to grasp the subject matter and later share how they applied it to personal situations.

The one thing that I love about my job is seeing my former students after they leave NOVA and hearing what they're up to. It's exciting finding out that they've completed a bachelor's degree and are continuing their education beyond that level. I recently invited a former student to be a guest lecturer in one of my classes. I feel proud when I see the growth that is exemplified throughout the students.

Matt Pinney

Best Teacher, Fine Arts
Most Helpful to Students, Fine Arts

Matt Pinney is an Artist and Educator who loves his community. His peers and students mean everything to him and have been instrumental in any success he has had. Matt is looking forward to an ongoing relationship with the Northern Virginia/Washington D.C. area for many years to come.

Alicia Tucker

Best Researcher/Scholar, Liberal Arts, Social Sciences, and Humanities

Cecilia Wiltshire

Best Teacher, English As A Second Language

Joesph Windham

Most Helpful to Students, History

"Joe really cares about his students and does everything he can to make sure that they do well in his classes. He is supportive and focuses on student success."

NORTHERN VIRGINIA COMMUNITY COLLEGE - LOUDOUN

Beverly Blois

Best Overall Faculty Member, History

BEVERLY BLOIS has been dean of the Humanities Division, Northern Virginia Community College-Loudoun Campus, since 1992. He has been affiliated with NOVA-Loudoun since 1974 as professor of history and as executive director of statewide curricular initiatives for both the Council of Higher Education of Virginia (1985–86) and the Virginia Community College System (1990–91). He has also served as director of planning studies for George Mason University's Prince William Campus (1987–88).

NORTHWEST MISSOURI STATE UNIVERSITY

Brenda Ryan

Most Helpful to Students, English

NORTHWEST VISTA COLLEGE

Julie Bajusz

Best Teacher, Liberal Arts, Social Sciences, and Humanities

Gary Bowling

Best Teacher, Liberal Arts, Social Sciences, and Humanities
Best Researcher/Scholar, Liberal Arts, Social Sciences, and Humanities

Gary Bowling served in the Air Force from 1982 to 2003 retiring as a Senior Master Sergeant. His permanent duty stations included Osan and Kunsan Air Bases, Republic of South Korea, Mather AFB, CA, Maxwell AFB, AL, and Lackland AFB, Tx. In addition to his duty stations, he was deployed many times throughout his career to various temporary duty stations including Kuwait, Saudi Arabia, Indonesia, Singapore, and Honduras. His final assignment was with the 7th Security Forces Squadron (SFS) at Dyess AFB, TX. At Dyess he served as the Operations Superintendent and Security Forces Manager (senior unit enlisted) for 300 person unit. He holds a BS in business administration from Golden Gate University and a MA in Adult Education from Troy University. His current position is Senior Coordinator for the Student Development discipline at Northwest Vista College. He has also taught as an adjunct history instructor at NVC since spring 2004.

Craig Coroneos

Best Researcher/Scholar, Liberal Arts, Social Sciences, and Humanities

Clint Dunagan

Best Overall Faculty Member, Philosophy
Best Teacher, Philosophy
Most Helpful to Students, Philosophy
Best Teacher, Liberal Arts, Social Sciences, and Humanities
Most Helpful to Students, Liberal Arts, Social Sciences, and Humanities
Best Overall Faculty Member, Liberal Arts, Social Sciences, and Humanities

Anival Gonzalez

Most Helpful to Students, English

Dana Goodrich

Best Researcher/Scholar, History

"Dana is always available to help his students and his colleagues. He is an expert in history as well as understanding and working with people."

Laura Lawrence

Best Overall Faculty Member, Liberal Arts, Social Sciences, and Humanities
Most Helpful to Students, Liberal Arts, Social Sciences, and Humanities

Crystal Manboard

Best Researcher/Scholar, English

Crystal Manboard has been teaching English at Northwest Vista since January 2010 and while still on active duty, she also attended Northwest Vista as a student. She retired after serving 23 years in the Air Force and began her educational journey immediately upon retirement. Her husband is also a veteran who proudly served and also continued his education upon retirement. Throughout their journey, much insight was gained but the biggest insight was that which came from other students/veterans who had similar backgrounds and experiences. She looks forward to offering any help or guidance needed to her fellow veterans.

Julie Moore-Felux

Best Teacher, English

Venetia Pedraza

Best Overall Faculty Member, English

Daniel Smith

Best Overall Faculty Member, Music
Best Researcher/Scholar, Music
Best Teacher, Music
Most Helpful to Students, Music

Michael Ward

Most Helpful to Students, Liberal Arts, Social Sciences, and Humanities

NORTHWESTERN STATE UNIVERSITY

Bill Brent

Most Helpful to Students, Liberal Arts, Social Sciences, and Humanities
Best Researcher/Scholar, Liberal Arts, Social Sciences, and Humanities

Helaine Razovsky

Best Overall Faculty Member, English
Best Researcher/Scholar, English
Best Teacher, English
Most Helpful to Students, English

Best Researcher/Scholar, Liberal Arts, Social Sciences, and Humanities
Best Teacher, Liberal Arts, Social Sciences, and Humanities
Most Helpful to Students, Liberal Arts, Social Sciences, and Humanities
Best Overall Faculty Member, Liberal Arts, Social Sciences, and Humanities

- Ph.D. Boston University
- M.A. Boston University • B.A. Kent State University
- Research and teaching interests:
- Renaissance and Reformation
- Early Modern Drama
- Gender Studies

Courses:

- Shakespeare
- Milton
- Early Modern English Drama
- British Literature
- World Literature
- Detective Fiction

Selected publications:

Rev. of Science, Religion and Politics in Restoration England: Richard Cumberland's De Legibus Naturae, by Jon Parkin. Seventeenth Century News 59 (Spring-Summer 2001): 74–77.

"Remaking the Bible: English Reformation Spiritual Conduct Books." Renaissance and Reformation/Renaissance et Réforme 22.4 (Autumn 1998): 5–25.

"Popular Hermeneutics: Monstrous Children in English Renaissance Broadside Ballads." Early Modern Literary Studies. 2.3 (1996): 1.1–34 <URL: http://www.shu.ac.uk/emls/02-3/razoball.html>.

NORTHWESTERN UNIVERSITY

Mira Balberg

Best Teacher, Religion

Sarah Jacoby

Best Researcher/Scholar, Religion

Sarah Jacoby studies South Asian Religions with a specialization in Tibetan Buddhism. She received her B.A. from Yale University, majoring in women's studies, and her M.A. and Ph.D. degrees from the University of Virginia's Department of Religious Studies. She joined Northwestern University in 2009 after completing a postdoctoral fellowship at the Society of

Fellows in the Humanities at Columbia University. Her research interests include Indo-Tibetan Buddhist doctrine and ritual in practice, studies in gender and sexuality, Tibetan literature, autobiography studies, Buddhist revelation, Buddhism in contemporary Tibet, and Eastern Tibetan area studies. She is the co-chair of the Tibetan and Himalayan Religions Group at the American Academy of Religion.

Professor Jacoby received an American Council of Learned Sciences (ACLS) Fellowship for the 2012–2013 academic year. Her research has also been funded by The Shelley & Donald Rubin Foundation, the Charlotte W. Newcombe Dissertation Writing Fellowship, the Fulbright Hays Dissertation Research Fellowship, and multiple Foreign Language and Area Studies Fellowships (FLAS).

She has recently published a monograph titled Love and Liberation: Autobiographical Writings of the Tibetan Buddhist Visionary Sera Khandro (Columbia University Press, 2014). This is the first study in any language of the autobiographical and biographical writings of one of the most prolific female authors in Tibetan history, Sera Khandro Künzang Dekyong Chönyi Wangmo (also called Dewé Dorjé, 1892–1940). She was extraordinary not only for achieving religious mastery as a Tibetan Buddhist visionary and guru to many lamas, monastics, and laity in the Golok region of eastern Tibet, but also for her candor. This book listens to Sera Khandro's conversations with land deities, dakinis, bodhisattvas, lamas, and fellow religious community members whose voices interweave with her own to narrate what is a story of both love between Sera Khandro and her guru, Drimé Özer, and spiritual liberation.

Her other books include a co-edited volume with Antonio Terrone entitled Buddhism Beyond the Monastery: Tantric Practices and their Performers in Tibet and the Himalayas (Brill, 2009) and a book she co-authored with Donald Mitchell titled Buddhism: Introducing the Buddhist Experience (Oxford University Press, 2014).

In 2014 Professor Jacoby was awarded a Searle Center for Advanced Learning and Teaching Innovation in Teaching Grant. In 2012 she was voted by Northwestern students onto the ASG Faculty Honor Roll and awarded a teaching excellence award from the Department of Religious Studies. Courses she teaches include Introduction to Buddhism, Buddhism and Gender, Buddhist Auto/biography, Tibetan Religion and Culture, Theory and Methods in the Study of Religion, South Asian Goddess Traditions, and Religion, Sexuality, and Celibacy.

Ken Seeskin

Best Overall Faculty Member, Religion

Kenneth Seeskin is the Philip M. and Ethel Klutznick Professor of Jewish Civilization. After receiving his PhD from Yale in 1972, he joined the faculty of Northwestern and has remained here ever since. He served for nearly 20 years as Chair of the Philosophy Department and with the start of the 2010–11 academic year, will serve as Chair of Religious Studies. He has won several teaching awards since coming to Northwestern and served as the Charles Deering McCormick Professor of Teaching Excellence from 1995–1998.

Seeskin is best known for his interpretation and defense of the rationalist tradition in Jewish Philosophy, including such figures as Maimonides, Spinoza, and Cohen. His most recent books include the Cambridge Companion to Maimonides (CUP, 2005), Maimonides on the Origin of the World (CUP, 2005), Autonomy in Jewish Philosophy (CUP, 2001), and Searching for a Distant God: The Legacy of Maimonides (OUP, 2000). The latter was awarded the Koret

Jewish Book Award in 2001. The Cambridge Guide to Jewish History, Religion, and Culture, co-edited with Judith Baskin, is scheduled to appear in 2010.

His published articles dealing with religious themes are wide-ranging including studies on the Book of Job, the meaning of the Holocaust, negative theology, the concept of holiness, and recent studies on the role of reason in Jewish philosophy, the role of miracles in Jewish philosophy, and the Greek background to Jewish Philosophy. He is currently working on a book on messianism that will take up issues pertaining to evil, rationality, and the philosophy of history.

NOTRE DAME DE NAMUR UNIVERSITY

Judy Buller

Best Overall Faculty Member, Communication

The Bachelor of Arts in Communication program provides students with hands-on experiences in journalism and communication as well as an understanding of the social and cultural impact of media. NDNU students are well prepared for positions in a vast range of fields such as journalism, public relations, corporate communication, and marketing. A strong internship program provides opportunities for practical application and access to job opportunities. The Argonaut is our award-winning newspaper and website that gives students an opportunity to dive into the process of news production from writing and editing to layout and advertising sales. Students can earn a general degree in Communication or choose a concentration in Public Relations or Media Studies. Students in other majors are also able to obtain a minor in Communication.

NOVA SOUTHEASTERN UNIVERSITY

Allison Brimmer

Most Helpful to Students, Liberal Arts, Social Sciences, and Humanities

Assistant Professor Division of Humanities Farquhar College of Arts and Sciences

Ph.D. University of South Florida M.A. Iowa State University M.A. University of South Florida B.A. Simpson College

David McNaron

Best Teacher, Liberal Arts, Social Sciences, and Humanities

Andrea Shaw

Best Overall Faculty Member, Liberal Arts, Social Sciences, and Humanities

Charles Zelden

Best Researcher/Scholar, Liberal Arts, Social Sciences, and Humanities

Charles L. Zelden is Professor of History at Nova Southeastern University in Ft. Lauderdale, Florida where he teaches courses on legal/constitutional history, Twentieth Century American history and the judicial process. He is the author of Bush v Gore: Exposing the Hidden Crisis of American Democracy (2008), The Battle for the Black Ballot: Smith v Allwright and the Defeat of the Texas All-White Primary (2004) and Voting Rights on Trial (2002).

OAKWOOD UNIVERSITY

Audley Chambers

Best Researcher/Scholar, Music

OCEAN COUNTY COLLEGE

David Bordelon

Most Helpful to Students, English

"Takes time with students in and out of the classroom."

Henry Jackson

Best Researcher/Scholar,
English Best Teacher, English

"Classes are very upbeat and stimulating."

John McGinley

Best Teacher, Philosophy

"I had doc McGinley @ Univ. of Scranton @1976 . . . he has only gotten better!"

Heidi Sheridan

Best Overall Faculty Member, English

"She is very caring, patient and knowledgeable."

Frank Siracusa

Best Teacher, Liberal Arts, Social Sciences, and Humanities

OHIO STATE UNIVERSITY

Frederick Luis Aldama

Best Researcher/Scholar, Liberal Arts, Social Sciences, and Humanities
Best Teacher, Liberal Arts, Social Sciences, and Humanities
Most Helpful to Students, Liberal Arts, Social Sciences, and Humanities
Best Overall Faculty Member, Liberal Arts, Social Sciences, and Humanities

Rederick Luis Aldama is Arts and Humanities Distinguished Professor of English. He teaches Latino/a and Latin American post-colonial literature, film, and comics, as well as narrative theory and cognitive science approaches to culture. He is also founder and director of LASER: Latino and Latin American Space for Enrichment Research.

Franco Barchiesi

Best Researcher/Scholar, African-American Studies

Professor Franco Barchiesi's latest book is Precarious Liberation: Workers, the State, and Contested Social Citizenship in Postapartheid South Africa (State University of New York Press and University of KwaZulu-Natal Press, 2011). The book is the winner of the 2012 CLR James Award of the Working Class Studies Association for best Published Book for Academic or General Audiences in working-class studies. Barchiesi has also edited (with Tom Bramble) the book Rethinking the Labour Movement in the 'New South Africa' (London: Ashgate, 2003).

In the 2014–15 academic year, Franco Barchiesi will be at Harvard University as a Sheila Biddle Ford Foundation Fellow at the W.E.B. Du Bois Research Institute.

Professor Barchiesi is the recipient of the 2010 Distinguished Undergraduate Research Mentor Award at the Ohio State University and in 2008 he has received the Distinguished Contribution to Scholarship (Article) Award from the American Sociological Association, Labor and Labor Movements Section.

Carmel Buckley

Best Overall Faculty Member, Fine Arts
Best Teacher, Fine Arts
Most Helpful to Students, Fine Arts

Danille Christensen

Best Researcher/Scholar, English
Best Teacher, English

"Dr. Christensen works tirelessly to educate her students with innovative lesson plans, intellectually rigorous activities and assignments, and thoughtful pacing to maximize their energy and the time available. She is patient and generous with their needs."

Danille Elise Christensen holds an MA and PhD in Folklore from Indiana University's Department of Folklore and Ethnomusicology, where she learned to pay close attention to the patterned things people say, make, and do in the course of daily life. Her work is broadly concerned with the ways individuals shape vernacular expressive culture in order to influence and persuade others.

Tiffani Clyburn

Best Overall Faculty Member, Liberal Arts, Social Sciences, and Humanities

Lesley Ferris

Best Researcher/Scholar, Theater

Lesley Ferris is a director and scholar. Her research interests are focused on gender and performance, carnival, and the use of masks. Her books include Acting Women: Images of Women in Theatre (Macmillan,1990) and Crossing the Stage: Controversies on Cross Dressing (Routledge, 1993). She has published numerous essays the most recent being "Cooking Up the Self': Booby Baker and Blondell Cummings 'Do' the Kitchen" in Interfaces: Women/Autobiography/Performance/Image, edited by Sidonie Smith and Julia Watson (University of Michigan Press, 2002). In July 2002 she presented a paper entitled "Performing New York: Drag Kings Memorialize 9–11" at the XIV World Congress of International Federation of Theatre Research in Amsterdam. In March 2003 she presented "On the Streets of Notting Hill: Carnival as/is Theatre" at the Mid America Theatre Conference in Indianapolis. From 1979–1984 she was the Artistic Director of the York and Albany Theatre in London. She served as the head of the drama program at Middlesex University in the late 1980's where she helped co-found the M.A. in Performance Art. In 1982 she received a playwriting award from the Arts Council of Great Britain for her script Subjugation of the Dragon, which was produced twice in London and once at the University of Essex. She has chaired departments of theatre at Lousiana State University and University of Memphis. Lesley has directed over 50 productions both in Britain and the U.S.A. These include the award-winning Portrait of Dora (London and Memphis), Angels in America, Part Two: Perestroika (Memphis and Columbus), Wit (Central Ohio premiere), and Bertolt Brecht's Saint Joan of the Stockyards (Columbus). Most recently she directed Adrienne and Adam P. Kennedy's Sleep Deprivation Chamber. She has served as the Resident Director for the London Theatre Program in summer of 2000 and 2001. The Office of International Education awarded her an Outstanding Faculty Award in 2002 for her contributions to international education.

Areas of Expertise Gender and Performance Theatre History, Dramatic Literature and Criticism Carnival Education B.A. Mount Union College M.A. San Diego State University Ph.D. University of Minnesota

Cami Freeman

Best Researcher/Scholar, English

"Cami Freeman uses research quite rigorously in her own creative work which contributes to the relevance of her fiction and essays. In turn, she instructs her students in such a way that not only teaches them how best to research and incorporate scholarly articles into their essays, but also excites them about the fundamental process."

Evonne Halasek

Best Teacher, Liberal Arts, Social Sciences, and Humanities
Best Overall Faculty Member, Liberal Arts, Social Sciences, and Humanities

Ann Hamilton

Best Researcher/Scholar, Fine Arts

Sarah Johnston

Best Overall Faculty Member, Classics
Best Researcher/Scholar, Classics

Tim Leasure

Most Helpful to Students, Music

"Works long hours without complaint helping both students and faculty. His students are outstanding, reflecting Tim's outstanding teaching."

Timothy Leasure received his BM from Ohio University in 1990 and his MM from The Ohio State University in 1993. Following post-master's work at Ohio State, he joined the United States Navy Band in Washington, D.C. in 1995. While a member of the Navy Band, he was chosen as the 1997 National Tour soloist, and as a featured soloist in the band's concerts in Stockholm, Sweden.

He has served as principal trumpet of the Springfield (Ohio) Symphony during its 1994–95 season, and as second trumpet in the ProMusica Chamber Orchestra of Columbus from 1992 to 1995. He has also performed as principal trumpet in the Arlington (Virginia) Symphony, the Annapolis (Maryland) Symphony, the West Virginia Symphony, the Columbus Symphony Orchestra, the ProMusica Chamber Orchestra of Columbus, and the United States Navy Band. He is currently second trumpet in the ProMusica Chamber Orchestra.

Mr. Leasure can be heard in over 40 recordings by professional orchestral, band, and chamber music ensembles. He has performed extensively in the United States and Europe, and in such prestigious venues as Carnegie Hall, the White House, the Kennedy Center for the Performing Arts, La Chapelle Notre-Dame de Rocamadour, France, and the Esterhazy Palace in Eisenstadt, Austria.

He is also the author of two software applications for the Apple iPhone, Tromba Musica and Mahler Translations, and is busy working on his third app, a 12,000 word music dictionary. These applications are currently being sold worldwide in the iTunes App Store.

In 2001 Mr. Leasure was awarded the Distinguished Alumni Award from Ohio University in Athens, Ohio. He was also the recipient of the 2004 Distinguished Teaching Award at The Ohio State University School of Music and was recently made an honorary member of Phi Mu Alpha Sinfonia.

Annie McGreevy

Best Teacher, English

James Morganstern

Best Overall Faculty Member, Art History

Julia Nelson-Hawkins

Best Teacher, Classics
Most Helpful to Students, Classics

I am preparing a monograph on medical imagery in Augustan poetry, titled The Poetics of Medicine in Augustan Rome: Therapoetics after Actium. In this study I attempt to account for why there are so many prominent scenes of plague and, more generally, disease and healing in the long, narrative poems of the Augustan period after Actium and, in particular, why the final books of the Georgics, Aeneid, and Metamorphoses all deal with significant medical themes. I see the proliferation of medical imagery in such poetry of this period as more than just a nod to Lucretius and Thucydides, who both helped to establish the plague narrative as a literary topos. I take a more socio-political approach and argue that these images are connected to broader shifts in medicine and its increasing association with politics in the late Republic and, in particular, to Augustus' creation of a "healing persona" after the Civil Wars.

Jennifer Patton

Most Helpful to Students, English

"Jenny Patton's office hours overlap with mine, so I've had the privilege of seeing first-hand how incredibly generous and patient she is with her students on all levels: intellectually, socially, mentally, and emotionally. She bends over backwards to meet them exactly where they are so that they can get the absolute most out of her class. One result of this is that former students are often in her office, continuing to benefit from her expertise and generosity."

Jenny Patton incorporates creative writing and narrative theory practices into her first-year composition and second-year writing classes. Her stories and essays have been published in Brevity, Prism Review, River Teeth online, 751 Magazine and Ohio Writer, for which her entry earned first place in the Best of Ohio Writers Contest sponsored by Poets' & Writers' League. She was a 2011 Peter Taylor Nonfiction Fellow at Kenyon Review Writers Workshop and a 2012 scholarship winner at New York Summer Writers Institute at Skidmore College.

Laura Podalsky

Best Overall Faculty Member, Spanish
Best Researcher/Scholar, Spanish
Best Teacher, Spanish
Most Helpful to Students, Spanish

Professor Podalsky specializes in Latin American film and cultural studies. Her research involves the relationship between Latin American culture, politics, and socio-historical formations, and she has developed projects on urban culture, questions of affect and, more recently, youth cultures. Her first book, Specular

City: Transforming Culture, Consumption, and Space in Buenos Aires, 1955–1973 Temple University Press, 2004), was a substantive, interdisciplinary, cultural studies project. Through the analysis of new films, literature, magazines, advertising, architecture, and car culture, Specular City discusses the material and discursive transformation of the Argentine capital in relation to contemporary struggles between middle-class and working-class sectors in the aftermath of the first Peronist administration. The Politics of Affect and Emotion in Contemporary Latin American Cinema: Argentina, Brazil, Cuba, and Mexico (Palgrave Macmillan, 2011) is a more medium-specific study, exploring the evocation of affect and the deployment of emotion in contemporary film as a response to the adjudicating tendencies of postdictatorial politics and the hyper-rationalist discourses of neoliberalism. She is currently working on a monographic study of youth, modernity, and film, comparing Argentina and Mexico in the 1950s–1960s and in the 1990s–present.

Prof. Podalsky also has published essays on a wide variety of topics, including landscapes of masculinity in contemporary Mexican cinema, Mexican youth films, the work of Brazilian director Ana Carolina, telenovelas and globalization, cosmopolitanism in tango films, and pre-revolutionary Cuban cinema in journals such as New Cinemas, Studies of Hispanic Cinemas, Framework, Screen, Cinemais (Brazil), and Archivos de la Filmoteca (Spain).

She is an Associate Editor of the Journal of Spanish and Latin American Cinemas

Daniel Reff

Best Teacher, Humanities

Daniel Reff is an anthropologist and ethnohistorian with a particular interest in the comparative study of the role of missionaries in "the rise of Christianity" in various cultural-historical contexts. His research has been made possible by a Weatherhead Fellowship from the School of American Research as well as Fellowships and grants from the Research Division of the National Endowment for the Humanities.

His first book, Disease, Depopulation, and Culture in Northwestern New Spain, 1518–1764 (University of Utah Press, 1991), explores the dynamics of Jesuit and Indian relations in what is today northern Mexico and the American Southwest. In 1992 Professor Reff headed-up an NEH-funded, interdisciplinary team, which prepared the first critical, English-language edition of Andrés de Ribas monumental history of the Jesuit missions of colonial Mexico (History of the Triumphs of Our Holy Faith Amongst the Most Fierce and Barbarous Peoples of the New World (1645), University of Arizona Press, 1999). Professor Reff subsequently completed a comparative study of missionary texts from early medieval Europe and colonial Latin America (Plagues, Priests and Demons, Cambridge University Press, 2005). More recently, again with funding from NEH, he assembled a team of scholars that produced the first English-language edition of Luis Frois' "Striking Contrasts in the Customs of Europe and Japan" (The First European Description of Japan, 1585, Routledge, 2014).

Professor Reff's research also has appeared in journals such as American Anthropologist, Ethnohistory, Romance Philolology, Colonial Latin American Review, The Journal of Anthropological Research, and the Journal of the Southwest, which awarded him the Spicer Prize for his 1996 article on Cabeza de Vaca's eight-year sojourn across North America ("Text and Context, Cures, Miracles and Fear in the Relación of Alvar Núñez Cabeza de Vaca")

Joy Reilly

Most Helpful to Students, Liberal Arts, Social Sciences, and Humanities

Joy Reilly, born in Dublin, Eire, and raised in England, joined the Department of Theatre faculty in 1985, after earning a doctoral degree from the department in 1984. She is a director, actress, historian and writer, who worked as a journalist and copywriter in the international corporate arena in London, and Frankfurt, Germany, before coming to America. Professor Reilly has written several introductory theatre texts, most recently Introducting Theatre 3rd ed (FORBES 1998) with M. Scott Phillips, and The Introduction to Theatre Workbook 4th ed (FORBES 1998). She has been honored with the 1995 Distinguished Teaching Award at OSU for her leadership of the Introduction to Theatre course, the largest program of its kind in the country. She is currently serving as Chair of Undergraduate Studies and conducted a tour of OSU Honors Undergraduates to London in 1997.

Professor Reilly also advises masters and doctoral students in history, literature and criticism, with a particular interest in the history of performance, and in the contributions of 19th Century women. She has taught seminars in acting and directing theory, contemporary literature, history of acting, creative writing for solo performance and, in Autumn 1998, is offering a seminar in 20th Century Irish Dramatic Literature and its Celtic influences.

Among the plays she has directed for OSU are Once A Catholic, Female Transport, Twelfth Night, Romeo and Juliet, Ourselves Alone, Our Country's Good, Mad Forest, and The Complete Works Of William Shakespeare (Abridged). She has performed at OSU in Lion In Winter, Private Lives, A Taste of Honey, Richard III, Auntie Mame, The Three Sisters, The Seagull, La Ronde, A Pagan Place, and Dancing at Lughnasa.

Professor Reilly currently serves as chair of Senior Adult Theatre for ATHE's focus group on Senior Theatre and helped to inaugurate the bi-annual national Senior Theatre USA convention for older actors. The next convention will be co-hosted by the Walt Disney organization in Orlando in January 1999. She is active nationally and internationally as a consultant for Theatre and Aging and will host the first international conference of older actors at OSU in August 2001. She founded Grandparents Living Theatre, (an internationally known company based in Columbus) in 1985 and served as its Artistic Director through March of 1998. Professor Reilly has conceived, written and directed eight original playscripts out of oral history including the Emmy-award-winning I Was Young, Now I'm Wonderful, and A Picket Fence, Two Kids, and a Dog Named Spot?

Professor Reilly is currently pursuing research in Irish and Celtic studies and in the creation of original scripts for solo work. She has also served as Theatre Critic for WOSU (PBS) Radio in Columbus since 1978, and provides opportunities for graduate students to write and broadcast radio reviews.

Robin Rice

Best Teacher, Liberal Arts, Social Sciences, and Humanities

Baritone Robin Rice, associate professor of voice, received his BM degree from Baylor University and his MM and DMA degrees from the University of Cincinnati College-Conservatory of Music. A specialist in the French repertoire, Rice became an honored participant under Elly Ameling, Gerard Souzay, and Dalton Baldwin at the Cleveland Art Song Festival. Dr. Rice also won the Carolyn Brice Award in the National Association of Teachers of Singing Artist Award Competition. He has worked with many luminaries such as Maestra Alberta Masiello, Theodore Uppmann, Barbara Honn, Joyce Farwell, Tom Jaber and Edward Bak, to name a few.

Rice's operatic repertoire spans such diverse roles as Mozart's Count Almaviva in The Marriage of Figaro and the joyous Papageno in The Magic Flute, to Lee Hoiby's Vernon in Summer and Smoke. He has also performed the world premieres of Reynold's The Hollow Men, Rytel's Blake Songs, and Dmitriev's R.O.S. In 1996, he was a recitalist with Michael Pilkington for the national convention of the Association of Teachers of Singing of the United Kingdom in Bournemouth, England. He also appeared as a soloist with the Northern Czech Philharmonic Orchestra in Teplice, Czech Republic with John Apeitos of the Athens, Greece Radio Symphony conducting.

Rice is a highly acclaimed and active international teacher. He has taught master classes internationally and in the United States. Venues include: Instituto Superior de Arte del Teatro Colón (Buenos Aires, Argentina), Castellina in Chianti Vocal Workshop (Tuscany, Italy), Cullasaja Vocal Workshop (North Carolina), University of Colorado at Boulder, Rice University, Saint Mary's College of Notre Dame, and Baylor University; Tung-hay University, Taipei Municipal University of Education and National Sun, Yet-sen University in Taiwan, R.O.C. Rice has been a guest vocal coach for Utah's Young Artist Apprentice Program, the San Diego Opera Ensemble and Florida Grand Opera's Young Artist Apprentice Programs. Dr. Rice has also taught at the Amalfi Coast Music Program in Vietri sul Mare, Italy and in 2009, he was a master teacher at the International Asian Opera Workshop in Taipei, Taiwan R.O.C. In July 2011, he was a master teacher at the International Asian Opera Project in Kunming, China.

Rice has mentored and worked with many remarkable singers of varied careers - artists like Taylor Stayton, Kerri Marcinko, Michele Angelini, Maureen McKay, David Hibbard, Robert Kerr and Daniel Neer, among others. Rice's students are performing on major stages in the United States, Europe and Asia which include the Metropolitan Opera, New York City Opera, Houston Grand Opera, San Diego Opera, Washington National Opera, Los Angeles Opera, Santa Fe Opera, The Kennedy Center, Carnegie Hall, Alice Tully Hall, Teatro Dell'Opera Roma, Deutsche Opera am Rhein, Komische Oper Berlin, Théatre de la Monnaie (Brussels), Teatro Comunale (Bologna), Varna Symphony Orchestra (Bulgaria), Royal Liverpool Philharmonic, Severoceska Filharmonie (Teplice, Czech Republic), and Festival Lyrique en Mer (France). Many of his students have also won many international vocal competitions. In summer 2012 he was part of the faculty of the Up North Vocal Institute in Boyne City, Michigan.

Agi Risko

Best Teacher, Slavic Languages

PhD (The Ohio State University, 1998) MA (The Ohio State University, 1991)
Areas of Expertise Swedish language and culture

Vincent Roscigno

Best Researcher/Scholar, Liberal Arts, Social Sciences, and Humanities

Social Stratification, Work, Education, and Culture. Current research includes historical and contemporary research regarding inequality, culture, and legitimation, using historical, qualitative and quantitative methods; contemporary workplace discrimination surrounding race, sex and age; and the dynamics of workplace bullying and abuse; and re-examining theories of power and their relevance in the contemporary world.

I was intrigued from my very first classes in the social and behavioral sciences as an undergraduate with ways in which inequality is reproduced and the interplay of structure and agency in that process. During my graduate training and as an assistant professor, I followed through on this interest by engaging in several projects pertaining to inequality in labor markets, politics and education. I also diversified methodologically, realizing that many aspects of the types of inequality I was interested in often lay under the surface of conventional research designs and analyses. Social inequality and explicating how it is enacted, reproduced, and legitimated by those in power remain core to my sociological passion, as reflected in recent and current multi-methodological projects on workplace discrimination and incivility, labor unrest and mobilization, the subordination of Native Americans historically, and educational inequalities and the ways they are recreated across generations and by institutional processes. Despite seemingly unique topics, the core sociological foci remain the same: In what ways do structure and action intersect to create such inequality? What role does culture and discourse play in the legitimating and making such inequalities seem acceptable; and, how do subordinated groups challenge, resist, or rectify the inequalities they face?

Jennifer Schlueter

Best Overall Faculty Member, Theater
Best Teacher, Theater
Most Helpful to Students, Theater

Andrea Sims

Best Overall Faculty Member, Slavic Languages
Best Researcher/Scholar, Slavic Languages
Most Helpful to Students, Slavic Languages

Jennifer Suchland

Best Researcher/Scholar, Liberal Arts, Social Sciences, and Humanities

Dr. Suchland works with an interdisciplinary approach to the study of rights, law and political discourses. She is particularly interested in how rights-based categories emerge and evolve and what is at stake in those articulations. Her current project is a genealogy of global political discourses on women's human rights looking specifically at sex trafficking and sexual harassment. She is interested in how rights norms developed at the global level and comparatively, looking specifically at Soviet and post-Soviet Russia.

Key Research Concepts: Genealogy; Sex Trafficking; Human Rights; Law; Gender; Geopolitics; Cold War; Feminist Methodologies; Postsocialist Cultural Studies; Theorizing the Global and Transnational.

Sarah Van Beurden

Most Helpful to Students, African-American Studies

Sarah Van Beurden joined the department in 2009. She is also affiliated with the departments of History and Art History. A Belgian national, she obtained her Ph.D. in History at the University of Pennsylvania. Dr. Van Beurden has been the recipient of, among others, the Annenberg fellowship at the University of Pennsylvania, a Rockefeller Archive research grant, a postdoctoral research fellowship at the Institute of Historical Studies University of Texas in Austin and a research fellowship at the Centre for Global Cooperation Research at the University of Duisburg-Essen (Germany.) She teaches African history, African art history and transnational cultural history. Her research and publications focus on colonial and postcolonial cultural politics, the history of the Congo and transnational cultural history. She is currently completing Authentically African: Arts and the Transnational Politics of Congolese Culture, a book that explores the role of arts and cultural politics in the Congo from the late colonial until the postcolonial era.

OHIO UNIVERSITY

Frederick Lewis

Best Teacher, Communication
Most Helpful to Students, Communication

Mary Manusos

Best Overall Faculty Member, Fine Arts

Mary Manusos has exhibited her work in over two hundred and fifty exhibitions in the United States and abroad including exhibitions at Franklin Furnace in New York, the Museum of Modern Art in New York, The Tweed Museum, and The San Francisco Museum of Modern Art. She has had one person exhibitions at Bajas Artes, Mexico; Casa de Cultura, Mexico and Museo de Architectura in Quito, Ecuador. Her work has been included in many juried exhibitions including the 2nd Kochi International Triennial Exhibition of Prints in Japan; eleven American Embassies abroad; as well as in Seoul, Korea and Lubiana. She continues to work in a manner reflective of the images she created in D'Art Objects, a collaboration with John Chakeres, published by Nuance Press in 1979.

A Professor at Ohio University for forty years, she is currently teaching during the Fall quarter in the Printmaking Department. Mary Manusos received her MFA from University of Wisconsin and her BA from San Diego State College.

Kasia Marciniak

Best Researcher/Scholar, English

"Works with great integrity."

2013–2016 Presidential Teacher; Co-Editor of Global Cinema Transnational Feminist Media Studies, Cinema and Media of Exile and Displacement, Visual Culture, Ethnic and Racial Studies, Migration Studies, Discourses of Nationalism and Globalization, Women and Literature, Postsocialist Cultures, Critical Pedagogy

Drew McDaniel

Best Researcher/Scholar, Media Arts

Dr. McDaniel is Professor and Director of the School of Media Arts & Studies at Ohio University. Previously, he held the posts of Director of the Center for International Studies, Director of Communication and Development Studies, and Director of Southeast Asian Studies. Prior to joining Ohio University, he worked as a broadcast journalist in the states of Washington and Colorado. He has held Visiting Professor positions at Nanyang Technological University in Singapore and Universiti Teknologi MARA in Malaysia. In 1990 he held a Fulbright Regional Research Fellowship in Southeast Asia, and more recently he received an appointment as Distinguished Scholar in Residence at the University of Queensland in Australia. He was twice named Ohio University's outstanding graduate faculty member.

For many years, Professor McDaniel served as Staff Consultant at the Asia-Pacific Institute for Broadcasting Development (AIBD) in Kuala Lumpur, Malaysia, a UN-chartered agency that provides research and training assistance across Asia. He has also organized training courses for the Asia-Pacific Broadcasting Union (ABU) and the Voice of America, as well as numerous other radio and television organizations.

Dr. McDaniel has written extensively on media in Asia and on technology in communication. Among his book publications are Electronic Tigers of Southeast Asia: The Politics of Media, Technology, and National Development; Broadcasting in the Malay World; Fundamentals of Audio Production; and Fundamentals of Communication Electronics.

OKLAHOMA CITY UNIVERSITY

Edward Knight

Most Helpful to Students, Music

Dr. Edward Knight is composer-in-residence and director of music composition at Oklahoma City University.

He was named Oklahoma's 2002 Musician of the Year and awarded fellowships to the Yaddo Arts Colony in Saratoga Springs, New York, in 2001 and 2002, and to the MacDowell Colony in Peterborough, New Hampshire in 2003. Since his arrival at OCU in 1997, he has premiered more than 30 compositions, including works for orchestra, symphonic band, percussion ensemble, piano trio; a clarinet sonata; cabaret songs; two fanfares; and two original full-length romantic musical comedies.

Knight has a background in jazz, classical, and theatrical composition. He toured Europe as a jazz trumpeter and earned his DMA from University of Texas at Austin. He studied privately with John Corigliano and was the first American to win the Sir Arthur Bliss Memorial award, for outstanding postgraduate composer at London's Royal College of Music.

Knight's orchestral works have been premiered by the New York Philharmonic, the Civic Orchestra of Chicago, and the Los Angeles Philharmonic Institute. John von Rhein of the Chicago Tribune called Knight's music "visceral in its excitement . . . The score (of the orchestra piece BIG SHOULDERS) seizes and holds your imagination and makes you want to hear it again."

Bernard Holland of The New York Times called Knight a "fresh, original voice, a young man to be encouraged. He hears the sounds of his time in his own way and is especially good at transferring bright, hard synthesized colors to the upper registers of acoustical instruments . . . Brightness and silence are major tools in this piece (CONCERTO FOR CLARINET) and Knight uses them with an inventive sense of humor."

Timothy Mangan of the Los Angeles Times cited "the composer's canny combination of steady meter with atonal lyricism, a waltz-like lilt with expressionist angst." After its Hollywood Bowl premiere, Mangan called Knight's orchestra piece GRANITE ISLAND "tightly unified, suave and sinister, confidently orchestrated."

Wayne Lee Gay of Knight-Ridder News Service called Knight's song cycle LIFE IS FINE "inventive and melodic" with "the sound of a work that may well be a major addition to the American vocal-orchestral repertoire."

Recent awards include fellowships to the arts colonies at Yaddo and MacDowell; the 2001 Oklahoma Music Teachers' Association Composer of the Year; 2002 Musician of the Year from the Oklahoma Federation of Music Clubs; ASCAP's Rudolf Nissim Award; first prize in the National Orchestra Association's New Music Orchestral Project; and special commendations from the Bergen Festival and Vienna Modern Masters. In addition, he has won ASCAP Standard Awards for each of the past 14 years.

His compositions are published exclusively by Subito Music, distributed nationally by Theodore Presser, and internationally by Boosey & Hawkes.

Matthew Mailman

Best Overall Faculty Member, Music
Best Researcher/Scholar, Music
Best Teacher, Music
Most Helpful to Students, Music
Best Researcher/Scholar, Liberal Arts, Social Sciences, and Humanities
Best Teacher, Liberal Arts, Social Sciences, and Humanities
Most Helpful to Students, Liberal Arts, Social Sciences, and Humanities
Best Overall Faculty Member, Liberal Arts, Social Sciences, and Humanities

David Steffens

Best Overall Faculty Member, Music
Best Teacher, Music

Dr. David Steffens is principal percussionist of the Oklahoma City Philharmonic. He is a recording artist, clinician, educator and champion of new music who has directed the percussion program at the Bass School since 1997.

As a founding member of the Robert Hohner Percussion Ensemble, Steffens is featured on four internationally released CDs on the DMP recording label – Different Strokes, The Gamut, The World Music Tour, and Lift-Off – and on Albany Records' The Percussion Music of David Maslanka. A Percussive Notes review of Different Strokes declared: "This might well be the most artistic balance and blend of percussion music ever recorded."

In 2005, Steffens recorded Lost Luggage – a five-movement work for solo marimba he commissioned from composer Edward Knight – for the CD Where The Sunsets Bleed, also released by Albany.

Steffens has performed with the Midland Symphony, West Shore Symphony Orchestra, Greater Lansing Symphony, Lawton Philharmonic Orchestra, and Schlossfestspiele in Heidelberg, Germany. He is a frequent guest artist and has presented master classes and percussion clinics on campuses nationwide.

Under Steffens' leadership, Oklahoma City University's Percussion Ensemble has performed works ranging from a Red Norvo tribute in Austin to a regional premiere of Stravinsky's Les Noces to a Maslanka work at the national SCI convention. He has built collaborative partnerships with local artists. In 2004, the Percussion Ensemble appeared with Canterbury Choral Society and Ballet Oklahoma in Tango Take Three, at Oklahoma City's historic Civic Center. Rick Rogers of The Daily Oklahoman singled out the percussionists for special praise: "An unexpected surprise was the OCU Percussion Ensemble, an accomplished group of instrumentalists ably directed by David Steffens. Many in the audience were no doubt unfamiliar with the myriad sounds capable of such a group, but multiple mallet keyboards produce impressive sonorities that turned their featured number into a showstopper." In 2005, the Percussion Ensemble reprised its collaboration with Canterbury to power the "World Harmonies" concert at the Civic Center.

He received his Performer's Certificate and his DMA from Eastman School of Music. He is a faculty member at the Oklahoma Arts Institute at Quartz Mountain and has taught at Michigan State, Central Michigan University, and the Eastman School of Music. He is president of the Oklahoma Chapter of the Percussive Arts Society.

He is an artist/clinician for Yamaha Percussion, Black Swamp Percussion, Bosphorus Cymbals, and Vic Firth Mallets.

Bosphorus Cymbals http://www.bosphoruscymbals.com.

Vic Firth Mallets http://www.vicfirth.com/

Black Swamp Percussion http://www.blackswamp.com

Yamaha Percussion http://usa.yamaha.com/products/musical-instruments/percussions/

Yamaha YM-5100A marimba Yamaha YG-2500 glockenspiel Yamaha 9000 series timpani Yamaha Live Custom Drums

OLD DOMINION UNIVERSITY

Megan McKittrick

Best Teacher, English

Konrad Winters

Most Helpful to Students, Theater

Mr. Winters is the resident designer for the Theatre Arts Program at ODU and teaches design and technical production courses in theatre and film for the Department of Communication & Theatre Arts. Mr. Winters is a member of the United States Institute for Theatre Technology (www.usitt.org).

Robert Wojtowicz

Best Researcher/Scholar, Fine Arts
Best Teacher, Fine Arts
Most Helpful to Students, Fine Arts
Best Overall Faculty Member, Fine Arts

Robert Wojtowicz is Professor of Art History and Associate Dean for Research and Graduate Studies in the College of Arts and Letters at Old Dominion University. He is the literary executor of the Estate of Lewis and Sophia Mumford and the author of Lewis Mumford and American Modernism: Eutopian Theories for Architecture and Urban Planning (1996); the editor of Sidewalk Critic: Lewis Mumford"'s Writings on New York (1998) and Mumford on Modern Art in the 1930s (2007); and the co-editor (with Bruce Brooks Pfeiffer) of Frank Lloyd Wright and Lewis Mumford: Thirty Years of Correspondence (2001).

ONONDAGA COMMUNITY COLLEGE

Mike Kaminski

Best Overall Faculty Member, Communication
Best Overall Faculty Member, Liberal Arts, Social Sciences, and Humanities

ORAL ROBERTS UNIVERSITY

Keith Gogan

Most Helpful to Students, English

Keith Gogan believes he is able to relate well to his students since he was once a student here himself. Although he has seen a lot of changes, he believes ORU has only gotten better. Because of ORU's unapologetic synthesis of academics and Christianity and whole-person educational philosophy, he believes ORU is unique among Christian educational institutions, as it addresses the totality of the human being, not simply the spiritual.

Gogan began his work at ORU in 1987 as supervisor of its new tutoring center, overseeing its operation and writing instructional programs for English composition students' use. Later, he taught English composition classes as an adjunct. In 1997, he attained full-time Instructor status, and in 2008, he advanced to Assistant Professor. He has been employed by ORU for 27 years.

"I have to admit that I love my work," he says. "I can say with all honesty that, when I head to work each day, I am excited to see what will happen with my students." Because his work has meaning, he would rather teach than do anything else, even if it paid five times more than his current salary!

Lori Kanitz

Best Researcher/Scholar, English
Best Overall Faculty Member, English
Best Teacher, English
Best Researcher/Scholar, Liberal Arts, Social Sciences, and Humanities
Best Teacher, Liberal Arts, Social Sciences, and Humanities
Best Overall Faculty Member, Liberal Arts, Social Sciences, and Humanities

Lori Kanitz loves hiking and being out in nature. She hikes all around Tulsa with her black lab Emma and has made several trips to the Colorado Rockies for mountain hiking, twice tackling "fourteeners."

She first got into hiking while in England for her graduate work at the University of Durham. Joining a hiking group, she tromped over much of the English countryside discovering the quiet pleasures of gardens, hedgerows and historic castles.

Lori spent her early years in Salina, Kansas, until her family moved to Tulsa when she was in junior high school. She earned her B.A. with highest honors from Oklahoma University in Norman. After teaching English for three years in an Oklahoma City high school, she went to the University of Durham in England on full scholarship for her M.A. in Literature and Theology. She is currently pursuing a Ph.D. in Literature and Theology at St. Andrews University in Scotland.

After completing her degree, Lori returned to Tulsa and applied for a teaching position at ORU. The rest is history. She has taught in the College of Arts and Cultural Studies for the past 16 years. "The students are the most rewarding part of my job," Lori remarked. "It is wonderful to sense God's calling on their lives and the purpose with which they approach their coursework. It is inspiring to see them connect what they are doing in class to more than just their career, but to their spiritual life as well."

Kay Meyers

Most Helpful to Students, Liberal Arts, Social Sciences, and Humanities

Dr. Kay Meyers has been at Oral Roberts University for more than twenty years. As a Professor of English, she teaches a number of literature classes, including Literary Criticism, American Literature, and the Modern Period. She serves as the sponsor for the English Honors Society, Sigma Tau Delta, and as the sponsor for Promethia, ORU's literary magazine. In 2001, she co-founded the Honors Program with Dr. John Korstad, where she serves as Associate Director. With this, she also co-teaches Artistic Expression, an honors interdisciplinary course, and in collaboration with Dr. Korstad, oversees the Freshmen Honors Seminar. She is also very involved in the Faculty Senate, and serves on the editorial board for the ORU academic journal.

For several years, Dr. Meyers has been working closely with a former student to transcribe and publish diaries belonging to a Civil War nurse. She also enjoys taking current students on department trips and to conferences. This allows her to get to spend time with her students and get to know them better. She applies her faith to her interactions with her students and tries to be understanding of any problems or troubles. She thoroughly enjoys being a part of her students' lives.

Dr. Meyers has also been a participant in the Let's Talk About It, Oklahoma! program since 1989. This book discussion program, which is sponsored by the Oklahoma Humanities Council and funded by the National Endowment for the Humanities, provides an opportunity for Dr. Meyers to not only serve her community, but her entire state by allowing her to travel around Oklahoma a few times a year for humanities presentations on various pieces of literature.

Dr. Meyers received her B.A. and M.A. in English from the University of Oklahoma, and her Ph.D. in English from the University of Tulsa. She has been a tenured professor at ORU since 1997. - See more at: http://web.oru.edu/academics/faculty_profile.php?id=101&k=#sthash.bVgSSnEF.dpuf

Paul Vickery

Best Researcher/Scholar, History

ORANGE COAST COLLEGE

Kat Carroll

Most Helpful to Students, Communication

OREGON INSTITUTE OF TECHNOLOGY

Robin Schwartz

Most Helpful to Students, Writing

"Robin Schwartz has always been student-oriented. While he certainly has classroom standards for writing, he goes above and beyond what is expected in helping students achieve the writing level needed to move to the next sequence."

OREGON STATE UNIVERSITY

David Robinson

Best Researcher/Scholar, English

OTTERBEIN UNIVERSITY

Nicholas Hill

Best Overall Faculty Member, Fine Arts
Best Researcher/Scholar, Fine Arts
Best Teacher, Fine Arts
Most Helpful to Students, Fine Arts
Best Researcher/Scholar, Fine Arts
Best Teacher, Fine Arts
Most Helpful to Students, Fine Arts
Best Overall Faculty Member, Fine Arts

"Sets high standards and maintains them himself and with students, garners student and colleague respect, compassionate and considerate in personal interactions with all students."

Nicholas Hill is Professor of Art at Otterbein University where he teaches printmaking and drawing. Widely exhibited, Nicholas is represented by Printworks Gallery in Chicago. He has exhibited his work in over 200 juried and invitational exhibitions in the United States and abroad. His work is represented in such public collections as the Library of Congress, the New-York Historical Society, the United States Department of State and the Nelson-Atkins Museum of Art. Nicholas has received grants from the Ohio Arts Council, the Greater Columbus Arts Council, the New York State Council on the Arts, the Kansas Arts Commission and Artists Space- New York City. He has been granted two international residencies in Dresden, Germany by OAC and GCAC. Additional residencies awarded include the MacDowell Colony, New Hampshire; the Virginia Center for the Visual Arts, and four residencies at the Ragdale

Foundation in Illinois. In 2012, Nicholas co-curated an exhibition of work by contemporary American and Indian printmakers in New Delhi. The exhibition toured India. His work has been included in group exhibitions in Chile, Germany and France in the past year. In 2013, he had a solo exhibition at Redlands University in California. In 2014, Nicholas presented solo exhibitions of his work in Santiago, Chile and Dresden, Germany. He also had an artist's residency at Taller 99 printmaking studio in Santiago, Chile.

OWENS COMMUNITY COLLEGE: TOLEDO

Tracy Darr

Most Helpful to Students, English

Denise Grupp-Verbon

Best Teacher, Music

Carey Marten

Best Overall Faculty Member, Graphic Arts
Best Teacher, Graphic Arts
Best Overall Faculty Member, Fine Arts

"Carey is the hardest working CRT teacher in the department. She takes responsibility for many extras that happen in the CRT program such as the annual CRT student exhibit, open lab assignments, and special programs. She is also a quality teacher that promotes excellence in the classroom."

Carey Marten has been teaching in the Commercial Art program since August 2003. She was hired full time in August 2006.

Marten graduated from Bowling Green State University with a Master of Education majoring in Career and Technology Education in August of 2004. She also is a 2002 graduate from BGSU with a Bachelor of Science in Technology, specializing in Visual Communication Technology. Like many of her students, she also started her education at a two-year community college, where she received an Associate of Applied Science degree specializing in Graphics from Terra Community College.

Marten does the majority of her freelance work for her hometown of Bellevue, Ohio, where she designs many different types of materials from print to multimedia. She designed the 2010 Northwest Ohio Aids Walk logo that was held at International Parkway in Toledo. Marten's freelance design work is an important part of how she keeps current with design trends and programs.

Eric Wallack

Best Researcher/Scholar, Music

PACE UNIVERSITY (ALL)

Eugene Richie

Best Overall Faculty Member, Liberal Arts, Social Sciences, and Humanities

"Eugene Richie understands the craft not only of creative writing but also of developing writers. His dedication to his students and his colleagues is phenomenal, bringing both a great intelligence and a kind heart–the perfect combination. He brings great wisdom and calm to the English Department–the intelligent calm in the storm."

Eugene Richie has published four books of poems, including Island Light (Painted Leaf, 1998) and Psyche and Amor (Factory Hollow, 2009) with Rosanne Wasserman, and a fifth collection, Curves and Constant Stars, is currently in progress. His translations include stories by Venezuelan author Matilde Daviu and Colombian author Jaime Manrique's My Night with Federico García Lorca (University of Wisconsin Press, 2003), a Lambda Literary Award finalist. He has edited John Ashbery's Selected Prose (University of Michigan Press; Carcanet, 2004) and, with Wasserman and Olivier Brossard, five volumes of John Ashbery's translations from the French—most recently, Pierre Martory's The Landscapist (2008), a London Poetry Book Society Recommended Translation and a National Book Critics Circle Award poetry finalist, and Ashbery's Collected French Translations (Farrar, Straus and Giroux, 2014), a London Poetry Book Society Recommended Translation and a finalist for the Poetry Foundation Pegasus Award for Poetry Criticism.

PARADISE VALLEY COMMUNITY COLLEGE

Marilyn Cristiano

Most Helpful to Students, Liberal Arts, Social Sciences, and Humanities

Brett Reed

Best Overall Faculty Member, Music
Best Researcher/Scholar, Music
Best Teacher, Music
Most Helpful to Students, Music

Brett Reed (composer, percussion) is a performer and composer specializing in contemporary and improvised music. He regularly performs as a solo percussionist, as a member of several ensembles, including Crossing 32nd Street, Skin & Bones, a percussion duo he co-founded with Terry Longshore, and as a jazz vibraphonist in solo and ensemble settings. He has performed at numerous festivals, including the Bang On A Can Marathon in New York City and the Los Angeles Philharmonic's Green Umbrella Series. Reed has also produced and performed on numerous recordings including Skin & Bones critically acclaimed first release Boom and their upcoming release Mixmaster. Other recording credits include the music of composers Anthony Davis, Peter Garland, and Iannis Xenakis. In addition, Reed has had performances of his own compositions "that's what it is, grape!" for septet, "Five Loops", for marimba and electronics, and "Translucent Divide" for solo vibraphone. Brett holds a Doctor of Musical Arts degree from the University of California, San Diego where he studied with renowned percussionist Steven Schick. Dr. Reed is currently the director of the commercial music program and the percussion program at Paradise Valley Community College.

Lois Roma-Deeley

Best Overall Faculty Member, Liberal Arts, Social Sciences, and Humanities
Best Teacher, Liberal Arts, Social Sciences, and Humanities

David Rubi

Best Researcher/Scholar, Liberal Arts, Social Sciences, and Humanities

PARK UNIVERSITY

Mark Noe

Most Helpful to Students, Communication

"He is always very nice and helpful. Seeks out answers when need be."

PARKLAND COLLEGE

George Johnston

Best Teacher, Speech

PELLISSIPPI STATE COMMUNITY COLLEGE

Luanne Dagley

Best Teacher, Liberal Arts, Social Sciences, and Humanities

"Ana Hansen is an excellent teacher. She has exciting activities for her students."

Roberto Jimenez

Best Teacher, Liberal Arts, Social Sciences, and Humanities
Most Helpful to Students, Liberal Arts, Social Sciences, and Humanities

Best Overall Faculty Member, Liberal Arts, Social Sciences, and Humanities

"He teaches in spirit"

PEPPERDINE UNIVERSITY

Graciela Boruszko

Best Researcher/Scholar, Internatl Studies & Languages

"Outstanding overall production and international scope."

Dr. Graciela Susana Boruszko is an associate professor of comparative literature and culture in the International Studies and Languages at Pepperdine University. She received her Ph.D.in French Philology from the Universidad Nacional de Educación a la Distancia in Madrid, Spain, summa cum laude, a DEA in French Philology from the same university, cum laude, and a second DEA, cum laude, and a master's degree (Maîtrise) in Modern Languages (Lettres et Langues), cum laude, from the Université de Bourgogne, Dijon, France, a B.A. (Licence et DEUG) in Modern Languages (Lettres et Langues), cum laude, from the Université de Bourgogne, Dijon, France.

Her areas of research include comparative literature, linguistics, and culture, world literature, Hispanic and French studies, literary theory, literary socio-politics, identity in literature, globalization, history and literature, hospitality, migration, ethnicity, cultural identity in literature, linguistic, cognitive and computational approaches to literary studies. The languages of her research are: Spanish, French, English, Galego, Català, Portuguese, Italian, and Latin. She conducted research in the United States, Europe and Central and South America as well as participated in international colloquia.

Dr. Boruszko received the medal for exceptional academic leadership at Pepperdine University in 2013 and a Values Award in 2005. She was the beneficiary of several recurrent research grants since 2007. Along with an active membership in many associations like ICLA, ACLA, SCLA, NEMLA, MLA, ATINER, ADFL, NAFSA, she holds several research leadership positions: Director of the International Research Group: Identidades en las literaturas de España (ILE) and she is an academic member of the Literature Research Unit in the Athens Institute for Education and Research (ATINER). She completed interdisciplinary and collaborative research in music and literature and she also served as editor on several occasions.

Lila Carlsen

Best Teacher, Languages

Gregory Daum

Most Helpful to Students, Communication

"Greg has been here a number of years and has always been seen as a mentor in both a spiritual sense and a communicative sense to all of the students in his class. Though a tough grader, Greg has been seen as an educator that tries to set his students up for success in life through the material that they learn in his class."

April Marshall

Most Helpful to Students, Languages

Ginger Rosenkrans

Most Helpful to Students, Communication

PHILADELPHIA UNIVERSITY

Barbara Kimmelman

Best Researcher/Scholar, History

Tom Schrand

Best Overall Faculty Member, History
Best Teacher, History
Most Helpful to Students, History

PIERCE COLLEGE

Cheryl L Goss

Best Overall Faculty Member, English

Jack Sterk

Best Overall Faculty Member, Liberal Arts, Social Sciences, and Humanities

PIERCE COLLEGE (ALL)

Jere Knudtsen

Most Helpful to Students, Music

"Jere has spent his 40 year career at Pierce fighting for what is best for music students, especially in times when budget cuts threatened to eliminate necessary curriculum."

PIKES PEAK COMMUNITY COLLEGE

Robert Olson

Most Helpful to Students, Graphic Arts
Best Teacher, Fine Arts

Glenn Rohlfing

Best Overall Faculty Member, History
Best Teacher, History
Most Helpful to Students, History
Best Teacher, Liberal Arts, Social Sciences, and Humanities
Most Helpful to Students, Liberal Arts, Social Sciences, and Humanities
Best Overall Faculty Member, Liberal Arts, Social Sciences, and Humanities

Glenn Rohlfing served honorably in the United States Army for eight years, where he was a 31 Uniform—Signal Support System Specialist. Under the Montgomery G.I. Bill, Glenn selected Pikes Peak Community College because of its flexibility and smaller class sizes. With the mentoring of Wayne Artis and Dr. Karen Wagner-who are now his colleagues, Glenn earned his Associates Degree in History, and transferred to the University of Colorado at Colorado Springs. At UCCS, he immersed himself in his studies by spending countless hours in the Writing Center and in the study rooms at the Kramer Family Library, which he fondly calls his monastic cell. Under the guidance of Dr. Wagner, Dr. Richard Wunderli and Jan Myers, Glenn fell in love with medieval history, studying medieval Scandinavia and the Icelandic Sagas while an undergraduate. In 2002, he entered graduate school and studied Jesuits in New France, Material Culture in the High Middle Ages, and the British Raj. After graduating Magna Cum Laude with High Distinction and earning the Outstanding Graduate in History for 2005, Glenn returned to teach at Pikes Peak Community College as an adjunct for three years, after which he was hired full-time. As a Pikes Peak Community College alum, Glenn is excited to teach at his alma mater because the students here are resourceful and overcome many personal challenges in an effort to improve their lives and the lives of the loved ones around them. In 2010, Glenn became the Co-advisor to Phi Theta Kappa, an organization Glenn joined while a student. Phi Theta Kappa is an international honors society for two-year college students, focusing on four hallmarks: Scholarship, Service, Fellowship and Leadership. Mentoring students is one of Glenn's passions; the Colorado Region recognized him as the Distinguished Chapter Advisor in Colorado for 2011. In 2012, Glenn was selected as one of Phi Theta Kappa's Faculty Scholars for 2013; in the same year, he was also recognized as one of Phi Theta Kappa's Distinguished Chapter Advisor internationally. Additionally the Communications, Humanities, Technical Services Division (CHTS) selected him as Pikes Peak Community College's Faculty of the Year, and consequently, the University of Texas at Austin and the National Institute of Staff and Organization Development honored him as one of NISOD's Faculty of the Year. Glenn still has ambitions to obtain his PhD, either in Medieval History or in Community College Leadership. Learning is a life-long endeavor; everyday Pikes Peak Community College gives him the opportunity to learn from his students and colleagues.

Jenna Shearn

Best Overall Faculty Member, Graphic Arts
Best Researcher/Scholar, Graphic Arts
Best Researcher/Scholar, Fine Arts
Most Helpful to Students, Fine Arts
Best Overall Faculty Member, Fine Arts

Carrie Spencer

Best Researcher/Scholar, History
Best Researcher/Scholar, Liberal Arts, Social Sciences, and Humanities

Carrie spent her early years as an Air Force dependent and later as an Air Force wife traveling around the world and learning about varied cultures and their histories. She now enjoys teaching the World History courses here at PPCC and is especially passionate about Eastern studies. Her educational background includes a Master's Degree in History from the University of Colorado and a Master's Degree in Education also from the University of Colorado. Carrie loves the beautiful state of Colorado and enjoys hiking in the mountains and walking every morning. She is the mother of five sons . . . four of whom are also now in college. Carrie thoroughly enjoys her new role as an advisor to AA students and welcomes the chance to assist students in any way that she can.

Betsy Tuma

Best Teacher, Graphic Arts

PIMA COMMUNITY COLLEGE

Howard Allen

Best Teacher, Fine Arts
Most Helpful to Students, Fine Arts
Best Overall Faculty Member, Fine Arts

Kathleen Bradley

Best Researcher/Scholar, French

Elena Grajeda

Best Teacher, Spanish

Hassan Hijazi

Best Teacher, Arabic

"Hassan Hijazi is an accomplished teacher. Throughout the years, he has produced remarkable results. His passion for the subject he teaches is a great inspiration for his students. He cares for them and motivates them."

Tommie Miller

Best Teacher, Social Work

Stephen Romaniello

Best Overall Faculty Member, Digital Arts
Best Researcher/Scholar, Digital Arts
Best Teacher, Digital Arts
Most Helpful to Students, Digital Arts
Best Researcher/Scholar, Fine Arts

Yosei Sugawara

Most Helpful to Students, Japanese

After earning an undergraduate degree in law at Meiji Gakuin University in Tokyo, Yosei Sugawara worked for a number of years for a multinational company then came to the United States for his graduate education. He earned his first master's degree in the interdisciplinary study of intercultural communication at California State University at San Bernardino then moved to the University of Arizona where he received a second master's in Japanese linguistics and a doctorate in language and culture education.

While in Arizona, he taught Japanese language, business and culture classes at the University of Arizona, IBM, and Pima Community College where he was recognized with two awards for teaching excellence. In 2005, he received the Japan-U.S. Friendship Commission Prize from the Donald Keene Foundation at Columbia University for his translation of Yoko Ogawa's Hakase no Aishita Sushiki. He was subsequently recognized by the Japanese Consulate in Los Angeles for his contributions to the advancement of Japanese language and culture education in Arizona in founding both the Tucson Japanese Speech Contest and the Southern Arizona Association of Japanese Education, and in acting for many years as an interviewer for the JET Program.

In 2014 he and his family (four dogs, three cats and, so far at least, one wife) came to UNCG. In addition to his teaching, his current professional interests include conducting a follow-up study of his earlier work on Japanese expatriate adjustment, publishing his research on the teaching and socio-cultural functions of academic language, correlating student motivations for second language acquisition, course content and retention rates, and tracing the influence of Shingon Buddhism on the Tale of Genji – not necessarily in that order.

PITTSBURG STATE UNIVERSITY

Susan Carlson

Best Teacher, English
Most Helpful to Students, English

Kathy Degrave

Best Overall Faculty Member, English

PLYMOUTH STATE UNIVERSITY

Liz Ahl

Best Teacher, English

Liz Ahl is the author of Luck (Pecan Grove Press, 2010). Her first chapbook, A Thirst That's Partly Mine, won the 2008 Slapering Hol Press chapbook contest. A third chapbook, Talking About the Weather, was published in a limited edition in 2012 by Seven Kitchens Press. A limited edition (30) collection of her poetry, On The Avenue of Eternal Peace, was designed and printed by Joe Ruffo (Lyra Press) and beautifully hand-bound by Denise Brady. Individual poems by Ahl have appeared in numerous literary journals and anthologies. She has been a writing resident at the Vermont Studio Center, the Jentel Artist Residency Program, and the Kimmel Harding Nelson Center for the Arts. Ahl's interests as a poet include performance, collaboration, and cross-disciplinary work (letterpress printing, working with musicians, dancers and visual artists). Her work has been set to original symphonic music, combined with dance choreography, and riffed to improvisational jazz. She has read in galleries in collaboration with thematic visual art shows and has published broadsides and chapbooks of both her own work and the work of others.

Rebecca Alosa

Most Helpful to Students, English

Rebecca Alosa has been teaching with the English Department at PSU for five years. During that time, she has been fortunate to work with many undergraduate students developing reading, writing, and critical thinking skills. She is a NH certified teacher for grades 5–12, and has worked with local middle school students as an instructor for the Writing Project. She has also been a member of the Editorial Board for the PSU Composition Journal.

Bruce Heald

Best Researcher/Scholar, Liberal Arts, Social Sciences, and Humanities

Karolyn Kinane

Best Overall Faculty Member, English
Best Overall Faculty Member, Liberal Arts, Social Sciences, and Humanities

Burrett McBee

Most Helpful to Students, Liberal Arts, Social Sciences, and Humanities

Ann McClellan

Best Researcher/Scholar, English

Anyone familiar with Ann McClellan knows she is proud of her Midwestern heritage (Go Blue!) and a dedicated Anglophile. Specializing in Twentieth century British literature, especially the novel and women writers, Ann's work explores the complex relationships between literature and culture, with research ranging from fictional representations of British women intellectuals to her current project on fan culture and the popularity of Sherlock Holmes. Her classes explore questions of literary value and canonicity and the inter-connection of history and text, with an emphasis on literary theory. Her goal: to lift the veil that often obscures the strategies and skills underlying literary analysis to teach students how to think, speak, and write about texts in interesting and sophisticated ways.

Robert Swift

Best Teacher, Liberal Arts, Social Sciences, and Humanities

POINT PARK UNIVERSITY

Thomas Baggerman

Best Researcher/Scholar, Communication

Thom Baggerman, Ph.D., comes to Point Park from Capital University, where he was an assistant professor in the Department of Communication. He was also previously a visiting instructor at the University of Pittsburgh and a part-time faculty member at Robert Morris University. Baggerman has worked at various television stations, including WTOV-TV, where he was operations manager, and WGGB-TV, where he was a director/producer.

Kim Bell

Best Overall Faculty Member, Liberal Arts, Social Sciences, and Humanities

Amy Kim Bell, M.A., is a professor of linguistics and modern languages in the Department of Humanities and Human Sciences. She has been a full-time faculty member at the University since 1985. Over this period, she has served in various administrative capacities including chair and interim chair as well as director and coordinator of the ESL program. While completing her undergraduate degree at West Virginia University, Bell participated in a study abroad program in Colombia at the Universidad de Antioquia. After graduation, she moved to Barcelona, Spain to teach English as a foreign language. When Bell returned to the United States, she began working as a professional translator with agencies in New York, Atlanta and Pittsburgh.

Helen Fallon

Most Helpful to Students, Liberal Arts, Social Sciences, and Humanities

Helen Fallon, professor of journalism and director of Point Park's Honors Program, has taught and served as an administrator at Point Park University since 1986. She was chair of the Department of Journalism and Mass Communication from 1999 to 2008 and was the acting dean of the School of Communication from 2008 to 2009. She then became director of the Honors Program.

Under Fallon's leadership, the School of Communication began annual student trips to Washington, D.C., and a different foreign country each year to focus on national and international media as well as broaden travel experiences for students. Past trips have included Germany, Paris, Italy, London, Toronto, Dublin and Prague.

Fallon remains an active journalist, working every Friday as a copy editor for the Pittsburgh Post-Gazette, where she's been gathering real-world materials for her classes since 1990. Earlier in her career, Fallon worked as a reporter, copy editor and editor for several papers in the Pittsburgh area for six years, and did public relations for a university and a community hospital for five years.

Fallon is involved in journalism education beyond Point Park, working with high school students by teaching apprenticeships for the Allegheny Intermediate Unit's gifted and talented program and visiting area high schools to talk with journalism and communication classes. She has been an active member and leader in the Pennsylvania School Press Association and has made presentations with the Journalism Education Association at its national conferences as well.

Fallon and her family, daughter Amy and son Stephen, established a travel fund at Point Park in memory of their late husband and father, John E. Fallon Jr. The fund helps School of Communication and Honors Program students in need travel nationally and abroad to complement their Point Park education.

Robert Lewis

Best Teacher, Liberal Arts, Social Sciences, and Humanities

Bill Moushey

Best Researcher/Scholar, Liberal Arts, Social Sciences, and Humanities

Bob O'Gara

Best Overall Faculty Member, Communication
Best Teacher, Communication
Most Helpful to Students, Communication

Bob O'Gara joined the University teaching ranks part time in 1983 and became a full-time professor in 1993. His teaching style blends extensive real-world experience with the best of contemporary theory spiced generously with informal professional insights and humor. Because of O'Gara's ties to the Pittsburgh public relations and advertising community, top communications professionals from some of the largest and best communication organizations regularly provide interaction with Point Park University students. O'Gara also serves the University as a member of the budget and finance committee, Honors Program advisors, faculty development committee, and the friends/alumni and scholarship committees of the School of Communication.

Prior to teaching, O'Gara was in corporate public relations and advertising for Pittsburgh Corning Corp. and Koppers Company Inc., where he served as manager of the Communications Services Department and Director of Public Relations. He then joined PNC Financial Corp. as head of its corporate and public affairs department. O'Gara has managed his own communications consulting practice covering employee, financial and marketing communications for a variety of industrial, commercial and nonprofit clients.

O'Gara is a board member of Mainstay Life Services (residential services and support for people with developmental disabilities and their families) and community adviser to Western Pennsylvania Conservancy for its Greenspace and Community Gardens program. O'Gara and his students regularly provide local non-profit organizations with communications counsel and services.

He lives in Mount Lebanon with his wife, Sally. They have two sons, Brian and Kevin.

PRATT INSTITUTE

Claudia Hernandez

Best Overall Faculty Member, Design

"Tireless dedication to the students and advancing interior design as a discipline."

Catherine Redmond

Best Teacher, Fine Arts

PRESBYTERIAN COLLEGE

Chad Helms

Best Overall Faculty Member, Languages
Best Researcher/Scholar, Languages
Best Teacher, Languages
Most Helpful to Students, Languages

PRINCETON UNIVERSITY

Joshua Katz

Best Overall Faculty Member, Liberal Arts, Social Sciences, and Humanities

I am a linguist by training, a classicist by profession, and a comparative philologist at heart. The recipient of degrees in linguistics from Yale (B.A. 1991, summa cum laude and with Exceptional Distinction in the major), Oxford (M.Phil. 1993; British Marshall Scholar), and Harvard (Ph.D. 1998), I had the good fortune to be able to reinvent myself as a classicist at Princeton. Now a Professor, I started my career here in the spring of 1998, when I was hired, in the first place as a one-semester Lecturer, to teach a class on the history of English while finishing a dissertation with the catchy title Topics in Indo-European Personal Pronouns. Widely published in the languages, literatures, and cultures of the ancient world, from India to Ireland via Greece, Rome, and the Near East, I am more or less equally interested in Greek and Latin. While there are recurrent themes in my work — for example, the reconstruction of Proto-Indo-European, poetics, and etymology, which I view as part of the history of ideas — I prefer in both my research and my teaching to prowl around topics rather than pursue one single line of inquiry: Hesiodic belly-prophecy and Horatian self-fashioning; Basque badgers and Roman testicles; the phonology of Tocharian monosyllables and the morphological peculiarities of Gothic pronouns; hieroglyphic Egyptian puns and modern English slang; etc. Recent articles have reinterpreted the opening of Vergil's Georgics, refined a phonological rule in Hittite, proposed a novel explanation for the form of the pluperfect in Ancient Greek, and taken a fragmentary verse of Archilochus about blind eels as a launching point for insights into Anatolian, Celtic, and Greek mythology; a paper in press makes a stab at coming to terms with the place of the curious "anagram notebooks" in the life and thought of Ferdinand de Saussure. One of my current preoccupations is wordplay, whose theory and practice across time and space I am investigating thanks to a John Simon Guggenheim Memorial Foundation Fellowship, a Loeb Classical Library Foundation Fellowship, and a

Visiting Fellowship at All Souls College, Oxford. Among the other institutions to which I am deeply grateful for awards over the years are the American Council of Learned Societies, the Andrew W. Mellon Foundation, the Institute for Advanced Study, the National Endowment for the Humanities, and the National Science Foundation. I was a Directeur d'études invité at the École pratique des Hautes Études in Paris in May 2011.

PURDUE UNIVERSITY - WEST LAFAYETTE

Raymond Olson

Best Researcher/Scholar, Liberal Arts, Social Sciences, and Humanities

Carolyn Perrucci

Best Overall Faculty Member, Sociology
Best Overall Faculty Member, Liberal Arts, Social Sciences, and Humanities

"Carolyn Perrucci has served the CLA - Sociology Department for 50 years, longer than any faculty member in University history. She is the 2014 Distinguished Woman Faculty Member for her scholarship. She also works closely with Sociology and Law and Society majors as faculty advisor of the local Gamma Chapter of the International Sociology Honor Society, Alpha Kappa Delta. She has won several awards, Violet Haas and Helen Schleman Awards, for service to women students on campus and elsewhere."

Robert Perrucci

Best Overall Faculty Member, Sociology

Professor of Sociology at Purdue University. Author of over 90 articles and chapters, and authored/edited 19 books. Research funded by National Science Foundation, National Institute of Mental Health, and Alfred P. Sloan Foundation. Former editor of The American Sociologist, Social Problems, and Contemporary Sociology, and Associate Editor of The American Sociological Review. Elected President of the Society for the Study of Social Problems and the North Central Sociological Association.

Robin Shay

Best Teacher, Speech & Hearing Sciences

QUINNIPIAC UNIVERSITY

Joseph Avitable

Best Teacher, Liberal Arts, Social Sciences, and Humanities

Anat Biletzki

Best Overall Faculty Member, Philosophy
Best Researcher/Scholar, Philosophy

Edward D'Angelo

Best Teacher, Philosophy

Timothy Dansdill

Best Researcher/Scholar, Liberal Arts, Social Sciences, and Humanities
Best Teacher, Liberal Arts, Social Sciences, and Humanities

Sharon Magnarelli

Best Teacher, Languages

Sharon Magnarelli, Ph.D. is an internationally acclaimed scholar who has published four books and nearly one hundred articles on Spanish American theatre and narrative. She is frequently invited to lecture at universities in the United States and abroad. She holds a Ph.D. from Cornell University and teaches advanced Spanish courses as well as literature courses and senior seminar.

Ronald Quirk

Best Overall Faculty Member, Languages

"Ron is an exceptional person and was a great department chair. He is always a gentleman and always professional. He is a gifted teacher and scholar."

I am a Professor of Modern Languages teaching Spanish at the undergraduate and graduate levels. My areas of specialization are nineteenth-century Spanish literature, the Hispanic Caribbean, and linguistics (the history of the Romance languages). I have published books and articles on all these areas.

Dave Valone

Best Teacher, History

RADFORD UNIVERSITY

John Knowles

Most Helpful to Students, Foreign Languages & Literature

Rita Martin

Best Researcher/Scholar, Foreign Languages & Literature

RAMAPO COLLEGE - NEW JERSEY

Bobby Deitch

Best Overall Faculty Member, Music

"Excellent musician and teacher"

RARITAN VALLEY COMMUNITY COLLEGE

Lauren Braun-Strumfels

Best Researcher/Scholar, Humanities

"Lauren was the first community college faculty member invited to participate in an International Humanities experience in Rome, Italy this past summer."

Brandyn Heppard

Best Overall Faculty Member, Philosophy

"Brandyn is just an incredible colleague who shows the students, faculty, staff and administrators alike why he makes RVCC a better place to work and learn!"

Joseph Kudless

Best Overall Faculty Member, History
Best Researcher/Scholar, History
Best Teacher, History
Most Helpful to Students, History

Education is a lifelong process. We can no longer be limited in our planning to the "traditional" college age student filling our classes. The strength of our enrollments will increasingly depend on our ability to provide the kind of learning experiences that part-time learners desire in their pursuit of personal and career satisfaction.

The central characters in the teaching-learning process remain the student and his/her teachers. The college administration and staff are there to facilitate student-faculty interaction so that effective learning happens and mutual satisfaction results. Leadership provides the suitable environment through strategic planning and careful utilization of budgeted funds.

Outside scrutiny has been on the increase and will continue to grow. The best response to such outside pressures will be to keep in mind the "bottom line": What will be the impact of a particular decision on our students and the community we serve? I wonder sometimes if we are too timid in our off-campus world in admitting that striving toward the goal of a cultured society remains most worthwhile in that when one uses ones talents to the utmost, the greater good of the community usually is served well.

Future stability will be best cultivated in an atmosphere of collegiality and trust. My approach to any leadership role on campus has been to give consistent evidence to those who take their education seriously that faculty and administrators take each other's ideas seriously. We should be no more timid and no less forthright in setting each other straight than in questioning our students about their progress or lack of it. If we are demanding of each other in setting reasonable goals and standards of performance, we are not kidding each other or the students whom we serve.

READING AREA COMMUNITY COLLEGE

Karen Jacobson

Best Overall Faculty Member, Liberal Arts, Social Sciences, and Humanities

"Karen is a very special caring professor and assistant dean and has gained the respect of faculty, staff, and students. Her compassion, love for RACC, and diligence are beyond reproach. She is a good listener so positive as well as a strong leader. I wish her well in her retirement."

Jorge Nouhra

Best Overall Faculty Member, Languages

"Jorge Nouhra is a capable, dedicated, and experienced language teacher. As a role model and leader, Professor Nouhra has impacted the lives of many of his students. His impeccable credentials and caring personality, makes Jorge Nouhra a perfect recipient for a Language Faculty Award recognition at Reading Area Community College. Congratulations Professor Nouhra!"

RIO HONDO COLLEGE

Ada Brown

Best Teacher, Fine Arts

Ada Pullini Brown, M.F.A., is a painter and art educator. Educated at the University of Houston (M.F.A.), Queens College of the City University of New York (B.A), Sarah Lawrence College and the High School of Art & Design, Ms Brown is currently an Associate Professor of Painting & Drawing at Rio Hondo College. She is also the director of ArTravel Workshops, a landscape painting summer program for artists which has taken groups of painters and water colorists to Europe for the last 18 summers. In the Los Angeles area, Ms. Brown's paintings have been exhibited at the Riverside Art Museum, Rancho Santiago College, Venice Art Walk, I–5 Gallery, Gallery Otra Vez, Avenue 50 Gallery and the Sabina Lee Gallery. Her work is held in collections in the United States, South Africa, Mexico and Europe. Ms. Brown is a former President of the Southern California Women's Caucus for Art and currently serves on the board of the Brewery Artwalk Association. She teaches courses in beginning, intermediate and advanced Freehand Drawing, Life Drawing and Painting.

Cynthia Lewis

Best Researcher/Scholar, Art

Cynthia Lewis received a B.A. in Art History from Pomona College, Claremont, and an M.A. in Art History from CSU Fullerton, specializing in the colonial art of the Americas. She teaches courses in the history of western art (Art 105, Art 106), The Art of Mexico, (Art 108), Modern Art (Art 112), Understanding Visual Art (Art 110), and History of Photography (Art 113, PreColumbian Art of the Americas (Art 109) and Art of the Americas: Colonial to the Present (Art 104). An art researcher for numerous documentaries produced by the History Channel, Discovery Channel, and A&E, her current writing and research is on the contemporary art of Mexico City and the art of the missions of Northern New Spain. She enjoys teaching honors courses, serving on several campus committees, and traveling to Spain, Mexico, Italy, and France with her art and art history students.

RIVERSIDE CITY COLLEGE - RIVERSIDE CCD

Joycee Beck

Most Helpful to Students, Graphic Arts

"She is always prepared and always ready to give her students what they need to succeed. She is patient and gives many more hours to her students than her part-time status requires."

Joycee Beck is an entertainer, instructor, and lecturer. She has traveled all over the world as a performing artist, entertaining and lecturing, including six armed forces tours to war-torn countries in the Middle East as well as the Balkans. She taught elementary school for ten years, taught beginning tax law at H&R Block for a couple of seasons as well as preparing taxes for many years, and is currently teaching Photoshop, After Effects, Illustrator, InDesign Flash and Designing Web Graphics at Riverside City College. Joycee has a Master of Arts degree in Education from Azusa Pacific University.

Kathleen Sell

Best Overall Faculty Member, Liberal Arts, Social Sciences, and Humanities

"Great reputation as a great teacher"

Takashi Suzuki

Best Overall Faculty Member, Languages
Best Researcher/Scholar, Liberal Arts, Social Sciences, and Humanities

ROANOKE COLLEGE

Dolores Flores-Silva

Best Researcher/Scholar, Languages

ROCK VALLEY COLLEGE

Joann Heck

Best Overall Faculty Member, English

"Professor Heck is as strong and supportive a colleague as she is a practicing English professor. She helps students selflessly and supports her colleagues in a professional and direct manner. Joann Heck is an honor to work with, and her students love her and tell everyone how much they appreciate her work."

A strong advocate for community college education, Joann earned her Associates of Arts degree from Rock Valley College. She earned a BA with a major in English at Northern Illinois University and went on to complete her Master of Arts in English at NIU. Joann likes to remind students of the impact that other community college students, like Gwendolyn Brooks, George Lucas, Eileen Collins, Tom Hanks, Walt Disney and others have made as she encourages each student's unique potential. Prior to entering the academic arena, Joann worked in the business world as a Marketing Director and Business Development Executive in the Managed Care, Health Care, and Electronic Security Industries. Her passions include teaching, shopping, traveling, (during the summer when she's not teaching and buried under a stack of papers) and spending time with her loud but very loving family

Donna Soltys

Best Overall Faculty Member, Liberal Arts, Social Sciences, and Humanities

"Donna consistently gives her all to help her students. She goes above and beyond helping the underprepared student and transforming them into the prepared student. She is a blessing and someone I try to emulate."

ROCKHURST UNIVERSITY

Curtis Hancock

Best Researcher/Scholar, Liberal Arts, Social Sciences, and Humanities

ROGER WILLIAMS UNIVERSITY

Sargon Donabed

Best Researcher/Scholar, History

Sargon George Donabed is Assistant Professor of History at Roger Williams University where he teaches Middle Eastern history and religious studies. Dr. Donabed completed his PhD at the department of Near and Middle Eastern Civilizations, University of Toronto in 2010. He serves on the advisory board of the journal Chronos, published by the University of Balamand in Lebanon as well as the Institute for Genocide Awareness and Applied Research (IGAAR) and the Modern Assyrian Research Archive (MARA). He is the co-organizer of the biannual Roger Williams University Conference on Religion and the State. He is a 2010 The American Academic Research Institute in Iraq (TAARII) fellow based on the project Documenting the Oral Folk Epic of Qatine Gabbara: Translation, Historical and Cultural Analysis, and

Transmission. Donabed is also co-editor of Religion and the State:Europe and North America in the 17th and 18th centuries (Lexington Books, 2012) and also co-editor and contributor to The Assyrian Heritage: Threads of Continuity and Influence (Uppsala University, 2012). His work takes him around the world regularly and has researched and presented at conferences most frequently in Europe and the Middle East. He is also an active member of a variety of conservation-based NGOs and has strong interest in fantasy and science fiction film and literature and wildlife/environmental conservation.

Autumn Quezada-Grant

Best Teacher, History

Autumn Quezada-Grant is an Assistant Professor of History. She received her Ph.D. from the University of Mississippi in 2010, in Latin American History. Her dissertation on Indian litigates in nineteenth-century Chiapas, Mexico examined the legal culture and patterns of negotiation and contestation between Highland indigenous groups with local Ladino elites. The study of litigation and attempts at redress in the past helps us understand the present engagement of present day Maya in politics and rebellion. Dr. Quezada-Grant has a broad research interest, which include modern Latin America, indigenous history, social justice, and Public Health history. She is currently working on a manuscript entitled Indigenas Rebeldes: Race, Power and Litigation in Nineteenth Century Chiapas, Mexico. She is the co-editor for Narratives and Negotiation: Agency, Religion and the State (Lexington Press, forth coming).

Professor Quezada-Grant loves to teach and travel. She had a study abroad course in the Dominican Republic in January 2015 titled "Social Justice in Hispaniola." In the past, she has taught a study abroad course in El Salvador.

She is also a co-advisor for the RWU Chapter of FIMRC (www.fimrc.org) and travels with students to do volunteer service-learning throughout Latin America.

Jennifer Stevens

Most Helpful to Students, History

Jennifer Stevens is an Associate Professor of History and American Studies specializing in American cultural studies and 20th century United States History. She received her Ph.D. in American Studies from Michigan State University in 2005. She also holds degrees from Bowling Green State University (M.A. in Popular Culture), National-Louis University (M.A.T. Early Childhood Education) and Carleton College (B.A. History).

ROGUE COMMUNITY COLLEGE

Michael Torguson

Best Teacher, Liberal Arts, Social Sciences, and Humanities

ROLLINS COLLEGE

Julian Chambliss

Best Overall Faculty Member, History
Best Teacher, History
Most Helpful to Students, History
Best Overall Faculty Member, Liberal Arts, Social Sciences, and Humanities
Most Helpful to Students, Liberal Arts, Social Sciences, and Humanities

"I have received numerous student comments about how inspiring he is."

Yusheng Yao

Best Researcher/Scholar, History

"His research is groundbreaking and informative."

Areas of Specialization: Twentieth century China, modern education reform and rural reconstruction

ROSE STATE COLLEGE

Claudia Buckmaster

Best Overall Faculty Member, English

"Dr. Buckmaster has always treated me with respect and fairness. When I have gone to her for information or about a student discipline issue, she has listened and responded immediately. Her educational background and experience are assets to the Rose State College community, and she is very deserving of this honor,"

Antoinette Castillo

Best Teacher, Humanities

Edmund Gert

Best Overall Faculty Member, Liberal Arts, Social Sciences, and Humanities

"Mr. Gert has always been extremely professional and fair with me. When my classes have not been made, he has worked to find replacement classes for me. He also listens when I have a problem and has shown lots of care and concern to myself and other colleagues through difficult situations. He also teaches, so he understands how difficult working with students can be at times. I feel as though I can take a problem to him at any time, and he will listen. He has supported me as an instructor and professional since I began teaching there. I have also heard many positive comments from his students about him and his teaching methods."

SACRED HEART UNIVERSITY

Kenneth Knies

Best Teacher, Philosophy

SADDLEBACK COLLEGE

Bob Cosgrove

Best Overall Faculty Member, English Best Teacher, English

Rich Goodman

Best Teacher, Social Science

Rich Goodman has been a professor in Human Services at S. C. C. for 38 years, 20 being full-time. He teaches courses in Alcohol and Drug Studies, namely HS 170- Alcohol and Other Drugs in Society. He was department chair for 7 years until retiring his full time position in 2010. That year he was honored as Teacher of the Year, an award he cherishes. His continued love of teaching is now as an adjunct faculty again. His Criminal Justice background resulted from working for the Los Angeles Co. Probation Dept. for 25 years. He worked primarily with substance abusers doing narcotic testing and was the Department Narcotics Trainer for 5 years. He also ran a mandated drug education program for minors sentenced to Juvenile Camp. Rich has been married to his high school sweetheart for the past 52 years. He is a father of 2 wonderful children,Scott and Kristin and loves to spoil his 4 grandchildren, Jake and Josh and Sara and Siena,his twin girl granddaughters! Sports and travel are his joys aside from family and education.

Carmenmara Hernandez-Bravo

Most Helpful to Students, Languages

Veronica Obermeyer

Most Helpful to Students, Art

Bill Stevenson

Most Helpful to Students, English

Richard White

Best Overall Faculty Member, Fine Arts

B.A. Arts and Crafts and Their History, honors, June 1974 Cowell College, University of California at Santa Cruz, Santa Cruz, California

M.F.A. Ceramics, June 1978 Alfred University, New York State College of Ceramics, Alfred, New York

SAINT LOUIS UNIVERSITY

Simone Bregni

Most Helpful to Students, Languages

Ellen Crowell

Best Researcher/Scholar, English

Ellen Crowell received her Ph.D in English from the University of Texas at Austin in 2004, with areas of emphasis in modernism, Irish Studies and the Literature of the American South. She joined the SLU English faculty in 2004.

EDUCATION M.A. University of Texas (1998) Ph.D. University of Texas (2004)

RESEARCH Ellen Crowell's major research areas coalesce around the cultural pre-history of modernism. Through research and teaching that seeks to link the late-Victorian intellectual and artistic movements of Aestheticism and Decadence to Modernist writing in English, her work illuminates the pre-history of modernism as a historical period heralding a transformative change in political, social, and aesthetic expression. Particular areas of focus include: Modernist writing in English; the cultural history of Decadence and Aestheticism; Queer Theory and the

History of Sexuality; Film Studies and Modernist Cinema; and Irish Studies, with particular emphasis on Oscar Wilde, W.B. Yeats, and Elizabeth Bowen. Her publications include Aristocratic Drag: The Dandy in Irish and American Southern Fiction (Edinburgh, 2007), "The Picture of Charles Bon: Oscar Wilde's Trip Through Faulkner's Yoknapatawpha" (2004), and "Scarlet Carsons, Men in Masks: The Wildean Contexts of V for Vendetta (2009). Her current book project, Oscar Wilde's Body, reconstructs forgotten subcultures of mourning, fandom, and queer self-fashioning to reimagine Oscar Wilde's presence in the literary and cultural landscapes of early modernism.

Kathleen Llewellyn

Best Teacher, Languages

Nil Santianez

Best Researcher/Scholar, Languages

Angela Smart

Best Overall Faculty Member, Languages

Kenneth Steinhauser

Best Researcher/Scholar, Theology Best Teacher, Theology

A native of Brooklyn, New York, I studied philosophy and theology in New York, Rome, Tübingen and Freiburg in Breisgau. From 1980 to 1990 I taught at Sacred Heart College and Belmont Abbey College in Belmont, North Carolina. Since 1990 I have been on the faculty of SLU, which has been and continues to be a wonderful experience for me. In the spring of 2007, I was Visiting Professor at the Istituto Patristico Augustinianum in Rome. It was great to return as a professor to the city where I had spent four years as a student. In 2010, I participated in an NEH Seminar on autobiography in antiquity in Tunis. In 2012,

I studied the oldest manuscript of Augustine's works at the National Library of Russia in Saint Petersburg. In 2013, I completed a term as president of the North American Patristics Society.

My specialization is patristics and early church history. First, I have a special interest in the influence of the Latin classics on the fathers of the church, particularly the influence of Cicero on Augustine. Second, I also have philological interests and have produced a critical edition of the heretofore unedited text of Anonymi in Iob commentarius, which was published in 2006 in the series CSEL (vol. 96) by the Austrian Academy of Sciences Press. Third, I work in the area of manuscript studies. I serve on the advisory board of Manuscripta: A Journal for Manuscript Research, published by SLU's Vatican Film Library. Currently I am working on the USA volume of The Manuscript Tradition of the Works of Saint Augustine. I periodically offer a graduate seminar on "Manuscripts and Texts." The Use of Textual Criticism for the Interpretation of Patristic Texts: Seventeen Case Studies, edited by me and my research assistant, Scott Dermer, and published in 2013, presents the research of seminar participants over several years.

I am also a priest of the Diocese of Jefferson City. I was ordained on December 18, 1970 in St. Peter's Basilica in Rome, and served three years in St. Matthias Church, Brooklyn, New York, before beginning my academic career in earnest. Occasionally I serve a rural parish, St. Patrick's Church in Jonesburg, Missouri.

Donald Stump

Most Helpful to Students, English

Dr. Stump received his Ph.D. from Cornell University in 1978. He is the Director of Saint Louis University's Micah House program and his research interests include the Bible, Renaissance Literature, and Edmund Spenser. He has recently finished editing a special journal issue of Explorations in Renaissance Culture.

EDUCATION Ph.D. Cornell University English (1978) M.A. Cornell University English (1974) B.A. Grinnell College English (1969)

TEACHING & RESEARCH INTERESTS Renaissance literature and its classical and biblical backgrounds; Elizabeth I and the writers of her court; the Bible and literature; fantasy.

BOOKS Primary author, Sir Philip Sidney: An Annotated Bibliography of Texts and Criticism (1554–1984). New York: G.K. Hall/Macmillan, 1994. Co-authors: Jerome S. Dees and C. Stuart Hunter.

Editor-in-Chief, 'Hamartia': The Concept of Error in the Western Tradition. Essays in Honor of John M. Crossett. Co-editors: J.A. Arieti, Lloyd Gerson, and Eleonore Stump. Texts and Studies in Religion, vol. 16. New York and Toronto: Edwin Mellen Press, 1983.

SPECIAL JOURNAL ISSUE Images of Elizabeth I: A Quadricentennial Celebration. Edited by Donald Stump and Carole Levin. Explorations in Renaissance Culture 30 (Summer 2004): 158 pp.

ARTICLES "Reforming the Greek Tragic Hero: Cross-Dressing in Sidney's Old Arcadia." In Teaching Early Modern Prose, edited by Margaret W. Ferguson and Susan Monta. Options for Teaching Series. New York: Modern Language Association. Forthcoming.

"Learning from Women: Sidney's Critique of Humanist Education." In Challenging Humanism, edited by Arthur F. Kinney and Ton Hoenselaars. Newark, Del.: University of Delaware Press. Forthcoming.

"Abandoning the Old Testament: Shifting Paradigms for Elizabeth, 1578–82." In Images of Elizabeth I: A Quadricentennial Celebration. Edited by Donald Stump and Carole Levin. Explorations in Renaissance Culture 30 (Summer 2004):89–109

"Fashioning Gender: Cross-Dressing in Spenser's Legend of Britomart and Artegall." Spenser Studies: A Renaissance Poetry Annual 15 (2001): 95–119.

"Marlowe's Travesty of Virgil: Dido and Elizabethan Dreams of Empire." Comparative Drama 34 (Spring 2000): 79–107.

Articles on "Prayer" and "Trinity." In A Dictionary of Biblical Tradition in English Literature, ed. David Lyle Jeffrey. Grand Rapids, MI: William B. Eerdmans, 1992, pp. 626–31, 785–88. Co-author for "Trinity": David L. Jeffrey.

"Edmund Spenser." In The Concise Dictionary of British Literary Biography, vol. 1: Writers of the Middle Ages and the Renaissance. Detroit: Bruccoli, Clark, Layman, Gale Research, 1992, pp. 359–87. Reprinted in a revised and enlarged version in Sixteenth-Century British Nondramatic Writers, 3rd Series, ed. David A. Richardson. Detroit, Washington, D.C., and London: Gale Research, 1996, pp. 228–63.

"The Two Deaths of Mary Stuart: Historical Allegory in Spenser's Book of Justice." Spenser Studies: A Renaissance Poetry Annual 9 (1991): 81–105.

"Pride" and "Tragedy." In The Spenser Encyclopedia, ed. A. C. Hamilton, Donald Cheny, W. F. Blissett, David Richardson, and William W. Barker. Toronto: University of Toronto Press, 1990, pp. 556–57, 697.

"A Slow Return to Eden: Spenser on Women's Rule." English Literary Renaissance 29 (Fall 1999): 401–21.

"The Sidney Database and Sir Philip Sidney: A Reference Guide." Sidney Newsletter 10 (1989): 75–79.

"Sidney's Astrophil, Vanishing." Renaissance Papers (1988), pp. 1–13.

"Hamlet, Cain and Abel, and the Pattern of Divine Providence." Renaissance Papers (1985), pp. 27–38.

"Spenser's Inferno: The Order of the Seven Deadly Sins at the Palace of Pride." Journal of Medieval and Renaissance Studies 14 (1984): 203–18. Co-author: John M. Crossett.

"Greek and Shakespearean Tragedy: Four Indirect Routes from Athens to London." In 'Hamartia': The Concept of Error in the Western Tradition. Essays in Honor of James M. Crossett. Texts and Studies in Religion, vol. 16. New York and Toronto: Edwin Mellen Press, 1983, pp. 211–46.

"Isis Versus Mercilla: The Allegorical Shrines in Spenser's Legend of Justice." Spenser Studies: A Renaissance Poetry Annual 3 (1982): 87–98.

"Sidney's Concept of Tragedy in the Apology and in the Arcadia." Studies in Philology 79 (1982): 41–61.

"Britomart's Mock-Heroic Quest." In Spenser and the Middle Ages, ed. David Richardson (Cleveland, 1976), pp. 157–66. Microfiche proceedings of the Spenser Sessions, International Congress on Medieval Studies, May 1976.

SALT LAKE COMMUNITY COLLEGE

Lisa Bickmore

Best Teacher, English

Lisa Bickmore attended Brigham Young University where she received a B.A. (1979) and an M.A. (1984) in English, and did graduate work at the University of Utah. An associate professor in the English Department, she teaches composition, creative writing, and writing studies courses.

Lisa's book of poems Haste was published by Signature Press, and her scholarly and creative work has been published in such journals as Teaching English in the Two-Year College, The Bedford Sourcebook on Multimodal Composition, Tar River Poetry, Caketrain, Hunger Mountain Review, Terrain and Quarterly West. She was awarded the Salt Lake City Mayor's Artist Award for the Literary Arts in 2008. Her book Flicker is the winner of the 2014 Anti-venom Prize (forthcoming from Elixir Press).

Her current interests include digital forms of composition, the constantly shifting landscape of digital and print publication, and the power of narrative in argument, as well as a deep and abiding commitment to seeing movies in theaters as often as is humanly possible.

Nathan Cole

Best Researcher/Scholar, Liberal Arts, Social Sciences, and Humanities
Best Teacher, Liberal Arts, Social Sciences, and Humanities
Most Helpful to Students, Liberal Arts, Social Sciences, and Humanities
Best Overall Faculty Member, Liberal Arts, Social Sciences, and Humanities

Nathan Cole is an Assistant Professor in the Department of English at Salt Lake Community College, in Salt Lake City, Utah. He is an enrolled member of the Akwesasne Mohawk Reservation (formerly known as the St. Regis Mohawk Reservation) in northern New York state. His research interests include Native American/American Indian history, culture, political, governmental and societal issues. As a graduate student, he taught introductory and advanced writing courses in English composition at Brigham Young University, and has taught in the past, SLCC's English 2100, Introduction to Technical Writing, and English 2600, Critical Introduction to Literature.

Jane Drexler

Best Overall Faculty Member, Philosophy
Best Researcher/Scholar, Philosophy
Best Teacher, Philosophy
Most Helpful to Students, Philosophy

Allison Fernley

Best Overall Faculty Member, English

Allison is associate professor of English who has been teaching at the college level since 1982 and at SLCC since 1989. Before joining the faculty at SLCC, she taught basic, introductory, and advanced composition at the University of Utah where she worked as a teaching assistant, teaching fellow, and associate instructor. She also worked for three years as a professional technical writer and editor for Shipley Associates, Learning Technologies Associates, and Eyring Research.

Her teaching and research interests are in writing studies, queer theory and criticism, and literary and rhetorical theory and criticism. She is especially interested in the tensions between identity politics and queer theory, media representations and queer theory/politics, post-queer politics, teaching for transfer, and genre theory.

She worked as a film and theater critic for two now-defunct Wasatch Front alternative papers, The Junction City News and The Event, from 1987–1994 and appeared as a weekly film critic/guest on the K-Talk radio with Joe Redburn from 1989–90. Her film criticism has been published in Film Quarterly.

Rick Graham

Best Teacher, Fine Arts

Whitney King

Most Helpful to Students, Fine Arts

Elisa Stone

Most Helpful to Students, English

Elisa is the Associate Director of the Community Writing Center and an Associate Professor of English at Salt Lake Community College, where she received the Utah Campus Compact's Engaged Scholar Award as the College's Service-Learning Faculty Mentor. She has volunteered alongside students at various non-profit organizations, and is a consultant to the State Office of Education in creating technical writing curriculum for Utah high schools. She adores animals, the ocean, and traveling.

SAN ANTONIO COLLEGE

Mariano Aguilar

Best Overall Faculty Member, Liberal Arts, Social Sciences, and Humanities

Brian Barwise

Most Helpful to Students, Languages
Most Helpful to Students, Liberal Arts, Social Sciences, and Humanities

Michael Connor

Best Overall Faculty Member, Architecture
Best Teacher, Architecture

Donna Duke-Koelfgen

Best Overall Faculty Member, English
Best Researcher/Scholar, English
Best Teacher, English

Juaquin Escamilla

Most Helpful to Students, Architecture

Jeff Hunt

Best Researcher/Scholar, Liberal Arts, Social Sciences, and Humanities

Alice Laffere

Most Helpful to Students, Reading

Laurie Metcalf

Best Researcher/Scholar, Languages

Jolinda Ramsey

Best Overall Faculty Member, Communication
Most Helpful to Students, Communication

Julie Razuri

Best Teacher, Languages
Best Teacher, Liberal Arts, Social Sciences, and Humanities

Cindy Sanchez

Most Helpful to Students, Music

"She always goes above and beyond to assist the students. Individual sessions are always going on behind the scenes in whatever they need."

Judy Scheinberg

Best Teacher, Liberal Arts, Social Sciences, and Humanities

SAN DIEGO CITY COLLEGE

Ella Sloan

Best Overall Faculty Member, Cosmetology
Best Researcher/Scholar, Cosmetology
Best Teacher, Cosmetology
Most Helpful to Students, Cosmetology

"Because of her years of expertise and the knowledge she has given the students. Her concern for the community has also been an inspiration to the students, which has enabled them to strive to reach their goals.

Dr. Sloan is, without a doubt, the best teacher in our department. She exemplifies what makes a great Instructor.

For many years after students have graduated, I have witnessed their gratitude for all the help they have been given by Dr. Sloan.

Dr. Sloan's contribution to our campus, the community as well as our department has shown students that continued education with hard work will grant the desires and goals of one's heart. Dr. Sloan is an example of one making dreams a reality."

SAN DIEGO MESA COLLEGE

Christie Allred

Most Helpful to Students, English

Jennifer Cost

Best Researcher/Scholar, English
Best Researcher/Scholar, Liberal Arts, Social Sciences, and Humanities

B.A., Humboldt State University, Arcata
M.A., San Diego State University

Ms Cost received her MA degree in English/American Literature from San Diego State University, San Diego, CA, and her BA degree in English, minor in Art History from Humboldt State University, Arcata, California. Personal interests include: literature, writing, photography, music, film, travelling and swimming.

Cooper Darius

Best Teacher, English

Chris Sullivan

Best Overall Faculty Member, English
Best Teacher, English

Chris holds the position of Professor in Mesa College's English Department. He was hired in 1996, and since then, he has assumed roles of increasingly greater responsibility to serve his department, the college, and even the State of California's educational programs, particularly in the areas of curriculum development and Student Learning Outcomes. Chris chaired the English Department from 2004–2009, and he guided faculty members through the creation of a rubric for learning outcomes in English 49 (formerly English 51). He also co-authored the so-called "Genesis Paper" on SLOs for the associate degree. He has served on the Academic Program Review Committee since 2003, and he co-chaired Standard 1 for the 2004 Accreditation Self-Study.

In a larger arena, Chris Sullivan served as a member of the Curriculum Committee for the Academic Senate for California Community Colleges. In this capacity, he was consulted for editing and finalizing language for the rubrics for English and reading. He also played a key role in developing the SLOs for English 1A throughout the California Community Colleges.

SAN FRANCISCO STATE UNIVERSITY

Yu-Charn Chen

Best Overall Faculty Member, Design

Jennifer Hammett

Most Helpful to Students, Cinema
Best Teacher, Fine Arts

"Jennifer is a wonderful, warm teacher and brilliant intellectual who both challenges students to do their very highest level of work, and supports them through the process with kindness, generosity and understanding of the challenges they face, academically, personally and socially. She sees what students need, collectively and individually, and gives unstintingly of her time, skills and connections to help them. She is unique in that that her warmth and nurturing do not undermine or overshadow her high expectations for excellence. She does not pat students on the back for merely trying, but pushes them to reach an objective level of achievement. Her ability to bring out the best in students – to instill in them a realistic sense of how much work goes into the production of truly outstanding scholarship or creative work, rests on a Socratic insistence on the pupil's role in the process of learning. At the same time, as teacher, she provides unstinting emotional support to keep them afloat as they persevere beyond the first draft of a paper, first cut of a film, or first version of a presentation – to reach a standard of excellence they would not have attempted without her urging.

Professor Hammett has consistently developed innovative classes and assessment methods, including an exemplary online course, an excellent GWAR course, and a very successful internship program. She is also sought out as a mentor by MFA students and is highly rated by undergraduates and graduates alike."

Julian Hoxter

Best Teacher, Cinema

Jeff Jacoby

Best Teacher, Broadcast Elec Comm Arts

"Jeff goes above and beyond mentoring students and being there for them as an advisor. He genuinely cares. He's a gem!"

Jacoby practices sound art and design, radio production, directing, producing, writing, and performing, and happily serves as Associate Professor of Audio & Radio in the Department of Broadcast & Electronic Communication Arts at San Francisco State University. He has served on the faculty of Quinnipiac University, Real Art Ways, The Media Arts Center, and the National Radio Project. Current projects include Into Sonic Space (sound art installations), The Traveling Radio Show (creative radio program), and collaborations with artists working in a variety of mediums.

Amy Kilgard

Best Teacher, Communication

Lucia Lachmayr

Most Helpful to Students, English

"she cares"

Yumi Wilson

Most Helpful to Students, Liberal Arts, Social Sciences, and Humanities

Yumi Wilson is an Associate Professor of Journalism at San Francisco State. Previously, she was a news writer for the Associated Press, a content producer at Yahoo! and an editor and reporter at the San Francisco Chronicle. She has a BA in Journalism from the University of Southern California and an MFA in Creative Nonfiction from the University of San Francisco.

SAN JACINTO COLLEGE - CENTRAL

Danielle Bible

Best Overall Faculty Member, History
Best Researcher/Scholar, History
Best Teacher, History
Most Helpful to Students, History

Joseph Schenck

Best Teacher, Music

"Is an excellent teacher and mentor and is responsible for our outstanding transfer rate to four-year colleges."

SAN JOAQUIN DELTA COLLEGE

Melanie Marshall

Best Teacher, Fine Arts

Mario Moreno

Best Overall Faculty Member, Fine Arts

"Mario Moreno has worked at Delta College for many years and has done fantastic work for the students. He fosters students to find their inner voices through the vehicle of art. He allows them to explore and expand their creative talents. Many of his students have won art contests at Delta and within our community. Some of his work in collaboration with his students are displayed on campus and beyond."

SAN JOSE STATE UNIVERSITY

Deanna Fassett

Best Researcher/Scholar, Communication Studies
Most Helpful to Students, Communication Studies

"She is awesome!"

Ph.D., Speech Communication (Communication Pedagogy), Southern Illinois University, Carbondale, IL, 2000; M.S., Speech Communication (Communication Pedagogy), Southern Illinois University, Carbondale, IL, 1996; B.A., Communication, University of California, San Diego, 1995

Alberto Gutierrez

Best Overall Faculty Member, Communication Studies

Rona Halualani

Best Overall Faculty Member, Communication
Best Researcher/Scholar, Communication
Best Teacher, Communication
Most Helpful to Students, Communication

Chris Jochim

Best Overall Faculty Member, Humanities

Christian Jochim is Professor of Comparative Religious Studies and Director of the Center for Asian Studies at San Jose State University. He received his Ph.D. in Religion (and East Asian Studies) at University of Southern California (1980). He teaches in the area of Asian religions and culture. He is the author of Chinese Religions: A Cultural Perspective in the Prentice-Hall Series in World Religions (1986). He has published many articles on Chinese religion and philosophy in Journal of Chinese Religions, Modern China, Philosophy East and West, and other journals and compilations. Currently, he is studying contemporary Confucianism in China and Taiwan. One more thing: He loves to swim!

J. Todd Ormsbee

Best Researcher/Scholar, Humanities

I'm an interdisciplinary scholar by training and temperament, but my research and thinking are most closely allied with sociology. My courses lie to the traditionally academic side, pushing students to develop ideas and thoughts grounded in facts, and to think analytically about cultural and social issues. I do this in the classroom by combining straight-forward transmission of knowledge with intellectual problem-solving, asking students to apply and evaluate historical and theoretical data and ideas by putting them "to work" solving problems the students observe in society today.

In my research, I am currently coding data from interviews conducted last winter of individuals who have left Mormonism and now consider themselves to no longer be Mormons and who are generally now in a state of unbelief (agnostic or atheist). I am working on two separate articles from the data. The first uses grounded theory methodology to create a theory of the psycho-social processes of leaving a tightly scripted religion, such as Mormonism. The second article will follow an unexpected vein in the data: the role that emotions and the interpretation of emotions play in ex-Mormons transformation from belief to unbelief. The sociology of emotions is new to me and an exciting and rich topic. As always, my analysis is through the lens of symbolic interaction.

My first book, out in March 2010 (after years of nail-biting and revisions), treats the development of gay male publicity and community formation in 1960's San Francisco through a symbolic interactionist lens.

By nature, teaching and researching in the social sciences and humanities touches on areas of deep personal and social concern; so I try to be as critical as possible of my own values and work to envision pragmatic, outcome-based solutions to real problems instead of being driven by ideology.

Jennifer Rycenga

Most Helpful to Students, Humanities

Jennifer Rycenga is Professor of Comparative Religious Studies at San José State University. Her areas of interest include American religious history, religion and music, feminist analyses of religion, lesbian history, religion and politics, and feminist analyses of music.

She has published numerous articles, and co-edited the volumes Frontline Feminisms with Marguerite Waller (Routledge, 2000), and Queering the Popular Pitch with Sheila Whiteley (Routledge, 2006). She is currently working on a cultural biography of white Abolitionist educator Prudence Crandall (1803–1890), and co-editing Sin Big: The Mary Daly Reader, with Linda Barufaldi. She is also working on two edited volumes: Amazon Freethinker with Emily Culpepper, and That Word is Liberty: Essays on Matilda Joslyn Gage.

She received the Ph.D. in the area of Religion and Music from the Graduate Theological Union and the University of California at Berkeley in 1992. She has been teaching at SJSU since 1995.

Prof. Rycenga is an avid birder, with a North American Life List of 701 species, and over 1300 species worldwide. Ask her about your backyard visitors and encounters with birds that go bump in the night! She serves as president of the Sequoia Audubon Society in San Mateo County.

Lisa Stenmark

Best Teacher, Humanities

David Terry

Best Teacher, Communication Studies

SANTA BARBARA CITY COLLEGE

Mihir Mahapatra

Most Helpful to Students, Multimedia

"Mihir Mahapatra teaches all the advanced Maya classes and prepares students to be able to create 3D assets for gaming engines. He spends endless hours in the lab helping anybody that asks for help, in his or any other media discipline. He is always up to date with any news or new technology in the field."

SANTA CLARA UNIVERSITY

Francis Smith

Best Teacher, Theology

"Long standing professor dedicated to his students and always available to make sure that students have understood the material and encourages them to engage in thought provoking discussion."

Fr. Smith is a San Francisco native. As an undergraduate at Santa Clara University in the 1950s, Smith earned a B.A. in Philosophy with a minor in History (1956), and followed with an M.A. in Philosophy from Gonzaga University ten years later. By then a member of the Jesuit community, Smith next went to the Graduate Theological Union in Berkeley, where he was awarded the S.T.M. from the Jesuit School of Theology (1972), and from there to Rome, where he received an S.T.D. in Fundamental Theology from the Pontifical Gregorian University in 1977.

Fr. Smith began to teach at Santa Clara University in 1974. His interests include Christology, ecclesiology, and liberation theology, and he offers a range of courses on these subjects at both the undergraduate and graduate levels.

SANTA MONICA COLLEGE

Maria Erickson

Best Overall Faculty Member, Modern Languages

"Professor Maria Erickson is ready to help other faculty members and share a new technical knowledge; such as Mango etc."

Donald Hartman

Best Overall Faculty Member, Fine Arts

"An excellent teacher who is loved dearly by his students for the last 40 years."

SAVANNAH TECHNICAL COLLEGE

Sarah Hunt

Best Teacher, English

"Sarah is extremely knowledgeable in her field. She is an excellent planner; therefore, she is always organized. She is very helpful and always cheerful."

SETON HALL UNIVERSITY

Dena Levine

Most Helpful to Students, Music

Gloria Thurmond

Best Overall Faculty Member, Music

SHIPPENSBURG UNIVERSITY OF PENNSYLVANIA

Sharon Harrow

Most Helpful to Students, English

SHORELINE COMMUNITY COLLEGE

Ernest Johnson

Best Teacher, Liberal Arts, Social Sciences, and Humanities

"Tireless, and always there for his students and his faculty colleagues"

SKYLINE COLLEGE

Katherine Harer

Best Teacher, Liberal Arts, Social Sciences, and Humanities

SLIPPERY ROCK UNIVERSITY

Myra Balok

Most Helpful to Students, English
Best Teacher, Liberal Arts, Social Sciences, and Humanities
Most Helpful to Students, Liberal Arts, Social Sciences, and Humanities

"Dr. Baylok loves each student like her own child. She approaches everyone as an individual and treats students with respect."

Robert Crafton

Best Researcher/Scholar, Liberal Arts, Social Sciences, and Humanities

Hunter Davis

Best Teacher, English

Danette Dimarco

Best Researcher/Scholar, English

Alan Levy

Best Teacher, History

"His breadth of knowledge in and beyond his fields of teaching and his commitment to helping his students perform up to the highest standards make him worthy of recognition."

Dr. Levy is the author of nine books including: Musical Nationalism, a study of the American composers who came to prominence amidst the expatriate era of the 1920s; Elite Education and the Private School, a discussion of the vagaries of private secondary education at one of the nation's premier academies; Radical Aesthetics and Music Criticism, which looks at the relationship of leftist political ideologies to the realms of music and aesthetics; Government and the Arts, a history of the debates over the subject of the Federal Government's support for the arts; a biography of the American composer Edward MacDowell; a biography of the zany, brilliant baseball player of the early 20th century, Rube Waddell; and Tackling Jim Crow, a study of the origins, breakdowns, and legacies of racial segregation in professional football. He has recently published a biography of Joe McCarthy, Architect of the Yankee Dynasty, and a bio of Floyd Patterson, a Boxer and a Gentleman, a look at the athletic, personal, and political travails of a heavyweight champion. He has also given legislative testimony and made conference presentations on the subject of academic freedom and its enemies in modern higher education.

Rachela Permenter

Best Overall Faculty Member, Liberal Arts, Social Sciences, and Humanities

Frederick White

Best Overall Faculty Member, English

SOKA UNIVERSITY

Xiao-Xing Liu

Best Teacher, Languages

"Dr. Liu is the best language professor at SUA as attested by students and her colleagues. She cares a lot about SUA's students. As the Language and Culture Program director, she is always fair and very concerned about the well-being of the program, the student body and her colleagues. She's a wonderful teacher and human being."

SOMERSET COMMUNITY COLLEGE

Brandy Brock

Best Teacher, Cosmetology

"After a tragic accident that left Brandy paralyzed from the chest down, she has returned to the classroom stronger than ever. Although, I am the program coordinator, I could not complete my task without Brandy's dedication to the students. She is now our classroom and lab manager. She teaches at least 28 credit hours with 32 student contact hours each week. She orders all supplies as well as all the other many duties she necessary including overseeing the clinic floor. She is a great role model to our students. She presents subject matters is a clear effective manner that enables all students learning styles are addressed. She is available to the needs of her students even during breaks. She is a great adviser and makes sure that students can achieve their goals. She is also very committed to our community by serving on the Kentucky Congress on Spinal Cord Injuries and acting as the facilitator for the Look Good Feel Better for Eastern Kentucky through the American Cancer Society. These are just a few reasons I feel that Brandy Brock should be considered for the Best Teacher for SCC."

SOUTH DAKOTA STATE UNIVERSITY

Janet Johnson

Best Teacher, Liberal Arts, Social Sciences, and Humanities

"Janet is an excellent teacher, and she helps to make students feel comfortable and accepted regardless of their ability level."

Ed.D. in Adult and HIgher Education (University of South Dakota); M.A. in Russian Linguistics and Literature (University of Iowa); M. A. in Education (Reading Specialist); B. A. in English (Augustana College); In the mid-1990's, Dr. E. Janet Johnson was a pioneer in the field of distance learning when she first taught Russian on interactive television at South Dakota State University, receiving United States Distance Learning Association (USDLA) awards for Best Live Interactive Television Course in both 1994 and 1995. Since then, she has continued to focus on active, student-centered learning in the blended classroom (face-to-face classes with online component). Positions held by Dr. Johnson on campus include Russian Culture & Language Instructor (1990–1996); the first Distance Ed Coordinator (1996–1997); English Instructor (1998–2003); the first Teaching/Learning Center Coordinator (2001–2002); College Reading Instructor 2003–to the present. Dr. Johnson holds a doctorate in Adult and Higher Education from the University of South Dakota, a master's degree in Russian Literature and Linguistics from the University of Iowa, and a master's degree in Education (Reading Specialist) from Virginia Tech.

Charles Woodard

Best Overall Faculty Member, English

SOUTH PIEDMONT COMMUNITY COLLEGE

Ryan Brown

Most Helpful to Students, English

Victoria Efird

Best Overall Faculty Member, English

Kamisha Kirby

Best Researcher/Scholar, English

SOUTH TEXAS COLLEGE (ALL CAMPUSES)

Liana Andreasen

Best Overall Faculty Member, English Best Researcher/Scholar, English

Robin Bell

Most Helpful to Students, English

Irma Garcia

Best Overall Faculty Member, Communication

Rufino Garcia

Best Teacher, College Success

Sharon O'Leary

Most Helpful to Students, Music

Michael Polinard

Most Helpful to Students, Liberal Arts, Social Sciences, and Humanities

Florinda Rodriguez

Best Overall Faculty Member, Reading

I received my Bachelor's in English with a minor in Reading in 1991 and a Master's in Reading in 1994 from UTPA. Prior to STC, I taught reading for 11 years in a junior high setting during which I also received my certification in Mid-Management.

Ronald F Schermerhorn

Best Overall Faculty Member, Music
Best Researcher/Scholar, Music
Best Teacher, Music

SOUTHEAST MISSOURI STATE UNIVERSITY

Mary Collier-Mims

Best Teacher, Music

"She is an outstanding teacher in the Music Ed classes!"

A graduate of the Alabama School of Fine Arts, Mary Collier Mims, Instructor of Music Education, returned to the Music Department in 2010. She was Assistant Professor of Music at Troy University in Alabama from 1982–1989, Instructor of Music at Southeast from 1989–1992, and taught public school music in Cape Girardeau (K–9) from 1992–2009. Mims holds the lifetime Missouri teaching certification in K–12 Instrumental and Vocal Music and serves the department as Coordinator of Field Experiences.

Chris Goeke

Best Overall Faculty Member, Music
Best Researcher/Scholar, Music

Christopher Goeke, tenor, is a Professor of Music at Southeast. Goeke holds both the D.M.A. in Vocal Performance and Pedagogy and M.M. in Vocal Performance degrees from the University of Iowa. Dr. Goeke is beginning his 22nd year of teaching voice at Southeast and is also very active in opera including conducting, directing, and serving as technical coordinator for the full operatic productions at Southeast. Dr. Goeke is an active member of the National Association of Teachers of Singing (NATS), and has served as the Missouri Governor for NATS.

Dr. Goeke's performance resume includes engagements with the Central City Opera in Colorado, the State Repertory Opera of New Jersey, Westchester Opera in New York, the Light Opera of Manhattan, Santa Fe Desert Choral, Cedar Rapids Symphony, and the Bach Society of St. Louis. Some of his operatic roles include Almaviva in Il Barbieri di Siviglia, Tamino in The Magic Flute, and Alfred and Eisenstein in Die Fledermaus. He has been featured as tenor soloist for Bach's Christmas Oratorio, St. Matthew Passion, and the Magnificat, The Messiah by Handel, and The Creation by Haydn. The tenor is an active recitalist and has presented a wide range of literature including lute songs, traditional art songs and song cycles in German, French, Italian, and English, and contemporary literature including many works by Benjamin Britten and popular songs of the Broadway stage. Dr. Goeke's hobbies include woodworking, bicycling, and learning to play the accordion.

SOUTHEASTERN BAPTIST THEOLOGICAL SEMINARY

Chip McDaniel

Best Overall Faculty Member, Languages

SOUTHEASTERN LOUISIANA UNIVERSITY

Bonnie Ahn

Best Researcher/Scholar, Social Work

SOUTHERN ADVENTIST UNIVERSITY

Michael Hasel

Best Researcher/Scholar, Languages

"Excellent researcher!"

Dr. Michael G. Hasel has taught Near Eastern studies, and archaeology at Southern since 1998. In the 1995–96 academic year, he was the Samuel H. Kress Fellow at the W.F. Albright Institute of Archaeological Research in Jerusalem. In 2005 he served as a Senior Fulbright Scholar at the Cyprus-American Archaeological Research Institute (CAARI) in Nicosia, Cyprus, funded by the U.S. Department of State. He is a named fellow of the Institute for Biblical Research.

SOUTHERN ILLINOIS UNIVERSITY - CARBONDALE

Lourdes Albuixech

Best Teacher, Languages

"Lourdes Albuixech is an innate teacher: responsible of all her duties, knowledgeable and a continuous source of information to the students and colleagues, interested in contributing to the improvement of the department, patient wirh students of all levels,"

Lourdes Albuixech earned a B.A. in Hispanic Philology from the Universidad Complutense de Madrid in 1992, an M.A. in Spanish Literature from University of California, Riverside, in 1994, and a Ph.D. in Spanish Literature from University of California, Riverside, in 1997. She has taught in Southern Illinois University since 1997.

SOUTHERN METHODIST UNIVERSITY

Robert Howell

Best Overall Faculty Member, Philosophy
Best Researcher/Scholar, Philosophy
Best Teacher, Philosophy

My work mostly concerns the nature of subjectivity, the self and the mind. As a result, I wind up working in metaphysics, epistemology, the philosophy of mind, and the philosophy of language. I have recently published a book, Consciousness and the Limits of Objectivity (Oxford University Press, 2013) that develops a view called subjective physicalism, which maintains that while everything is physical, some physical states cannot be fully grasped unless they are occupied. I am currently at work on a book on the self, and am beginning to dip my toes into the emerging field of neuro-ethics.

Matthew Lockard

Most Helpful to Students, Philosophy

My current research focuses on various issues in epistemology and philosophy of mind, including the nature of epistemic normativity, the individuation of perceptual states, and the role of inference in justifying belief. I also have interests in ethics, philosophy of language, history of modern philosophy, and history of early analytic philosophy.

SOUTHERN POLYTECHNIC STATE UNIVERSITY

Kathryn Bedette

Best Researcher/Scholar, Architecture

Kathryn Bedette, AIA, is an Associate Professor in the Department of Architecture at Southern Polytechnic State University where she coordinates second-year architecture studios and the history and theory sequence, "Architecture Culture". Kathryn is a registered architect and serves on the Board of Directors for AIA Georgia.

Durham Crout

Best Overall Faculty Member, Architecture
Best Teacher, Architecture
Best Teacher, Fine Arts

Howard Itskowitz

Best Overall Faculty Member, Fine Arts

Elizabeth (Liz) Martin

Best Researcher/Scholar, Fine Arts

Liz Martin-Malikian is a tenured faculty at Kennesaw State University (formerly SPSU), where she coordinates thesis prep, research & studio and also teaches materials & methods in the environmental tech sequence (2006–present). Prior to that Liz served as the Paul Rudolph Visiting Professor at Auburn University's School of Architecture (2004–06). She has also taught and lectured in Los Angeles, Beirut and has been an invited juror at the Southern California Institute of Architecture (SCI-Arc), University of Virginia, Tulane University, Clemson University, Woodbury University, University of Utah, and Rice University.

While on research leave in 2014, Liz conducted fieldwork exploring postwar reconstruction in Beirut, Lebanon. While there, she also curated a Pop-up Studio-X Beirut for Columbia University's global networking program creating an educational forum to explore the future of the city. Made-up of local practitioners, academic scholars and neighborhood activists, the forum works as a global network for sharing ideas and projects about the built environment in postwar Beirut. Conceived by former Dean Mark Wigley, Studio-X is organized as a think-tank with labs all over the world and is considered by many to be a possible new model for design education, addressing the most important questions facing our shared world.

Liz is currently a PhD Candidate at the University of Edinburgh exploring the architecture of alterity. She received a M.Arch. from SCI-Arc and a B.Arch. from Tulane University.

Tony Rizzuto

Most Helpful to Students, Architecture

Professor Rizzuto's Ph.D. explored the relationship between philosophy and architecture. Examining three distinct moments in the history of western philosophy, the work traces the impact of their epistemological structures on the development of western architectural theory, resulting in three distinct approaches to tectonics; Classical Tectonics, Rational Tectonics and Poetic Tectonics.

He is also interested in urban theory and design and has been actively involved in planning and zoning in the City of Atlanta for the past five years, chairing the Midtown Land Use Committee and serving on the Development Review Committee for SPI-16 and 17.

Ermal Shpuza

Best Researcher/Scholar, Fine Arts

Associate Professor of Architecture at Southern Polytechnic State University. His research interests span across complex systems in cities, architectural and urban morphology, urban history and design computing. His work is focused on the study of evolution of Adriatic and Ionian coastal cities, interaction between boundary shape and circulation network in buildings and cities, development of descriptive measures of shape, impact of physiography on street networks and effects of building floorplate on daylight and acoustics. He teaches design studios focusing on urban design and environmental systems and courses on research methods, morphology and design, urban design theory and environmental technology. He received his architectural professional degree from Polytechnic University of Tirana, MSc in Architecture from the Bartlett, University College London, and PhD in Architecture from Georgia Institute of Technology.

Chris Welty

Most Helpful to Students, Architecture
Most Helpful to Students, Fine Arts

SOUTHERN UTAH UNIVERSITY

Mark Miller

Best Researcher/Scholar, History

"Mark has published two major books on Native American tribal identity in the last 5 years. That is a remarkable accomplishment for anyone dealing with the heavy teaching load and administrative responsibilities at this institution."

Professor Mark E. Miller (Ph.D. University of Arizona, 2001) teaches general United States history, the history of the American West, Native Americans, Utah, and modern World History. His published works include articles in the Journal of Arizona History, Journal of Mormon History, Religious Educator, and Journal of the Southwest, and a chapter in the forthcoming edition of the textbook Utah's History. He has two books,

Forgotten Tribes (University of Nebraska Press) and Claiming Cherokee (University of Oklahoma Press), published since 2004, and has been quoted in the Washington Post, Wall Street Journal, and other newspapers in the last few years. He is married with three children.

Virginia Stitt

Best Teacher, Music

SOUTHWEST TENNESSEE COMMUNITY COLLEGE

Ken Dunn

Best Researcher/Scholar, Graphic Arts
Best Teacher, Graphic Arts
Best Researcher/Scholar, Fine Arts
Best Teacher, Fine Arts
Most Helpful to Students, Fine Arts
Best Overall Faculty Member, Fine Arts

As the program coordinator of the Graphic Arts Technology program here at Southwest Tennessee Community College in Memphis, Tennessee, my goal and that of all of our faculty members is to help students learn to use graphic arts to communicate their own ideas and those of their clients or employers in order to attain their personal, career and business goals. Our students learn to use software tools to solve complex and difficult business problems in an efficient and cost-effective manner. What does that mean? In a recent New York Times column, Pulitzer Prize winner Thomas Friedman sketched a scenario that precisely characterizes our graduates.

Patsy Fancher

Best Overall Faculty Member, Graphic Arts
Most Helpful to Students, Graphic Arts

Jerome Wilson

Best Teacher, English

SPELMAN COLLEGE

Beverly Guy-Sheftall

Best Researcher/Scholar, English

"Dr. Guy-Sheftall is a member of the English Department and the founder of the Women's Research and Resource Center, and the Women's Studies Program at Spelman College. She holds the Anna Julia Cooper Endowed Chair. The Spelman College Women's Studies Program is still the only one that grounds its study of women and feminist politics in activism. She is perhaps the most influential scholar at the College, and her work here continues to distinguished the College among its peer institutions."

Beverly Guy-Sheftall, Ph.D., is the founding director of the Women's Research and Resource Center and the Anna Julia Cooper Professor of Women's Studies. She is also adjunct professor at Emory University's Institute for Women's Studies where she teaches graduate courses.

At the age of 16, she entered Spelman College where she majored in English and minored in secondary education. After graduation with honors, she attended Wellesley College for a fifth year of study in English. In 1968, she entered Atlanta University to pursue a master's degree in English; her thesis was titled, "Faulkner's Treatment of Women in His Major Novels." A year later, she began her first teaching job in the Department of English at Alabama State University in Montgomery, Ala. In 1971, she returned to her alma mater Spelman College and joined the English department.

She has published a number of texts within African-American and women's studies, which have been noted as seminal works by other scholars, including the first anthology on Black women's literature, Sturdy Black Bridges: Visions of Black Women in Literature (Doubleday, 1980), which she co-edited with Roseann P. Bell and Bettye Parker Smith; her dissertation, Daughters of Sorrow: Attitudes Toward Black Women, 1880–1920 (Carlson, 1991); Words of Fire: An Anthology of African American Feminist Thought (New Press, 1995); and an anthology she co-edited with Rudolph Byrd titled Traps: African American Men on Gender and Sexuality (Indiana University Press, 2001).

Her most recent publication is a book co-authored with Johnnetta Betsch Cole, "Gender Talk: The Struggle for Women's Equality in African American Communities" (Random House, 2003). In 1983, she became founding co-editor of Sage: A Scholarly Journal of Black Women that was devoted exclusively to the experiences of women of African descent.

SPRING ARBOR UNIVERSITY

Bruce Brown

Best Researcher/Scholar, Music

Dr. Jonathan Bruce Brown is a professor of music and chair of the music department at Spring Arbor University. He is an active composer and serves as composer in residence with the Jackson Symphony Orchestra. He also hosts the JSO's pre-concert lecture series known as Backstage Glimpses.

Dr. Brown's modern yet lyrical compositional style has made him a frequent recipient of commissions for new works. In recent years, he has completed commissions for the Jackson Symphony Orchestra, Brass Band of Battle Creek, District VIII of the Michigan School Band and Orchestra Association and the Jackson Chorale. He was also commissioned to write festive works for the 15th anniversary celebration of the South Carolina School for the Arts, the opening of the new music building at Central Michigan University, and the dedication of a new library at Spring Arbor University. Performances of his works have taken him across the country.

His composition, Legacy of Vision: Martin Luther King, has been performed in many venues including Constitution Hall in Washington D.C. In 1992, it was included in a nationally-syndicated radio program produced by the University of Illinois. His Symphony For String Orchestra has been recorded by the Prague Radio Symphony Orchestra, and is available on compact disc through MMC Music in Boston, Mass. After receiving a bachelor's degree from Central Michigan University, he studied composition with Neil McKay at the University of Hawaii, where his thesis composition, Three Pieces for Orchestra, was performed by the Honolulu Symphony.

Dr. Brown completed his doctoral studies under Martin Mailman at The University of North Texas, where his works won top honors in competitions in New York, Texas and California. At SAU, he was honored to receive the 1991 Sears-Roebuck Foundation Award for Teaching Excellence and Campus Leadership, and Faculty Merit Awards in 1988 and 1995. His compositions have also been recognized by annual awards from the American Society for Composers, Authors and Publishers (ASCAP) since 1992.

In addition to his teaching duties in music, Dr. Brown has led SAU cross cultural programs in the Czech Republic and the Republic of South Africa.

ST. AMBROSE UNIVERSITY

Debbie Sanborne

Best Teacher, Music

ST. CHARLES COMMUNITY COLLEGE

Boxerman

Best Teacher, History

"Dedicated to teaching students the history that he believes is important for them to learn."

Barak Cook

Best Overall Faculty Member, History

Kirk Hanser

Best Overall Faculty Member, Music
Most Helpful to Students, Music

Erin Haupt

Best Teacher, Music

Grace Moser

Best Researcher/Scholar, History

Daniel Rezny

Best Teacher, History

Mary Sweetin

Best Researcher/Scholar, Music

Mary Sweetin is a classically trained violinist, with Bachelors and Masters Degrees in Music from SIUE. Mrs. Sweetin performs with numerous ensembles in the St. Louis area, including the Bach Society Orchestra, St. Louis Ballet Orchestra and the American Chamber Chorale Orchestra, to name a few. She has been the first violinist with the Landolfi String Quartet for over 15 years, and has written over 75 string arrangements for this ensemble. Mrs. Sweetin recently accepted a post as Concertmaster of the University City Orchestra, and is excited to begin her first season with them.

As an Adjunct Faculty Member SCC, Sweetin serves as Orchestra Director, Violin Instructor, and teaches several sections of Music Appreciation. Since accepting the position at SCC in 2008, the SCC Symphonic Orchestra has performed a major symphonic work every semester, including such works as Scheherazade, by Rimsky-Korsakov, Symphonie Fantastique, by Hector Berlioz, and Holst's The Planets.

Mrs. Sweetin enjoys private teaching at SCC, and offers lessons to music majors and beginners. alike. Sweetin performs an annual Faculty Violin Recital at SCC, performing many of the standards of the violin repertoire, including Bloch's Baal Shem, Suite Italienne, by Igor Stravinsky, Violin Sonata in G minor by Claude Debussy, and The Lark Ascending by Ralph Vaughan-Williams, to name a few.

In May 2012 Sweetin was named Adjunct Faculty of the Year in the Humanities Division at St. Charles Community College.

ST. CLOUD STATE UNIVERSITY

Brad Chisholm

Best Overall Faculty Member, Fine Arts

Sharon Cogdill

Best Researcher/Scholar, English

1983, Michigan State University; Director, Dr. Richard Benvenuto
Dissertation: "Thomas Hardy's Conception of Consciousness and the Narration of his Novels"

Areas of Comprehensive Examinations: Victorian Literature, 1814–1914; 19th-Century American Literature; Writing and Composition Theory
B.A.
1972, State University of New York at Fredonia, summa cum laude
Texas Christian University, Honors Program (freshman, sophomore)
Additional Coursework and Formal Professional Development
Digital Humanities Summer Institute, University of Victoria, Victoria, Canada
Brown University, Rhode Island (with NEH-supported scholarship) Digital Humanities Observatory, King's College, Dublin (with NEH-supported scholarship) William Wilmot, advanced mediation workshop in large groups; long-term, persistent conflict; intake (Spring 2005); SCSU Mediation Program Anti-racism Organizing for Institutional Change (2004–2005, about 10 full days); MCARI, Minnesota Collaborative Anti-racism Initiative Leah Wing, training for mediation program in critical race theory (Spring 2003); Janelle Soderquist, training for mediation program (Spring 2003); Bureau of Mediation Services, State of Minnesota Private Mediation Training Group: Interest-based Mediation Model in a Collective-bargaining Environment (1997–99) Management Analysis Division, Minnesota Department of Administration: Performance Measurement Training (Outcome-based Performance Measures) (1998) Council of Colleges of Arts and Sciences (CCAS): Seminar for New Deans (1998) Harvard Extension: Data Structures (1984), Pascal Programming (1984) Bookbuilders Guild Technical Seminars: Paper Manufacturing (1988), Four-Color Film Preparation (1984), Introduction to Paper (1984)

Philippe Costaglioli

Best Teacher, Fine Arts

Philippe has been teaching at St Cloud State since 1991. Raised and educated in Europe [France and Catalunya (Barcelona)] Philippe is a native speaker of Catalan, French and Spanish. As a scholar, his interests, passion and expertise have been nourished by four parallel fields: Sociolinguistics, philosophy of language (and translation), comparative literature study and last but not least film studies.

Philippe did his Masters (1989), doctoral work (1991) and post doctoral work (1998) at the University of Toulouse, France as well as in collaboration with the University of Barcelona (Catalunya). As a scholar, Philippe has explored connections between spirituality and sensuality, problems of symbolic and visual representations in contemporary discourse and film. International cinema is a crucial component of his teaching material.

Philippe is also a published poet as well as a performer. He constantly pursues collaborations with composers, musicians, painters, photographers and of course filmmakers. His work has been shown in the states, France, Spain and the Czech Republic.

As a teacher, Philippe wants to instill in his students a passion for the aesthetics and ethics of cinema through developing a taste and ability for analytical and symbolic interpretation of a film. Philippe also encourages his students to creatively explore what they have learned throughout the semester (through film, writing, photography and even music).

Philippe is currently working for our Program as a Liaison with the University of Lyon (France) to build for our students an extraordinary opportunity to go study film in the city of the Lumiere brothers.

Glenn Davis

Best Overall Faculty Member, English

Judith Dorn

Best Teacher, English

Rebecca Nelson-Crowell

Best Overall Faculty Member, Communication Disorders

"The students love her teaching style. The faculty appreciate her leadership and diplomacy as the chair of the department and graduate directors. The clients love her clinical skills and advocacy on their behalf in the public sector. The administration and Dean love her because she is very successful in gaining grants and funding for the CSD program. She is the all-around winner."

Shannon Olson

Most Helpful to Students, English

Shannon Olson, associate professor of English, has been applying her unique wit and writing style to the "Last Word" essays on the last page of Minnesota Public Radio's Minnesota Monthly magazine since May 2011. Olson, author of the sardonic semi-autobiographical novels "Welcome to My Planet" and "Children of God Go Bowling," has been teaching English at St. Cloud State for 10 years.

Olson's relationship with Minnesota Monthly began when editor Joel Hoekstra was seeking a writer to do a "Last Word" column for Mother's Day. He and others on the magazine's editorial staff thought of Olson, whose novels and other writings frequently include her mother, Flo. Olson's family stories have drawn praise from many critics and writers, including Garrison Keillor, who referred to Flo as "one of the great moms of American fiction."

ST. EDWARD'S UNIVERSITY

Stacey Borasky

Best Overall Faculty Member, Behavioral & Social Sciences

ST. JOHN'S UNIVERSITY

Roderick Bush

Best Overall Faculty Member, Social Science
Best Researcher/Scholar, Social Science
Most Helpful to Students, Social Science

Roderick D. Bush, (Ph.D. SUNY Binghamton) Associate Professor of Sociology specializes in race and ethnicity, the black experience, social movements, world-systems studies, globalization, social inequality, social change, urban sociology, community organizing and political sociology.

Elaine Carey

Best Overall Faculty Member, History
Best Researcher/Scholar, History

Elaine Carey is an Associate Professor and Chairperson of the Department History at St. John's University in Queens, New York. She also holds the Lloyd Sealy Research Fellowship at CUNY's John Jay College of Criminal Justice. She is the author of Plaza of Sacrifices: Gender, Power, and Terror in 1968 Mexico (2005), co-editor with Andrae Marak of Smugglers, Brothels, and Twine: Transnational Flows of Contraband and Vice in North America (2011), and Women Drug Traffickers: Mules, Bosses, and Organized Crime (2014) and the forthcoming Protests in the Streets: 1968 Across the Globe.

As a historian who researches crime and human rights, she has served as an expert witness in courts across the United States, and she has consulted for radio, film, television, archives, libraries, and museums. In January 2013, she began a three-year term as the Vice-President for the Teaching Division of the American Historical Association (AHA).

Richard Thomas

Most Helpful to Students, Communication

Richard Thomas is Chair and an Assistant Professor in the Division of Mass Communications, Journalism, Television and Film earned his MFA Degree from the University of London. His experiences include positions as a filmmaker, playwright, actor, reporter and photojournalist. For the past two years, Mr. Thomas has taught as an adjunct at St. John's University.

ST. JOHN'S UNIVERSITY - STATEN ISLAND

Chiara Cillerai

Best Researcher/Scholar, English

ST. LOUIS COMMUNITY COLLEGE AT FLORISSANT VALLEY

Sharon Fox

Most Helpful to Students, Liberal Arts, Social Sciences, and Humanities

ST. LOUIS COMMUNITY COLLEGE AT MERAMEC

Sophia Pressman

Best Teacher, Humanities

ST. MARY'S COLLEGE OF CALIFORNIA

Cynthia Ganote

Best Teacher, Sociology

ST. PETERSBURG COLLEGE (ALL CAMPUSES)

Anthony Ballard

Best Researcher/Scholar, Communication
Best Teacher, Communication

It's been my pleasure and an honor to be a part of the Communications Department at St. Petersburg College since 1992. I received a B.A. in English from Elizabeth City State University in North Carolina, and a M.A. in Speech-Language Pathology from South Carolina State University. Prior to joining the SPC faculty, I was a speech-language pathologist in the Pinellas County School District. In addition to teaching the basic speech course for the college, I also teach Interpersonal Communication. My philosophy of education is based on the belief that education is a basic human right. The society that fails to educate its people undermines its own progress and security in what should be an ever advancing world civilization. In one holy text it is written, "Regard man as a mine rich in gems of inestimable value. Education can, alone, cause it to reveal its treasures, and enable mankind to benefit therefrom." This view undergirds my belief that the educator should act as a facilitator of growth and discovery in a cooperative process that enables the realization of the potential that lies latent in each student. My interests include spending time with my family, personal spiritual growth, writing songs based on social and spiritual concepts, playing guitar, cycling on the trails of Florida, and the amazing sport of racquetball.

Matt Bowen

Most Helpful to Students, Liberal Arts, Social Sciences, and Humanities

Evelyn Finklea

Most Helpful to Students, Communication

Sheryl Peterson

Best Overall Faculty Member, Social Science
Best Teacher, Social Science

Sheryl S. Peterson received her Bachelor's degree from University of Florida, her Master's degree in Counseling Psychology from Anna Maria College, and is currently working on her Doctorate from the University of South Florida. She has been associated with St. Petersburg College since 1981, when she left teaching Psychology to become the counselor of Project Support. She then became the coordinator of Distance Learning until August 1989. At this time she became a member of St. Petersburg College Social Sciences faculty until 1996, when she became Program Director of Social and Behavioral Science on St. Pete/Gibbs Campus. In 2003 she moved to Seminole's eCampus to become a fulltime online instructor and is currently the Academic Chair of Social Science on Seminole Campus.

Julia Rawa

Best Overall Faculty Member, Communication
Best Researcher/Scholar, Liberal Arts, Social Sciences, and Humanities
Best Teacher, Liberal Arts, Social Sciences, and Humanities
Best Overall Faculty Member, Liberal Arts, Social Sciences, and Humanities

My primary research focuses on global history as a theoretical approach to aesthetics and literary studies. In 2008, I was honored to win a National Endowment for the Humanities grant to pursue study of W. B. Yeats at the National University of Ireland and Trinity College I am committed to innovative, interdisciplinary, and student-centered teaching. To that end, I encourage students to read assiduously and to appreciate, as Aldous Huxley rightly observes, that "every [individual] who knows how to read has it in his power to magnify himself, to multiply the ways in which he exists, to make his life full, significant and interesting."

Carmen Simpson

Most Helpful to Students, Liberal Arts, Social Sciences, and Humanities

Pamela Wilkins

Most Helpful to Students, Social Science

BA - University of South Florida, Criminology MS - Troy State University, Counseling and Psychology Professional Certification: Guidance Counselor K–12, State of Florida

ST. THOMAS UNIVERSITY

Joe Holland

Best Overall Faculty Member, Philosophy
Best Researcher/Scholar, Philosophy
Best Teacher, Philosophy
Most Helpful to Students, Philosophy

Michael Mulvey

Best Overall Faculty Member, History

I am a European and Global historian interested in housing, aging, finance, the environment, and laughter. My research looks at these issues in post-1789 France with a focus on the postwar and postcolonial period. The challenges of the last two centuries were global; France reminds us that cultures responded to them differently. As a teacher and scholar, I trace the intellectual sources of responses to real or perceived social problems that have biopolitical implications. In other words, how do historical actors arrive at solutions to social questions and how do those solutions bear on sexuality, procreation, health, human relations, life and death?

STEPHEN F. AUSTIN STATE UNIVERSITY

Jennifer Dalmas

Best Overall Faculty Member, Music
Best Teacher, Music
Most Helpful to Students, Music

Dr. Jennifer Dalmas, a native of Lynchburg, Virginia, joined the music faculty of Stephen F. Austin University in fall 2003. Dr. Dalmas graduated from the Virginia School of the Arts in 1991 and received her Bachelor's music degree from Mary Washington College in 1995. For the next six years, she held a teaching and performing assistantship at Florida State University, where she studied with Professor Karen Clarke and Dr. Gary Kosloski. After receiving the Doctor of Music Degree in 2001, she served as a Visiting Violin Instructor at Florida State University the following year.

While living in Florida, she regularly performed with the Tallahassee, Jacksonville, and New World Symphonies. Dr. Dalmas has also played with several period orchestras, including the Atlanta Baroque Orchestra and the Apollo Ensemble. She has performed solo repertoire such as Zigeunerweisen and various concerti with orchestras in the United States and Europe, and as part of the Promising Artists of the 21st Century program has given recitals and master classes both in the United States and Central America. She won the doctoral concerto competition at Florida State University in 1999, performing the Glazunov violin concerto with the FSU orchestra and was awarded second prize in the National MTNA competition the following year. As a member of the Hachidori Piano Trio, she has been a semi-finalist in the Fischoff Chamber Music Competition, and has performed the Beethoven Triple Concerto with orchestras in the United States, the Czech Republic, and Bulgaria.

In Texas, Dr. Dalmas has been an active orchestral musician, and has served as concertmaster for the Orchestra of the Pines, in addition to performing with other regional symphonies. An avid chamber music player, she has been a member of both the Alazan Piano Trio and East Texas Baroque, and also serves as the music director of the Chamber Music in the Pines summer festival. She is active as an adjudicator and clinician, and in 2009 was invited to present at the national A.S.TA convention.

Her string students have won numerous awards and competitions, and have enjoyed summer studies at prestigious music festivals such as Brevard, Eastern, and Meadowmount.

Jack Heifner

Best Overall Faculty Member, Theater

"Jack Heifner is the most knowledgable, experienced, passionate and creative faculty member in the School of Theatre. He sincerely cares about the students and their growth as artists and individuals."

Jack Heifner is best known for his play Vanities, which ran for five years in New York and became one of the longest running plays in off-Broadway history. He is also the author of Patio/Porch, Natural Disasters, Running on Empty, Bargains, Boys' Play, Home Fires, Heartbreak, Comfort and Joy, The Lemon Cookie, Dwarf Tossing and over thirty other plays produced in New York, Los Angeles and theaters around the world. Mr. Heifner has written the books to six musicals, including Leader of the Pack on Broadway and Vanities—A New Musical, which opened in New York in 2009. He has also written for television and film. Mr. Heifner was born in Corsicana, Texas and lives in New York City. Since 1997, he has been playwright-in-residence at SFA where he teaches play, screen and television writing one semester each academic year. He is the director of The Festival of New American Plays at SFA featuring works by major American playwrights. At SFA, he has also directed Home Fires, The Seagull, The Member of the Wedding and How to Succeed in Business Without Really Trying.

STETSON UNIVERSITY

Tom Farrell

Best Researcher/Scholar, English

"Recognized by Southeast Medieval Studies Association as top scholar in the region. Brings breathtaking erudition to every conversation, form the hallway to the classroom – even to meetings."

STONEHILL COLLEGE

Allyson Sheckler

Best Teacher, Fine Arts

Allyson Sheckler joined the faculty of the Fine Arts and Theatre Arts Department in 1995. She teaches Art History period courses including Greek and Roman Art, the Middle Ages, the Renaissance and the Seventeenth Century

SUFFOLK COUNTY COMMUNITY COLLEGE

Joseph Gansrow

Best Teacher, English

SUFFOLK UNIVERSITY

Paul Andrade

Most Helpful to Students, Art & Design

"Paul Andrade is simply one of the most helpful teachers at NESAD. He comes to NESAD 5 days a week to teach or monitor the wood shop lab. After class hours, I have overheard some of his very excellent critiques he has given to his students. Also I had the opportunity to view many of his student projects which are often both innovative and well thought out."

1990 BFA, Rhode Island School of Design 1997 MFA, Mason Gross School of the Arts Rutgers University Foundation Faculty & Fabrication Lab manager, New England School of Art and Design @ Suffolk University

Charles Cramer

Best Teacher, Liberal Arts, Social Sciences, and Humanities

His primary area of specialization is the theory of European Classicism from Polykleitos to Postmodernism, with a primary historical focus on late-eighteenth and early-nineteenth century France and England, though he also has interests in late-nineteenth century French Impressionism and Symbolism, in the early-twentieth century Cubist milieu, and in semiotics, the philosophy of mind and cognitive science.

Jennifer Fuchel

Most Helpful to Students, Fine Arts

"Jennifer Fuchel is so incredibly helpful to students. During finals week, I have seen her assist students into the wee hours of the night and into the weekend. She is so considerate and kind and really wants her students to understand graphic design."

Concept is king! Although I teach many technically oriented classes, I stress developing strong concepts, and then learning the skills necessary to execute the design. I try to expose students to as many areas of design as possible - to me, they are all interrelate. I help students see connections and become more comfortable with designing multi-disciplinary and multi-dimensional design works.

Micky Lee

Best Overall Faculty Member, Liberal Arts, Social Sciences, and Humanities

"A good researcher with a steady stream of academic articles published. A good teacher with a dry sense of humor and a well-organized lecture. Guides her students with a careful and caring hand."

Micky Lee is a feminist, a Hong Kong citizen, a British national, and a US resident.

Wallace Marosek

Best Researcher/Scholar, Fine Arts

"Wallace Marosek gives his entire mind, body and soul to his students. His research skills and knowledge in Graphic Design and Illustration are to be commended."

I enjoy the process of learning, the thrill of being presented with new problems, experimenting to find solutions, conquering new challenges, celebrating accomplishments. As well I enjoy the structure of learning – being given the freedom to discover my own way while at the same time being guided, the give-and-take between mentor and apprentice, the luxury of time dedicated to exploring new concepts, talents, and abilities. This is precisely my approach to teaching – giving each student the freedom to find his/her own way while at the same time imparting the rigors of classic design elements, ultimately guiding each student to surpass his/her own limit.

Steve Novick

Best Overall Faculty Member, Fine Arts
Best Teacher, Fine Arts

"Steve Novick is an innovative and all around amazing teacher. He really assists students in their projects and really gets them to think about creativity and what it all entails.

Steve Novick is one of the finest teachers I have ever known. He really assists students in tying art, design and fine arts all together."

Sean Solley

Best Researcher/Scholar, Art & Design

Randall Thurston

Best Overall Faculty Member, Art & Design
Best Teacher, Art & Design

"Randal is an incredible teacher. Just by viewing many of the innovative student projects he has developed for his classes, I am always amazed at how much I have also learned from him.

Randall Thurston simply is one of the finest teachers at NESAD/Suffolk University. Patient, kind, articulate, intelligent and really gets students to think on all aspects of art, design and fine arts. I have never heard a student speak negatively about his teaching methods. All praises."

SUNY DELHI

William Campbell

Best Overall Faculty Member, Humanities
Best Researcher/Scholar, Humanities
Best Teacher, Humanities

Most Helpful to Students, Humanities
Best Teacher, Liberal Arts, Social Sciences, and Humanities

Professor, Liberal Arts and Sciences Division
A.A.S. (Plant Science) SUNY Delhi
B.S. (Liberal Arts) SUNY Oneonta
M.F.A. (Dramatic Arts) The University of Connecticut

Terry Hamblin

Most Helpful to Students, Liberal Arts, Social Sciences, and Humanities

Alicia Thompson

Best Researcher/Scholar, Liberal Arts, Social Sciences, and Humanities
Best Overall Faculty Member, Liberal Arts, Social Sciences, and Humanities

SUNY FREDONIA

John Arnold

Best Researcher/Scholar, History
Best Teacher, History
Best Researcher/Scholar, Liberal Arts, Social Sciences, and Humanities
Best Teacher, Liberal Arts, Social Sciences, and Humanities

James Davis

Best Researcher/Scholar, Music

Jim Davis is Professor of Musicology and Chair of the Musicology Area at The State University of New York at Fredonia School of Music. His primary research focuses on the music and musicians of the American Civil War. He has also worked in the areas of music history pedagogy, American popular music of the 19th and 20th centuries, and the history of bands.

Davis received a BS in Music from the University of Colorado at Denver with an emphasis in scoring and arranging; an MM in composition from the University of Texas; and a PhD in music history and theory from Boston University.

Davis has presented papers at regional and national meetings of the American Musicological Society, Society for American Music, College Music Society, Society for Ethnomusicology, Society for Military History, Music Library Association, and American Culture Association. His research has been published in American Music, College Music Symposium, Journal of Military History, Nineteenth Century Studies, Journal of the History of Ideas, Journal of Band Research, and the Journal of American Culture. His articles on pedagogy have appeared in International Journal of Teaching and Learning in Higher Education, Journal of Aesthetic Education, and Philosophy of Music Education Review.

He is the author of Music Along the Rapidan: Civil War Soldiers, Music and Community During Winter Quarters, Virginia (University of Nebraska Press, 2014) and Bully for the Band! The Letters and Diary of Four Brothers in the 10th Vermont Infantry Band (MacFarland, 2012). He also edited the collections The Music History Classroom (Ashgate, 2012) and Art and the American Civil War, both from Ashgate Publishing. He is currently working on a manuscript that examines the creation and tumultuous reception of the song "Maryland, My Maryland" during the Civil War.

Jennifer Hildebrand

Most Helpful to Students, History
Most Helpful to Students, Liberal Arts, Social Sciences, and Humanities

Barry Kilpatrick

Most Helpful to Students, Music

Barry Kilpatrick is Professor of Trombone/Euphonium at The State University of New York at Fredonia School of Music, where he has been a member of the faculty since 1979. As Assistant Director of the School of Music, he coordinates the music admission process. He also serves as head of the Brass Area.

Principal trombonist of the Erie Philharmonic Orchestra from 1987–2008, and of the Western New York Chamber Orchestra since 1981, Kilpatrick has toured Jamaica and New Zealand with the Fredonia Faculty Brass. He has performed and toured, as a member of the trombone and euphonium sections, with Keith Brion's "New Sousa Band."

Kilpatrick earned degrees in performance at the University of Wisconsin-Madison in the 1970s, where he became acquainted with composers who heard him play euphonium. He played the first performance of Alec Wilder's Euphonium Concerto in 1971, as well as of works by Hilmar Luckhardt, Walter Hartley, and John Davison. These were included in his recording American Music for Euphonium (Mark Records), of which one critic wrote, "Kilpatrick ... plays with a beautiful tone, expressive phrasing, and sure, solid technique. This is an essential disc."

In 1989, Kilpatrick was invited to become a music critic for American Record Guide, an internationally distributed magazine that has since published almost 2000 of his reviews of classical recordings. From 1991–2008, he devoted his summers to the Pennsylvania Governor's School for the Arts, teaching low brass and chairing the Music Department.

Julie Newell

Best Overall Faculty Member, Music Best Teacher, Music

Soprano Julie Newell is a recipient of the SUNY Chancellor's Award for Excellence in Scholarship and the Arts, as well as a recipient of the SUNY Research Foundation Award for Excellence. Ms. Newell has performed major operatic roles such as Mimi, Desdemona, Alice Ford, Countess Almaviva, and Pamina, amongst many others throughout the U.S., with companies including: Arizona, Indianapolis, Memphis, Orlando, Kansas City, Long Beach, Syracuse, Chautauqua, to name a few. She has premiered works by Antonio Vivaldi and Gian-Carlo Menotti to critical acclaim at New York's Lincoln Center. As a concert singer, she has collaborated with numerous conductors including Kazuyoshi Akiyama, David Zinman, Fabio Mechetti and Maximiano Valdes.

Ms. Newell has received support grants from the William Matheus Sullivan Foundation and OPERA America. She was a Metropolitan Opera District winner and has served as a judge for that same internationally recognized competition.

Ms. Newell serves as the Coordinator of Opera at the SUNY Fredonia School of Music in addition to administrating Lyric Arts International, which oversees an operatic exchange with the professional musicians of the Orchestra da camera delle Marché in Italy.

Mary Sievens

Best Overall Faculty Member, Liberal Arts, Social Sciences, and Humanities

John Staples

Best Overall Faculty Member, History

SUNY ONEONTA

Dora Dumont

Best Researcher/Scholar, History
Most Helpful to Students, History

Andrew Kahl

Best Overall Faculty Member, Theater
Best Teacher, Liberal Arts, Social Sciences, and Humanities

Ralph Watkins

Best Overall Faculty Member, History
Best Teacher, History

SUNY PLATTSBURGH

Genie Babb

Best Researcher/Scholar, English

Vincent Carey

Best Researcher/Scholar, Liberal Arts, Social Sciences, and Humanities

Elizabeth Cohen

Best Teacher, English

Rick Davies

Best Overall Faculty Member, Music

Rick Davies is a composer/trombonist who works primarily in the fields of Latin jazz, salsa, and Afro-Caribbean music. He is professor of music and was chair of the music department at SUNY Plattsburgh.

Davies has had an extensive career with major artists in New York City and around the world. Artists he has worked with include Wayne Gorbea, Tito Puente, Charlie Palmieri, Wyclef Jean, Arrow, Manny Oquendo's Libre, Blondie, and many others. He has traveled extensively both in the United States and around the world. Among the places he has appeared are Tokyo, Sydney, Budapest, Amsterdam, Paris, Great Britain, Mexico, Colombia, and throughout the Caribbean. Salsa Norteña is his third CD release and contains eight high-energy salsa and Latin jazz songs.

Davies received a Ph.D. from New York University in 1999 (his dissertation was on Cuban brass performance). In 2009, Rick Davies received the SUNY Chancellor's Award for Scholarship and Creative Activity. As a scholar, Davies has written several monographs, chapters, and articles primarily about Afro-Cuban music and salsa. His book, Trompeta: Chappottín, Chocolate, and the Afro-Cuban Trumpet Style, was released in May 2003 by Scarecrow Press. He contributed a chapter on Alfredo "Chocolate" Armenteros to Music From Cuba (Praeger Press, 2001). Other articles he has written include "The Cuban Sonero: Beny Moré and Miguelito Cuní," "Miguelito Valdés, the First Salsa Superstar," "The Latin Jazz Continuum," and "Americanizing the Eurocentric Music Curriculum".

Thomas Friedrich

Best Overall Faculty Member, English

Tom Friedrich is Assistant Professor of English and Director of Freshman Composition at the State University of New York College at Plattsburgh. He holds a Ph.D. in Literacy Education with a Minor in Literacy and Rhetorical Studies from the University of Minnesota.

Pat Hoffmann

Most Helpful to Students, English

SUSQUEHANNA UNIVERSITY

Maria Munoz

Best Teacher, History
Best Teacher, Liberal Arts, Social Sciences, and Humanities

Muñoz completed her doctoral degree at the University of Arizona in Latin American History with a minor in Cultural Anthropology. Her dissertation was entitled, "We Speak for Ourselves": The First National Congress of Indigenous Peoples and the Politics of Indigenismo in Mexico, 1968–1982." She also co-edited the volume, Men of the People: The Presidencies of Lazaro Cardenas and Luis Echeverria, (University of Arizona Press, 2010). She just finished a chapter titled "De Pie y en Lucha! Indigenous Mobilizations After 1940" for an edited volume as well as a contributor on the encyclopedia project, Native Peoples of the World. Currently she is working on her book manuscript. Muñoz's research interests include social movements, race, ethnicity, identity construction and nation-state building. She teaches Latin America (colonial and modern surveys), Mexico Since Independence, Race and Nation in Modern Latin America, and Latin American Revolutions, among others.

Edward Slavishak

Best Overall Faculty Member, History
Best Overall Faculty Member, Liberal Arts, Social Sciences, and Humanities

Karol Weaver

Best Researcher/Scholar, History
Most Helpful to Students, History
Best Researcher/Scholar, Liberal Arts, Social Sciences, and Humanities
Most Helpful to Students, Liberal Arts, Social Sciences, and Humanities

Weaver offers courses on American history, women's history and the history of medicine. Her research interest focuses on the history of healers. Her first book, Medical Revolutionaries: The Enslaved Healers of Eighteenth-Century Saint Domingue, was published by the University of Illinois Press in 2006. In 2011, her book Medical Caregiving and Identity in Pennsylvania's Anthracite Region, 1880–2000, was published by Penn State Press. Weaver also serves as the coordinator of the Women's Studies Program.

SYRACUSE UNIVERSITY

Steven Davis

Best Teacher, Communication

Steve Davis has worked in newsrooms of all kinds and of all sizes as a manager, editor, writer and page designer. He was executive editor at the community newspaper in Chambersburg, Pa., and also served as both national editor and Washington editor at USA TODAY. His time there included one year as the lead editor directing the newspaper's coverage of the Persian Gulf War.

He also worked in Jackson, Miss. and Rochester, N.Y. on the sports, local and national desks. In the mid-1980s, he worked for Gannett Co., Inc. in Washington, D.C. on a research and development project to create some of the earliest prototypes for personalized, online news.

He is the co-author of "Click on Democracy: The Power of the Internet to Change Political Apathy into Civic Action" (Westview Press).

Notable projects include The Stand, a community newspaper and companion website serving Syracuse's South Side which Davis created in 2009, and his blog, Think Like an Editor, which he runs with his wife, Emilie, an adjunct at Newhouse.

In 2011, Davis took a group of students to Grahamstown, South Africa where they created the multimedia World Journalism Project. He was also one of three Newhouse faculty members who led the News 21 student team that produced Apart from War in 2010

Dennis Harrod

Best Teacher, Liberal Arts, Social Sciences, and Humanities

Mary Ann Stanley

Best Teacher, French

TACOMA COMMUNITY COLLEGE

Andrew Cho

Best Overall Faculty Member, Sociology

TARRANT COUNTY COLLEGE (ALL)

Rebecca Balcarcel

Best Researcher/Scholar, English
Most Helpful to Students, English

Anthony Burns

Most Helpful to Students, English

I am an instructor with a diverse background in composition, first year and developmental writing programs. Direct experience with traditional and non-traditional classrooms settings and students. Focus on critical thinking, analysis, persuasion, organization, form and content.

Mary Cinatl

Best Researcher/Scholar, Liberal Arts, Social Sciences, and Humanities
Most Helpful to Students, Liberal Arts, Social Sciences, and Humanities
Best Overall Faculty Member, Liberal Arts, Social Sciences, and Humanities

Lee Cowan

Best Teacher, History
Best Researcher/Scholar, Liberal Arts, Social Sciences, and Humanities

Eric Devlin

Best Overall Faculty Member, English

Mike Downs

Best Researcher/Scholar, History

Angela Drummonds-Mathews

Most Helpful to Students, English

Candace Eldridge

Best Teacher, Speech Communication

Robin Fletche

Best Teacher, Liberal Arts, Social Sciences, and Humanities

Cheryl Hamilton

Best Overall Faculty Member, Liberal Arts, Social Sciences, and Humanities

Gregory Kosc

Best Overall Faculty Member, History
Best Teacher, History
Most Helpful to Students, History
Best Researcher/Scholar, Liberal Arts, Social Sciences, and Humanities
Best Teacher, Liberal Arts, Social Sciences, and Humanities
Most Helpful to Students, Liberal Arts, Social Sciences, and Humanities
Best Overall Faculty Member, Liberal Arts, Social Sciences, and Humanities

"Greg could very well be the brightest individual we have working for us at TCC. Not only is he profoundly intelligent, he has a powerful work ethic to go along with it. I can honestly say that I would not have been able to accomplish nearly as much as I have at TCC so far without his help. We often spend hours each week discussing plans to improve our history department and make history more meaningful, engaging, and interesting for our students. He works tirelessly trying to get outside scholars from out-of-state universities to come give talks at our campus, and the events he has helped organize have truly changed many of our students' lives."

Paulinda Krug

Best Researcher/Scholar, Liberal Arts, Social Sciences, and Humanities

Cristina Martinez

Best Overall Faculty Member, Speech Communication
Best Researcher/Scholar, Speech Communication
Most Helpful to Students, Speech Communication
Most Helpful to Students, Liberal Arts, Social Sciences, and Humanities
Best Overall Faculty Member, Liberal Arts, Social Sciences, and Humanities

Hector Menchaca

Best Teacher, Liberal Arts, Social Sciences, and Humanities

A.A. Tarrant County College B.A Texas Wesleyan University M.Ed. University of North Texas Additional coursework at Texas Woman's University, Tarleton, Texas A&M Commerce Faculty member/Administrator

Charles Overstreet

Most Helpful to Students, Sociology

"Experience and personality!"

Karen Parsons

Best Teacher, Music

Rick Stitzel

Best Overall Faculty Member, Music
Best Teacher, Music
Most Helpful to Students, Music

"Outstanding colleague, kind, compassionate and most professional."

Natalia Verjat

Best Teacher, Liberal Arts, Social Sciences, and Humanities

Chad Wooley

Best Researcher/Scholar, Liberal Arts, Social Sciences, and Humanities

TAYLOR UNIVERSITY

Faye Chechowich

Most Helpful to Students, Liberal Arts, Social Sciences, and Humanities
Best Overall Faculty Member, Liberal Arts, Social Sciences, and Humanities

"terrific encourager and mentor; a champion of faculty and students alike; wonderful administrator too"

Phil Collins

Best Overall Faculty Member, Liberal Arts, Social Sciences, and Humanities

PERSONAL INFORMATION
Born: Madison, Wisconsin
Wife: Terri Collins; Married October 1, 1983
Children: Luke and Katelyn
Ordination: Ordained 1994 Castleview Baptist Church Indianapolis, Indiana
Licensed: Licensed 1988 First Baptist Church of Geneva Geneva, Illinois

Linda Deneau-Manganello

Best Teacher, Liberal Arts, Social Sciences, and Humanities

Kevin Diller

Best Researcher/Scholar, Philosophy
Most Helpful to Students, Philosophy

I can still recall the moment in room 212 of Bergwall Hall when it dawned on me that I would vastly prefer studying and teaching theology and philosophy to becoming one of the world's elite computer scientists. I had, up until that point, been a computer science major, but soon switched to philosophy. I doubt that the world was deprived that day of any great future advances in computer science, but I am quite confident that it was a very good decision—after all, it is a decision that 19 years later would bring me back to Taylor University.

I have a deep sense of gratitude as I reflect on the journey of my life to this point. After graduating from Taylor I enjoyed the luxury of spending time in colleges and universities pursuing the philosophical aspects of the theological questions that have been most significant for me personally. At the first of those destinations—Princeton, NJ—I met and was married to a phenomenal partner in life's journey, who is of course my wife, Gwen. We went on to study in Grand Rapids, MI; Marburg, Germany; St Andrews, Scotland; and finally, Notre Dame, IN.

Along the way, I always said that I thought I would be best suited to teach in an environment similar to that of my own undergraduate experience. That we would actually come back to Taylor would have been too good to have hoped for. And yet, by God's grace, here we are.

I find myself deeply appreciating three aspects of the position I hold at TU. The first is the opportunity to work with Taylor students, who—by and large—are hungry to learn, grow and serve. I can get fairly excited about the possibility of serving as a catalyst for propelling that learning and growth in students who really desire it. The second is the opportunity to work in a very supportive yet highly energized collegiate environment. The esprit de corps we enjoy at this institution across administration, faculty and staff is a great gift. The third aspect of my position that I deeply value is the opportunity to give and receive in an inclusive, welcoming, tight-knit community. We see here uncommon manifestations of what Jesus called the Kingdom of God, living together with others in grace, forgiveness, celebration and exhortation which presents enormous benefits for the quality of life we and our children can experience.

I'm thrilled to be teaching at Taylor. My prayer is that God, who is made known in Christ, would by His Spirit use my courses in the lives of students to refine their thoughts, hearts and lives for fuller relationship with God and each other.

Al Harrison

Most Helpful to Students, Music
Most Helpful to Students, Liberal Arts, Social Sciences, and Humanities
Best Overall Faculty Member, Liberal Arts, Social Sciences, and Humanities

"Dr. Harrison is the consummate professional and servant leader; he always puts students first and goes above and beyond on their behalf"

I have been a trombonist with several symphony orchestras and bands and am currently playing with the Marion Philharmonic Orchestra and the Mississinewa Valley Band. I also enjoy playing at Church and singing in the Church Choir at Upland Community Church. Over the years I have taken the Taylor Wind Ensemble and Jazz Ensemble on a variety of tours presenting concerts in both domestic and overseas locations. Each year, I serve the Indiana School Music Association and the Indiana Bandmasters Association as both an adjudicator and guest clinician for honors bands. In 1995 I was presented with the Distinguished Professor citation by the Taylor University Alumni Association.

Dennis Hensley

Best Teacher, Liberal Arts, Social Sciences, and Humanities

"known as a master motivator and communicator; industrious and a champion of his students"

Aaron Housholder

Most Helpful to Students, English

Lorne Mook

Best Overall Faculty Member, English
Best Teacher, English

Joseph Ricke

Best Researcher/Scholar, English

Patricia Robertson

Best Researcher/Scholar, Music
Best Researcher/Scholar, Liberal Arts, Social Sciences, and Humanities

I love teaching at Taylor University for numerous reasons, but primarily because I enjoy the day to day interaction with the students that I teach, my colleagues, the administration and support staff here, as well as the deep and enduring friendships these relationships have produced.

Bradley Seeman

Best Overall Faculty Member, Philosophy
Best Teacher, Philosophy

Why I came to Taylor: Dr. Brad Seeman joined the Taylor community in 2006, after teaching at Judson University for a number of years, and has found an intellectual and spiritual home among colleagues who are also passionate about exploring issues at the intersection of philosophy and theology.

Hobbies: Perhaps Brad's favorite activity is taking walks on country roads with Kiersten and the family dog, Nestle. He also enjoys doing nearly anything outside with the kids, cycling, volleyball, singing, playing guitar, reading Scripture on the front deck, having friends over, and pizza/movie nights with Kiersten and the kids. Indeed, having been born and raised in the Chicago area, Brad considers himself something of a deep dish pizza connoisseur, and in his considered judgment, Kiersten makes the best pizza in Grant County, Indiana.

Family

Brad met his wife, Kiersten, at Trinity Evangelical Divinity School in Deerfield in 1993. They were married in 1994 and have been blessed with five children: Grace, Emma, Luke, Elias, and Josiah. They live on a 20 acre farmette three miles from the Taylor campus.

Jim Spiegel

Best Researcher/Scholar, Liberal Arts, Social Sciences, and Humanities

TEMPLE UNIVERSITY

Ayala Guy

Best Researcher/Scholar, Judaic Studies

TENNESSEE TECH UNIVERSITY

Tony Baker

Best Teacher, English

Bob Coogan

Best Overall Faculty Member, Fine Arts

Kristen Deiter

Best Overall Faculty Member, Liberal Arts, Social Sciences, and Humanities

George Webb

Best Researcher/Scholar, History

Kimberly Winkle

Best Researcher/Scholar, Fine Arts Best Teacher, Fine Arts

"Kim is very active as a fine craft artist and a wonderful Foundation professor."

TEXARKANA COLLEGE

Lisa Jones

Best Researcher/Scholar, Liberal Arts, Social Sciences, and Humanities
Best Teacher, Liberal Arts, Social Sciences, and Humanities

Doug Saffel

Best Overall Faculty Member, Social Science
Best Teacher, Social Science
Most Helpful to Students, Social Science
Best Overall Faculty Member, Liberal Arts, Social Sciences, and Humanities

Candy Smith

Best Researcher/Scholar, Social Science

Mary Ellen Young

Most Helpful to Students, Liberal Arts, Social Sciences, and Humanities

TEXAS A&M UNIVERSITY AT COLLEGE STATION

R.J.Q. Adams

Best Researcher/Scholar, History
Best Teacher, History

R.J.Q. Adams received the Ph.D. degree from the University of California, Santa Barbara in 1972 and joined the A&M History Department in 1974. His professional interest is the history of Britain in the 20th century. He became professor of history in 1987, and from 2001 until 2004 was Claudius M. Easley, Jr., Faculty Fellow in Liberal Arts. In 2004, he was named Patricia and Bookman Peters Professor of History. Professor Adams teaches graduate and undergraduate lecture and seminar classes in modern British history; and since 1993 he has directed twelve M.A. and eleven Ph.D. students to completion of their degrees and currently chairs eight doctoral committees. Widely published here and abroad, among Professor Adams's recent books are (with Robin W. Winks) Europe, Crisis and Conflict: 1890–1945 (New York and London: Oxford University Press, 2003); Bonar Law (London, John Murray; Stanford, CA, Stanford University Press 1999); British Politics and Foreign Policy in the Age of Appeasement, 1935–39 (London, Macmillan Ltd.; Stanford, CA., Stanford University Press, 1993 and many reprints) and Balfour: The Last Grandee (London, John Murray, 2007 and 2008). He is currently writing a book about the reign of King George V, 1910–1936.

Terry Anderson

Best Researcher/Scholar, Liberal Arts, Social Sciences, and Humanities

James Bradford

Best Teacher, Liberal Arts, Social Sciences, and Humanities

Bradford grew up in northern Michigan, received degrees from Michigan State University and the University of Virginia, and specializes in naval, maritime, and early American history.

Olga Dror

Most Helpful to Students, Liberal Arts, Social Sciences, and Humanities

Chester Dunning

Best Overall Faculty Member, History
Most Helpful to Students, History

Professor Chester Dunning, winner of four distinguished teaching awards, conducts research in Russian history and early modern European history. He also has a strong interest in comparative history and the impact of censorship on Russian literature. Professor Dunning has received several research grants, including two from the National Endowment for Humanities, as well as a research fellowship from the Harvard University Russian Research Center. His books include Russia's First Civil War: The Time of Troubles and the Founding of the Romanov Dynasty, a History Book Club Selection during Summer 2001; A Short History of Russia's First Civil War; Jacques Margeret's The Russian Empire and Grand

Duchy of Muscovy: A Seventeenth Century French Account; and (with Caryl Emerson and Sergei Fomichev) The Uncensored Boris Godunov: The Case for Pushkin's Original Comedy (forthcoming in Spring 2005). Professor Dunning has also published many refereed articles in leading scholarly journals, including Slavic Review, The Russian Review, The Slavonic and East European Review, Comparative Studies in Society and History, Revue Historique, Albion, Voprosy istorii, Russian History/Histoire Russe, The Sixteenth Century Journal, Harvard Ukrainian Studies, Forschungen zur osteuropäischen Geschichte, and The Pushkin Review.

Adam Seipp

Best Overall Faculty Member, Liberal Arts, Social Sciences, and Humanities

Dr. Adam R. Seipp's research and teaching focus on war and social change in 20th century Europe. He is the author, most recently, of Strangers in the Wild Place: Refugees, Americans, and a German Town, 1945–1952, published in 2013 by Indiana University Press. This book is an "international history of a very small place" - the rural Bavarian community of Wildflecken. In 1936, the Nazi state created a massive military training site near the town. During the Second World War, this base housed an industrial facility that drew forced laborers from all over conquered Europe. At war's end, the base became Europe's largest Displaced Persons camp, housing thousands of Polish refugees and German civilians fleeing Eastern Europe. As the Cold War intensified, the US Army occupied the base, removed the remaining refugees, and stayed until 1994. Strangers in the Wild Place tells the story of these tumultuous years through the eyes of these very different groups, who were forced to find ways to live together and form a functional society out of the ruins of Hitler's Reich. He is also the author of The Ordeal of Peace: Demobilization and the Urban Experience in Britain and Germany, 1917–21, along with articles and book chapters on the First World War, American basing policy in Europe, and re-armament debates in West Germany. He is interested in supervising undergraduate and graduate research in European and Military History, along with the history of sports.

TEXAS A&M UNIVERSITY AT CORPUS CHRISTI

Ruth Aipperspach

Best Teacher, Liberal Arts, Social Sciences, and Humanities

Diana Ivy

Best Researcher/Scholar, Liberal Arts, Social Sciences, and Humanities
Most Helpful to Students, Liberal Arts, Social Sciences, and Humanities
Best Overall Faculty Member, Liberal Arts, Social Sciences, and Humanities

Terry Lewis

Best Overall Faculty Member, Theater
Best Researcher/Scholar, Theater
Best Teacher, Theater

Kelly Russell

Most Helpful to Students, Theater

For the last four years Kelly Russell's "Camp on the Coast" summer theatre workshop has attracted aspiring actors from across Texas and other states to the University campus. Around 100 high school students perform in one-act plays and receive instruction from University faculty and the finest high school theatre teachers in Texas during the two-week camp. The assistant professor of theatre, who is president-elect of the Texas Educational Theatre Association (TETA), will become the organization's president in January 2009. Russell is a frequent workshop presenter for TETA and Texas Speech Communication Association conventions. Last spring his production of "Anima Mundi" won the Directors' Choice Award for outstanding play at the Region VI American College Theatre Festival. Prior to arriving at the University in 2004, he taught theatre arts and speech communication in various Texas high schools and coached seven state champions and two national finalists.

Ethan Thompson

Best Overall Faculty Member, Liberal Arts, Social Sciences, and Humanities

"-Author of a book -Dedicated professor teaching Film Criticism, Television Criticism, Advertising Criticism both in-class and some on-line classes Served as interim Chair of Communication-Media Dept."

TEXAS CHRISTIAN UNIVERSITY

Richard Enos

Best Teacher, English

Donald Frischmann

Best Researcher/Scholar, Liberal Arts, Social Sciences, and Humanities

Donald Frischmann (Ph.D., University of Arizona; M.A., Ohio State University; B.A., University of Missouri) teaches courses in Spanish language and Latin American literature, culture and civilization. His areas of additional specialization include Mexican theatre, culture, and literature. His twenty-five years of field research throughout Mexico have resulted in an in-depth understanding of the dynamics of Mexican culture and civilization. In 1994, he established the TCU Summer Study in Mexico Program, which will take place in Mérida, Yucatán, in May–June, 2009.

His specialty course, Contemporary Mexican Indigenous Literatures, is designed to broaden cultural perspectives, demonstrating how Mexico can claim a four-millennia tradition as a multicultural, multilingual land. In his recent series of critical anthologies Words of the True Peoples/Palabras de los Seres Verdaderos (U.T. Press) he analyzes and translates the work of thirty-three contemporary writers in Spanish and thirteen living Mexican languages. For this decade-long project, he held several grants, including a Senior Fulbright Lecture-Research Award at the Universidad de las Américas-Puebla and a National Endowment for the Humanities Fellowship.

Dr. Frischmann's books have been formally presented by leading experts and writers at the America's Society and the National Museum of the American Indian (NYC), the National Palace of Fine Arts and the National Museum of Art (Mexico City), and the Latino Cultural Center (Dallas), among others. He is presently working on a new book on emerging writers in Mexican languages. This project focuses upon the fresh perspectives a new generation of writers is bringing to world literature.

Linda Hughes

Best Overall Faculty Member, English
Best Researcher/Scholar, English

Carrie Leverenz

Most Helpful to Students, English

Claire Sanders

Most Helpful to Students, Liberal Arts, Social Sciences, and Humanities

Partnering with TCU Career Services, Dr. Sanders teaches a course that helps Liberal Arts students prepare for the job market and market themselves successfully to potential employers

Tamas Ungar

Best Teacher, Liberal Arts, Social Sciences, and Humanities

Keith Whitworth

Best Overall Faculty Member, Liberal Arts, Social Sciences, and Humanities

Dr. Whitworth is a Sociology Instructor at Texas Christian University specializing in the two areas; sustainability and the social impacts of technology.

TEXAS SOUTHERN UNIVERSITY

Arbolina Llamas Jennings

Most Helpful to Students, English

Arbolina L. Jennings Associate Director, Quality Enhancement Plan (QEP) Co-Chair General Education Subcommittee Tenured Assistant Professor, English 713-313-7661/6764 jennings_al@tsu.edu Texas Southern University Houston, Texas 77004

Iris Lancaster

Best Researcher/Scholar, English
Best Teacher, Liberal Arts, Social Sciences, and Humanities

Iris Lancaster is Assistant Professor of English. She holds a Ph.D. in Critical Literacy (Texas A & M) and a Master of Arts in English (Texas Southern University). Professor Lancaster teaches Composition and Rhetoric courses as well as American Literature. Her specialization area is American Literature, 1899–1946. Professor Lancaster's publications include work in the collection, The Street: A Critical Response to Ann Petry and an article on composition, literature, and classical rhetoric in the Louisiana English Journal. Her memberships and affiliations include the Modern Language Association, National Council of Black Studies, and College Language Association.

TEXAS STATE UNIVERSITY - SAN MARCOS

Michael Conner

Best Teacher, German

TEXAS TECH UNIVERSITY

Stefano D'Amico

Best Researcher/Scholar, History

Stefano D'Amico joined the department in 1999 after three years of teaching at the University of Colorado at Boulder. He is a specialist in early modern socio-economic European history and urban history, with a focus on Italy.

Curtis Bauer

Best Teacher, English

Bauer specializes in creative writing (poetry) and Spanish translation. His areas of interest are American and world poetry, poetry and fiction in translation and chapbook publishing. His collection of poems, Fence Line, won the 2003 John Ciardi Poetry Prize. His poems, prose and translations have appeared in The American Poetry Review, Iowa Review, Rivendell, and Ninth Letter, among others. He has been a fellow at the Vermont Studio Center and a Lannan Writer in Residence at IAIA in Santa Fe. He is the publisher of Q Ave Press Chapbooks.

Erika Brooks-Hurst

Most Helpful to Students, Human Dev & Family Services

Genevieve Durham DeCesaro

Best Researcher/Scholar, Theater

Janis Elliot

Most Helpful to Students, Art History

Dr. Elliott's research has been awarded international fellowships including two Social Sciences and Humanities Research Council of Canada Fellowships and the British School at Rome Grant in Aid of Research (1994). She writes on 14th-century Italian church decoration. She co-edited the highly-acclaimed book, The Church of Santa Maria Donna Regina: Art, Iconography and Patronage in Fourteenth-Century Naples, 2004.

Sybil Hart

Best Overall Faculty Member, Human Dev & Family Services
Best Researcher/Scholar, Human Dev & Family Services
Best Teacher, Human Dev & Family Services
Best Researcher/Scholar, Liberal Arts, Social Sciences, and Humanities

I specialize in development during infancy and early childhood and conduct research in two general areas. The first pertains to the emergence of jealousy. These studies explore factors that underlie the ontogenesis of sensitivity to exclusivity as well as those responsible for individual differences in jealousy's presentation and differentiation along typical and atypical trajectories. The second pertains to the breastfeeding dyad. These entail multidisciplinary approaches involving teams of researchers from TTU and TTUHSC toward understanding and enhancing health and cognitive benefits for the breastfed infant during the first year of life.

Francisco Ortega-Grimaldo

Best Teacher, Design
Best Teacher, Fine Arts

"Dr. Ortega gets the best evaluations in his area. Students love him not just because of how he teaches but because of the ethical standards he has. He devotes hours to advise his students and is very tough when he needs to be. In spite of how strict he is, as I said, his gets very good evaluations."

Francisco Ortega-Grimaldo began his professional career as a graphic designer working for the Model Institutions for Excellence (MIE) Program, an initiative created by the National Science Foundation. At the same time, he also freelanced for many education programs in the US and in México, and taught at the New Mexico State University. As part of the first generation (2000) of SWEAT (http://www.sudor.net/) a collaborative dedicated to making socially conscious video games, he worked in partnership on the design of the digital games Crosser and La Migra, which have received international acclaim. His game Crossing the Bridge has been presented at the UTEP library, Mexic-Arte Museum in Austin, TX, and has a permanent display at the Inter-American and Border Studies Center (UTEP). In addition, he is working on his doctoral degree and presenting papers on the importance of games as social discourse. He has been awarded with the SBC Chancellor Scholarship, the Walt Disney's Imagineering award, and two national literature awards in México. Please visit the web sites: fog-site.com and ludoztli.com.

Carolyn Tate

Best Overall Faculty Member, Art History
Best Researcher/Scholar, Art History

"Dr, Tate is an outstanding researcher and teacher. her service is outstanding. As the only full faculty member, she has been part of many search committees and has literally shaped the art history area. She wrote the proposal for the MA in art history. She is a very good colleague too."

Dr. Tate has expressed her fascination with the ancient art of the Americas through museum exhibitions, teaching, and academic publications. Two major themes in her research involve why Mesoamericans created their elaborate ceremonial cities and how individuals perceived of themselves in relation to social groups and to the symbol-laden cities they built. Her latest book, Re-Considering Olmec Visual Culture: The Unborn, Women, and Creation (University of Texas Press, 2012), addresses the development of visual culture in Mesoamerica and its roles in defining belief systems, creating ritual centers, and in the evolution of phonetic and pictorial writing.

Tate's perceptions of the expressive capacity of Pre-Columbian objects were honed during five years during which she worked as Associate Curator of Pre-Columbian Art at the Dallas Museum of Art. There, with the support of a grant from the National Endowment of the Arts, she revamped the installation of the Museum's permanent collections of Andean and Mesoamerican art, making the works more accessible though audio tours, a video of ancient sites, and maps. She curated Chocolate Pots and Tomb Guardians, an exhibition of Mesoamerican works from collections in Dallas-Fort Worth and collaborated on several exhibitions, including the blockbuster Images of Mexico (twentieth century art) and Gold of Three Continents. With Kent Reilly and David Grove, she organized the first major exhibitions of Olmec art to be shown in the US, The Olmec World: Ritual and Rulership, (Art Museum, Princeton University and Houston Museum of Fine Arts), and Olmec Art of Ancient Mexico, (National Gallery of Art). Other exhibitions include Human Body, Human Spirit: A Portrait of Ancient Mexico (Emory University and Royal Ontario Museum) and 1519/The One Reed Year: Wonders of Aztec Mexico (Lubbock, TX).

In 1993, Carolyn Tate joined the faculty of Texas Tech University, where she teaches courses on the arts and cultures of Mesoamerica and Native North America. She creates special topics that reflect and intensify her own research interests, so that students participate in the refinement of her ideas and interpretations. Courses such as "The Body and the City in Mesoamerica" stem from her book on the Maya city of Yaxchilan, which is populated by over 100 monumental figures and 90 buildings. Working with students in her courses on "Gender in Mesoamerican Art" and "Creation Narratives in Mesoamerican Art" helped her develop her latest book, Re-Considering Olmec Visual Culture: The Unborn, Women, and Creation (2012). She also enjoys teaching Art History Survey 1. Each semester she devises special projects and assignments that engage students in aspects of ancient art history - such as curating exhibitions that explore how ancient people conceived of time.

Dr. Tate's research has been supported by fellowships at Dumbarton Oaks (Harvard University; Washington DC), the Harvard Center for the Study of World Religions (Cambridge, MA), the Clark Art Institute (Williamstown, MA), and the Instituto de Investigaciones Estéticas at the Universidad Nacional Autónoma de México. These fellowships have facilitated her publication of over 30 articles or book chapters on the ancient Maya, the Olmec, and on issues of gender, narrative, creation stories, spirituality, astronomy, and city planning in ancient Mesoamerica.

Among her awards are Texas Tech Integrated Scholar (2011), Texas Tech Barney Rushing Research Excellence Award (2006), Appreciation Award from the Consulado General de México (Dallas 1990), and Fulbright-Hayes Dissertation Fellowship (1984).

Bill Wenthe

Most Helpful to Students, English

TEXAS WOMAN'S UNIVERSITY

Anno Simpoon

Best Overall Faculty Member, Reading

"Dr. Simpson's work for Reading Recovery and our department has been instrumental in moving our department forward and also getting our TWU Reading Dept. known nationally. Due to all of her work with the i3 Grant, she has impacted the lives and literacy success stores for thousands of young struggling readers. She inspires me on a daily basis in her work effort, her ability to teach, serve, and research, and she is just THE BEST!"

THE CATHOLIC UNIVERSITY OF AMERICA

Joshua Benson

Best Teacher, Religion

Tim Noone

Best Researcher/Scholar, Liberal Arts, Social Sciences, and Humanities

THE COLLEGE AT BROCKPORT (SUNY BROCKPORT)

Joseph Chesebro

Best Researcher/Scholar, Communication
Most Helpful to Students, Communication

I actually graduated from Brockport in the early 1990's and, after completing my doctoral program at WVU, I found myself back at Brockport in 1999 as a faculty member. Since then, I've thoroughly enjoyed having the opportunity to teach undergraduate and graduate courses in interpersonal communication, workplace communication, and training and development. My research has examined student-teacher communication, and more recently I've begun to look at how experts in the workplace communicate their hard-to-describe expertise to others (it's not easy!). My goal in the classroom and in my research is to identify practical and useful ways we can apply what we are learning.

When I'm not busy at work, I'm enjoying time with my family which consists of my wife and I trying to keep up with our two young kids! I fill whatever time is left playing tennis & drums, learning to play piano, going to concerts, and collecting live recordings of concerts.

Donna Kowal

Best Teacher, Communication

"Donna personally taught me as a graduate student, for three courses in my program. I have gone to her for advice repeatedly. On at least two occasions, students of mine have taken a course for honors credit. She is creative and inspiring."

Education: M.A. and Ph.D., University of Pittsburgh
Graduate Certificate, Women's Studies Program, University of Pittsburgh
B.S., The College at Brockport

Area(s) of Specialty: Rhetorical criticism; political communication.

Recent Publications: "Anarchist Women and the Feminine Ideal: Voltairine de Claire, Emma Goldman and Lucy Parsons." Co-authored with Catherine H. Palczewski and Linda Horwitz. The Rhetoric of Nineteenth Century Reform and the Perfecting of American Society, Vol. 5 of Rhetorical History of the United States. Ed. Martha Watson and Thomas Burkholder. Michigan State UP. 2008

"Digitizing and Globalizing Indigenous Voices: The Zapatista Movement." Critical Perspectives on the Internet, Ed. Greg Elmer, Lanham, MD: Rowman Littlefield: 2002. 105–126.

"Academic/Activist, Theory/Practice: Academic Studies on Ecofeminist Activism." Book Rev. Women's Studies Quarterly. Vol. 14.1&2 (Spring/Summer 2001): 279–283.

"One Cause, Two Paths: Militant vs. Adjustive Strategies in the British and American Suffrage Movements." Communication Quarterly. 48.3 (Summer 2000): 240–255.

In Press: Emma Goldman, Public Womanhood, and the 'Sex Question,' Albany, NY: SUNY Press, forthcoming. Awards: Donald H. Ecroyd and Caroline Drummond Ecroyd Teaching Excellence Award, Eastern Communication Association, 2005. State University of New York Chancellor's Award for Excellence in Teaching, 2000. Research Interests: Rhetoric of social movements, focusing on representations of gender and culture; History and social influence of American feminism.

Administrative Roles: Director of the Honors College, The College at Brockport, 2009–present. Past chair and program planner for the following Eastern Communication Association Interest Groups: Kenneth Burke (2004–06), Voices of Diversity (1999–2000), and Rhetoric and Public Address (1997–98).

THE COLLEGE OF NEW JERSEY

Pierre LeMorvan

Best Teacher, Philosophy

TIDEWATER COMMUNITY COLLEGE

Kevin Brady

Best Researcher/Scholar, Liberal Arts, Social Sciences, and Humanities

Robert Pappas

Most Helpful to Students, Design

TOURO COLLEGE

Shakira Kennedy

Best Overall Faculty Member, Social Work

"best teacher"

TOWSON UNIVERSITY

Jenifer Ballengee

Best Researcher/Scholar, Liberal Arts, Social Sciences, and Humanities

Jennifer Ballengee received her B.A. from Emory University, M.A. from University of Georgia and Ph.D. in Comparative Literature from Emory University. Her expertise is in Visual Culture, Cultural Studies, Literary Theory, Rhetoric, and Mythology. Her monograph, The Wound and the Witness: The Rhetoric of Torture explores the rhetorical functions of torture and the witnessing of torture in both classical texts and contemporary contexts. Her current book project focuses on the use of ancient ruins in the modern construction of national identity.

Carl Behm

Most Helpful to Students, Liberal Arts, Social Sciences, and Humanities

Carl Behm III (Ph.D., Maryland) specializes in Victorian and early 20th-century literature. He regularly teaches courses on the Victorian Age, the 19th-century British novel, and fiction writing, as well as special studies of authors such as Charles Dickens, George Eliot, Thomas Hardy, E.M. Forster, and D.H. Lawrence. He is also the coordinator of the English internship and practicum. The co-author of two books, he has presented papers on Lawrence and Forster to international conferences in England, France, Canada, and the United States

David Bergman

Best Researcher/Scholar, English
Best Researcher/Scholar, Liberal Arts, Social Sciences, and Humanities

David Bergman received his B.A. from Kenyon College, and his M.A. and Ph.D. from The Johns Hopkins University. His areas of expertise are Victorian literature, modern American poetry, and gay literature. He is the author or editor of more than a dozen books, including poetry, literary and cultural criticism, children's books, textbooks and collections of essays, short fiction and autobiography. He is the author of Gaiety Transfigured: Gay Self-Representation and American Literature, chosen as an Outstanding Book of the Year by Choice, an of The Violet Hour: The Violet Quill and the Making of Gay Culture. He is winner of the Lambda Book Award and the George Elliston Prize. His latest book of poetry is Heroic Measures. He is currently editing an anthology Gay American Autobiography and translating with Katia Sainson of the Foreign Languages Department, The Selected Poems of Jean Senac.

Frances Botkin

Best Teacher, Liberal Arts, Social Sciences, and Humanities

Frances Botkin received her B.A. from the University of Illinois at Urbana-Champaign, and her M.A. and Ph.D. from the University of Illinois at Chicago. Her areas of expertise are British Romanticism, Caribbean literature, and gender studies. She has presented papers and published articles on Maria Edgeworth, Sydney Owenson, William Wordsworth, Jane Austen, and Jack Mansong (a nineteenth-century Jamaican slave rebel). She is currently working on her book project: Rewriting the Colonial Caribbean: A Cultural History of Obi, or "Three-Finger'd Jack."

Tom Casciero

Best Overall Faculty Member, Fine Arts
Best Overall Faculty Member, Fine Arts

"Well rounded; expert; caring; committed"

Tom Casciero is a Professor in the Department of Theatre Arts at Towson University in the USA. He trains graduate and undergraduate actors in movement; voice/movement integration; movement theatre techniques, improvisation, and devised theatre.

Tom received his Ph.D. in Theatre (1998) from The Union Institute Graduate School. His thesis, Laban Movement Studies and Actor Training, is a contextual essay and teaching manual for training actors in physical awareness and expressivity. It is widely consulted throughout the world. He is a Certified Laban Movement Analyst (CMA) and a Research Associate for the Laban/Bartenieff Institute of Movement Studies (LIMS). He has taught in the LIMS Certification Program and is currently on the editorial board of the Journal of Laban Movement Studies.

He was awarded an Overseas Fellowship and Visiting Scholarship at the University of Pretoria and Pretoria Technikon University, both in South Africa. He was a Visiting Artist at the National University of Costa Rica, teaching in both the Theatre and Dance departments. Tom has taught as a guest artist at Teatro Abya Yala, the Kentucky Shakespeare Festival, Webster Movement Institute, the University of Texas at Austin, Southern Illinois University, Calvin College, and in various colleges, theaters, and festivals.

His research has been presented at International Laban Conferences in Rio de Janeiro, Bratislava (Slovakia), Amherst, Minneapolis, and Baltimore. He has also presented at the conferences of the South African Performer's Voice and Movement Educators, the Association of Theatre in Higher Education, the Association of Theatre Movement Educators, the Voice & Speech Trainers Association, and at the Southeast Theatre Conference.

Tom is currently a primary researcher on the MERGE project, an investigation into the foundational elements of movement, their uses in actor training, and the intersection of common elements of movement seen across diverse methods of movement-based actor training. Co-researchers include Wendell Beavers and Erika Berland (Naropa University), Judith Chaffee (Boston University), Sara Romersberger (SMU), and Rebecca VerNooy (Ohio University).

He has served as movement director/consultant for over 40 University Main Stage and studio productions. He was the liaison, co-producer, and on site director for the Mid-Atlantic Movement Theatre Festival and the past vice-president of ATME. He is also a performer and director, having toured his solo performances of comedy and physical theatre nationally and internationally, and having directed regional and university theatre. He also conceptualized, co-created and produced the multi-media production PostMarks, a devised theatre piece based on the first three books of Nick Bantock's Griffin and Sabine series.

Rita Costa-Gomes

Most Helpful to Students, History

Rita Costa-Gomes joined the History Department in 2005. She earned her PhD in Medieval History at the Universidade Nova de Lisboa, Portugal with a dissertation titled, "A Corte dos Reis Portugueses no final da Idade Média." Her research focuses on the history of Portugal, Iberia, the Mediterranean, and the Early Atlantic, 1000–1500 AD, particularly social, cultural, urban, and environmental history, the history of court societies, the history of landscapes, the history of the late medieval nobility, and the history of historiography.

Dr. Costa-Gomes's curiosity about the Middles Ages was piqued as an undergaduate. As she recalls, "I was working as a research assistant for the Portuguese historian Joel Serrão on his new edition of the essays of the nineteenth-century author and medievalist Alexandre Herculano (1810–1877). More than three decades later, my curiosity remains, and I keep trying to answer all my questions. I would like to provide to Towson students similar opportunities to cultivate their curiosity." Her motto is "sapiens ubicumque peregrinatur."

Dr. Costa-Gomes published the first study of the royal court of Portugal in a book entitled The Making of a Court Society: Kings and Nobles in Late Medieval Portugal (Cambridge University Press, 2003). Two volumes of her ongoing series about medieval frontier castles, A Ring of Castles, have now been published in English. She is currently writing a book-length study of the impact of royal itinerance in the emergence of political territories and in the landscapes of Iberia between the thirteenth and the sixteenth centuries.

Dr. Costa-Gomes has developed a webpage called "Medieval Baltimore" resulting from a collaborative writing project of Towson students that explores how the Middle Ages is visible and alive in the Baltimore metropolitan area in buildings, museum artifacts and the lives and work of people who devote themselves to the study and appreciation of all things medieval. Medieval Baltimore can be found at www.medievalbaltimore.net

Nicole Dombrowski

Best Researcher/Scholar, History

Nicole Dombrowski-Risser earned her degrees from University of Wisonsin (BA) and New York University (Ph.D.,1995). Her dissertation was titled "Beyond the Battlefield: Civilian Flight in the Exodus of May–June 1940 in France" and was overseen by Professors Tony Judt and Mary Nolan. She is the author of France under Fire: German Invasion, Civilian Flight and Family Survival during World War II which is being published by Cambridge University Press, and the editor of Women and War in the Twentieth Century: Enlisted with or without Consent (Routledge 2004). Her research insterests are in France and Europe, 1789–present, especially social history, history of women and the family, history of gender and war, and rural history.

Dr. Dombrowski-Risser's current project is "A French Family Farm: Two Hundred Years of Survival against War, Weather and Market Change on an Olive Farm in Provence." This is a book-length study which takes as a focus the olive farm La Blachette located in Nyons, France in the Rhones-Alpes region. Using family archives, land records, regional agricultural journals and personal testimonies, I tell the story of survival of one family farm which survived against the obstacles of political turmoil, phyolloxera epidemic, climate crisis and shifting markets to carve out a niche for a family and a way of life swiftly disappearing in European and around the globe. A French Family Farm applies the tools of social and women's history to write the biography of one farm which survived against the odds by combining entrepreneurialism, collectivist agricultural strategies and market innovation. In 2010 she was made a Knight of the Order of the Black Olive of Nyons, France, her research on French olive agriculture.

Dr. Dombrowski-Risser is the coordinator of the Paul Gantt Nuremberg Papers Digital Archive Collection at Towson university, an on-going project to digitize and annotate the Towson University holdings of Nuremberg Papers from the Special Crimes Division which prosecuted German Industrialists for using slave labor, contributing to genocide of European Jewry, contributing to the waging of aggressive war and violating the Peace.

Elizabeth Gray

Best Teacher, History

Elizabeth Gray earned her PhD from the College of William and Mary in 2002 for her dissertation, "American Attitudes Toward British Imperialism, 1815–1860." She is currently writing a book-length study of drug addiction in early America. Her research interests are American history in an international context in the early national and antebellum eras, focusing on cultural history, and diplomatic history, with particular interest in the history of addiction and American foreign relations, broadly defined.

Diane Hall

Best Researcher/Scholar, Family Studies

Lisa Martinelli

Best Overall Faculty Member, Family Studies

Benjamin Zajicek

Best Overall Faculty Member, History

"Dynamic."

TRIDENT TECHNICAL COLLEGE

Rebecca Pryce

Best Teacher, Fine Arts

TROY UNIVERSITY

Ben Robertson

Best Researcher/Scholar, English

Ben P. Robertson, Ph.D. (University of Tulsa), Associate Professor of English. British Romantic Literature, Trans-Atlantic/Transnational Literature, Cultural Studies, Religion in Literature, Women's Studies. Bibliographer, Keats-Shelley Journal. Folger Fellow (2006, 2005). Editor, The Diaries of Elizabeth Inchbald, 3 volumes (Pickering & Chatto, 2007); General Editor, Themes of Conflict in the Nineteenth- and Twentieth-Century Literature of the American South: The Proceedings of the 2004 Conflict in Southern Writing Conference (Mellen, 2007) and Conflict in Southern Writing (ATSP, 2006). Publications in Keats-Shelley Journal, The Catholic Church and Unruly Women Writers (Palgrave, 2007), James Joyce Quarterly, South Asian Review, Names: A Journal of Onomastics, South Central Review, South Atlantic Review, Companion to the British Short Story and Short Fiction (Facts on File, 2006), The University of Tulsa Graduate Review, The Learning, Wit, and Wisdom of Shakespeare's Renaissance Women (Mellen, 1997), Alpha Chi Recorder, and The Rectangle; forthcoming in Students' Companion to American Literary Characters, Companion to Twentieth-Century British Poetry, and Keats-Shelley Journal. Contributing Editor, Annotated Bibliography of English Studies. Member, Modern Language Association, North American Society for the Study of Romanticism, Keats-Shelley Association, Byron Society of America, American Name Society, British Association for Romantic Studies, Wordsworth-Coleridge Association, South Central Modern Language Association, South Atlantic Modern Language Association. Current research, Inchbald, Hawthorne and the Romantic Moral Romance: Little Histories and Neutral Territories (Pickering & Chatto).

TUFTS UNIVERSITY

Elizabeth Ammons

Best Overall Faculty Member, Liberal Arts, Social Sciences, and Humanities

"A committed teacher, researcher, scholar, admired and respected by students and colleagues alike. Elizabeth Ammons has worn many hats and advocated for those who needed a voice. Her dedication to her students, to the English Department, and to the School of Arts and Sciences should continue to be recognized."

B.A., M.A., University of Cincinnati, 1964, 1966; Ph.D., University of Illinois, 1974; part-time and adjunct teaching positions at various institutions before coming to Tufts in 1976 as an Assistant Professor; Dean for Arts and Humanities at Tufts 1994–96; Harriet H. Fay Professor of Literature at Tufts, 1996 to the present; Chair of the English Department, 1991–1994; Director of the First-Year Writing Program at Tufts, 1998–2003; numerous departmental and university committees, including Tenure and Promotion and co-chair of the Equal Educational Opportunity Committee; recipient of Mellon, National Endowment for the Humanities, and American Association of University Women faculty fellowships; Phi Beta Kappa; recipient of Tufts University Multicultural Service Award, 1996; Fulbright Lecturer in Japan, summer 2003, giving lectures in Tokyo and Kyoto on American literature; WISE Speaker, England, summer 2006; Outstanding Graduate Teaching and Mentoring Award 2011.

Anne Cantu

Best Teacher, Languages

Eva Hoffman

Best Researcher/Scholar, Fine Arts
Best Researcher/Scholar, Fine Arts

Ronna Johnson

Best Teacher, English

"Ronna's enthusiasm is infectious. Her knowledge of her field is vast, deep and ever expanding. Her ability to animate the era is well known and much sought after by students."

Ph.D. Tufts University. 1984 Tufts University Nominee University Microfilms International Distinguished Dissertation Award. Doctoral Thesis: Jack Kerouac's Art: The Artist as Literary Hero in The Duluoz Legend M.A. University of Michigan, Ann Arbor Horace H. Rackham School of Graduate Studies. Master's Thesis: Social Commentary in the Fiction of Mary McCarthy B.A. Boston University, College of Liberal Arts, Division of General Education (honors college) magna cum laude, Distinction in Literature. Senior Thesis: Rites of Passage in Ernest Hemingway and Thomas Wolfe

Jeanne Penvenne

Best Teacher, History

I am a social and labor historian of colonial Mozambique. Although much of my work focuses on ordinary workers, and specifically explores the many ways they framed their experiences in oral histories, I also have engaged the development and contradictions of Mozambique's black intellectual elites largely through their newspapers. My early research documented the extent to which forced labor in Southern Mozambique fueled men's migration to South Africa's mines and Mozambique's towns and cities. As I recorded scores of men's work and life histories, I was struck, first, by the fact that, although most of these men had women in their lives, it was difficult to get a sense of their families, and second, that migration and waged labor choices played out very differently for men and women. That led me to conduct an oral history project among the women workers in Mozambique's colonial era cashew-shelling industry, and to seek gendered perspectives on households, waged labor and migration. That project resulted in a book manuscript, Mozambique's Factory for Women. It may actually be published someday.

I taught on a visiting basis at several universities in the greater Boston area between 1985 and 1990, and was hired full time at Tufts in 1993. I have also taught and developed courses on African Historiography, Gender and Oral History at Universidade Eduardo Mondlane in Maputo Mozambique, most recently in 2004–2005. Generous funding from Fulbright, Fulbright-Hays, Social Science Research Council, Gulbenkian and Mellon Foundations supported my research and teaching in Mozambique since 1976. During my sabbatical in 2014–15, I pursued two ongoing projects, an exploration of Mozambican colonial intellectuals in the late colonial era and a Southern Africa textbook and reader.

Mingquan Wang

Best Overall Faculty Member, Chinese

"is undoubtedly one of the most productive researchers at Tufts. Mingquan is the most dedicated and hardworking colleague in our department."

Carol Wilkinson

Best Overall Faculty Member, English

"A gifted teacher, beloved by her students, generous to her colleagues, a bright light in the English Department. She has long deserved recognition!"

TULANE UNIVERSITY

Michaela Cannon

Best Researcher/Scholar, Dance

Aaron Colier

Best Teacher, Fine Arts

Aaron has taught drawing and painting at Tulane since the fall of 2006. His classes are aimed at providing the drawing and painting student with an expansive vocabulary in communicating their personal concerns, posing formal elements as the gateway to conceptual considerations of the work. Aaron's own work is aimed at marrying the crude and wonderful in a way that mimics life experience and personal makeup. His paintings and drawings also explore the limitations of artwork, the artist, and the viewer

Jeffrey Gunshol

Best Overall Faculty Member, Dance
Best Teacher, Dance
Most Helpful to Students, Dance

Adam Mysock

Most Helpful to Students, Art Studio

Adam Mysock, Professor of Practice, Drawing & Painting Painter Adam Mysock was born in Cincinnati, Ohio in 1983 - the son of an elementary school English teacher and a lab technician who specializes in the manufacturing of pigments. On account of a steady stream of folk tales from his mother, his father's vividly dyed work clothes, and a solid Midwestern work ethic, he developed an interest in painting and drawing all things Americana from a very early age. Mysock earned a Bachelor of Fine Arts degree in Painting and Art History in 2004 from Tulane University and an MFA from Southern Illinois University at Carbondale in 2007. After his studies, he served as the coordinator for the city of Cincinnati's MuralWorks program and worked as an adjunct drawing professor at Sinclair Community College in Dayton. In the summer of 2008, Mysock became a Professor of Practice at Tulane University where he currently teaches and maintains a studio.

Mysock's work has been exhibited both nationally and internationally and is in private collections throughout the world, including those of Thomas Rusche, Thomas Coleman, and Michael Wilkinson. In 2012, he was awarded first prize "Best in Show" in the Ogden Museum's first annual Louisiana Contemporary Juried Exhibition, and he was included in the 2013 South edition of New American Paintings this past summer.

Most recently, Mysock has exhibited at Pulse Miami, Volta9 in Basel, Switzerland, and White Linen Night at Jonathan Ferrara Gallery. He is scheduled to participate in a group exhibition, "Crescent City Connection," at the Dishman Art Museum in Beaumont, Texas in late September.

TULSA COMMUNITY COLLEGE

Jon Boyne

Best Researcher/Scholar, Liberal Arts, Social Sciences, and Humanities
Best Teacher, Liberal Arts, Social Sciences, and Humanities
Most Helpful to Students, Liberal Arts, Social Sciences, and Humanities
Best Overall Faculty Member, Liberal Arts, Social Sciences, and Humanities

Tamara Britton

Best Overall Faculty Member, English
Best Teacher, English
Most Helpful to Students, English
Most Helpful to Students, Liberal Arts, Social Sciences, and Humanities

Tamara R. Britton, a member of Phi Theta Kappa, received her A.S. in Biology in 2002 from Tulsa Community College. She transferred to Oregon, where she attended Linn-Benton Community College, majoring in English with a minor in Creative Writing. She worked for 2 years in LBCC Student Life and Leadership where she donated many hours to community service, working with Habitat for Humanity and local food banks. She graduated from Oregon State University in 2007, receiving a B.A. in English, with a concentration in Medieval Literature, and a minor in Creative Writing, and tutored for 1 year in the Writing Center before returning to Oklahoma. She earned her graduate degree at Oklahoma State University, originally working on an M.F.A. in Creative Writing before switching her major to earn an M.S. in Teaching, Learning, and Leadership in Reading/Literacy. She has worked at TCC's West Campus for over seven years, teaching Reading I and II, Writing I and II, Academic Strategies, and Comp I. In addition to teaching and working in the Reading & Writing Lab, she also works in the West Campus Writing Center.

Mary Cantrell

Best Overall Faculty Member, English
Best Teacher, English
Most Helpful to Students, English
Best Overall Faculty Member, Liberal Arts, Social Sciences, and Humanities

David Charlson

Best Researcher/Scholar, English

Ryan Craig

Best Teacher, Liberal Arts, Social Sciences, and Humanities
Best Researcher/Scholar, Liberal Arts, Social Sciences, and Humanities

Thomas Culpepper

Most Helpful to Students, English
Best Overall Faculty Member, Liberal Arts, Social Sciences, and Humanities

Miguel Da Corte

Best Overall Faculty Member, Languages
Best Researcher/Scholar, Languages
Best Teacher, Languages
Most Helpful to Students, Languages

Kasty France

Best Teacher, Liberal Arts, Social Sciences, and Humanities

Jon Herring

Most Helpful to Students, Liberal Arts, Social Sciences, and Humanities

Mary Larson

Best Teacher, English

Kara Ryan-Johnson

Best Researcher/Scholar, English
Best Researcher/Scholar, Liberal Arts, Social Sciences, and Humanities

UNION COUNTY COLLEGE

Pat Gallagher

Best Overall Faculty Member, Communication
Best Teacher, Communication
Most Helpful to Students, Communication

Derek McConnell

Best Teacher, Liberal Arts, Social Sciences, and Humanities

"Derek is a strong advocate for experiential learning. He supports interactive learning across multiple disciplines."

Mary Meeks

Best Researcher/Scholar, Liberal Arts, Social Sciences, and Humanities

Dr. Mary Meeks is an Associate Professor of English. In addition to fostering the love of literature and writing in her students, she actively encourages her students to be engaged in the life of the college community outside the classroom. Dr. Meeks has served as an advisor to the Student Government Association and is the advisor to the Catholic Student Organization and the Christian Fellowship Club. She has also served on various faculty committees in support of student activities and student success. Dr. Meeks has a bachelor's in English from Brooklyn College, a master's of Philosophy from Drew University and a Ph.D. in English from Drew University.

Michael Z Murphy

Best Teacher, English

"Michael unselfishly provides many hours to work closely with the students who stage the wonderful plays performed at UCC."

UNIVERSITY AT ALBANY (SUNY ALBANY)

Charles Shepherdson

Best Researcher/Scholar, English
Best Researcher/Scholar, Liberal Arts, Social Sciences, and Humanities

"Who is the person in this picture? The box has my name, but this is not me."

UNIVERSITY AT BUFFALO (SUNY BUFFALO)

Rachel Ablow

Best Researcher/Scholar, English
Best Teacher, English

Linda Dinger

Best Teacher, Social Work

Robert Knopf

Best Teacher, Theater

A theater director, dramaturg, and scholar, Robert Knopf is the author of The Theater and Cinema of Buster Keaton (Princeton University Press, 1999). For Yale University Press, he edited Theater of the Avant-Garde, 1890–1950 (2001), Theater and Film (2004), and Theatre of the Avant-Garde, 1950–2000 (2011). Prior to coming to UB in 2004 to serve as chair, he taught at Purdue University, Connecticut College, and the University of Michigan.

For the stage, he has directed at Circle-in-the-Square Downtown, Cherry Lane Studio, Paradise Factory, Circle Rep Lab, and historic Town Hall, all in New York City. He served as dramaturg for the National Public Radio series The Archeology of Lost Voices, for which he adapted and directed the docudrama Hidden Dragon. He is the author of a textbook for directing, The Director as Collaborator (Allyn & Bacon, 2005).

Lynne Koscielniak

Best Overall Faculty Member, Theater
Most Helpful to Students, Theater

Based in Buffalo, NY, she is an Associate Professor of Scenography and the Director of Design and Technology in the Department of Theatre and Dance at the University at Buffalo. She has held office in the USITT Upstate New York Regional Section and is a past Chair of Design and Technology for Region II, Kennedy Center American College Theater Festival. She holds an MFA in Stage Design from Northwestern University.

Her work has earned her an Emerging Designer Residency (Steppenwolf), a Design Fellowship at the Kennedy Center (Intensive in Scenic & Costume Design: The Collaborative Process with Ming Cho Lee and Constance Hoffman), a Joseph Jefferson Citation Nomination, The Michael Merritt Scholarship for Excellence in Design and Collaboration and the UB Faculty Award for Excellence in Mentoring Undergraduate Research and Creative Activity. Lynne is a member of United Scenic Artists – Local 829.

Barbara Tedlock

Best Researcher/Scholar, English

Tom Tucker

Best Researcher/Scholar, Theater

UNIVERSITY OF AKRON

Janet Bean

Most Helpful to Students, Liberal Arts, Social Sciences, and Humanities

She is co-director of the UA Summer Writing Institute, a professional development program for high school English teachers. Her research interests include writing and identity, teacher education, and writing across the curriculum.

Julie Drew

Best Teacher, Liberal Arts, Social Sciences, and Humanities

Julie Drew is Professor of English. She teaches writing, film, and cultural studies, and her research interests include contemporary discourse theory, rhetoric and public policy, and the politics of education.

J. Thomas Dukes

Best Overall Faculty Member, English
Best Overall Faculty Member, Liberal Arts, Social Sciences, and Humanities

Thomas Dukes is Professor of English and teaches business and professional writing, modern literature, the Bible as literature, and other courses. His poetry collection Baptist Confidential (Bluestem/Emporia State U) came out in 2007, and he is the author of a memoir Sugar Blood Jesus: A Memoir of Faith, Madness, and Cream Gravy (OSL/Upper Room Books, 2009). He has published in a variety of journals including Poetry, New Orleans Review, Women's Studies, Bulletin of the Association for Business Communication, The Journal of Florida Literature, and others. He is a member of the Northeast Ohio Romance Writers Association and the Marjorie Kinnan Rawlings Society. He enjoys maintaining elitist cultural standards while sampling the vulgarity of modern culture on an almost daily basis.

Andrea Ferraro

Best Overall Faculty Member, Liberal Arts, Social Sciences, and Humanities

Antonia Forster

Best Researcher/Scholar, English
Best Researcher/Scholar, Liberal Arts, Social Sciences, and Humanities

Antonia Forster, Professor. Her Index to Book Reviews in England 1749–1774 was awarded the 1992 Rose Mary Crawshay Prize by the British Academy, and the second volume to take the Index to 1800 was published by the British Library in 1997. She spent 1996 in Cambridge and Oxford working with James Raven on English Novels 1770–1799: a Bibliographical Survey of Prose Fiction Published in the British Isles, published by Oxford University Press in 2000. She is also preparing an edition of the correspondence of Ralph Griffiths, founding editor of the Monthly Review, and working on a study of Griffiths and his reviewers, and smaller projects concerned with the reviewing of women writers. She has published articles in such journals as

The Age of Johnson, Shakespeare Quarterly, and Studies in Newspaper and Periodical History and has been awarded fellowships for research in the Newberry, Houghton, Beinecke and British Libraries. Her teaching interests focus mainly on writers of the period from 1590 to 1850, especially women and Shakespeare.

Stephen Harp

Best Overall Faculty Member, Liberal Arts, Social Sciences, and Humanities

Steve Harp has BA and MA degrees in French and History, and a PhD in European history from Indiana University. Though his home base is in History (CAS 204c), he is professor of history and professor of French, teaching courses and doing research in French and European history. He is the author of books on the history of primary schooling and nation building in Alsace-Lorraine, Michelin and cultural identity in modern France, and nudism and European tourism in twentieth-century France. He is now at work on a study of the post-World War II cultural reinvention of Alsace as "European"–and not just French, German, and/or Alsatian. He encourages every student he meets to study abroad in a non-English-speaking country.

Elizabeth Kennedy

Best Researcher/Scholar, Liberal Arts, Social Sciences, and Humanities

Young Lin

Most Helpful to Students, Communication

Therese Lueck

Best Researcher/Scholar, Communication

Hillary Nunn

Most Helpful to Students, English

Hillary Nunn teaches courses in Shakespeare and early modern literature Her current research focuses on intersections between Renaissance literary culture and the era's domestic medical texts and cookery books. She is the author of Staging Anatomies: Dissection and Tragedy in the Early Stuart Era (Ashgate, 2005), as well as "On Vegetating Virgins: Greensickness and the Plant Realm in Early Modern Literature" in the collection The Indistinct Human in Renaissance Literature (Palgrave Macmillan, 2012) and "Home Bodies: Matters of Weight in Renaissance Women's Medical Manuals" in the volume The Body in Medical Culture (SUNY, 2009).

Andrew Rancer

Best Overall Faculty Member, Communication
Best Teacher, Communication

Mark Rittenour

Best Teacher, Communication

Lance Svehla

Most Helpful to Students, Liberal Arts, Social Sciences, and Humanities

Lance Svehla teaches undergraduate and graduate courses in composition, writing, and rhetoric. Co-director of the Akron Writing Institute and Director of the Graduate Studies Certificate in Composition, his academic interests include composition pedagogy, rhetorical theory, scriptwriting, interactive narrative, postmodernism, pragmatism, grant writing, and the politics of literacy. His current research centers on trying to enact a pragmatic sense of self, ethics, and history in the writing classrooms of postmodernity. He has published essays in such journals as Teaching English in the Two-Year College, Dialogues: A Journal for Writing Specialists, and The Journal of Dracula Studies. He is currently working on a contracted book (co-authored with Drs. Julie Drew and William Lyons) entitled Sound-Bite Saboteurs: Public Discourse, Education, and the State of Democratic Deliberation. His non-academic interests include cooking and playing electric blues guitar.

William Vail

Most Helpful to Students, Communication

Eric Wasserman

Best Teacher, English
Best Teacher, Liberal Arts, Social Sciences, and Humanities

"Eric is a favorite of students who gives his time willingly and with genuine care. He has followed our Northeast Ohio Master of Fine Arts students through their program and beyond, always counseling them with their interests at heart. He has been inventive in the creation of classes both at the undergraduate and graduate levels, and developed our program in ways that benefit students and colleagues."

Biography Eric Wasserman is the author of a collection of short stories, The Temporary Life, and a novel, Celluloid Strangers. His fiction has won the David Dornstein Memorial Creative Writing Contest and the Cervená Barva Press Fiction Chapbook Prize and has most recently been featured in Confrontation, Journal of Experimental Fiction, Soundings East, and Great Lakes Review. He specializes in teaching courses in film, literature and fiction writing and advances the English Department's minor in Popular Literature and Film. He is the founder and faculty advisor for the university's graduate student literary journal, Rubbertop Review, and its undergraduate student literary journal, Ashbelt. You can visit him at www.ericwasserman.com.

UNIVERSITY OF ALABAMA

Jacqueline Morgan

Most Helpful to Students, Honors

UNIVERSITY OF ALABAMA BIRMINGHAM

Doug Baulos

Most Helpful to Students, Art

Douglas Pierre Baulos received his MFA from the University of New Orleans and a Bachelor of Fine Arts from the University of Alabama at Birmingham.

He regularly teaches workshops and lectures on his research in book arts, drawing and mixed media. In 2009, Baulos won the President's Award For Excellence In Teaching at UAB.

UNIVERSITY OF ALASKA ANCHORAGE

Sudarsan Rangarajan

Best Researcher/Scholar, Languages

Dr. Sudarsan Rangarajan is the author of a 2012 book entitled "Critical Essays on Michel Butor's L'Emploi du temps," a book published from the Peter Lang publisher in NY. He has published eight papers and presented 18 conference papers since he joined UAA in 2003. His research interest is Twentieth Century French and Quebecois literature, and Critical theory, and is the most productive researcher in the Department of Languages.

UNIVERSITY OF ALASKA FAIRBANKS

Eric Heyne

Best Researcher/Scholar, English

UNIVERSITY OF ARIZONA

Rosati Connie

Most Helpful to Students, Philosophy

Richard Eaton

Best Overall Faculty Member, History
Best Researcher/Scholar, History

My research interests focus on the social and cultural history of pre-modern India (1000–1800), and especially on the range of historical interactions between Iran and India, and on Islam in South Asia. I have published monographs on the social roles of Sufis (Muslim mystics) in the Indian sultanate of Bijapur (1300–1700), on the growth of Islam in Bengal (1204–1760), and on the social history of the Deccan from 1300 to 1761. Most recently, I have co-authored a monograph entitled Power, Memory, Architecture: Contested Sites on India's Deccan Plateau, 1300–16. These four historical monographs employ as analytical tools, respectively, Weberian social thought, Annales School methodology, biography, and architecture.

I am also active in the growing subfield of world history, as well as comparative history. My regularly taught courses include History of Medieval India, History of Modern India and Pakistan, Comparative History, and World History.

Martha Few

Best Teacher, History
Best Teacher, Liberal Arts, Social Sciences, and Humanities

Melissa Fitch

Best Overall Faculty Member, Languages
Best Researcher/Scholar, Languages
Best Teacher, Languages
Most Helpful to Students, Languages
Most Helpful to Students, Liberal Arts, Social Sciences, and Humanities

Dr. Melissa A. Fitch (Ph.D. 1995 ASU), a second generation Chicana, was born in Los Angeles and raised in the City of San Francisco. Her research is currently focused on transnational representations of Latin American popular culture, film, theater and narrative. She is author of the book Side Dishes: Latin/a American Women, Sex and Cultural Production (Rutgers UP, 2009). In 2013 she received the UA Creative Teaching Award and in 2008, she received the University of Arizona's Five Star Teaching Award, the institution's highest teaching honor. In 2004 she received the UA General Education Teaching Award. Since 2002 she has been editor-in-chief of the academic journal Studies in Latin American Popular Culture (University of Texas Press). She was named Outstanding University Educator by the Arizona Languages Association in 1997. Her essays have been published in Latin American Theater Review; Gestos: Teoría y práctica del teatro hispánico; Chasqui: Revista de literatura latinoamericana; ADFL Bulletin; Luso-Brazilian Review; Romance Languages Annual, ADE Bulletin and in the books Dale Nomás! Dale que va! (Buenos Aires: Editorial Nueva Generación, 2006); Latino/a Popular Culture (NYU Press, 2002) and Interventions: Feminist Dialogues on Third World Women's Literature and Film (Garland, 1997). She is co-author of the book Culture and Customs of Argentina (Greenwood, 1998). Professor Fitch is past president of the Arizona Languages Association and the statewide Languages Articulation Task Force. She directed the UA Study Abroad program in Fortaleza, Brazil in 2001 and the UA programs in Alcala de Henares, Spain in 2004 and in Segovia, Spain in 2007. She will direct the study abroad program in Chile in fall of 2014. She has traveled to 33 countries. These include Peru, where she studied Quechua in Cuzco in 1998; Turkey, where she studied Turkish in Istanbul in 2004 and 2005, China, where she studied Mandarin in Shanghai in 2007 and 2008 and Paris, France, where she studied French while on sabbatical in 2010. In 2011 she was named a Fulbright scholar to the Chinese University of Hong Kong for 2011–12. Her free time revolves around travel, learning new languages, working as a volunteer with Casa de la Luz hospice and dancing Argentine tango.

Kevin Gosner

Best Overall Faculty Member, History

Andres Guzman

Best Researcher/Scholar, Liberal Arts, Social Sciences, and Humanities

Katherine Morrissey

Most Helpful to Students, History

Judith Nantell

Best Overall Faculty Member, Spanish & Portuguese

Judith Nantell received her M.A. and Ph.D. from Indiana University and taught at the University of Wisconsin-Madison and Florida State University before coming to the University of Arizona in 1987. Her publications center largely on Spanish Peninsular Poetry, Modern and Contemporary. Her recent research explores Spain's newest poets of the Generation of 2000. She has published two books: The Poetry of Francisco Brines: The Deconstructive Effects of Language, (Lewiston, Pa.: Bucknell University Press) 1994 and Rafael Alberti's Poetry of the Thirties: The Poet's Public Voice, (Athens, Georgia: University of Georgia Press) 1986, the Winner of the 1984 South Atlantic Modern Language Association (SAMLA) Studies Award. In fall 2011 she delivered the Merle E. Simmons Distinguished Alumni (MESDA) Lecture at Indiana University–Bloomington. In that same year she received the College of Humanities Distinguished Faculty Teaching Award. In spring 2013 she organized and participated in the bilingual reading of selected poems by the Spanish contemporary poet, Luis Muñoz, when he visited the Department.

Most Recent Courses Taught: Spanish 350 Readings in the Literary Genres; Spanish 449A Topics in Spanish Literature and Cultural Studies. Topic: La poesía de la Generación del 1927 en España: La Modernidad, los Ismos y el Cine Mudo; Spanish 449A Topics in Spanish Literature and Cultural Studies. Topic: Visualizar La Modernidad 1907–1937 en el Cine Experimental Internacional, el Arte Moderno Español, la Poesía Española: Lo Novísimo sin Fronteras. New Course for spring 2014: Spanish 521 Topics in 18th, 19th, 20th and 21st Century Spanish Literature. The Topic of this course is: Los poetas de "aquí y ahora" en España de cambio de siglo.

R. Ivet Ore Giron

Best Teacher, Spanish & Portuguese

Charles Tatum

Best Overall Faculty Member, Spanish & Portuguese
Best Researcher/Scholar, Spanish & Portuguese
Most Helpful to Students, Spanish & Portuguese
Best Overall Faculty Member, Liberal Arts, Social Sciences, and Humanities

Dr. Charles Tatum is Professor of Spanish and Dean of the College of Humanities at the University of Arizona. He was born in El Paso, Texas and raised in Parral, Chihuahua, Mexico. His mother, Eloísa Aínsa, a Mexican-American, was born and raised in El Paso. Tatum received his B.A. from the University of Notre Dame, his M.A. from Stanford University, and his Ph.D. from the University of New Mexico. Tatum is the author of a monographic study Chicano Literature (1982)–published in translation in Mexico in 1986-and co-author of Not Just for Children: The Mexican Comic Book in the Late 1960s and 1970s (1992). He is co-founder and senior co-editor of the journal, Studies in Latin American Popular Culture. He is editor of three volumes of New Chicana/Chicano Writing (1991–1993) for the University of Arizona Press and co-editor of a volume of essays, Recovering the U.S. Hispanic Literary Heritage, Vol. II. His book, Chicano Popular Culture (The University of Arizona Press, 2001), was selected as a "Best of the Best of the University Presses" book by the American Association of American Presses. His published book chapters and articles include studies on Latin American prose fiction, Chicano/a literature, and Mexican popular culture. Tatum serves on the advisory board of the Recovering the U.S. Hispanic Literary Heritage Project. He is a member of an editorial group that produced an anthology of U.S. Hispanic literature that was published in 2001 by Oxford University Press and of a Spanish-language anthology, En otra voz. Antología de la literature hispana en los Estados Unidos (Arte Público Press, 2002). His study of the Chicano poet Gary Soto was published in Scribner's Encyclopedia of Latino/a Writers (2004). Tatum's most recent book, Chicano and Chicana Literature: Otra voz del pueblo, has recently been published by The University of Arizona Press. Tatum has been very active in advancing diversity issues at the University of Arizona where he is a member of the Diversity Coalition and various committees and task forces that deal with diversity issues. He has advanced the College of Humanities as a model for recruiting and retaining a diverse faculty as well as undergraduate and graduate students. In 2005, Tatum authored a 160-page report, Charting a University of Arizona Course Toward Becoming an Hispanic Serving Institution: Recommendations, at the request of Peter Likins, former President of The University of Arizona. The document deals broadly with the recruitment and retention of Latina/o undergraduate and graduate students, faculty, and administrators.

Jeremy Vetter

Best Researcher/Scholar, History

Jeremy Vetter's research is at the intersection of environmental history and the history of science and technology in the American West. His current projects focus on the Great Plains and Rocky Mountains between 1860 and 1920, including (1) the history of scientific practices in the field and environmental sciences, and (2) the relationship between the field sciences, economic development, and environmental transformation. He is also interested in agriculture and the history of the food system, the history of human interactions with grasses, the involvement of lay people in the field sciences, and the history of science in the national parks.

UNIVERSITY OF ARKANSAS FAYETTEVILLE

Kameri Christy-Mcmullin

Best Researcher/Scholar, Social Work

Glenda House

Best Teacher, Social Work

Prior to joining the School of Social Work faculty, Professor House was a post-MSW clinical social worker providing community-based mental health services for adults with serious and persistent mental illness. She also provided individual and group psychotherapy for adult survivors of childhood sexual abuse. Previous administrative experience includes: Director of the NWA Adult Day Care Center (services for frail elders and caregivers) and Co-Director of the NWA Rape Crisis Center. Professor House teaches primarily in the MSW Program. Currently she teaches a graduate foundation course, Practice II (4343), in the fall and spring, and undergraduate Field Seminar (4422) in the summer.

Yvette Murphy

Most Helpful to Students, Social Work

Dr. Yvette Murphy-Erby, a Professor of Social Work, is the Associate Dean of Social Sciences in the J. William Fulbright College of Arts and Sciences. Former roles within the university include Bachelor of Social Work Program Director, Associate Director of the School of Social Work and Director of the School of Social Work. Dr. Murphy-Erby received her BA from the University of North Carolina – Charlotte (1985), MSW from UNC-Chapel Hill (1991) and her PhD from UNC-Greensboro (2004). She joined the University of Arkansas faculty in 2004, left in 2008 to serve as a Program Officer for the Duke Endowment in North Carolina and returned to the University of Arkansas in 2009. In addition to serving as an administrator, Dr. Murphy-Erby maintains an extensive research agenda. In collaboration with colleagues, she has served as a PI, Co-PI or Investigator on over 11.0 million dollars in external funding and published numerous articles and a book. Committed to service, mentorship and leadership, Dr. Murphy-Erby has received numerous awards and recognitions including, The 2012 Pacesetter Award presented by the College of Health and Human Services, University of North Carolina-Greensboro, the 2102 Outstanding Faculty Leader Award presented by the University of Arkansas Academic Affairs: Center for Multicultural and Diversity Education and the 2002 University of North Carolina - Chapel Hill School of Social Work Distinguished Alumna Award.

UNIVERSITY OF CALIFORNIA DAVIS

Laura Grindstaff

Best Overall Faculty Member, Sociology

Laura Grindstaff came to UC Davis 12 years ago from the University of Pennsylvania. She teaches in the areas of popular culture, cultural sociology, gender and society, and field methods. Her research focuses broadly on American popular culture and its role in constructing gender, race, and class relations. Her book, The Money Shot, is an ethnographic account of daytime television talkshows. Based on interviews and participant observation behind-the-scenes, it explores how the production process works to transform the private experiences of "ordinary" people into extraordinary public performances, and what these performances tell us about class inequality in the US. Currently she is writing a series of articles on cheerleading based on more than ten years of ethnographic research with colleague Emily West. The authors use cheerleading as a vehicle to examine how and why young people embrace particular cultural scripts regarding gender and sexuality, why young people participate in cheerleading when other sports are open to them, and current tensions between the "feminine" realm of cheerleading and the "masculine" realm of sport. Dr. Grindstaff is also publishing a series of articles on reality television, focusing on the constituency most neglected in current studies of reality tv: character-participants. Drawing on in-depth interviews, she explores why people desire television exposure, the contradictions between expectation and reality that participants experience, and the larger sociological implications of the rise of ordinary celebrity. She argues that reality programs, like daytime talk shows before them, are a form of self-service media, and that such media point to an important transformation in public discourse; specifically, they indicate a movement away from the concept of public sphere per se toward more a dramaturgical notion of public stage.

Bill McCarthy

Best Overall Faculty Member, Sociology

UNIVERSITY OF CALIFORNIA RIVERSIDE

Paul Beehler

Best Overall Faculty Member, English
Best Researcher/Scholar, English
Best Teacher, English
Most Helpful to Students, English

Alexandra Maryanski

Best Researcher/Scholar, Sociology

Alexandra Maryanski is a sociologist who was also trained in biological and cultural anthropology. She is one of the founding figures in a new area of sociological inquiry labeled evolutionary sociology. She is the founding chair of the Biology, Evolution, and Society section of the American Sociological Association. Her areas of specialization are the evolution of social institutions, the neurology of the hominin, primate, and human brain, the sociology of religion and kinship, as well as the history of sociology and, more generally, history of ideas. She is best known for her work on what present-day primates, especially great apes can tell us about human nature and social organization. Her research methods include cladistic analysis and social network analysis.

Ivan Strenski

Best Researcher/Scholar, Religion

"Ivan Strenski is an international authority in the field of religious studies."

UNIVERSITY OF CALIFORNIA SAN DIEGO

Ricardo Dominguez

Best Researcher/Scholar, Fine Arts

"He does original, provocative and socially conscious research"

Ricardo Dominguez is a co-founder of The Electronic Disturbance Theater (EDT), a group who developed virtual sit-in technologies in solidarity with the Zapatistas communities in Chiapas, Mexico, in 1998. His recent Electronic Disturbance Theater 2.0/b.a.n.g. lab project (http://bang.transreal.org/) with Brett Stalbaum, Micha Cardenas, Amy Sara Carroll, and Elle Mehrmand, the Transborder Immigrant Tool (a GPS cell phone safety net tool for crossing the Mexico/US border) was the winner of "Transnational Communities Award" (2008), an award funded by Cultural Contact, Endowment for Culture Mexico–US and handed out by the US Embassy in Mexico. It also was funded by CALIT2 and the UCSD Center for the Humanities. The Transborder Immigrant Tool has been exhibited at the 2010 California Biennial (OCMA), Toronto Free Gallery, Canada (2011), The Van Abbemuseum, Netherlands (2013), ZKM, Germany (2013), as well as a number of other national and international venues. The project was also under investigation by the US Congress in 2009–2010 and was reviewed by Glenn Beck in 2010 as a gesture that potentially "dissolved" the U.S. border with its poetry. Dominguez is an associate professor at the University of California, San Diego, in the Visual Arts Department, a Hellman Fellow, and Principal/Principle Investigator at CALIT2 and the Performative Nano-Robotics Lab at SME, UCSD. He also is co-founder of *particle group*, with artists Diane Ludin, Nina Waisman, Amy Sara Carroll, whose art project about nano-toxicology entitled *Particles of Interest: Tales of the Matter Market* has been presented at the House of World Cultures, Berlin (2007), the San Diego Museum of Art (2008), Oi Futuro, Brazil (2008), CAL NanoSystems Institute, UCLA (2009), Medialab-Prado, Madrid (2009), E-Poetry Festival, Barcelona, Spain (2009), Nanosférica, NYU (2010), and SOMA, Mexico City, Mexico (2012): http://hemisphericinstitute.org/hemi/en/particle-group-intro.

Dima Saab

Most Helpful to Students, Linguistics

"Dima has an outstanding track record as an instructor and lecturer. Her evaluations are perfect. She gives her all to the students. She presents them the language and the culture. She organizes activities that require lots of preparation, time and expertise on her part and often times she share them selflessly with her colleagues. She accommodates her students as much as she can, seeing them outside office hours, explaining how to cook original authentic food and beverages and having them taste it. She has a positive attitude and an encouraging smile that makes a difficult language fun and interesting to learn. Often time I hear students and colleagues alike comment on her abilities."

UNIVERSITY OF CALIFORNIA SANTA BARBARA

Catherine Albanese

Best Overall Faculty Member, Religion

Tony Brueckner

Best Overall Faculty Member, Philosophy
Most Helpful to Students, Philosophy

"Anthony Brueckner"

Dorothy M. Chun

Best Overall Faculty Member, Languages

Howard Giles

Best Overall Faculty Member, Communication

"Howie is a prolific researcher who always has time for building strong relationships with his students. His guidance is wise, patient, and constructively critical."

John Mohr

Best Overall Faculty Member, Sociology
Best Researcher/Scholar, Sociology
Best Teacher, Sociology
Most Helpful to Students, Sociology
Best Researcher/Scholar, Liberal Arts, Social Sciences, and Humanities
Best Teacher, Liberal Arts, Social Sciences, and Humanities
Most Helpful to Students, Liberal Arts, Social Sciences, and Humanities
Best Overall Faculty Member, Liberal Arts, Social Sciences, and Humanities

Bill Prizer

Best Researcher/Scholar, Music

Geoffrey Raymond

Best Researcher/Scholar, Sociology

"Geoff is one of the best if not the best researcher in his field. He has a ready grasp of developments in his field and contributes important work to the discipline."

Dwight Reynolds

Best Researcher/Scholar, Liberal Arts, Social Sciences, and Humanities

Nathan Salmon

Best Researcher/Scholar, Philosophy

I investigate issues from a philosophical point of view, and try to achieve a deeper understanding using philosophical methods. My primary tool is reason, my primary criterion for success truth. There will always be those who condemn reason as somehow excessively confining. There will also be those who hold that truth is over-rated, that it is somehow subjective, or even nonexistent, and that therefore it should not be the principal aim of rational inquiry. These people typically replace reason or truth with some favored practical, political, or social agenda, implicitly suggesting that the reasoned search for truth is immoral. Often, they misleadingly apply the word 'true' to any proposition that promotes or supports their substitute for genuine truth. These people are not merely mistaken; naturally understood, their stance is inconsistent. Worse, they devalue humanity's greatest intellectual achievements. They soil the noblest intellectual pursuit of which humanity is capable. Their stance is also dangerous. It is advisable to keep those who place little or no value on genuine reason and truth at a safe distance, at least intellectually (even though they frequently occupy positions of power and status). It is better to seek truth and miss than to aim elsewhere and hit the bull's eye. There is no sin in having erroneous beliefs, provided one endeavors not to. My principal area is the philosophy of language. What is the meaning or content of a sentence—what is that which the sentence says, and which someone may believe or disbelieve? And how do the meanings of the individual words contribute toward forming the meaning of the sentence? I defend the theory of direct reference. According to direct-reference theory, the content of a name like 'Simone Salmon' is simply the person it stands for. This is direct reference because the name means directly what it stands for, rather than meaning some third entity that intervenes between the name and its bearer. Direct-reference theory is saddled with serious philosophical problems. One set of problems concerns identifications. If direct-reference theory is correct, then a statement like 'Mark Twain is Samuel Clemens' simply says about Mark Twain that he is himself. So the sentence would mean the same thing as 'Mark Twain is Mark Twain'. But these two sentences do not appear to be synonymous. Everyone knows the information contained in the second sentence, but some people evidently don't know the information contained in the first: that Mark Twain and Samuel Clemens are one and the very same. This problem is called "Frege's puzzle," after the great philosopher who first used it to argue against direct reference. Frege held instead that the meaning of a word or expression is a concept of the thing referred to. Although both names stand for the same person, the name 'Mark Twain' means or expresses one concept, the name 'Samuel Clemens' another. I have become known for my defense of direct-reference theory against Frege's puzzle. The other set of problems with direct-reference theory concerns names for individuals that do not exist. If the content of a name is just the thing for which it stands, as direct-reference theory holds, then a name like 'Harry Potter' should mean nothing at all, since it stands for a completely fictional character. If the name means nothing at all, then the sentences that make up the Harry Potter stories should be meaningless. But they are clearly meaningful. They make some sense, we understand them, and they entertain. I have defended direct-reference theory against this problem. I argue that although Harry Potter is fictional, he is also every bit as real as you or me. What distinguishes a fictional character like Harry Potter is that he is not a real person. That is, Harry Potter is not actually a person. He is only a fictional person. A fictional person is a real thing, although not a real person. Harry Potter is an object—a real object—created by author J. K. Rowling. He is every bit as real as the novels themselves. In fact, he is a component part of those novels. This sort of consideration does not lay the problem to rest finally. One can devise names that, unlike names from fiction, really do stand for nothing at all. The problem thus has a good deal of force and resilience. I have done much to bring the remaining problem into sharp focus, to pave the way for a full defense of direct reference. My defense of direct reference points back to a theological theme. God, it turns out, is every bit as real as you or me. On the other hand, as an atheist I hold that God is also no more real than Harry Potter. God is depicted in modern mythology as an omnipotent, omniscient, benevolent, and divinely perfect being. In reality, He is an entirely mythical object, no more capable of real thought, action, intelligence, or even consciousness than any purely fictional character. Those who are offended by this simple observation ought to look inside themselves and dispassionately ask "Why?". Most will not. Let us hope some will.

Cynthia Stohl

Best Researcher/Scholar, Communication
Best Teacher, Communication

Cynthia Stohl has been a member of the Department of Communication since January 2002. Prior to joining the UCSB faculty, she was the Margaret Church Distinguished Professor and Head of the Department of Communication at Purdue University. Professor Stohl's work connects several areas in organizational and group studies. She is concerned with the relationships among internal and external communication processes as they are manifest in global collaborations. Her early research focused upon communication networks and issues of quality in global manufacturing groups and has extended to exploring the changing communication partnerships among workers, management, communities, and civic and multinational organizations. Her most recent work addresses a diversity of network and collective action organizations in the global context focusing specifically on the role of new communication technologies in contemporary organizing. She is a Co-Principal Investigator on a National Science Foundation Grant titled "Technological Change and Collective Association: Changing Relationships Among Technology, Organizations, Society, and the Citizenry" and an co-investigator on a Marsden Foundation grant studying new media and the Global Social Justice Movement in New Zealand.

Professor Stohl has been the featured speaker at universities and international conferences throughout the United States, Europe, Asia, Australia and New Zealand. Her research and teaching in organizational communication have been recognized for excellence by university and national associations. The author of over 75 books and articles, Professor Stohl has been the recipient of several research awards, including the 1995 National Communication Association's Best Book Award, and over 15 top paper awards. Last year she became the first communication scholar to receive the International Communication Association's Outstanding Article Award two years in a row, 2007 and 2008. Also the recipient of several outstanding teaching awards, her excellence in teaching was most recently recognized in 2006 when she became one of UCSB's "Primo Professors" and in 2007 when Professor Stohl was awarded the UCSB Distinguished Teaching Award.

Molly Talcott

Best Overall Faculty Member, Sociology

Hsiao-Jung (Sharon) Yu

Best Overall Faculty Member, Languages

Hsiao-jung Yu's research in Chinese linguistics focuses primarily on two areas: historical linguistics and applied linguistics. In historical linguistics, she has conducted research in applying linguistic analysis to solving authorship problems, identifying translators of Buddhist sutras, and dating Chinese literary works. Her current research interest has concentrated primarily on exploring external influences on the changes of the Chinese language, such as the influence of translated Chinese Buddhist sutras and the language policy when China was under the Mongol and Manchu control. In applied linguistics, her primary interests are in language-teaching theory and methodology, context and culture in language learning, and second-language acquisition.

Professor Yu feels that she has learned tremendously while working on these above mentioned areas and topics and wholeheartedly welcomes everyone who shares similar interests to come and join her to explore these areas.

UNIVERSITY OF CENTRAL ARKANSAS

Kevin Browne

Best Overall Faculty Member, Theater
Best Researcher/Scholar, Theater
Best Teacher, Theater

"The only Ph.D. in the Theatre Program. Continues research, writes papers, presents at conferences, and still acts!

Understands material, is passionate and enthusiastic about his teaching, demands the best from his students."

Jann Bryant

Most Helpful to Students, Music

"Mrs. Bryant is perhaps the most conscientious, caring individual in the entire School of Music. She spends countless hours assisting students in planning course loads and schedules. She is known for her patience and all-encompassing knowledge of the music degree at UCA."

Lynn Burley

Best Overall Faculty Member, African/African-American Studies
Best Teacher, African/African-American Studies

I teach a myriad of courses including Semantics, Sociolinguistics, Linguistics for Educators, Introduction to Linguistics, World Languages, Composition I and II, and the new course, Teaching Writing in the Mid-level Grades. I have written numerous reviews for Language and The Linguist List since I spend a lot of time reading books for my classes. I have published several book chapters, written the entry on Dakota and Siouan Languages in the Encyclopedia of Linguistics and appeared in several conference proceedings.

Teri Colaianni

Most Helpful to Students, Liberal Arts, Social Sciences, and Humanities

"Teri cares so much for her students and helps them develop their abilities to speak and use their voice to let others know what they feel. She is a very fun and nurturing person and her students know she is looking out for them."

Shauna Meador

Most Helpful to Students, Theater

Brent Shires

Best Teacher, Music

"Dr. Shires teaches individual horn lessons, advises the student horn club, attends workshops (sometimes travelling with the entire studio), directs brass ensembles, teaches high brass methods, music fundamentals and ear training. He expends tremendous amounts of energy

when teaching, in order to produce well-rounded, highly knowledgeable future educators. He challenges his students to not just do "well enough" but to be excellent in whatever they attempt."

Tiffanie Snyder

Best Teacher, Family & Consumer Science

Registered and Licensed Dietitian. Adjunct Faculty Member for the past 11 years. Instructor of Nutrition, Sports Nutrition, and Resource Management.

UNIVERSITY OF CENTRAL FLORIDA

Jason Burrell

Best Overall Faculty Member, Fine Arts

In front of us is industry – the world that was built, with I-beams, iron, brick, cinder blocks, paper, glass, plexiglas, fiberglass, steel, rebar, plastic, disposability, calogen and silicone. This imposition of bureaucracy and policy over contact is the largest of nightmares and most tragic of fates. Drawing is the surrogate that leads to the transmutation of this vision.

Balancing oppositional forces is the energy of my artwork. The immediacy of black and white on paper can be one of the most informal means of communication; yet its precedence and pedigree has witnessed our construction from shopping lists, plumber advertisements, ransom notes and DaVinci doodles. The decaying factory is our reality. Its antithesis and correct response is hope and aspiration for the future. This dream occurs in the realm of the aesthetic. Finally haunting alienation is vanquished when the humanless, nameless, exclusively functional is transformed into an aesthetic experience.

Drawing is the open medium that alleviates the weight of the subject. It is the embodying of optimism. It overwhelms and evades the strict geometry of structure. The drawing is an escape from regimented severity to livelier hope; ugly places in a moment of beauty. The restricted palette and open brushwork renders calculated construction with the passion of a gesture, rigid structure revealed through light; linear perspective drawn with organic paint.

Ida Cook

Most Helpful to Students, Liberal Arts, Social Sciences, and Humanities

Matt Dombrowski

Best Teacher, Fine Arts
Most Helpful to Students, Fine Arts

"I share an office with Matt Dombrowski. He is incredibly patient and giving of his time with students. His critiques are straight forward and fair."

My research has been continually coupled with my teaching philosophy and experience at UCF. My artwork focuses on simulated environments, as art is a very interdisciplinary field. Most recently, I worked with a fellow colleague to produce Cu Na Mara a time arts work that will be featured in the Mikro Makro Festival in Poland. As well as university research, I freelance for Disney, Orlando Magic, T. G. Lee, and Mizuno Golf, to name a few. In my personal, fine art related works, my realities are simulated; I want the viewer to be absorbed in my art, as he would be absorbed in his own memories. I believe the computer has become an extension of our minds. I strongly feel that digital art is meant for so much more than merely industry work, but can be developed and utilized in the international world of contemporary art. Please, enjoy my art and feel free to contact me if you have any further questions

Jill Fjelstul

Best Overall Faculty Member, Hospitality
Best Researcher/Scholar, Hospitality
Best Teacher, Hospitality
Most Helpful to Students, Hospitality

Dr. Fjelstul is an Associate Professor at the UCF Rosen College of Hospitality Management. She joined the faculty in August, 2005, completing her doctorate in Curriculum and Instruction from the UCF College of Education in January, 2006. Her research interests are aligned with the RV industry, golf and club management, and online education.

Dr. Fjelstul is an avid RVer, affording her a true understanding of the RV lifestyle. She is an author and speaker of RV trends and patterns and is actively involved with the leading RV industry associations.

Dr. Fjelstul also has extensive background in developing successful programs, both academically and industry based. At Rosen College, Dr. Fjelstul wrote and successfully defended a Bachelor of Science degree program in golf and club management. She also designed and implemented the first online education courses for the LPGA Teaching and Club Professionals Division of the LPGA. Fjelstul, with grant monies received from her UCF Toni Jennings special initiative award, created and conducted two unique golf academies for the exceptional population.

Her passion for golf has led to widespread accomplishments, including two state team championships, numerous individual championships, and one collegiate individual title. Fjelstul was honored as an inductee into the University of Northern Iowa's Athletic Hall of Fame in 2001. She is also a Class A LPGA Teaching and Club Professional Member, spending

13 years as a Division I head women's golf coach. Fjelstul was twice named Conference Coach of the Year, led two teams and one individual to three NCAA Regional appearances, and has coached two teams to conference titles. Dr. Fjelstul began her teaching, coaching, and research career at her alma mater, the University of Northern Iowa.

Andrew Gay

Best Teacher, Film

Andrew Kenneth Gay is an educator, innovator, and digital cinema entrepreneur. His work aims to equip the next generation of emerging filmmakers to control the means of production and leverage social media tools to distribute their output directly to a micro-targeted audience. His theories were recently published in the peer-reviewed Journal of Screenwriting.

As a writer/director of shorts and features he has received multiple awards and over 50 official selections at festivals across the globe. His first feature-length motion picture, A Beautiful Belly (2011), recently received worldwide distribution online, and his new documentary feature, Highly Sought After (currently in production), successfully raised more than $48,000 through Kickstarter with the support of over 1,000 backers and has over 5,500 fans on Facebook.

Andrew has worked as a production coordinator, production manager, and assistant director on numerous commercials, reality shows, and features, including media for Spike TV, VH1, the Discovery Channel, and Disney. He also consults with other independent filmmakers, writing feature budgets, business plans, and offering script notes for producers in search of financing.

He is the Festival Director of the StarLite Film Festival in Winter Garden, FL, a competitive exhibition focused on programming student shorts and microbudget features produced for $200,000 or less.

Kevin Haran

Most Helpful to Students, Fine Arts

Kevin Haran has a B.F.A in Studio Art and a B.A. degree in Art History from the University of Central Florida. He earned an M.F.A. degree in Painting/Drawing from the University of Tennessee, Knoxville in 1991. Haran was Associate Curator of Education at the Knoxville Museum of Art and currently teaches studio art at the University of Central Florida.

Ana Leon

Best Overall Faculty Member, Social Work

"Ana is always there to assist students and provides outstanding lectures."

Dr. Leon's research, training and clinical work focuses on child health and mental health with an emphasis on Infant Mental Health. She is passionate about improving the health and mental health well-being of very young children and their families. Her passion for this population is reflected in her teaching, community training, research and community service. Since 2000, her research grants/contracts have totaled over $3,190,443. Dr. Leon's scholarly work includes over 23 articles, co-author of one book, and national and international presentations. She is Chair of the NASW Child Welfare Specialty Practice Section, and is a member of the Editorial Board for the Journal of Social Service Research and a Consulting Editor for Families in Society. She developed a guidebook, Babies & toddlers in a new country: A developmental guide for parents new to the United States for immigrant and refugee families served by Catholic Charities. In 2010, she received the University of Central Florida Women Making History: The Faculty Women Prominence Project award.

For the past 36 years, Dr. Leon has specialized in working with children and families in diverse agency settings and roles. She has worked as a clinician, has held numerous administrative jobs, served as Executive Director of Psychiatry at Winter Park Memorial Hospital, has been in private practice, served as a program evaluator and program consultant. She also serves as a qualified supervisor for state licensure for both Mental Health Counselors and Clinical Social Workers. She is currently a Professor of Social Work at the University of Central Florida where she has taught mostly clinical, children's courses and research courses for the past 20 and years. Dr. Leon also served as the Vice Chair for the university's Institutional Review Board which reviews and approves all research studies at UCF.

Dr. Leon's primary areas of training, clinical and research interests are Infant and Child Mental Health and documentation in social work and the human services. Her other professional and research interests include: Latino children and families, child development, and community based family services. For seven years she developed and received funding for a successful university-community collaborative project that provided community based mental health services to children and families. As part of that grant funding, she developed an assessment tool (the CALF) for assessing areas of life functioning in various client systems. An article on the CALF was published in Child & Family Social Work. She has also published in the areas of child mental health, family systems theory, assessment tools, challenges of Latina mothers, the impact of migration, clinical documentation and psycho-pharmacology.

Ann Miller

Best Overall Faculty Member, Communication

"Dr. Ann Miller's research is outstanding and she gets good reviews from both her undergraduate and graduate students."

Ann Neville Miller has actively taught, researched and served in the field of intercultural communication for more than 10 years. Graduating from the University of Georgia in 2005, she won the Gerald R. Miller dissertation award for her research on self-disclosure of positive diagnosis of HIV in Kenya. She has since established research programs on intercultural factors in health communication, instructional communication, and scholarly productivity of communication faculty in sub-Saharan Africa. She has published more than 35 peer-reviewed articles in journals like Journal of International and Intercultural Communication, Health Communication, Journal of Health Communication, AIDS and Behavior, and Communication Education, and co-edited a scholarly volume. Miller teaches courses in intercultural communication, persuasion, research methods and health communication. She is a member of the editorial board for the Journal of International and Intercultural Communication and Communication Education.

Olga Molina

Best Overall Faculty Member, Liberal Arts, Social Sciences, and Humanities

Jennifer Navarra

Best Researcher/Scholar, Liberal Arts, Social Sciences, and Humanities

Estelli Ramos

Best Teacher, Liberal Arts, Social Sciences, and Humanities

Master of Social Work Coordinator Estelli Ramos is an expert in gangs, gang life and gang violence.

He has an extensive background working with grassroots groups providing a host of training from cultural competency to street gang intervention. He can speak to a host of social issues including youth at risk, teen pregnancy and the role families play in their children's lives.

Ramos has degrees from the University of Chicago, School of Social Service Administration and Northeastern Illinois University. He joined UCF in 2006.

When he is not teaching at UCF he provides private consulting and therapy to families. Before moving to Florida he worked extensively with the Hispanic community in the Chicago area. Ramos is founder and director of Teatro Juvenil, a youth based theater program for at risk youth in the inner city of Chicago. He also was the director of youth services in the Chicago area. Previously he worked in the Chicago Public Schools as a school social worker.

Ramos has spoken at various conferences on gangs, social work and cultural competency. He has appeared in local and regional media including the Orlando Sentinel and Univision. Like "Estelli Ramos" on Facebook Tweet "Estelli Ramos" on Twitter Share "Estelli Ramos" on Google+

Stephen Schlow

Best Teacher, Fine Arts

Maria Grazia Spina

Best Teacher, Languages

M.A. in Modern Foreign Languages and Literatures from University of Genova, Italy
Italian, French and German languages and cultures. Second language acquisition. My online persona for my fully online courses is a world traveler.

Joshua Truitt

Best Overall Faculty Member, Leadership

Joshua Truitt, Ph.D. is a Postdoctoral Associate and Special Assistant to the Dean of the UCF College of Education and Human Performance. He is an instructor for the UCF LEAD Scholars program, and is active in several charitable and civic organizations. Dr. Truitt received his B.S. and Ph.D. from UCF, and MBA from Rollins College.

Barbara Turnage

Best Researcher/Scholar, Social Work
Best Teacher, Social Work
Most Helpful to Students, Social Work

"Barbara has the strongest record of research."

Barbara is outstanding in her delivery of educational materials. She is creative and innovative in her approach to education.

UNIVERSITY OF CENTRAL OKLAHOMA

Eva Dadlez

Best Researcher/Scholar, Liberal Arts, Social Sciences, and Humanities
Best Teacher, Liberal Arts, Social Sciences, and Humanities

Dr. Eva M. Dadlez is Professor of Humanities and Philosophy, Dr. Eva Dadlez, Athena College of Liberal Arts, University of Central Oklahoma. She received her Ph.D. from Syracuse University. She writes on issues at the intersection (often at the collision) of aesthetics, ethics and epistemology. She has written two books on the preceding: What's Hecuba to Him? Fictional Events and Actual Emotions (Penn State University Press, 1997) and Mirrors to One Another: Emotion and Value in Jane Austen and David Hume (Wiley-Blackwell, 2009). She is also a feminist ethics dilettante.

Aida Dean

Best Researcher/Scholar, Foreign Languages
Most Helpful to Students, Foreign Languages

Darian Debolt

Most Helpful to Students, Liberal Arts, Social Sciences, and Humanities
Best Overall Faculty Member, Liberal Arts, Social Sciences, and Humanities

Darian C. De Bolt teaches moral philosophy and ancient Greek at the University of Central Oklahoma. He received a B.A. magna cum laude (1968) in philosophy and Greek from the University of Oklahoma. From that same institution, he also received an M.A. (1984) and Ph.D. (1993) in philosophy. Dr. De Bolt pursued a Master of Theological Studies degree at the Lutheran School of Theology at Chicago. He is also a graduate of the Federal Bureau of Investigation National Academy at the University of Virginia. Dr. De Bolt specializes in ethics, social and political philosophy, and epistemology. However, he has a wide range of interests including comparative religion and Greek literature. He also pursued a career in law enforcement at the Norman (OK) Police Department. He retired from police work in 1993 at the rank of Captain. He was subsequently elected to the city council of the City of Norman, Oklahoma.

David Forbat

Most Helpful to Students, Music

A member of the UCO piano faculty since 2005, David Forbat teaches applied piano, piano ensemble, keyboard functional skills and class piano. He holds a DMA from the Peabody Institute of the Johns Hopkins University, a MM from the University of Cincinnati, College-Conservatory of Music and a BM from the University of Southern California. His teachers included John Perry, Frank Weinstock, Claude Frank and Ann Schein. Prior to 2005, he held full-time teaching appointments at Geneva College (PA) and William Carey College (MS) and part-time appointments at the University of Pittsburgh and Carnegie Mellon University (PA).

Dr. Forbat maintains an active performing schedule as a solo recitalist and collaborative pianist. During the 2012–13 season, he performed solo recitals at universities in Mississippi, Texas and Oklahoma. He then traveled to China to perform and teach at four universities. At the invitation of Beijing Normal University, he returned to China in May/June 2013 for a 2-week teaching residency. His recital performances have been aired on Ohio, Mississippi, and Oklahoma public radio stations on programs such as "Mississippi Concert Hall" and "Performance Oklahoma."

In addition to his chamber and vocal collaborations with members of the UCO music faculty and guest artists, he has participated in large ensemble programs (such as the UCO Cantilena, Concert Chorale and Wind Symphony). He has at times presented informative and cross-disciplinary programs (Liszt and the Italian Poets, The French Connection and Schumann's Carnaval, Op.9), and has combined creative ideas with UCO dance professor Robyn Pasternack. Together, they were co-recipients of the 2011 College of Fine Arts & Design Collaborative Team Award.

During the spring 2007 semester, he was awarded a Faculty Pedagogical Research Grant to explore the integration of keyboard learning with moveable-Do solfege. On this topic, he was the featured presenter for the College Faculty Forum at the 2007 State Convention of the Oklahoma Music Teachers Association. With Jacquelyn Karash, NCTM, he co-authored Key Awareness: Building Confidence Playing the Piano in all 30 Keys (2012). He continues to investigate ways to foster tonal understanding at the keyboard and enjoys composing pieces directed toward achieving this end.

In recent years, Dr. Forbat has been an advocate of digital keyboards and the many pedagogically-relevant ways in which they can be used to build keyboard musicianship. In 2011, with the assistance of a UCO Foundation Grant, he traveled to Scottsdale and Tucson, AZ with 15 students and 3 faculty members, giving keyboard ensemble concerts and presentations. A project for the spring semester of 2014 featured piano concertos by Chopin and Mozart with digital keyboard orchestra accompaniment.

Marilyn Govich

Best Overall Faculty Member, Music

Dr. Marilyn Govich is a professor of applied voice, and previously served as Division Head of Music Theatre and Opera, and Coordinator of Vocal Arts at UCO. She holds a DMA in Vocal

Performance from the University of Oklahoma where she received the Provost's Outstanding Dissertation Award. Her dissertation also won the Biennial Dissertation Award of the National Opera Association. At UCO, Dr. Govich has been awarded the Hauptman Fellowship Award for Distinguished Achievements in Research, Creative and Scholarly Activity, as well as Faculty Merit Award for Outstanding Teaching. Two of her students recently won national Outstanding Performance Awards at the Kennedy Center of the American College Theatre Festival. Dr. Govich was recently seen as The Grand Duchess Sophie in Call Me Madame with Tony Award winner Beth Leavel. Other performance roles include Mother Abbess in The Sound of Music, Widow Douglas/Sally Phelps in Big River, Mother in Ragtime, Lady Beaconsfield in Jekyll and Hyde, Older Emma in Timeless, Ethel McCormack in Footloose, Aldonza in Man of La Mancha, Nancy in Oliver, Margot Bonvalet in The Desert Song, Winnifred in Once Upon a Mattress and Domina in A Funny Thing Happened on the Way to the Forum. In addition, she has performed Rosalinde in Die Fledermaus, Ann Putnam in The Crucible, Dorabella in Cosi Fan Tutte, Constanze in Abduction from the Seraglio, Madame Flora in The Medium, Lola in Gallantry and Miss Baggott in The Little Sweep. She has musical directed shows at Oklahoma City Repertory Theatre and The Palace Players in Lorain, Ohio. In addition Dr. Govich has operated a private voice studio in New York City, and previously taught at Oklahoma Baptist University. Dr. Govich has performed as a soloist with the newly created Center for Historical Performance Practice, and previously was a soloist with the Oklahoma Collegium Musicum. She continues as an active recitalist, oratorio soloist, adjudicator and clinician. She has twice been selected to perform as a soloist at the National Conference of the National Opera Association in New York City and has twice presented Artist Recitals at the Texoma Regional Conference of the National Association of Teachers of Singing, as well as performed for regional conferences of The College Music Society. Dr. Govich has recorded a scene from the David Yeagley opera "Jacek" with the Polish National Radio Symphony in Katowice, Poland for Opus One Records. She is extremely proud of her students who have become outstanding teachers and those who have performed in Europe, China, Japan, Canada, on Broadway (including her daughter and son), Broadway national tours, cruise ships and numerous regional theaters.

Keith Higa

Best Researcher/Scholar, Curriculum & Instruction

Ted Honea

Best Researcher/Scholar, Music

Dr. Ted Honea combines expertise in scholarly research and practical performance in a way that is rare among university music faculty. Well-known in the U.S. for his articles on horn pedagogy and performance practice, he also maintains an active performance career, performing repertoire from all periods of literature for the horn. Returning home to teach at his alma mater, Dr. Honea holds a BM from UCO, a MA and is ABD in Musicology from the Eastman School of Music and a second MA and PhD degree in Classics from the State University of New York-Buffalo. He has also studied horn with Verne Reynolds,

Paul Ingraham and Melvin Lee and natural horn with Richard Seraphinoff. Dr. Honea has authored numerous scholarly and pedagogical publications, including articles in "The Horn Call," "Music Educators Journal," "Music Reference Services Quarterly," "MLA Notes," "American Music Teacher," "Ad Parnassum," "Journal of Religious History," "Classical World," "Journal of Ritual Studies," "Journal of Academic Librarianship," "Journal of Library Administration" and "New Library Scene." He has served as a member of the preservation committee research library group and music library association, treasurer of the American Liszt Society and current archivist for the International Horn Society. Dr. Honea is a member of the American Musicological Society, the Society for American Music, the International Horn Society, the College Music Society and the National Association of Schools of Music. He received the UCO "Vanderford Distinguished Teacher Award" for the College of Fine Arts and Design in 2006 and the University's "Modeling the Way Award" in 2008. Dr. Honea has recently focused his efforts in scholarship on the history of wind literature, its performance practice and performing on the natural horn. Previous staff employment included service at Sibley Music Library, Eastman School of Music (1978–1998) where he was conservator, head of rare books and special collections and archivist. Previous faculty appointments include the University of Rochester (1986–88), Eastman School of Music (1991–98) and George Eastman House Institute for Conservation (1992–98).

Diana Pardo

Best Teacher, Languages

"Is very helpful not only to students, but to other faculty and staff. Has led the summer study abroad for advanced students to Spain for last 2 summers. Is a wonderful part of the Department."

Tess Remy-Schumacher

Best Teacher, Music

Allen Rice

Best Teacher, English

Susan Spencer

Best Overall Faculty Member, English
Best Teacher, Liberal Arts, Social Sciences, and Humanities
Most Helpful to Students, Liberal Arts, Social Sciences, and Humanities

Wayne Stein

Best Researcher/Scholar, English
Most Helpful to Students, English
Best Researcher/Scholar, Liberal Arts, Social Sciences, and Humanities
Best Overall Faculty Member, Liberal Arts, Social Sciences, and Humanities

Wayne Stein, Ph.D., is a Professor in the English Department

A chapter called "Enter the Dracula: Asian Cinematic Vampires" will be included in Draculas, Vampires, and Other Undead Forms by Scarecrow Press. Five entries in Greenwood Asian American Encyclopedia have been released. A new edition of Languaging Force X (LFX) – the role playing game that promotes literacy, critical thinking skills and fun – has been published. Stratagems, a new edition, was just published by Fountainhead Press. Fresh Takes, a composition reader, was also published by McGraw Hill. A chapter on anime vampires found in Vampire Hunter D, "The Western Eastern: De Coding Hybridity and CyberZen Goth(ic)" in Asian Gothic, is now out.

Developing a Service Learning Composition Class, online composition classes, and other online classes (Kung Fu Films, History of Rhetoric 2, and Cyberpunk Films and Literature), he tries to create classes that make a difference.

He was awarded the 2008 DaVinci Fellow from the "Oklahoma's Creativity Think Tank." He taught EFL (English as a Foreign Language) in Seoul, Korea, the place of his birth. He was the past president and past webmaster of OKTESOL. He has chaired and helped to chair five conferences: three OKTESOL Conferences, and a couple of Oklahoma Buddhist Conferences.

He was awarded the ESL Professional of the Year Award from OKTESOL (2003), the Service to the Sanga (Community) Award from the Stillwater OSU Buddhist Society (2003), Faculty Merit for Service (2004), and LA Creative Award (2004).

Dr. Stein has presented papers at international, national, and local conferences. He teaches Asian American literature, modern world literature, and cyberpunk fiction and is interested in the magical realism of Gabriel Garcia Marquez, the paranoia of Thomas Pynchon, and the postmodernity of Theresa Hak Kyung Cha. He visited Vietnam (2002) as he prepared to teach his Vietnam War literature course.

He grew up visiting Chinatown in Los Angeles, watching Kung Fu films, and practicing martial arts, which were his ways of trying to act Asian. Thus, he brings much enthusiasm to his Kung Fu film class. Since Kung Fu films were influenced by the samurai films of Kurosawa, he also teaches a Kurosawa film class. Practicing the way of the sword (kendo), he is one of the sponsors of the Budo Society, the way of the warrior, and tells everyone to watch Red Beard, Toshio Mifune's last film with Kurosawa. He has also taught a course on graphic novels and on Anime (Japanese animation) and Manga (comics).

Believing in excellence, he has completed training under the Oklahoma Quality Awards Foundation. Promoting the benefits of technology, Dr. Stein has been the Liberal Arts Web-maintainer/master, the English Department webmaster, and the technical advisor of the UCO OWL, Online Writing Lab. He has been English Department Manager of Turnitin.com, Fall 2003–Spring 2005. He has promoted the idea of Lectureships for Composition Instructors.

Amalia Wilkinson

Best Overall Faculty Member, Foreign Languages
Best Teacher, Foreign Languages

UNIVERSITY OF CINCINNATI

Mark Lause

Best Researcher/Scholar, History

Professor Lause grew up in a small blue collar community and worked his way through college during the 1960s and 70s. His academic interests seemed to center naturally on the history of class and social movements in the United States.

Lause has done extensive work in nineteenth century labor and social history, including numerous articles in academic journals and reference material. His initial work focused on early printers to discuss the origins of an American labor movement: "Some Degree of Power": From Hired Hand to Union Craftsman in the Preindustrial American Printing Trades, 1778–1815. (Fayetteville: University of Arkansas Press, 1991) documented the first generation of unionists in that craft.

Lause's subsequent work has sought new ways of examining and understanding the sectional crisis and the Civil War "from the bottom up." He argued for the complexity of the Republican and Unionist coalition—before and after—in Young America: Land, Labor, and the Republican Community (Urbana IL: University of Illinois Press, 2005) on the antebellum land reform movement and The Civil War's Last Campaign: James B. Weaver, the Greenback-Labor Party & the Politics of Race & Section (Lanpham, Md.: University Press of America, 2001). His Race & Radicalism in the Union Army (Urbana IL: University of Illinois Press, 2009) explores the wartime collaboration of blacks, Indians and whites in the Transmississippi under the leadership of those abolitionists, land reformers, socialists and others who had been associated with John Brown before the Civil War. The Antebellum Political Crisis & the First American Bohemians (Kent, OH: Kent State University Press, 2009) discusses the cultural impact of escalating sectional and electoral pressures on antebellum radicalism. His Price's Lost Campaign: the 1864 Invasion of Missouri (Columbia: University of Missouri Press, 2011) uses social and institutional history to cast light on the neglected Civil War expedition that largely closed the conflict west of the Mississippi River. A Secret Society History of the Civil War (Urbana: University of Illinois Press, 2011) examines the importance of several clandestine, fraternal traditions as a means of understanding how ordinary citizens, including African Americans, struggled to shape their history.

In addition, Lause is completing a monograph on the politics of mid-nineteenth century American spiritualism and finishing a major work on the labor movement during the Civil War, as well as composing a second book on the close of 1864 Missouri campaign. Afterwards, he aspires to pursue several projects in nineteenth century European and comparative fields.

Dean Mogle

Most Helpful to Students, Theater
Best Teacher, Liberal Arts, Social Sciences, and Humanities

"Most dedicated professor on Campus
Extremely dedicated Professor. Students are always first as he educates."

Professor and Head of the Costume Design and Technology Program, Dean Mogle has designed numerous productions of drama, musical theatre, opera and dance during the past 40 years. Costume designs for CCM include Les Miserables, Into the Woods, The Magic Flute, Falstaff, Werther, Little Women, Hansel and Gretel, Rusalka, Pelleas et Melisande, Grand Hotel, Il Viaggio A Reims, The Secret Garden, The Rake's Progress, Lend Me a Tenor, Babes in Arms, Blithe Spirit, The Voyage of Edgar Allan Poe, Chicago, The Cunning Little Vixen, A Flea in Her Ear, La Bohéme, The Wild Party, Zaide (world premiere), Cendrillon, A Midsummer Night's Dream, Dracula: The Game of Love, Lady Be Good and Don Giovanni, among others.

Previously, he was resident costume designer for 30 productions and professor at the University of South Dakota. He was resident and guest costume designer for the Black Hills Playhouse for more than 35 productions and resident costume designer for the Fargo-Moorhead Community Theatre for 30 productions. As a guest designer he has worked for the F-M Civic Opera and Modern Dance companies, Nebraska and New Mexico Repertory Theatres, the Utah Shakespeare Festival and the Cincinnati Ballet, among others. He has been costume and fashion designer for the Don-Lu modeling institute, the Miss USA pageant, theme restaurants and mascots.

National Opera Association Awards include Falstaff, Rake's Progress, Seasons in Hell, Vixen, La Bohéme and A Midsummer Night's Dream. USITT-Ohio Peggy Ezekiel Awards include Falstaff, Vixen, Chicago, A Flea in Her Ear and A Midsummer Night's Dream for CCM; Othello for the New Mexico Repertory; King Lear and Macbeth for the Utah Shakespeare; and Nutcracker for the Cincinnati Ballet, and awards from the League of Cincinnati Theatres.

He is curator for CCM's Historic Costume Collection including the Italo Tajo Costume Archive. He participates locally, regionally and nationally as a lecturer, panelist, workshop presenter and adjudicator. He is a fiber artist, painter and sculptor and is a member of the United Scenic Artists-829.

UNIVERSITY OF CINCINNATI BLUE ASH COLLEGE

Diana Becket

Most Helpful to Students, English

Ruth Benander

Best Researcher/Scholar, English

Brenda Refaei

Best Overall Faculty Member, English

UNIVERSITY OF COLORADO

Ann Schmiesing

Best Teacher, Languages

Ann Schmiesing is an associate professor of German and Nordic Studies with research interests in 18th and 19th-century German and Norwegian literature and culture. She received an M.A. in comparative literature from the University of Washington and a Ph.D. in German from Cambridge University (U.K.). Her book Norway's Christiania Theatre, 1827–1867: From Danish Showhouse to National Stage was published by Fairleigh Dickinson University Press in 2006. Her many publications in nationally and internationally peer-reviewed journals include articles on Gotthold Ephraim Lessing, Daniel Chodowiecki, Theodor Gottlieb von Hippel, Bjørnstjerne Bjørnson, and Johan Falkberget, and on issues pertaining to drama, narrative fiction, theatre history, and book illustration. She has also published and/or presented papers on several issues in humanities pedagogy, including service learning, curriculum development, and the utilization of rare books collections in humanities teaching.

David Shneer

Best Overall Faculty Member, Liberal Arts, Social Sciences, and Humanities

"DAVID IS DIRECTOR OF THE PROGRAM OF WHICH I AM PART. HE IS A SUPERB ADMINISTRATOR, GREAT WITH STUDENTS, AN EXCELLENT RESEARCHER, A CREATIVE THINKER, A WONDERFUL COLLEAGUE, AND OVERALL MOST DESERVING OF THIS NOMINATION. I HAVE BEEN AT CU FOR 10 YEARS NOW, AND HE IS THE BEST FACULTY MEMBER I HAVE ENCOUNTERED."

Called a "pathbreaking" scholar by the Frankfurter Allgemeine Zeitung, Shneer's research focuses on 20th century European, Russian, and Jewish history and culture. His newest book, Through Soviet Jewish Eyes: Photography, War, and the Holocaust (Rutgers University Press, 2011), finalist for the National Jewish Book Award and winner of the 2013 Association for Jewish Studies Jordan Schnitzer Prize, looks at the lives and works of two dozen Soviet Jewish World War II military photographers to examine what kinds of photographs they took when they encountered evidence of Nazi genocide on the Eastern Front. In fall 2011, the traveling museum exhibit Through Soviet Jewish Eyes debuted at the CU Art Museum in Boulder, Colorado. Through Soviet Jewish Eyes was on display at the Holocaust Museum Houston in Texas in October 2013.

His other books include Queer Jews, finalist for the Lambda Literary award, Yiddish and the Creation of Soviet Jewish Culture, finalist for the National Jewish Book Award, and New Jews: The End of the Jewish Diaspora, which has sparked discussion in publications like the Economist and The Jerusalem Post. His new project, Not On Their Last Road, examines Yiddish musical culture's role in the clash between fascism and communism through the life and work of Lin Jaldati, a Dutch-Jewish Yiddish-singing cabaret singer, who survived the Holocaust and was the last person to see Anne Frank alive. After the war, she moved to East Germany and became the Yiddish diva of the Communist world until her death in 1988.

UNIVERSITY OF COLORADO COLORADO SPRINGS

Christopher Bell

Most Helpful to Students, Communication

Dr. Bell has served as the Director of the Center for Excellence in Communication and an Assistant Professor of Communication at the University of Colorado Colorado Springs since the summer of 2010. He specializes in the study of popular culture, focusing on the ways in which race, class and gender intersect in different forms of media. Dr. Bell's primary research areas are young adult culture, comic books, celebrity culture and reality television. He serves as the chair of the Southwest/Texas Popular Culture Association's Harry Potter Studies division. He teaches both theory and methodology courses in critical analysis of popular culture, rhetorical theory, representation theory and mass media. Dr. Bell is a nationally-touring speaker on issues of race, class and gender in the media, and has presented work at regional, national and international conventions and conferences. In what little spare time is left over, Dr. Bell is a competitive gamer, competing on regional and international circuits, and travels with his wife and daughter.

Ken Eurek

Best Teacher, Languages
Best Teacher, Liberal Arts, Social Sciences, and Humanities

Rieko McAdams

Most Helpful to Students, Languages
Most Helpful to Students, Liberal Arts, Social Sciences, and Humanities

Margaret Mistry

Best Overall Faculty Member, Languages
Best Researcher/Scholar, Languages

Harriet Napierkowski

Best Researcher/Scholar, Liberal Arts, Social Sciences, and Humanities

Dave Nelson

Best Teacher, Liberal Arts, Social Sciences, and Humanities

Rex Welshon

Best Overall Faculty Member, Liberal Arts, Social Sciences, and Humanities

UNIVERSITY OF CONNECTICUT

Anne Berthelot

Best Overall Faculty Member, Languages

An agrégée des lettres and a graduate of the Ecole Normale Supérieure, Anne Berthelot is now professor of French and Medieval Studies at the University of Connecticut. Following her dissertation for the French Doctorat dEtat on the writer in 13th century French literature, she has written numerous books and articles on Medieval literature, focusing especially on the Arthurian legend with a comparatist approach. She is part of the team who is making the so-called Lancelot-Grail Cycle accessible to a large audience in the prestigious series of La Pléiade (Gallimard). Her most recent book is a synthetic presentation of the Arthurian legend for the Editions du Chêne, La Légende du roi Arthur (Fall 2004), which has now been translated into German. She is working on a book-length study of the enunciation problems in the Roman de Perceforest, and at the same time is preparing an edition of a little-known Arthurian romance that may be considered as the source for the Perceforest, the Roman des fils du roi Constant.

Joel Blatt

Best Overall Faculty Member, History

Professor Blatt received his Ph.D. from the University of Rochester and an undergraduate degree from Cornell University. He has edited a book with an Introduction on The French Defeat of 1940: Reassessments (1998). He has also published a number of articles on France and Italy between the two world wars and is presently working on two books. He has received a number of grants from the UConn Research Foundation, and has served a Provost's Fellowship. He is twice the recipient of the UConn Stamford Campus Outstanding Teacher Award.

Alexis Bolan

Best Researcher/Scholar, Art

Alexis L. Boylan received her B.A. in history from Bryn Mawr College and her Ph.D. in art history from Rutgers University. Her research focus is on American visual culture from the colonial to the contemporary periods with particular emphasis on race and gender.

Boylan is editor of Thomas Kinkade, The Artist in the Mall (Duke University Press, 2011). Her current book project, entitled Man on the Street: Masculinity, Modernism, and Ashcan Art examines the Ashcan circle of artists and their visual commentaries on urban masculinity. Boylan's next book project, The Case of the Two Mrs. Sloans, uses the extraordinary lives of the two wives of artist John Sloan to critically consider the cultural roles and meanings of different stages of identity for heterosexual women, namely the positions of wife, mistress, and widow. Boylan has published in Rethinking Marxism, Woman's Art Journal, and Prospects and has contributed essays to recent museum exhibition catalogues including Seeing the City: John Sloan's New York (Yale University Press and the Delaware Art Museum, 2007); Landscape of Slavery: The Plantation in American Art, (University of South Carolina Press and the Gibbes Museum of Art, 2008); and The Life and Art of Alfred Hutty: Woodstock to Charleston (University of South Carolina Press 2012).

Rosa Helena Chinchilla

Most Helpful to Students, Languages

Cora Lynn Deibler

Best Overall Faculty Member, Art

Thomas Jambeck (Posthumously)

Best Overall Faculty Member, English

Frederic Lardinois

Best Researcher/Scholar, English

Sam Pickering

Most Helpful to Students, English

B.A. 1963 University of the South (Sewanee) B.A. 1965 M.A. 1970 Cambridge University M.A. 1985 Ph.D. 1970 Princeton University D Litt (Honorary) 2002 Oglethorpe University PROFESSIONAL HISTORY 1970–78 Assistant Professor, Dartmouth College 1978–84 Associate Professor, University of Connecticut 1984–2013 Professor, University of Connecticut 2013 - Emeritus, University of Connecticut Fellowships and Awards: Fellow, Institute of Modern Letters at the University of Tulsa, 1970. NEH "Special" Fellowship, 1974. ACLS Grant - in - Aid, 1976. Fulbright Lectureship, University of Jordan, 1975–76. Fulbright Lectureship, Tishreen University (Latakia, Syria), 1980–81. Bread Loaf Writers' Conference Fellowship, 1987. Fellowship of Southern Writers, Nonfiction Prize, 1993. Member of the Connecticut Academy of Arts and Sciences. Research Associate, University of Western Australia, 1993–94 and 2000–01. Robert B. Heilman Award for Reviewing, 1999 and 2005. Montgomery Bell Academy, Distinguished Alumnus, 2002. Writers' Colony at Dairy Hollow, Moondancer Fellowship, 2003. Fellow of the Institute for Advanced Studies in the Humanities at the University of Edinburgh, 2005. Member of the Fellowship of Southern Writers, 2005. McGinnis - Ritchie Award for Nonfiction, 2005. Honorary Research Fellow, University of Western Australia, 2008. Distinguished Alumni Professor, University of Connecticut, 2013. Miscellaneous Teaching: Montgomery Bell Academy, Nashville, Tennessee, 1965–66. NEH Institute in Children's Literature, University of Connecticut, 1983, 1985. Rare Book School, Columbia University, 1985. American Studies Seminar, American Antiquarian Society, 1994. Sewanee School of Letters, 2010.

Janet Pritchard

Best Teacher, Art

Janet L. Pritchard photographs place with a special interest in the intersection of nature and culture. She describes her method as the pursuit of artistic vision through historical empathy. Her current creative research project, Yellowstone Dreams: An American Love Story, explores our national love affair with America's first national park through the lenses of nature, culture and history. Her previous work, Dwelling: Expressions of Time, explored the layers of time found in the landscape of southern New England.

Pritchard's exhibition venues include: Fruitlands Museum, Massachusetts; International Center for Photography, New York; Martha Schneider Gallery, Chicago; Photographic Resource Center, Boston; RISD Museum of Art, Providence, RI; Sam Francis Gallery, Santa Monica, California; and the National Trust for Historic Preservation travelling exhibition America's Uncommon Places. She is the recipient of numerous awards, including: Artist-in-Residence Fellowship, Ucross Foundation, Clearmont, Wyo.; Jay and Deborah Last Fellowship, American Antiquarian Society, Worcester, Mass.; Fellowship, University of Connecticut Humanities Institute; Artist-in-Residence Fellowship, Institute for Electronic Arts at Alfred; and Honors General Education Curriculum Development Grant, University of Connecticut.

Pritchard has been on the faculty since 2001 and serves as the Area Coordinator for Photography. Before coming to the University of Connecticut, she taught at Tyler School of Art in Philadelphia and the University of Colorado. Pritchard received her BA in art history and classics from the University of Colorado, and her MA and MFA from University of New Mexico. Before pursuing a career in photography, she worked as an outdoor education instructor and spent her youth travelling between the Northeast and Rocky Mountain West. She makes her home in New England but considers herself geographically bilingual.

Tom Recchio

Best Teacher, English

Laurie Sloan

Most Helpful to Students, Art

Laurie Sloan is an Associate Professor of Art at the University of Connecticut. She received her M.F.A. degree from Tyler School of Art, Philadelphia, PA and Rome, Italy and her B.F.A. degree from Miami University, Oxford, OH. Sloan's artwork has been exhibited widely in the United States and abroad, including recent exhibitions at the Manhattan Graphic Arts Center,

Northern Arizona University, the International Print Center, NY, and Real Art Ways in Hartford, CT. She has won several grants and honors including a Connecticut Commission on the Arts Individual Artist fellowship and numerous juried exhibition awards. Sloan has been invited to participate in numerous exchange portfolios and her prints are in over twenty important collections including the DeCordova Museum, Lincoln, MA, Bucknell University, Lewisburg, PA and the Santa Barbara Museum of Art, CA. She has lectured on her work at various institutions including the University of Alberta, Edmonton, Massachusetts College of Art, Boston, MA, and the University of Tennessee, Knoxville, TN.

UNIVERSITY OF DELAWARE

Wendy Bellion

Best Researcher/Scholar, Art History
Most Helpful to Students, Fine Arts

Professor Wendy Bellion teaches American art history. She holds a B.A. in art history from Wesleyan University and an M.A. and Ph.D. from Northwestern University. Professor Bellion's scholarship takes an interdisciplinary approach to American visual and material culture, focusing on the late colonial and early national United States and exploring American art within the cultural geographies of the British Atlantic world and early modern Americas. Her book, Citizen Spectator: Art, Illusion, and Visual Perception in Early National America (2011) examines the exhibition of trompe l'oeil paintings, optical devices, and illusionistic spectacles within post-revolutionary cultures of sensory discernment and undeceiving. She is also co-editor (with Professor Mónica Domínguez Torres) of Objects in Motion: Art and Material Culture across Colonial North America (2011), a special issue of the journal Winterthur Portfolio that developed out of a University of Delaware symposium in 2008. Her new book project – What Statues Remember: Iconoclasm and Ritual in Early New York – explores the performative dimensions of sculptural monuments in lower Manhattan from the late eighteenth century to the Occupy Wall Street movement. Her publications include essays on Charles Willson Peale's use of perspective devices, the transatlantic art of waxwork sculptor Patience Wright, theatrical illusion in early Philadelphia, and art-historical methodologies.

Professor Bellion taught at Rutgers University and the College of William and Mary before joining the University of Delaware in 2004. An elected member of the American Antiquarian Society, she has been awarded grants and fellowships from the American Antiquarian Society, Center for Advanced Study in the Visual Arts (National Gallery of Art), Henry Luce Foundation, Library Company of Philadelphia, National Endowment for the Humanities, Omohundro Institute of Early American History and Culture, Terra Foundation for American Art, and Winterthur Museum. She has contributed to exhibitions at the National Gallery of Art, the Art Institute of Chicago, the Newberry Library, and the National Museum of Women in the Arts.

She currently serves on editorial boards for the University of Delaware Press and the scholarly journals American Art and J19.

At the University of Delaware, Professor Bellion actively advises graduate students in American art history and serves on the Executive Committee of the Winterthur Program in American Material Culture. Her undergraduate courses include surveys of American art history from colonial settlement to the present as well as field-based seminars exploring early photography, Surrealism in the United States, and the artistic cultures of early Philadelphia. Her graduate seminars include courses on methodology and historiography, the Peales and Philadelphia, visual culture and the late Enlightenment British world, and American sculpture. She has also enthusiastically collaborated with colleagues in art history, English, and American history to co-teach seminars exploring the transcultural arts of the colonial Americas, the painting and literature of the American Revolution, and English design history.

Lindsay Hoffman

Most Helpful to Students, Liberal Arts, Social Sciences, and Humanities

Larry Nees

Best Teacher, Art History

Professor Lawrence Nees received his B.A. from the University of Chicago and his M.A. and Ph.D. from Harvard. His primary area of interest is the art of the early Middle Ages, focusing on northwestern Europe but also considering the Mediterranean and Byzantine world. He also teaches Northern Renaissance art and is interested in non-Western artistic traditions, especially in Islam and East Asia. He has written The Gundohinus Gospels; From Justinian to Charlemagne: European Art A.D. 565–787; A Tainted Mantle: Hercules and the Classical Tradition at the Carolingian Court; and Early Medieval Art 300–1000. He is currently preparing two books: Frankish Manuscripts 7th–10th Centuries, and The Coronation Throne from Saint-Remi: Art and Political Legitimacy in the Twelfth Century. Professor Nees has received research fellowships from the Mellon Foundation, the American Council of Learned Societies, the National Endowment for the Humanities, the Center of Advanced Study in the Visual Arts (National Gallery of Art), the Institute for Advanced Study in Princeton, and the Guggenheim Foundation.

Linda Pellechia

Best Teacher, Fine Arts

Nancy Signorielli

Best Researcher/Scholar, Communication

David Stone

Most Helpful to Students, Art History

Professor David Stone, Professor of Art History and Director of the Curatorial Track Ph.D. in Art History, is a specialist in Italian Baroque art. He received his B.A. from the University of California, Berkeley, and his M.A. and Ph.D. from Harvard University. His method seeks to integrate technical art history, connoisseurship, and stylistic analysis with close study of seventeenth-century literary and artistic theory, religious and political culture, and patterns of patronage and art collecting. Many of his seminars involve working with original objects at local museums.

Professor Stone is best known for his work on Caravaggio. Much of his recent research is devoted to the artist's 15-month sojourn on the island of Malta, where Caravaggio served the Catholic military order of the Knights Hospitaller of St. John of Jerusalem, eventually becoming an honorary member of the order himself. For research on Caravaggio and the Knights of Malta, Stone received a 1995–1996 Andrew W. Mellon Senior Fellowship in the Department of European Painting at the Metropolitan Museum of Art, New York, and a 1997–1998 Andrew W. Mellon Post-Doctoral Rome Prize in Art History from the American Academy in Rome. For 2002–2003, he was awarded a two-term membership in the School of Historical Studies at the Institute for Advanced Study, Princeton, New Jersey. In 2004–2005, Stone participated in the groundbreaking exhibition, Caravaggio: The Final Years, organized by the Museo di Capodimonte in Naples and the National Gallery in London. He recently published (with K. Sciberras) Caravaggio: Art, Knighthood, and Malta (Malta: Midsea, 2006); "Self and Myth in Caravaggio's David and Goliath" (rev. version), in Caravaggio: Realism, Rebellion, Reception, ed. G. Warwick (Newark, Delaware: University of Delaware Press, 2006); and "Signature Killer: Caravaggio and the Poetics of Blood," Art Bulletin 94 (December 2012). In 2009, Stone co-chaired an historic six sessions on Caravaggio and his followers for the Annual Meeting of the Renaissance Society of America in Los Angeles. A selection of papers from this conference (and others) is now gathered in L. Pericolo and D. M. Stone (eds.), Caravaggio: Reflections and Refractions (Farnham, UK: Ashgate, 2014), which includes Stone's essay, "Caravaggio Betrayals: The Lost Painter and the 'Great Swindle.'"

A leading international authority on the Bolognese artist Guercino, Professor Stone organized Guercino, Master Draftsman: Works from North American Collections, an exhibition which traveled to the Harvard University Art Museums, the National Gallery of Canada, and the Cleveland Museum of Art in 1991. He also wrote the accompanying catalogue (Bologna: Nuova Alfa Editoriale, 1991). In the same year he published Guercino: Catalogo complete dei dipinti (Florence: Cantini Editore, 1991), a complete catalogue of the artist's paintings. While consultant to the National Gallery of Art, Washington, D.C., for their 1992 Guercino exhibition, Professor Stone organized and chaired "Guercino: Nature and Idea; A Quadricentennial Symposium." Sponsored by the National Gallery of Art, the Delaware Art Museum, and the Department of Art History at the University of Delaware, the two-day symposium was accompanied by an exhibition at the Delaware Art Museum curated by Stone: Mostly Baroque: Italian Paintings and Drawings from the Collection of Carlo Croce. More recently he co-curated (with B. Buscaroli) a small exhibition at the Fondazione Cassa di Risparmio in Bologna: Il Gesto Trattenuto: Torna a Bologna un alfresco del Guercino; "La Madonna che offer un bocciolo di rosa al Bambino" (exh. cat., Bologna: Bononia University Press, 2006). He also contributed an essay to Guercino: Early Paintings from Cento and Bologna, ed. F. Gozzi and M. Naldi (Bologna: MGP, 2009), which accompanied an exhibition of the same title held in New York and Bologna: "Il 'vero' Guercino e il vero" [and in English: "The 'Real' Guercino and Reality"]. Stone also contributed entries on Guercino to several recent exhibition catalogues, including Il Cavalier Calabrese: Mattia Preti, tra Caravaggio e Luca Giordano, ed. V. Sgarbi and K. Sciberras (Milan: Silvana, 2013), and Italian Master Drawings from the Princeton University Art Museum, ed. Laura M. Giles et al. (Princeton, NJ: Princeton University Art Museum; New Haven, CT: Yale University Press, 2014).

Stone teaches courses on a wide variety of topics, including seminars on Caravaggio: Methods and Mythologies; Authorship and Authenticity in Italian Renaissance and Baroque Drawings; Originality, Virtuosity, Creativity in Italian Art and Theory; Bernini and Roman Baroque Sculpture; Orazio and Artemisia Gentileschi; Guercino: Painter and Draftsman; Caravaggio's Followers, from Manfredi to Ribera; Bolognese Painting from the Carracci to Crespi; and Baroque Poetics and Imagery. He also regularly teaches an undergraduate survey of European Baroque Art and Architecture; and an introductory course on Western Art: Renaissance to Modern.

Stone is a Trustee of the American Academy in Rome and a member of the AAR Advisory Council to the Committee on the School of Classical Studies. He is also Chief External Examiner in the Department of History of Art, University of Malta.

Dannagal Young

Best Teacher, Communication

UNIVERSITY OF DENVER

Ricardo Iznaola

Best Overall Faculty Member, Music

"Ricardo's a fantastic teacher, mentor, leader, and colleague. His contributions to both the music school and the university are innumerable. And he's a great friend!"

One of the most attractive personalities of the guitar world, Ricardo Iznaola pursues a brilliant, multi-faceted musical career. An American citizen, he was born in Havana, Cuba, in 1949, and trained in Caracas under maestros Manuel Pérez Díaz and Alirio Díaz, and in Madrid under the eminent master Regino Sainz de la Maza, while pursuing studies in Theory and Composition at Madrid's Royal Conservatory. Critics, colleagues and audiences in four continents have acclaimed his path-breaking work as concert and recording artist, composer, teacher, lecturer, and writer. Mr. Iznaola has won top prizes in ten international competitions, including the Munich International Performance Competition, the Francisco Tárrega Competition, the Alirio Díaz Competition, the Stroud International Composer's Competition, the NATS Art Song Composition Competition and the Realize Music Challenge Orchestral Composition Competition, among others.

Susan Schulten

Best Researcher/Scholar, History

Richard Slavich

Best Researcher/Scholar, Liberal Arts, Social Sciences, and Humanities

Cellist Richard Slavich is a concert artist of vast experience, having appeared as soloist with many orchestras, including the Colorado Symphony and National Repertory Orchestras. He has performed in recital throughout the US, in venues ranging from Chicago's Orchestra Hall and Washington D.C.'s Kennedy Center to rural elementary schools in South Carolina

UNIVERSITY OF FLORIDA

Archie Chip Birkner

Best Teacher, Music

Dr. Chip Birkner serves as the Assistant Director of Bands at the University of Florida, where he directs the University Concert Bands, Gator Pep Bands, is Associate Director of the Pride of the Sunshine Marching Band, teaches courses in music education, and observes student teachers. Prior to his appointment at the University of Florida, Dr. Birkner held positions as the Assistant Director of Bands at College Park HS and Knox JH in The Woodlands, TX, as well as similar positions in Tomball, TX and La Porte, TX.

John Griffith

Most Helpful to Students, Liberal Arts, Social Sciences, and Humanities

Susan Hegeman

Best Overall Faculty Member, English

Susan Hegeman (AB, Harvard 1986; PhD in Literature, Duke 1992) teaches courses in American literature, cultural studies, American Indian literature, popular culture, and critical theory. Her research addresses the intersections of American literature and the social sciences, especially anthropology. She has written two books on the concept of "culture" in the U.S.: Patterns for America: Modernism and the Concept of Culture (Princeton, 1999) and The Cultural Return (California, 2012). The first examines the joint modernist project of anthropologists and literary critics at the turn of the century to define and deploy the idea of "culture"; and the second traces the history of the "cultural turn" of the 1980s and 1990s to offer an account of culture's ascendancy as a central interdisciplinary concept, its more recent failing fortunes, and argues for "culture's" continued importance to our understanding of present society. Prof. Hegeman has also published work in American Indian studies, museum studies, literary history and theory. She is an associate editor of the multi-volume Blackwell Encyclopedia of the Novel (2011). Prof. Hegeman is currently working on a project on indigeneity in a global context, and another on the history of the popular social sciences in the mid-twentieth century U.S.

Prof. Hegeman has taught at the University of Florida since 1995. In 1996, she was the William S. Vaughn Visiting Fellow at Vanderbilt University's Robert Penn Warren Center for the Humanities, and in 2012, she was the Fulbright Distinguished Scholar in American Studies at Uppsala University in Sweden.

Kenneth Kidd

Best Teacher, English

Kenneth Kidd received his PhD from the University of Texas at Austin. He came to UF in 1998 after teaching for several years at Eastern Michigan University. Professor Kidd works primarily in children's literature studies but his research interests span nineteenth- and twentieth-century American literature, psychoanalysis, queer theory, and cultural studies. He is the author of two books: Making American Boys: Boyology and the Feral Tale (University of Minnesota, 2004), and Freud in Oz: At the Intersections of Psychoanalysis and Children's Culture (University of Minnesota Press, 2011). With Sidney I. Dobrin he coedited the anthology Wild Things: Children's Culture and Ecocriticism (Wayne State University Press, 2004), and with Michelle Ann Abate he coedited Over the Rainbow: Queer Children's and Young Adult Literature (University of Michigan Press, 2011). He has published in a number of children's literature journals as well as American Imago and PMLA. Since 2004 he has served as Associate Editor of Children's Literature Association Quarterly. He is current at work on a new book about the children's literary classic.

Professor Kidd teaches undergraduate courses in children's and young adult literature, as well as graduate seminars such as "Into the Archive," "Theory's Child," "Childhood Studies," and "Disney and Its Discontents." He also serves as Associate Director of UF's Center for the Study of Children's Literature & Culture.

Barbara Mennel

Most Helpful to Students, English

Barbara Mennel received her PhD in German Studies from Cornell University in 1998. She holds a joint appointment in English and in the German section of the Department of Languages, Literatures, and Cultures.

She is author of The Representation of Masochism and Queer Desire in Film and Literature (Palgrave, 2007), Cities and Cinema (Routledge, 2008), and Queer Cinema: Schoolgirls, Vampires, and Gay Cowboys (Wallflower Press, 2012). With Jaimey Fisher she has edited the volume Spatial Turns: Space, Place, and Mobility in German Literature and Visual Culture (Rodopi, 2010) and with Sabine Hake she edited the collection Turkish German Cinema in the New Millennium: Sites, Sounds, and Screens (Berghahn, 2012). She has published in Camera Obscura, Germanic Review, Modern Austrian Literature, New German Critique, Quarterly Review of Film and Video, Studies in Twentieth Century Literature, Women in German Yearbook, and TRANSIT: A Journal of Travel, Migration and Multiculturalism in the German-Speaking World. In the summer of 2013, she was a Paul Mellon Visiting Senior Fellow at the Center for Advanced Study in Visual Arts at the National Gallery of Art in Washington, D.C., where she worked on her book manuscript tentatively titled "Women and Work in the Contemporary European Cinema."

Her research interests include transnational cinematic practices, feminist and queer theory, minority cultural production in Germany, European cinema, and the intersection of urban studies and cinema studies.

Christine Overdevest

Best Overall Faculty Member, Sociology

Chuck Peek

Best Teacher, Sociology

Stephen Perz

Best Researcher/Scholar, Sociology

Stephen Perz received his PhD in sociology with a specialization in demography from the University of Texas at Austin in 1997. Perz conducts research on demographic processes and environmental change, land use practices and land cover dynamics, and the socio-spatial processes of road-building in the Amazon. He has received funding from NASA and NSF for work on roads and forest fragmentation in the Brazilian Amazon. He is involved in fieldwork along the Transamazon Highway that crosses northern Brazil, as well as along the Transoceanic Highway that connects southern Brazil to Peru and Bolivia. This work emphasizes interdisciplinary and international collaboration and has resulted in roughly 30 publications in numerous journals and books.

Peter Rudnytsky

Best Researcher/Scholar, English

"Peter has accomplished an enormous amount of important publishing on various topics in his field. He offers visibility in Western Europe as well, where he has a high profile among psychoanalytic researchers."

Peter L. Rudnytsky received his undergraduate degrees from Columbia and Cambridge, and his PhD from Yale. He joined the University of Florida faculty in 1992. He has been a Mellon Faculty Fellow at Harvard in 1983–84, a Fulbright Western Europe Regional Research Scholar in 1988–89, an Eccles Fellow at the University of Utah Humanities Center in 1993–94, and a Fulbright/Freud Society Scholar of Psychoanalysis in Vienna in 2004. In 2008–09 and 2010–11, he was a Visiting Scholar in the Psychoanalytic Studies Program at Emory University.

Trained as a Renaissance scholar, Professor Rudnytsky's current areas of specialization include Freud, the history and theory of psychoanalysis, and psychoanalytic approaches to literature. He is the author of Freud and Oedipus (Columbia, 1987); The Psychoanalytic Vocation: Rank, Winnicott, and the Legacy of Freud (Yale, 1991); Psychoanalytic Conversations: Interviews with Clinicians, Commentators, and Critics (Analytic, 2000); and Reading Psychoanalysis: Freud, Rank, Ferenczi, Groddeck (Cornell, 2002). His most recent book is Rescuing Psychoanalysis from Freud and Other Essays in Re-Vision (Karnac, 2011). Among his numerous edited and co-edited books are Transitional Objects and Potential Spaces: Literary

Uses of D. W. Winnicott (Columbia, 1993), Freud and Forbidden Knowledge (NYU, 1994), Ferenczi's Turn in Psychoanalysis (NYU, 1996), Psychoanalyses/Feminisms (SUNY, 2000), Psychoanalysis and Narrative Medicine (SUNY, 2008), and Her Hour Come Round at Last: A Garland for Nina Coltart (Karnac, 2011).

In 2011, Professor Rudnytsky stepped down after ten years as editor of American Imago, founded by Sigmund Freud and Hanns Sachs in 1939 and published by The Johns Hopkins University Press. Concurrently, he has launched two new book series: The History of Psychoanalysis, with Professor Brett Kahr, published by Karnac Books; and Psychoanalytic Interventions, with Professor Esther Rashkin, published by Fordham University Press. Professor Rudnytsky serves on the editorial boards of The Psychoanalytic Quarterly, The Psychoanalytic Review, Psychoanalytic Psychology, American Journal of Psychoanalysis, and Psychoanalysis and History.

Professor Rudnytsky is an Honorary Member of the American Psychoanalytic Association as well as a Corresponding Member of the Institute of Contemporary Psychoanalysis in Los Angeles. In 2001, he received the Local Educators' Award from the International Forum for Psychoanalytic Education; and in 2003, he received the Gradiva Award for Reading Psychoanalysis from the National Association for the Advancement of Psychoanalysis. In 2011, an initiative led by Professor Rudnytsky resulted in the designation of the American Psychoanalytic Association as an Affiliated Organization of the Modern Language Association of America. He is the 2012 honoree of the Committee on Research and Special Training of the American Psychoanalytic Association. Professor Rudnytsky is co-chair of the Committee on Psychoanalysis and the Academy of the American Psychoanalytic Association and a member of the Committee on Research of the International Psychoanalytic Association.

Committed both personally and professionally to integrating the academic and clinical worlds of psychoanalysis, Professor Rudnytsky is an analyst in training with the Florida Psychoanalytic Institute. He received an MSW from Florida State University in 2008 and has since become a Licensed Clinical Social Worker and a Qualified Supervisor in Social Work in the State of Florida. He maintains a part-time private practice in psychoanalysis and psychoanalytic psychotherapy in Gainesville.

David Waybright

Best Overall Faculty Member, Music

Dr. David A. Waybright received his Bachelor of Arts and Master of Arts degrees at Marshall University and the Doctor of Musical Arts degree in orchestral conducting from the Cincinnati College-Conservatory of Music. He taught initially at Wahama High School in Mason, West Virginia and since that time has served as Director of Bands at Ferrum College, Plymouth State College, and Director of Bands and Orchestra at McNeese State University. Dr. Waybright is currently Director of Bands at the University of Florida, where he holds the rank of professor and is the head of the conducting area. He directs the wind symphony and supervises the band program and the graduate and undergraduate conducting curricula.

Barbara Zsembik

Most Helpful to Students, Sociology

Barbara Zsembik received her Ph.D. in Sociology and Demography from the University of Texas in Austin, 1988. She teaches and writes on health disparities in the United States, and Latino issues and experiences. Dr. Zsembik has received recognition as Provost Administrative Fellow (2000), T. Lynn Smith Term Professor (1999–2000), and University Teacher of the Year (1995).

UNIVERSITY OF GEORGIA

Maurice Daniels

Best Teacher, Liberal Arts, Social Sciences, and Humanities

Brian Allan Drake

Best Teacher, History
Best Teacher, Liberal Arts, Social Sciences, and Humanities

Brian Allen Drake specializes in environmental history, a discipline which explores the intricate ways in which nature has affected human history and vice versa. His recent research focuses on the postwar American environmental movement, particularly its relationship to postwar politics and ideology. The University of Washington Press published his book, Loving Nature, Fearing the State: Environmentalism and Antigovernment Politics Before Reagan, in August 2013 in its Weyerhaeuser Environmental Series. He is also the editor of The Blue, the Gray, and the Green: Toward an Environmental History of the Civil War, forthcoming from UGA Press in January 2015. Drake has published articles in Great Plains Quarterly, the Georgia Historical Quarterly, and Environmental History, many book reviews, and also contributed a chapter to Barry Goldwater and the Remaking of the American Political Landscape, published by the University of Arizona Press. In 2005 he received the Harry S. Truman Good Neighbor Award Foundation's Eddie Jacobson Scholarship to support his dissertation research. Prior to arriving at UGA, Drake taught for two years in the Humanities and Western Civilization program at the University of Kansas. He earned his B.A. in history at the University of Michigan in Ann Arbor, and his M.A. in U.S. history from the University of Georgia.

Kristen Kundert-Gibb

Best Teacher, Theater

Professionally, Professor Kundert-Gibbs has worked as a producer, director, vocal coach, and actress. In academia, she has taught in theatre programs across the country from Duke, to William and Mary to Indiana State University, where she headed the Graduate and Undergraduate Acting Programs, to Los Angeles and finally, happily embracing UGA. Ms. Kundert-Gibbs experiments in the integration of media and theatre and has directed many productions using varying forms of media including a production of Pinter's New World Order which was presented at the International Pinter Festival in London, a production of Cabaret with Theatre X from Manheim, Germany, and Death of Salesman at The Warehouse Theatre. Her most recent acting work includes playing Emer in the world premiere of Sword Against the Sea, which toured to The International Yeats Festival in Sligo, Ireland. While in L.A, she was instrumental in founding the Kingsmen Shakespeare Festival serving as vocal coach, pre-show director and actor. She has been extensively involved in KC/ACTF as a member of the Region III Executive Board. After spending a year acting and teaching with Karamu House Theatre, Ms. Kundert-Gibbs has been drawn to multicultural theatre work and has directed African American productions for The Warehouse Theatre, UNC-Asheville and Clemson University. When teaching, Ms. Kundert-Gibbs focuses both directing and acting on the work of Sanford Meisner. As a believer in the interaction of the breath, voice, body and emotions, she incorporates work such as Alba Emoting, Transactional Analysis and Bioenergetics into the classroom. Her voice classes focus on the work of Kristin Linklater and Edith Skinner. She is also interested in Eastern bodywork such as yoga and Tae Kwon Do, is a certified black belt and teaches kickboxing. In all disciplines, she hopes to teach students to value their instincts, trust their impulses, and celebrate their humanity.

Larry Nackerud

Most Helpful to Students, Social Work

The mix between the formulation of social welfare policies and the applied practice of social work is Larry Nackerud's primary teaching and research interests. He became interested in the mix of policy and practice while working as a child protection worker on a reservation in northern Montana. His interest intensified during his years working as a clinical social worker in the state psychiatric hospital of Alaska. The specific public policy area of U.S. immigration and refugee policy became more prominent after the completion of his Ph.D. studies in the Texas-Mexico border region. Nackerud includes in his teaching and research a great deal of attention to the elements of research methodology, particularly the use of an interpretive inquiry paradigm, different sampling strategies, and a combining of quantitative and qualitative methods.

Evgeny Rivkin

Best Overall Faculty Member, Music
Best Researcher/Scholar, Music
Best Teacher, Music
Most Helpful to Students, Music

"One of the best piano teachers in this country"

Fran Teague

Best Overall Faculty Member, English
Best Researcher/Scholar, English
Best Teacher, English
Best Researcher/Scholar, Liberal Arts, Social Sciences, and Humanities
Best Teacher, Liberal Arts, Social Sciences, and Humanities
Most Helpful to Students, Liberal Arts, Social Sciences, and Humanities
Best Overall Faculty Member, Liberal Arts, Social Sciences, and Humanities

Fran Teague is a University Professor and a Josiah Meigs Distinguished Teaching Professor. She teaches performance history and theory. Her special areas of interest are Renaissance drama, dramaturgy, and early women writers. She has written or edited seven books, including The Curious History of "Bartholomew Fair" (Bucknell, 1985), One Touch of Shakespeare (Folger, 1986), Shakespeare's Speaking Properties (Bucknell, 1991), and Shakespeare and the American Popular Stage (Cambridge, 2006); her articles have appeared in many journals including Comparative Drama, Shakespeare Quarterly, Shakespeare Studies, and Theatre Journal. Teague has a joint appointment with the Department of English and is an affiliate of the Women's Studies Program.

UNIVERSITY OF HARTFORD

Anton Miller

Best Overall Faculty Member, Music

Since giving his Carnegie Hall concerto debut in 1992 with the New Chamber Orchestra of New York, American violinist Anton Miller has appeared throughout the United States and abroad as a soloist, chamber musician, recitalist, and pedagogue. Miller gave the world premier performance in Beijing, China of Xiaogang Ye's Last Paradise for violin and orchestra, which was taped for broadcast within Asia and subsequently released on CD. He has made several apperances in Brazil to tour Tchaikovsky, Sibelius, and Brahms Concertos with the Orchestra Sinfonica de Campinas, and has performed recital tours in Germany, Austria, Switzerland, England, and China. His complete discography can be found on the Dorian, Jericho, Full House, Hugo, and Klavier lables.

Miller received his Bachelor of Music degree from Indiana University as a student of Franco Gulli and was awarded the prestigious Performer's Certificate. He completed his Master of Music at The Juilliard School as a scholarship violin student of Dorothy DeLay and studied chamber music with Felix Galimir and the members of the Juilliard Quartet.

UNIVERSITY OF HAWAII AT MANOA

R. Kelly Aune

Best Teacher, Speech

Bobby Tokunaga

Best Researcher/Scholar, Liberal Arts, Social Sciences, and Humanities

Bobby Tokunaga studies the social and psychological implications of using communication technologies. He is currently engaged in research that examines the deficient self-regulation of Internet use, the use of Internet-based technologies for the sexual solicitation of minors, and interpersonal factors promoting electronic monitoring and surveillance in relationships. He is also part of a group of international scholars charged with refining the definitions and methods used in cyberbullying research. His other research interests include investigating factors that moderate the aggression and hostility people experience following violent video game play and statistical methods used in meta-analysis. Bobby teaches various undergraduate courses in the Communicology Department, including theories of communicology, research methods, and the evoluation of public discourse in the digital era.

Bobby is active in high school speech and debate activities in Hawai'i. He advised speech and debate teams for eight years and has served in various capacities on the Executive Board of the Hawai'i Speech League.

UNIVERSITY OF HOUSTON

Mabel Cuesta

Best Researcher/Scholar, Languages

UNIVERSITY OF HOUSTON DOWNTOWN

Jillian Hill

Most Helpful to Students, English

Dr. Hill is an Assistant Professor of English at the University of Houston-Downtown. She teaches courses both undergraduate and graduate professional and technical writing courses.

Thomas Lyttle

Best Overall Faculty Member, Theater
Most Helpful to Students, Theater

Bill Pogue

Best Overall Faculty Member, Arts & Humanities

Bill Pogue is a lecturer in communication studies. Mr. Pogue is in his third year of a full-time appointment after thirty years of adjunct service. His undergraduate work was at Northwestern University and University of Denver, with graduate study at University of Minnesota and Jones International University. Apart from academia, his professional career has included broadcasting, arts administration and operating his own marketing firm serving a wide range of clients. Personal life can be summarized as "husband of one, father of two and grandfather of three." His wife is Kate Pogue, also a UHD faculty member. Mr. Pogue is "Retention Officer" for the College of Humanities and Social Sciences; as such he strives to assist students who experience course difficulty and helps put them back on track. This year he also serves as a faculty academic adviser for declared CHSS majors.

UNIVERSITY OF ILLINOIS AT CHICAGO

Mandy Farretta

Best Researcher/Scholar, Spanish

Rosilie Hernandez-Pecoraro

Best Overall Faculty Member, Spanish

Luis Lopez-Carretero

Most Helpful to Students, Spanish

Justyna Makowska

Best Teacher, Spanish

Gary Minnix

Best Overall Faculty Member, Fine Arts

Gary Minnix holds a BFA from Bowling Green State University and an MFA from Tyler School of Art, Temple University. Originally from Ohio, he has lived in Chicago and taught at UIC since 1981. His work is based primarily in photography but has included sculptural installations and an interactive virtual reality piece. Minnix received a National Endowment for the Arts photography fellowship in 1980 and a research award from the University of Illinois at Chicago in 2001. His photographic and digital work has been widely shown, including in Photography: Invention and Innovation at the Art Institute of Chicago; Form as Content at the Visual Studies Workshop, Rochester, NY; American Color at New Mexico State University (and national tour); Light/Color at Ithaca College, NY; Independent American Photography at the Union of Polish Art Photographers, Warsaw; Virtual Reality at Ars Electronica, Linz, Austria and the Museum of Contemporary Photography, Chicago.

Silvia Rosman

Best Overall Faculty Member, Spanish

"She is an excellent teacher, very didactic and creative who had many students following her courses. Moreover, as a scholar her work is original and rigorous."

Clinical Assistant Professor Education PhD Princeton University

Major Interests:

19th and 20th Century Latin American Literatures, Trans-Atlantic Studies (Mexico and Brazil), Visual Cultures, Cultural and Critical Theory

Selected Publications:

Books

Being in Common: Nation, Subject and Community in Latin America Literature and Culture. London/Lewisburg: Bucknell University Press, 2003. Dislocaciones culturales: Nación, sujeto y comunidad en América Latina. Rosario: Beatriz Viterbo Editora, 2003.

Articles

"Escrever a revolução: sujeito e acontecimento em El reino de este mundo de Alejo Carpentier," Ed. Elena Palmero González, Homenaje por los sesenta años de El reino de este mundo. Editora da FURG, Rio Grande (Brazil), 2009. "The Father, the Law, Desire: Perversions of the Letter in Luis Gusmán's Villa" Romance Studies, Vol. 26, No. 1. (January 2008). "La comunidad por-venir," Nación y literatura en América Latina. Ed.

Ramón Maíz Suárez. Buenos Aires: Prometeo Libros Editorial, 2007. "Entre el deseo y la memoria: los relatos de viaje en el Río de la Plata", Colorado Review of Hispanic Studies, 3:1 (Fall 2005).

UNIVERSITY OF ILLINOIS AT URBANA-CHAMPAIGN

Lynne Dearborn

Best Researcher/Scholar, Architecture

"This individual is a multidisciplinary scholar who engages undergraduate and graduate students in her research."

Lynne M. Dearborn, Ph.D. joined the faculty at the Illinois School of Architecture in 2001 and is now Associate Professor of Architecture and Urban Planning. She currently serves as Chair of Architecture's Health and Well-being Program and as Chair of the Ph.D. programs in Architecture and Landscape Architecture. Her research highlights adverse environmental and living conditions of low-income and other marginalized groups in the US and internationally and addresses questions of power and agency in seeking to advocate for more supportive environments for these groups. Through the use of community-based participatory processes and evidence-based practices, her design studios seek to ameliorate these conditions in underserved communities around the world.

Mary Eamon

Most Helpful to Students, Social Work

Rhonda Frank

Most Helpful to Students, Architecture

"Always pleasant and helpful"

Julie Gunn

Best Overall Faculty Member, Music

"Exceptionally committed to excellent education, community outreach, and high level performance"

Julie Gunn is a pianist, music director, vocal coach, and song arranger. She has appeared on many prestigious recital series, including the Carnegie Hall Pure Voice Series, Lincoln Center Great Performers, Boston's Jordan Hall, Brussels' Theatre de la Monnaie, San Francisco's Herz Hall, the 92nd Street Y, Toronto's Roy Thompson Hall, University of Chicago Presents, San Francisco Performances, Oberlin College, Cincinnati Conservatory of Music, the Krannert Center for the Performing Arts, the Ravinia Festival, Manhattan's legendary Café Carlyle, the Sydney Opera House, and the United States Supreme Court. She has been heard in recital with William Burden, Richard Croft, Elizabeth Futral, Isabel Leonard, Stefan Milenkovich, Mandy Patinkin, Yvonne Gonzales Redman, Michelle De Young, the Pacifica Quartet, and Nathan Gunn. This season she looks forward to concerts with Kelli O'Hara in Evanston, IL, and with Nathan Gunn in Denver, Salt Lake City, and Minneapolis.

In her faculty appointment at the University of Illinois, she works with singers, pianists, chamber musicians and songwriters as the Director of Lyric Theatre Studies. She has served on the music staff at the Metropolitan Opera Young Artist Program, Wolf Trap Opera, St. Louis Opera Theatre, Southern Methodist University, Opera North, Highlands Opera Studio,

Theaterworks, Chicago Opera Theater, and Illinois Opera Theater, and given master classes at universities and young artists' programs all over the United States, including the Ryan Young Artists' Program, Houston Grand Opera Studio, and Cincinnati Conservatory of Music, Santa Fe Opera, the Aspen Festival, the Interlochen Center for the Arts, and the Metropolitan Opera Guild, as well as tenures as artist-in-residence at Cincinnati Opera and the Glimmerglass Festival.

Dr. Gunn serves as the Assistant Director for Development and Public Engagement at the School of Music. She works as an advocate for state and local arts organizations in the development of projects ranging from constructing arts buildings to mentoring inner-city high-schoolers interested in the arts. She is the founder and director of the Illinois School of Music Academy, a program for talented pre-college chamber musicians and composers.

On a national level, she solicits and programs the work of emerging and established American songwriters on recitals, and is proud to have been the music director of her husband Nathan Gunn's solo disc, Just before Sunrise, released by Sony/BMG records, which included arrangements of songs by Gene Scheer, Ben Moore, Joe Thalken, Billy Joel, Sting, and Charles Ives. Her orchestral arrangements of traditional American songs and standards have been heard at the Kennedy Center, Chicago's Symphony Center, and London's Queen Elizabeth Hall.

Kevin Hinders

Best Teacher, Architecture

"He cares about his students and their learning."

Professor Hinders leads graduate seminars and studios that examines design on the urban scale. He is also the principal in charge at Champaign-based PREP A.R.E. He is well versed with projects large and small. As a professional, Professor Hinders has drafted new master plans for the University Campus, Research Parks, and Campustown 2000. His work truly spans the scales of architecture, urban design, and master planning.

Sally Jackson

Best Overall Faculty Member, Communication
Best Researcher/Scholar, Communication

Sally Jackson earned three degrees in Communication from the University of Illinois. She returned to Illinois as a member of the faculty in 2007, concurrent with appointment as the campus Chief Information Officer and Associate Provost, a position she held until 2011. Professor Jackson held prior faculty appointments at the University of Nebraska-Lincoln (1979–1982), Michigan State University (1982–1985), Oklahoma University (1985–1990), and the University of Arizona (1991–2007). While at Arizona, she held a series of academic leadership positions, rising to Vice President for Learning and Information Technologies and becoming the university's first Chief Information Officer.

Christopher Larrison

Best Researcher/Scholar, Liberal Arts, Social Sciences, and Humanities

"A phenomenal mentor for doctoral students"

Christopher R. Larrison earned a B.S. degree from Cornell University (1988), and a MSW (1996) and PhD (1999) from the University of Georgia. He has over 10 years of professional and research experience with community mental health agencies.

Dr. Larrison is interested in community based service agencies and how they interact with clients and policies to create outcomes for clients. Employing primary data, he has studied community based mental health services in the U.S., community development in rural Mexico, and the impact of welfare reform in Georgia. His use of field research techniques has led to extensive experience in training and supervising research staff, collecting and analyzing primary data, and using mixed methods. He has concentrated on two sets of factors related to community based service outcomes: 1) Client-level factors with a particular consideration to socioeconomic status, gender, level of health, and race, culture and ethnicity; and 2) agency-level factors with a particular consideration to program model, level of innovation, and organizational climate. Several of Dr. Larrison's peer reviewed articles have been placed on suggested reading lists by the U.S. Department of Housing and Urban Development, the U.S. Government Accounting Office, and The World Bank.

Dr. Larrison was the PI for a three year NIMH funded field project that is examining the relationship between organizational climate and treatment outcomes for African American adults receiving services at 12 community mental health agencies in the Upper Mississippi River Basin (1R34 MH074640-02A, Climate, diversity, and outcomes in rural mental health).

Debra Richtmeyer

Best Teacher, Music

Debra Richtmeyer, an internationally renowned saxophonist and pedagogue, has been Professor of Saxophone at the University of Illinois at Urbana-Champaign since 1991 and has performed as a soloist and clinician in North America, Europe, Slovakia, the Czech Republic, Thailand and China. She has performed or recorded as a concerto soloist with numerous bands and orchestras, including the Dallas Symphony Orchestra,

Siobhan Somerville

Best Overall Faculty Member, Liberal Arts, Social Sciences, and Humanities

Stephen Taylor

Most Helpful to Students, Music

Stephen Andrew Taylor's music often explores boundaries between art and science. His first orchestra commission, Unapproachable Light, inspired by images from the Hubble Space Telescope and the New Testament, was premiered by the American Composers Orchestra in 1996 in Carnegie Hall. Other works include the chamber quartet Quark Shadows, commissioned by the Chicago Symphony and premiered in 2001; and Seven Memorials, a 32-minute cycle for piano inspired by the work of Maya Lin and premiered by Gloria Cheng in Los Angeles, 2004; she also performed the work at Tanglewood in 2006. The Machine Awakes, a CD of his orchestra, chamber and electronic music was released in 2010 on Albany Records; and Paradises Lost, a new opera based on a novella by Ursula K. Le Guin, was premiered in Portland, Oregon and at the University of Illinois in 2012.

Besides composing for traditional instruments, Taylor also works with live electronics in pieces such as Agoraphobia for flute, harp and electronics, premiered by Jonathan Keeble and Ann Yeung in Montreal in 2009. He is also active as a conductor with the Illinois Modern Ensemble, and as a theorist, writing and lecturing on György Ligeti, Björk and Radiohead. He also collaborates with the band Pink Martini, and rock singer Storm Large.

Born in 1965, he grew up in Illinois and studied at Northwestern and Cornell Universities, and the California Institute of the Arts; his teachers include Steven Stucky, Karel Husa, Mel Powell, Bill Karlins and Alan Stout. His music has won awards from Northwestern, Cornell, the Conservatoire Américain de Fontainebleau, the American Academy of Arts and Letters, Composers, Inc., the Debussy Trio, the Howard Foundation, the College Band Directors National Association, the New York State Federation of Music Clubs, the Illinois Arts Council, the American Music Center, and ASCAP. Among his commissions are works for Northwestern University, University of Illinois, the Syracuse Society for New Music, Pink Martini and the Oregon Symphony, the Quad City Symphony, the Chicago Symphony, Quartet New Generation and the New Philharmonic, Piano Spheres, and the American Composers Orchestra.

UNIVERSITY OF INDIANAPOLIS

Jim Ream

Most Helpful to Students, Theater

Jim has worked with numerous Indianapolis theatres, including Indianapolis Civic, Edyvean Repertory Theatre, and the Phoenix Theatre. His designs have received three Corbin Patrick Award nominations and the best set design award for Phoenix Theatre's production of Fences.

Other scenic design credits include: One Voice, Ten Ten Theatre, New York; Noises Off, Brainerd, MN; Et Ta Soeur, Indiana Repertory Theatre Upper Stage; Translations and G.R. Point at Purdue University; Tale of Two Cities and Young Black Beauty for Stage One of Louisville. His most recent designs in the Indianapolis area include Joseph and the Amazing Technicolor Dream Coat and Talley's Folley at Edyvean Repertory Theatre, and numerous productions at the University of Indianapolis including Hamlet and God's Favorite. Jim received his bachelor of arts degree from Culver-Stockton College and his master of fine arts degree in design from Purdue University. He specializes in design and theatre technology.

UNIVERSITY OF IOWA

Florence Boos

Best Overall Faculty Member, English

Mary Cohen

Best Overall Faculty Member, Music
Best Teacher, Music

"Phenomenal teacher, great energy, great motivator, brings a freshness of spirit alloyed with high quality content. A master at what she does in every respect."

Miriam Gilbert

Best Teacher, English
Best Overall Faculty Member, English

My work–teaching courses in drama, editing a major drama anthology, creating CD-ROMs for use in classes, and writing about Shakespeare in performance–all grows out of a long-term immersion in both literature and theatre. I teach courses for both undergraduates and graduates in dramatic literature, stressing the notion of plays as performance-texts. The CD-ROMs highlight the variety of interpretive possibilities in Shakespeare's plays by juxtaposing 2–6 versions of selected moments, chosen from commercially available videotapes. My most recent book also stresses a range of interpretation, by focusing on a single (and highly controversial) play The Merchant of Venice, in a series of post-WWII productions at Stratford-upon-Avon.

Tammy Ho

Best Researcher/Scholar, Liberal Arts, Social Sciences, and Humanities
Best Teacher, Liberal Arts, Social Sciences, and Humanities
Most Helpful to Students, Liberal Arts, Social Sciences, and Humanities

Irina Kostina

Best Overall Faculty Member, Liberal Arts, Social Sciences, and Humanities

Irina Kostina, Ph.D., currently holds a lecturer's position at The University of Iowa where she has been working for over 20 years. Prior to coming to the U.S., Dr. Kostina worked at the Moscow State University as the Head of Russian as a Foreign Language Department. She has authored and co-authored a number of books and scholarly articles that were published in Russia, the U.S. and Europe, including her book "Perspectiva" which is used in Russia, Spain, Holland and many other countries. She is also a renowned specialist in intensive Russian language teaching methodology, and she has taught various levels of Russian language proficiency and methodology courses in Russia, the U.S. Cuba, Italy, North Korea, Estonia and Canada. Throughout her academic career, Dr. Kostina has been awarded and participated in several major educational grants. She served as Instructional Lead on two federally funded grants with a total budget of over $400,000. In 2011, The University of Iowa awarded Dr. Kostina with a highly prestigious university-wide President and Provost Award for Teaching Excellence in recognition of her unique expertise and extraordinary dedication to her profession. As a Program Director, she is working with STARTALK since 2012 receiving 5 grants from this organization.

John Manning

Most Helpful to Students, Music

Nathan Platte

Best Researcher/Scholar, Music

Nathan Platte's research and teaching interests include American film music, opera, Russian music, collaborative creativity, and musical adaptations across media. He has presented papers at national and international conferences, including the Society for American Music, Society for Cinema and Media Studies, and American Musicological Society. His articles and projects have received recognition from the University of Michigan (Louise E. Cuyler Prize in Musicology), Harry Ransom Humanities Research Center (Dissertation Fellowship), and Society for American Music (Mark Tucker Award and Wiley Housewright Dissertation Award).

Platte's publications explore film music of Hollywood's studio era and, in particular, the collaborative process of film scoring. His work has appeared in The Journal of Musicology, 19th-Century Music, Music and the Moving Image, and The Journal of Film Music. Other recent publications include The Routledge Film Music Sourcebook (Routledge, 2012; coedited with James Wierzbicki and Colin Roust), Franz Waxman's "Rebecca": A Film Score Guide (Scarecrow Press, 2012; coauthored with David Neumeyer), and chapters in Music, Sound, and Filmmakers: Sonic Style in Cinema (Routledge, 2012) and The Oxford Handbook of Film Music Studies (Oxford University Press, 2014). He has forthcoming work in The Journal of the Society for American Music and the Behind the Silver Screen (Rutgers University Press) series. He is currently writing Hearing Selznick (under contract with Oxford University Press), a book that investigates the scores for films like Gone With the Wind, Since You Went Away, and Spellbound.

Platte received his Ph.D. from the University of Michigan, where he also completed bachelor's degrees in history and trombone performance. Before joining the faculty at the University of Iowa in 2011, he taught at Michigan and Bowling Green State University.

Claire Sponsler

Best Researcher/Scholar, English

Claire Sponsler's work centers on medieval literature, with a special interest in the overlapping areas of book history, performance, and cultural studies. Her books include The Queen's Dumbshows: John Lydgate and the Making of Early Theater (University of Pennsylvania Press, 2014); John Lydgate: Mummings and Entertainments (Middle English Texts Series, 2010); Ritual Imports: Performing Medieval Drama in America (Cornell University Press, 2004), which won the Barnard Hewitt prize from the American Society for Theatre Research; East of West: Crosscultural Performance and the Staging of Difference (Palgrave Macmillan, 2000; co-edited with Xiaomei Chen); and Drama and Resistance (Minnesota University Press, 1997). She has also written numerous articles and reviews. She directed an NEH Institute on "Ritual and Ceremony from Late Medieval Europe to Early America" at the Folger Shakespeare Library in the summer of 2010. Her research has been supported by an NEH Fellowship and other grants and awards, including a 2013–14 National Humanities Center Fellowship. She has taught at the George Washington University and the Bread Loaf School of English and is a former chair of the Department of English at Iowa.

Anne Stapleton

Most Helpful to Students, English

My teaching, research, and writing explore the relationship between literature and culture of late eighteenth- and nineteenth-century Great Britain, with a particular focus on Scotland. I am especially intrigued by literary representations of national and regional identities in poetry, songs, fiction, and autobiographical works. Archival research in England and Scotland, a vital aspect of my work, uncovered a number of significant texts which led to my book entitled Pointed Encounters: Dance in Post-Culloden Scottish Literature (2014, in the SCROLL series from Rodopi Press). By tracing the central motif of Highland dance as an outwardly conforming but covertly subversive expression of Scottish identity, I investigate Scottish musical traditions and literary forms developed between 1750 and 1830.

More recently, I have been exploring Sir Walter Scott's remarkable legacy in the United States, particularly towns named Waverly. Published anonymously in 1814 and known as the first historical novel, Waverley; Or, 'Tis Sixty Years Since inaugurated a staggeringly successful series of books that became known as the Waverley novels, since Scott signed subsequent novels "by the Author of Waverley" until he publically acknowledged authorship in 1827. Waverly, Iowa, is one of many U.S. towns named for these popular novels as citizens sought to remember the past, articulate contemporary values, and define their identities in changing settings. I am eager to learn more about this fascinating phenomenon.

UNIVERSITY OF KANSAS

Robert Antonio

Best Teacher, Sociology

"Professor Antonio is the most demanding teacher in our department–and probably the most popular among students. Students (both grads and undergraduates) marvel at the depth of his knowledge and his passion for the topics he teaches. He is nearly always in his office with the door open, available to students (and faculty) and mentors ways beyond his teaching duties. When students do well in his classes, they feel that it is a real accomplishment."

Professor Antonio (PhD Notre Dame) specializes in social theory, macroscopic sociology, and economy and society. His writings have focused on Marx, the Frankfurt School, Weber, Dewey, Habermas, and others in the classical and continental tradition. Among his publications are Marx & Modernity (2002), "After Postmodernism: Reactionary Tribalism" (2000), "A New Global Capitalism? From 'Americanism and Fordism' to 'Americanization-Globalization'" (with A. Bonanno, 2000), "Karl Marx" (1999), "Mapping Postmodern Social Theory" (1998), "Nietzsche's Antisociology: Subjectified Culture and the End of History" (1995). Areas: Social Theory, Comparative and Historical Sociology, and Political and Economic Sociology.

Barbara Barnett

Best Researcher/Scholar, Journalism

Barbara Barnett is an associate professor in the William Allen White School of Journalism and Mass Communications. She previously worked as a print journalist, covering health, politics and higher education. She also worked as a public relations professional in health communications and has organized communication projects in Africa, Asia and Latin America. At KU, she teaches courses in reporting, research methods, media and diversity, media and popular culture and media and the military. She helps coordinate KU's Media and the Military Project, funded by the McCormick Foundation. Her research focuses on media coverage of women who are victims and perpetrators of violence. She also has conducted research on women's health and on gender and sports.

Muriel Cohan

Best Overall Faculty Member, Dance

"Over 25 years of producing high quality original works each semester for the University Dance Company, nurturing the dance students in these works and ongoing excellent teaching."

Mugur Geana

Best Overall Faculty Member, Journalism

Mugur V. Geana, MD, PhD is Associate Professor of Strategic Communication at the William Allen White School of Journalism and Mass Communications; he is also the Director of the Center for Excellence in Health Communication to Under-served Populations and holds an honorary faculty appointment with the Research Division, Department of Family Medicine, KUMC. Geana's interest in applied research and strategic communication campaigns allowed him to contribute to projects for the Kansas Department of Health and Environment and the Kansas Department of Transportation, as well as in other grants and projects totaling more than $5 million over the last five years. Geana founded the Experimental Research Media Laboratory at the School of Journalism and Mass Communications, a major research resource for our graduate program. Geana is the Principal Investigator for multidisciplinary research collaboration between KU and the University of Costa Rica addressing public health interventions to under-served communities in Central America, as well as for a grant from the Washington University in St. Louis focused on finding optimized ways of delivering updated health information about diabetes to American Indians. Geana teaches strategic communication campaigns at the undergraduate level and research theory and methods at graduate level. He is the recipient of the prestigious 2011 Baskett-Mosse Award from the Association for Education in Journalism and Mass Communication.

Elena Ghezzo

Best Teacher, Liberal Arts, Social Sciences, and Humanities

"She is truly a great teacher!"

ChangHwan Kim

Best Researcher/Scholar, Liberal Arts, Social Sciences, and Humanities

"Professor Kim is one of the most prolific scholars I know. He publishes in top tier journals and his work addresses one of the most important issues in our nation: growing economic inequality."

Professor Kim (PhD University of Texas at Austin) is specialized in the areas of work and organizations, race and ethnicity, Korea studies, and quantitative methodology. The common concern of his research is to contribute to the generation of the critical knowledge and information that will ultimately help policy makers to understand and eventually ameliorate the undesirable sources of increasing socioeconomic polarization in our society. Methodologically, he is interested in panel models and diverse statistical decompositions. His recent research appear in American Sociological Review, Work and Occupations, Social Science Research and Korean Journal of Sociology.

Charles Marsh

Best Teacher, Journalism

Charles Marsh is the Oscar Stauffer Professor of Journalism and Mass Communications at the University of Kansas. His primary areas of research involve public relations, ethics and classical rhetoric. His research has appeared in Public Relations Inquiry, Journal of Public Relations Research, Public Relations Review, Journalism and Mass Communication Quarterly, Journal of Mass Media Ethics and other journals. He is the author of Classical Rhetoric and Modern Public Relations: An Isocratean Model and co-author of the textbooks Public Relations: A Values-Driven Approach and Strategic Writing: Multimedia Writing for Public Relations, Advertising and More.

His current research interests involve examining the "fully functioning society theory" of public relations. Pioneered by Robert L. Heath, that theory holds that a core function of public relations involves working to build humane and effective societies. Toward that end, he is examining public relations through the lenses of evolutionary biology, social justice and classical rhetoric.

Patty Noland

Most Helpful to Students, Journalism

Patty Noland worked in the broadcast news and video production industries for 15 years before returning to her alma mater, The University of Kansas. Noland now serves as the career development coordinator for the School of Journalism and Mass Communications. In that capacity, she assists students with résumés, internships and job-search strategies, and conducts workshops on career-related topics. She also teaches journalism students in reporting, writing, broadcast production and professional development.

Among Patty's accomplishments:

Managing the KU Journalism Career Center. Hosting 15–20 recruiters each year. Producing, scripting and directing more than 70 video programs. Assisting with media relations for the Baltimore venues of Pope John Paul II, 1995. Producing and supervising edits for "Your New Home." Serving as script supervisor for the test-market of "The Tim and Daphne Reid Show." Teaching a broadcast news class and supervising the newsroom at Texas A&M University. While working at Proffer Productions in Kansas City, Patty worked for many clients, including American Century Investments, Hallmark, Sprint, HOK Sport, Burns & McDonnell, and Western Auto. During her time at SRB Productions in Washington, D.C., Patty worked on projects for the U.S. Department of Treasury, the U.S. Navy, the U.S. Postal Service, the U.S. Department of Labor, the AFL-CIO, the Smithsonian, Loyola College, the Maryland Lottery and the United Negro College Fund. At both firms, she wrote technical and cost proposals for prospective clients, developed marketing leads for new projects and handled client relations.

Patty has won awards from The Public Relations Society of America/Maryland Chapter, The Kansas City International Association of Business Communicators, Sigma Delta Chi/The Society of Professional Journalists, The Kansas Association of Broadcasters and United Press International. Patty was one of 38 instructors on staff when the KU Center for Teaching Excellence gave its Departmental Teaching Award to the School of Journalism in 2002.

Her experience in news includes working as an anchor and reporter for KANU-FM in Lawrence, Kansas, and serving as the assistant news director at KAMU-TV in College Station, Texas. She also worked for KOMU-TV, KFRU-AM and KBIA-FM in Columbia, Missouri.

At KU, Patty earned a Master of Science in Radio, Television and Film in 1987. She earned a Bachelor of Journalism from the University of Missouri-Columbia 1983. Patty is a member of the Public Relations Society of America, the Kansas City Ad Club and Phi Beta Delta, an international honor society. She serves as the adviser for the Public Relations Student Society of America, University of Kansas chapter, and the KU Ad Club. She also is the adviser for KU's chapter of Kappa Tau Alpha, a journalism honor society that recognized the top 10 percent of journalism students. Patty became a member of Kappa Tau Alpha while she was a student at the University of Missouri-Columbia.

Dennis Sander

Best Teacher, Architecture

Margot Versteeg

Best Overall Faculty Member, Spanish & Portuguese
Best Researcher/Scholar, Spanish & Portuguese
Best Teacher, Spanish & Portuguese
Most Helpful to Students, Spanish & Portuguese

Margot Versteeg holds two MAs, one in Translation Studies and one in Spanish Literature, from the University of Amsterdam and received her Ph.D. from the University of Groningen, The Netherlands. Her area of specialization is contemporary peninsular cultural studies, focusing mainly on the 19th and early 20th century. Before she came to KU, she taught at the University of Amsterdam and Utrecht University in The Netherlands.

Versteeg's scholarly articles have appeared in, among others, Diálogos hispánicos, Foro Hispánico, Gestos, Excavatio, España Contemporánea, New Comparison, cover of Journalerso de la pluma Bulletin of Spanish Studies, Hispania, Journal of Iberian and Latinamerican Studies, ALEC, Hecho teatral, Moenia, and Siglo Diecinueve. She also authored several book chapters, among others in Au natural: Rereading Hispanic Naturalism, edited by J. P. Spicer-Escalante and Lara Anderson, and in Perfiles del heroismo en la literatura hispanica de entresiglos (XIX–XX), edited by Luis Alvarez-Castro and Denise DuPont (2013).

Currently, she is working on two projects: a book on the theater of Emilia Pardo Bazán titled Postcolonial proposals: Emilia Pardo Bazan's Theatrical Vision of a New Spain, and Divas and Chorus Girls, a study of representations of female performers in 19th and early 20th century Spain. Versteeg is also editing a volume on Pardo Bazan in the MLA's Approaches to Teaching World Literatures series (with Susan Walter, U Denver). She has been awarded a Cramer Award and a Cramer professorship (2012–2014) for excellence in research and teaching. Previously, she received a Grotius research fellowship from the University of Amsterdam.

In 2009 she developed, with María García Otero (Francis Marion University), a summer course for European Studies that took a group of KU students along the Camino de Santiago in Spain. Versteeg has frequently (co)directed KU's Summer Language Institutes in Barcelona, Spain (2007, 2010, 2012), Puebla, Mexico (2008) and Buenos Aires, Argentina (2011, 2013).

Dissertators working with Margot Versteeg have recently worked or are currently working on a series of topics related to performance studies (adultery in 19th and early-20th century plays, female performances of violence, and theatre under Franco's dictatorship).

UNIVERSITY OF KENTUCKY

Ruth Adams

Most Helpful to Students, Fine Arts

Ruth Adams, a nationally recognized artist, is the Associate Director of the School of Art and Visual Studies and an Associate Professor of Art at the University of Kentucky. Ruth's photography deals with issues of intimacy, voyeurism, mindfulness, death and rebirth. She is best known for her series unremarkable: a journey through cancer, chemotherapy, radiation and recovery and Organic Studies: Scanner as Camera. She loves combining new imaging technologies with the traditional sensibilities of a photographer and is one of the pioneers in the field of Scanner Art. She has exhibited internationally, won numerous awards and grants, and her photographs hang in numerous private and public collections. Ruth holds an MFA in Photography and Digital Art from the University Of Miami, a BFA in Photography from Rochester Institute of Technology, and a BS in Computer Science from Syracuse University. An experienced photographer, digital artist, and educator, Ruth has developed a reputation as a dynamic instructor and an innovative artist, and enjoys introducing students and patrons of the arts to both the digital and analog techniques available to today's photographer.

Anna Brzyski

Most Helpful to Students, Fine Arts

Anna Brzyski joined the faculty of the Art Department at the University of Kentucky in 2003. She teaches courses in the 18th and 19th century European art, as well as theory, methodology, and criticism. She has taught courses on a broad range of topics, including visual culture, aesthetics, 19th century art, classicism, postmodernism, contemporary art, abstraction, and landscape.

Her research interests focus broadly on how value and legitimacy have been negotiated within particular artworlds and across artword meshworks and how particular paradigms of knowledge emerge and why they are maintained. Her dissertation ("Modern Art and Nationalism in Fin de Siècle Poland") and the majority of her publications have dealt so far with Central/Eastern Europe and in particular Poland during the late 19th and early 20th century. Her work has appeared in Art Criticism, Centropa, 19th Century Art Worldwide, an anthology Art and National Identity at the Turn of the Century, edited by Michelle Facos and Sharon Hirsh (Cambridge 2003), and an anthology Local Strategies - International Ambitions. Modern Art and Central Europe, 1918–1968, edited by Vojtech Lehoda (Czech Academy of Sciences, forthcoming in 2005). She co-edited with Peter Chametzky a special issue of Centropa (September 2001) entitled "Modernism and Nationalism, Postmodernism and Postnationalism?" and is currently working on two book projects, an anthology Partisan Canons (Duke University Press, forthcoming) and Art in the Age of Art History. Modernism, Nationalism and Legitimacy in 19th Century Europe.

Prof. Brzyski is also the project director and compiler of the Polish Art Archive, a digital database of primary source materials pertaining to Polish art, a project funded by Southern Illinois University and the US Department of Educations. She is also the designer and the site manager of the HGCEA website (Historians of German & Central European Art & Architecture). She is also the founder and editor-in-chief of the ArtWorlds Press.

Rae Goodwin

Best Overall Faculty Member, Fine Arts
Best Overall Faculty Member, Fine Arts

Hui Chi Lee

Best Teacher, Fine Arts

As a visual artist, my strength lies in both the formation of my thought processes and the manipulation of non-traditional painting materials. I have printed and drawn on Plexiglas and vinyl, and layered images with three-dimensional materials. The representation of my conviction is about the dehumanization of excessive materialism. I am not interested in reproducing the visible. Rather, I aim to make the invisible visible. While I seek to capture the viewer's attention, my deeper goal is to inspire the viewer to consider the implications of the work. By my creative strategies, every element of a work takes on a metaphorical meaning.

Tom Marksbury

Most Helpful to Students, English

Ebony Patterson

Best Researcher/Scholar, Fine Arts

Ebony G. Patterson (Jamaican, born Kingston 1981) completed her undergraduate studies at the Edna Manley College for the Visual and Performing Arts in Kingston and earned her M.F.A. in 2006 from the Sam Fox College of Design & Visual Arts at

Washington University, St. Louis. She currently serves as an Associate Professor in Painting and Mixed Media at the School of Visual Arts and Visual Studies at the University of Kentucky, Lexington. Patterson has been featured in publications such as The New York Times, Frieze Magazine, Huffington Post, Art Papers, Art Nexus and The International Review of African American Art. Her work has been included in notable group exhibitions such as Visual Art and Music in a Post-Hip-Hop Era at the Museum of Contemporary African Diasporan Arts (MoCADA), Brooklyn, 2013; Caribbean: Crossroads of the World at the Studio Museum in Harlem/El Museo del Barrio/Queens Museum of Art, 2012; Out and Bad, the National Gallery of Bermuda, 2012 the Jamaica Biennial and Aruba Biennial, 2012; Wrestling the Image: Caribbean Interventions at Art Museum of the Americas, Washington D.C., 2011; Young Talent V, National Gallery of Jamaica 2012; Ghetto Biennale, Haiti, 2009; and Infinite Island: Contemporary Caribbean Art at the Brooklyn Museum of Art, 2007. Patterson recently had a two person exhibit Six Degrees of Separate Nations, Patricia and Philip Frost Museum and solo show 'until you see them', Monique Meloche Gallery in 2013. Patterson is currently having her first US museum show at the Nerman Museum of Contemporary Art 'dy/nas/ty'. Looking ahead, Patterson will be included in the Caribbean: Crossroads of the World will travel to the new Perez Art Museum in Miami and her work will be included in Gold at the Bass Museum. Patterson is represented by Monique Meloche Gallery and divides her time between Kingston, Jamaica and Lexington, KY.

Jane Peters

Best Teacher, Fine Arts

Ellen Rosenman

Best Teacher, English

Dmitry Strakovsky

Best Researcher/Scholar, Fine Arts

He has been able to parlay this experience into a series of classes that deal with electro-mechanical fabrication and software development in the arts. Dima's work spans diverse media and conceptual interests. Collaborative performances, media installations, drawing and sculptural works are just some of the examples of different modalities that define his output.

UNIVERSITY OF LA VERNE

Claire Angelici

Best Teacher, Liberal Arts, Social Sciences, and Humanities
Most Helpful to Students, Liberal Arts, Social Sciences, and Humanities

Dan Campana

Best Researcher/Scholar, Liberal Arts, Social Sciences, and Humanities

Department Chair, and directs degree programs in minority communities. Interests are in the philosophy of religion, epistemology, social ethics, and interdisciplinary studies. Professional paper and presentations are in religion and ethics, and religion and ecology.

George Keeler

Best Overall Faculty Member, Communications
Best Researcher/Scholar, Communications

For 35 years at the University of La Verne, Dr. Keeler, Communications Department chair, has led his publication students to top awards in national and regional contests that have brought honor and acclaim to the individual student journalists and helped launch impressive careers. Those honors include best in the state California Intercollegiate Press Association Sweepstakes Awards, first place in Society of Professional Journalists Region 11 contests (California, Arizona, Nevada, Hawaii) for newspaper and magazine, 14 All-American Awards through the associated Collegiate Press as newspaper adviser, and many silver crown awards (top 3 percent in the nation) for excellence in newspaper and magazine from the Columbia Scholastic Press Association, among many others. With a professional background in national and regional magazines (The Thalians Magazine and Messenger Magazine), Dr. Keeler presently is chief adviser to the University's La Verne Magazine in addition to heading the Communications Department Internship Program. Dr. Keeler is a writing coach to industry, a national magazine photographer, a contributing writer to regional newspapers and a professional media workshop leader. He serves as adviser to the campus chapter of the Society of Professional Journalists and has for many years provided leadership on the Board of Directors of the Los Angeles Professional Chapter. National Society of Professional Journalism leadership selected Dr. Keeler to serve as academic chair and news room supervisor at the October 1998 SPJ National Conference, where he advised a scholarship student staff from Stanford, Northwestern, Towson, UCLA, ULV, University of Connecticut, Cal State Fullerton, Cal State Northridge and Eastern Kentucky University in covering the convention utilizing the web, video, and a daily 12–16 page newspaper. Dr. Keeler is active in faculty governance, presently serving as chair of the Faculty Assembly and as a member of the Faculty Senate. During the past three years, he has led as chair of the University's Undergraduate Academic Policies Committee, as President of the American Association of

University Professors, ULV Chapter, and as executive director of the California Journalism Conference. Dr. Keeler has led ULV student study trips to Brazil, Costa Rica, Cuba, Vietnam, Thailand and Hawaii.

Shane Rodrigues

Best Teacher, Communications
Most Helpful to Students, Communications

Shane is an appointed 2006 board member for the Alliance of Community West Region. He is an award winning videographer and a current Independent Producer, Director & freelance videographer of Happy Hawaiian Boy Productions in Claremont, CA. He is also a Producer and Director of LVTV-3, La Verne Community Television in La Verne, CA. Shane was the Video Specialist for LVTV-3 and worked as Production Assistant for the daytime soap "The Young & the Restless" in Hollywood, CA. Finally, Shane was the Adviser of the Maui High School Video Club in Kahului, Hawaii, a photography assistant for Photography by Irvin in Wailuku, Hawaii, and a Director for Maui Community College televised classes in Kahului, Hawaii. He also won awards from the Alliance for Community Media, California Intercollegiate Press Association, College Broadcasters Inc., National Association of College Broadcasters and the Broadcast Education Association.

UNIVERSITY OF LOUISIANA AT LAFAYETTE

Skip Fox

Best Researcher/Scholar, English

"He never stops–he is involved with so many different projects and manages to move forward with all of them!"

Skip Fox has written several books and chapbooks of poetry and mixed-genre work as well as a lengthy bibliography. He is currently working on a nine volume text: Dream of a Book, four of which have been published: What Of (Potes & Poets), At That (Ahadada), For To (BlazeVox Books) and Delta Blues (Ahadada). He occasionally performs with a group of electronic musicians as Live Capture Remix, a jazz quartet for voice. His work has recently appeared in Ambit, Poetic Inhalation, Tarpaulin Sky, Big Bridge, eratio, Gestalten, Word for/Word, moria, Talisman, Hambone, and Exquisite Corpse.

Marcia Gaudet

Best Overall Faculty Member, English

Marcia Gaudet is Dr. Doris Meriwether/BORSF Professor of English and Department Head of English at the University of Louisiana at Lafayette. She has a B.S. from Louisiana State University, M.Ed. from Auburn University, and Ph.D. in English from University of Southwestern Louisiana (now UL Lafayette). She is the author of Tales From the Levee: The Folklore of St. John the Baptist Parish and co-author of Porch Talk with Ernest Gaines: Conversations on the Writer's Craft. Recent publications include editing and compiling, with Reggie Young, Mozart and Leadbelly: Stories and Essays by Ernest J. Gaines (2005), Carville: Remembering Leprosy in America (2004), and co-editing (with James McDonald) a new edition of their collection of readings on Louisiana folklore and culture entitled Mardi Gras, Gumbo, and Zydeco.

Mark Honegger

Best Teacher, English

"He is the most organized of any professor I've had. He knows exactly where the class is every day, every week of the semester. But this doesn't mean he's a slave to the syllabus–not only does he accept but encourages students to go deeper or even off topic if it is tangentially related to the discussion. He is brilliant at reformulating student remarks or questions to the rest of the class so that it not only makes more sense but makes the student seem thoughtful as well. He is one of the toughest not because he's too "hard" but because he holds his student's work to high expectation, pushing them further than they believed they could go."

Mark Honegger's book English Grammar for Writing (Houghton Mifflin 2005) teaches grammar from the perspective of comparing speech with writing. Recent articles include "No Grammar and Grammar Appreciation: Two Approaches to the G-Word" and "ESL and Dialects in the Writing Classroom" in Winning Ways of Coaching Writing: A Practical Guide for Teaching Writing, Grades 6–12 (Allyn & Bacon, 2001) and reviews of three books: Interrogative Investigations: The Form, Meaning and Use of English Interrogatives, Lexical Categories: Verbs, Nouns and Adjectives, and Predicting New Words: The Secrets of Their Success (Journal of English Linguistics).

Clai Rice

Most Helpful to Students, English

UNIVERSITY OF LOUISIANA AT MONROE

Ralph Brown

Best Overall Faculty Member, Liberal Arts, Social Sciences, and Humanities

UNIVERSITY OF LOUISVILLE

Margaret D'Silva

Best Researcher/Scholar, Communication

Muffy Sinclair

Most Helpful to Students, Communication

Christine Steineck

Best Teacher, Communication

Katherine Taylor

Best Overall Faculty Member, Communication

UNIVERSITY OF MAINE ORONO

Carla Billitteri

Best Teacher, English
Best Teacher, Liberal Arts, Social Sciences, and Humanities

Academic Positions

Current

Associate Professor, University of Maine, English Department, 2008–

Associate Faculty, National Poetry Foundation, University of Maine, 2008–

External Faculty, Doctoral Program in Anglo-American Studies, University of Catania, Italy, 2003–

Examiner, Critical Languages Program, University of Maine, 2000–

Previous

Assistant Professor, University of Maine, English Department, 2002–2008

Visiting Assistant Professor, University of Maine, English Department, 2001–2002

Lecturer, University of Maine, English Department, 1999–2001

Lecturer, D'Youville College, English Department, 1999

Education

Ph.D., English, State University at Buffalo, 2001

M.A., English, State University at Buffalo, 1995

Laurea, magna cum laude, Modern Languages and Literatures, University of Catania, Italy, 1989

Foreign Languages

Italian (native speaker) and French

Current Book Project

I am currently working on two books: a study of politics and aesthetics among the modernist poets and a monograph on Delmore Schwartz for European scholars.

Tony Brinkley

Best Researcher/Scholar, English

Tony Brinkley has taught at the University of Maine since 1983. He is a graduate of Yale University (BA) and the University of Massachusetts, Amherst (PhD). He teaches English Romantic Poetry, English Renaissance Poetry, Critical Theory, Fascist Studies (particularly in the National Socialist and Bolshevik contexts), Poetry and Poetics, Translation Studies, and Film. Tony Brinkley's poetry has appeared in Another Chicago Magazine, Beloit Poetry Journal, The New Review of Literature, Cerise Press, Drunken Boat, Otoliths, Hungarian Review, and Poetry Salzburg Review. His translations from Russian, German, French, and Hungarian have appeared in Shofar, Beloit Poetry Journal, The New Review of Literature, Cerise Press, May Day, World Literature Today, and Hungarian Review. He is the author of Stalin's Eyes (Puckerbrush Press) and the co-editor with Keith Hanley of Romantic Revisions (Cambridge University Press).

Richard Brucher

Best Overall Faculty Member, Liberal Arts, Social Sciences, and Humanities

Areas of specialization include Shakespeare and English Renaissance drama, modern American and British drama.

Dylan Dryer

Most Helpful to Students, Liberal Arts, Social Sciences, and Humanities

Ben Friedlander

Best Overall Faculty Member, English
Best Researcher/Scholar, English

Benjamin Friedlander joined our faculty in 1999. He was Coordinator of Graduate Studies from 2002 to 2005. His areas of expertise include Poetry and Poetics, Nineteenth- and Twentieth-Century American Literature, and Critical Theory. He is currently completing a book on Emily Dickinson and the Civil War. He is the author of Simulcast: Four Experiments in Criticism (U of Alabama P, 2004) and coeditor of Charles Olson's Collected Prose (U of California P, 1997). In addition to his scholarly endeavors, he writes poetry. His most recent collection is A Knot Is Not a Tangle.

Judith Hakola

Most Helpful to Students, English

Maureen Smith

Best Researcher/Scholar, Liberal Arts, Social Sciences, and Humanities

Maureen E. Smith is an enrolled member of the Oneida Tribe of Wisconsin. She currently serves as the Director of Native American Studies and Associate Professor of Interdisciplinary Studies at the University of Maine at Orono. She received her Ph.D. in Urban Education from the University of Wisconsin at Milwaukee. Her dissertation entitled, "Effects of U.S. Policy on American Indian Identity", looked at the psychological effects of the boarding school experiences for Native people from 1820 to 1934. She obtained a Master's Degree in Counseling from the University of Wisconsin at Oshkosh.

Prior to coming to Maine, Maureen worked as a professor of Education at the University of Wisconsin-Oshkosh from 1993 to 1997. She was the acting Associate Vice Chancellor for Multicultural Affairs. She worked as an advisor to minority students from 1986–1990 and an advisor to American Indian students from 1982–1986. Maureen also served as an alcohol counselor at the Oneida Tribal Health Center, Oneida, Wisconsin and Fond du Lac County Mental Health Center.

UNIVERSITY OF MARY WASHINGTON

Claudine Ferrell

Most Helpful to Students, History

Claudine Ferrell holds a Ph.D. from Rice University where her doctoral research focused on lynching in the South. A member of the Mary Washington history faculty since 1984, she has offered a wide range of courses in American history including legal, constitutional, and African-American history. Among her advanced courses are the Gilded Age, the Progressive Era, and the Vietnam War. Her newest book is The Abolitionist Movement (2005).

Jeffery McClurken

Best Overall Faculty Member, History
Best Teacher, History

Dr. McClurken's research areas include the history of the Civil War, veterans, families, the Pinkertons, mental institutions, the 19th-Century American South, and the digital humanities. He teaches classes on a wide array of US History topics, including Civil War and Reconstruction, American technology and culture, digital history, women's history, and history & film.

The Princeton Review named him one of The Best 300 Professors in 2012 and his lectures and classes have been featured on C-SPAN and NPR's With Good Reason. He was the 2014 recipient of the Teaching with Technology Outstanding Faculty Award from the State Council of Higher Education of Virginia, the highest award the state gives higher ed faculty. He has written numerous essays for the Chronicle of Higher Education's ProfHacker column and has articles in several publications related to teaching with technology. His book, Take Care of the Living: Reconstructing the Confederate Veteran Family in Virginia, was published by the University of Virginia Press in 2009. Information about his teaching and research can be found at http://mcclurken.org/

Bruce O'Brien

Best Researcher/Scholar, History

Bruce O'Brien joined the Mary Washington faculty in 1990 from Yale University where he earned his Ph.D. He is primarily responsible for the Department's offerings in ancient Greek and Roman history and medieval history. His special interest is England from the tenth to the twelfth centuries. His book God's Peace and King's Peace: The Laws of Edward the Confessor (1999) examined the impact of the Norman conquest on English law. His new book, Reversing Babel: Translation among the English during an Age of Conquests, c. 850 to c. 1200, which evaluates medieval England's multilingualism and culture of translation, is due out in 2010 from the University of Delaware Press. He is chair of the board and intellectual lead for Early English Laws, an international project to re-edit and translate all English legal texts written before Magna Carta. The project is a collaboration between the Institute of Historical Research (London) and the Centre for Computing in the Humanities at King's College London. For more information on the project, see this link. Education Ph.D., Yale University, 1990. History. M.A., Yale University, 1984. History. B.A., St. Olaf College, 1982. Creative Writing and Medieval History. Courses FSEM 201: European History HIST 121: Western Civilization I HIST 331: History of Ancient Greece HIST 341: Europe in the Middle Ages I HIST 342: Europe in the Middle Ages II HIST 451: History of Anglo-Saxon England

UNIVERSITY OF MARYLAND

Ronit Eisenbach

Most Helpful to Students, Fine Arts

Ronit Eisenbach is an architect, artist, curator, and educator whose scholarship and multi-disciplinary spatial practice aims to engage others in dialogue about the world we make for ourselves. Combining art, design and architecture, she explores how the perception of subjective, invisible and ephemeral objects affects understanding and experience of place. An interest in thinking through making and refining perception has led her to teach a series of situation-based, design-build studios that frame elements of architecture such as light, color, space, and shadow.

At Maryland she teaches undergraduate and graduate architecture studios, contemporary theory and an introductory course about the built environment. She has also designed interdisciplinary courses that she has co-taught with faculty in dance and art. A new course "Making Place: Public Art and Design" developed with Sculpture Professor John Ruppert and Urban Planning Prof. Gerrit Knapp has been selected by the Academy of Innovation & Entrepeneurship as a new "Fearless Ideas Course."

Venues for Eisenbach's work include: the Detroit Institute of Arts, the Graham Foundation, the Cranbrook Art Museum, Princeton University, the Clarice Smith Performing Arts Center, and the streets of Tel Aviv. Collaborations with choreographers, Peter Sparling, Dana Reitz, Bebe Miller, and Sharon Mansur, have resulted in a series of performance/installation works that explore relationships between gesture, inhabitation, and place. These include: Detroit 300: Fast Forward, Play Back, Placing Space, Memory House/Desire House and Out of Place. Her current research and creative work focuses on the potential for a catalytic relationship between ephemeral works and sites-in-flux.

Eisenbach's design work is recognized with awards from the AIA, ACSA and ID magazine and has been supported by the Andrew W. Mellon Foundation, Clarice Smith Performing Arts Center, Graham Foundation, Detroit Institute of Arts, University of Michigan Arts of Citizenship program, Maryland State Arts Council and others. Reviews have appeared in the Journal of Architectural Education, Public Art Review, The Washington Post, Washington Times, Sculpture Magazine, + Metropolis.

Eisenbach co-authored, Installation by Architects: Experiments in Building and Design, co-produced the documentary film, The Radiant Sun: Designer Ruth Adler Schnee, and curated Ruth Adler Schnee: A Passion for Color and Design for the Palazzo Mocenigo in Venice, Italy and UMD's Kibel Gallery, which she directs. A graduate of the Cranbrook Academy of Art and Rhode Island School of Design, Eisenbach is a former member of the ACSA Board of Directors and a current member of the National Building Museum's Education Committee. She is faculty fellow of NYU's Center for Creative Research, a Beverly Willis Architecture Fellow, the Virginia Center for the Creative Arts and a MacDowell Fellow.

Joanna Findlay

Best Researcher/Scholar, English
Best Teacher, English

Marybeth Shea

Best Overall Faculty Member, English
Best Teacher, Liberal Arts, Social Sciences, and Humanities

"Mary Beth Shea has great insight in teaching and in language use. She is always willing to share her insights with colleagues. By all reports, her students find her to be an inspiring and engaging teacher.

Mary Beth Shea is a brilliant teacher and scholar of rhetoric. She inspires both students and her fellow faculty members."

UNIVERSITY OF MARYLAND - EASTERN SHORE

Amy Hagenrater-Gooding

Best Researcher/Scholar, English

UNIVERSITY OF MASSACHUSETTS

Nancy Noble

Most Helpful to Students, Art History

UNIVERSITY OF MASSACHUSETTS - BOSTON

Ann Blum

Best Overall Faculty Member, Languages
Best Researcher/Scholar, Languages
Best Teacher, Languages
Most Helpful to Students, Languages

Michael Fennimore

Best Teacher, Theater

"Michael is an excellent acting teacher and actor. He engages with his students, and his practical, no-nonsense approach to the craft makes him a valuable and rare teacher."

Daniel Gidron

Best Teacher, Theater

"Danny's depth of knowledge and experience along with his eagerness to share with his students makes him an outstanding teacher."

Peter Kiang

Best Overall Faculty Member, Ethnic Studies
Best Researcher/Scholar, Ethnic Studies

Professor of Cultural Studies, Social Studies, Multicultural Education, School for Global Inclusion and Social Development

Ginger Lazarus

Most Helpful to Students, Theater

"Ginger's work with the students on their playwrighting and screenwriting is of the highest caliber. She brings her expertise to the students and gives her all to their progress."

Ginger Lazarus is an award-winning playwright whose work has been frequently produced in her native Boston and beyond. Her most recent project was A Blessing and a Curse: A Duet of Plays on Motherhood, an evening of two one-act plays presented by Spiced Wine Productions. Her play Matter Familias received an IRNE nomination for Best New Play of 2004; other honors include the 1999 John Gassner Memorial Playwriting Award for MOCKBA: A Play About Moscow and selection as a ten-minute play finalist in the 2002 Kennedy Center American College Theater Festival for Shooting Sparks. Her plays have been produced locally by Boston Playwrights' Theatre, Centastage, Queer Soup, Playwrights' Platform, and the Boston Theater Marathon, and have been featured nationally in Untitled Theater's 24/7 Festival (New York), Philadelphia Gay and Lesbian Theater Festival, Pan Theater Ten Minute Play Festival (San Francisco), Bloody Unicorn Theater Company's Lesbian Shorts (Tucson), and the Last Frontier Theatre Conference Play Lab (Valdez, Alaska). Two of her short plays also appeared at the Warehouse Theatre and Canal Café Theatre in London. Ginger holds a master's degree in playwriting from Boston University and also teaches in Emerson College's Professional Studies division. She is a member of the Dramatists Guild and StageSource.

Robert Lublin

Best Researcher/Scholar, Theater

"Robert is an excellent scholar whose enthusiasm for his subject is infectious. His scholarship translates well to his work with the students. He is demanding and helpful, which is a great combination."

Robert Lublin specializes in theatre history and historiography and has published articles and reviews on English and American theatre in numerous journals, including Theatre Survey, Theatre Notebook, American Drama, and others. Additionally, he has written chapters in books on feminist pedagogy, Sarah Kane's radical dramaturgy, and the sacred and the profane in early modern English literature. He is the author of Costuming the Shakespearean Stage: Visual Codes of Representation in Early Modern Theatre and Culture. This monograph examines costuming practices in early modern London and considers the visual semiotics that determined how apparel was understood when it was first seen on the Shakespearean stage. He is currently co-editing a collection titled Reinventing the Renaissance: Shakespeare and His Contemporaries in Adaptation and Performance. In addition to teaching courses in theatre history and dramatic literature, Lublin also teaches acting, directs plays, and has experience as a fight choreographer.

Louise Penner

Best Overall Faculty Member, Liberal Arts, Social Sciences, and Humanities

Aminah Pilgrim

Best Teacher, Liberal Arts, Social Sciences, and Humanities
Most Helpful to Students, Liberal Arts, Social Sciences, and Humanities

Rajini Srikanth

Best Researcher/Scholar, Liberal Arts, Social Sciences, and Humanities
Best Overall Faculty Member, Liberal Arts, Social Sciences, and Humanities

Human Rights and Literature, American Literature (including Asian American literature, Native American literature, and literature of the American South), Interdisciplinary Approaches to Literature, Literature in the Context of Comparative Race and Ethnicities, Pedagogy of Literature, Literatures of the Middle East

Tony Van Der Meer

Best Teacher, Ethnic Studies

UNIVERSITY OF MASSACHUSETTS - DARTMOUTH

Robert Waxler

Best Overall Faculty Member, English

"I nominate Prof. Waxler for his teaching, scholarship and innovativeness in devising "Changing Lives Through Literature"–an international program to help felons convicted of non-violent (usually drug-related) crimes."

Robert Waxler is an English professor at the University of Massachusetts Dartmouth. Waxler co-founded the internationally known Changing Lives Through Literature program in 1991, and is the co-founder of the UMass Dartmouth Center for Jewish Culture. He has served as Chairman of the English Department, Associate Dean of the College of Arts and Sciences, and Dean of Continuing Education. He has authored or co-authored several books, including Changing Lives Through Literature, Finding a Voice, Transforming Literacy, Losing Jonathan, and Courage to Walk. He is currently at work on a new book, Story and Real Life, exploring the relationship between linguistic narrative and the contingent experience of everyday life.

UNIVERSITY OF MASSACHUSETTS - LOWELL

Jehanne-Marie Gavarini

Most Helpful to Students, Fine Arts

Jehanne-Marie Gavarini is a visual artist whose work has been exhibited extensively nationally and internationally. In addition to her artistic work, Gavarini writes about art and visual culture. She is co-translator of Tomboy (University of Nebraska Press, 2007), an autobiographical novel by acclaimed Franco-Algerian writer Nina Bouraoui. Recent publications include "Rewind: The Will to Remember, the Will to Forget in Michael Haneke's Caché" in Millennial Cinema Memory in Global Film (New York: Columbia University Press, 2012) and "Permeable Borders in Notre Musique" in Zoom in, Zoom out: Crossing Borders in Contemporary European Cinema (Cambridge Scholars Press, 2007). Gavarini is currently Professor of Art at the University of Massachusetts Lowell and Resident Scholar at the Women's Studies Research Center at Brandeis University. She is Co-Director of UML's Center for Arts and Ideas.

Arno Minkkinen

Best Researcher/Scholar, Fine Arts
Best Researcher/Scholar, Fine Arts

"Outstanding research. International gallery and museum shows, talks, lectures and workshops."

Arno Rafael Minkkinen is a Finnish American photographer, educator, curator, and writer with over two hundred one-person and group exhibitions at galleries and museums worldwide.

He was born in Helsinki in 1945 and emigrated to the United States in 1951. A graduate of Wagner College with a BA in English Literature, he began taking self-portraits in 1970 while working as an advertising copywriter on Madison Avenue in New York. Studying later with Harry Callahan and Aaron Siskind at Rhode Island School of Design, he earned his MFA degree in photography in 1974. Over the past four decades, Minkkinen has been primarily engaged as a teacher, curator, and writer while continuing to devote his photographic research and energies to the self-portrait: unmanipulated images of the naked human figure in natural and urban landscapes.

Professor of Art at the University of Massachusetts Lowell, Minkkinen also serves as Docent at Aalto University School of Art, Design, and Architecture, Helsinki. Earlier in his teaching career he served as Assistant Professor at MIT, Visiting Artist at Philadelphia College of Art (now University of the Arts), and the École d'Arts Appliqués in Vevey, Switzerland. He has conducted workshops in Maine, Tuscany, Colorado, New Mexico, China, Finland, France, Italy, Lithuania and Norway. Spirit Level Workshops have been offered periodically to UMass Lowell students together with Finnish, French, Italian, and Swiss students in Russia, Eastern Europe, Italy and Mexico.

UNIVERSITY OF MEMPHIS

Antonio De Velasco

Best Overall Faculty Member, Communication
Best Researcher/Scholar, Communication
Best Teacher, Communication
Most Helpful to Students, Communication

"Very invested in students, devoted to his objectives in transforming students into rhetorical agents in the community, committed to strengthening the department in unique and innovative ways, exudes a passion for the field that is contagious."

A teacher of rhetoric, Tony de Velasco works within a scholarly tradition more than 2,000 years in the making. Bringing this established tradition into dialogue with classical and contemporary social theory, de Velasco designs and teaches courses across the department's rhetoric curriculum.

He has received both the College of Communication and Fine Arts Dean's Outstanding Research Award and the University of Memphis Alumni Association Distinguished Teaching Award, a campus-wide distinction that recognizes excellence in teaching at the undergraduate and graduate levels. In 2012, de Velasco directed On Civic Learning: Rhetoric, Public Address, Political Division, a meeting of the Thirteenth Biennial Public Address Conference.

Professor de Velasco has published on a range of topics, including the rhetoric of current political controversies, argumentation, the history of rhetoric, and pedagogy. His first book,

Centrist Rhetoric: The Production of Political Transcendence in the Clinton Presidency, was named a "highly recommended" title of 2010 by Choice magazine. In 2011, de Velasco, with Melody Lehn, published Rhetoric: Concord and Controversy, an edited volume featuring top papers delivered at the Rhetoric Society of America's 2010 conference.

Currently, de Velasco is co-editing an anthology of essays by the late rhetorician Michael C. Leff, writing a book on contemporary rhetorical theory, and serving as Graduate Coordinator for the Department of Communication.

Leroy Dorsey

Best Researcher/Scholar, Liberal Arts, Social Sciences, and Humanities

PhD, 1993, Indiana University MA, 1988, California State University, Hayward BS, 1980, California State University, Hayward

Academic Summary

Dr. Dorsey's research examines the symbols used by political figures to promote their legislative agendas, shape their identities as morally sound advocates, and transform their audiences into seemingly active agents poised to support particular agendas. He studies the public speech of the presidency, most notably the rhetoric of Theodore Roosevelt and, to a lesser extent, Woodrow Wilson, John F. Kennedy and Ronald Reagan. His most recent book, We Are All Americans, Pure and Simple: Theodore Roosevelt and the Myth of Americanism, examines how Roosevelt used the frontier myth of national origin to create standards for non-whites and immigrants to achieve before they could be identified as Americans. The book won the 2008 National Communication Association Marie Hochmuth Nichols Award for the top book in public address studies.

Dr. Dorsey recently began exploring the intersection of politics with popular culture and how the resulting dialogue shapes notions of race, ethnicity, and American identity.

His research has been published in such journals as Quarterly Journal of Speech, Presidential Studies Quarterly and Rhetoric & Public Affairs. Dr. Dorsey is a contributor to African American Orators and edited The Presidency and Rhetorical Leadership.

Courses Taught

Communication and Popular Culture Women, Minorities & Mass Media American Public Discourse Since 1865 Political Communication

Selected Publications

Leroy G. Dorsey, "Narrating the Presidential 'Race': Barack Obama and the American Dream," in Communicating Marginalized Masculinities: Identity Politics in TV, Film, and New Media, edited by Ronald L. Jackson II & Jamie E. Moshin. Routledge, 2012.

Leroy G. Dorsey, "We Are All Americans, Pure and Simple": Theodore Roosevelt and the Myth of Americanism. Tuscaloosa: University of Alabama Press, 2007.

Leroy G. Dorsey, ed., The Presidency and Rhetorical Leadership. Texas A&M University Press, 2002.

Leroy G. Dorsey and Rachel M. Harlow, "'We Want Americans Pure and Simple': Theodore Roosevelt and the Myth of Americanism," Rhetoric & Public Affairs, 6 (2003): 55–78.

Leroy G. Dorsey, "Rereading The X-Files: The Trickster in Contemporary Conspiracy Myth," Western Journal of Communication 4 (2002): 448–468.

Honors and Awards

Marie Hochmuth Nichols Award for the top book in 2007 for Public Address Studies, Public Address Division, National Communication Association, 2008

Texas A&M Association of Former Students College-level Distinguished Teaching Award, 2003 and 2010

B. Aubrey Fisher Award for outstanding article published in the Western Journal of Communication, 1995

Professional Activity

Editorial Board Member Voices of Democracy, 2009–present Southern Communication Journal, 2007–2010 Communication Studies, 2005 Rhetoric and Public Affairs, 2000–05 Western Journal of Communication, 2000–02

Aram Goudsouzian

Best Overall Faculty Member, History

"outstanding dedication to students, colleagues, and the study of history."

Kenneth Kreitner

Most Helpful to Students, Music

UNIVERSITY OF MISSOURI - ST. LOUIS

Andrew Hurley

Best Overall Faculty Member, History

"Andrew works hard at all aspects of his job–his teaching, his research, and his service to the department. He's helpful to everyone."

Diane Touliatos

Best Researcher/Scholar, Music

"Dr. Touliatos is a world-known scholar in recovering the music manuscripts of Medieval and Byzantine women composers from the year 800. It is a a wonderful service she has done for the humanities, music history, women in the arts, and correcting the long-standing belief that there were no women composers before Hildegard von Bingen."

The doctoral work in historical musicology was completed by Diane Touliatos at The Ohio State University's School of Music under the supervision of her dissertation advisors, Richard Hoppin and Milos Velimirovic.

As a faculty member in the Department of Music at University of Missouri-St. Louis since 1979 and as a Research Fellow at the Center for International Studies at UMSL since 1982, Diane Touliatos has focused her research efforts in Eastern Medieval Chant, Ancient Greek Music, and Women Composers. In the area of musicology, she has achieved a world-wide reputation as an internationally published scholar. More specifically in the area of Medieval Byzantine Musicology and Women Composers, she has become a leading world-wide expert and consequently an invited speaker at conferences and university seminars all over Western Europe, Greece, Poland, Russia, and the former Yugoslavia.

As of January, 1997, Diane Touliatos was elected by the Humanities Faculty of the College of Arts and Sciences of the University of Missouri-St. Louis as Director of the Center for the Humanities.

Laura Westhoff

Best Researcher/Scholar, History
Best Teacher, History
Most Helpful to Students, History

"Laura is smart and creative, and definitely one of the most imaginative scholars in our department."

Laura Westhoff is an associate professor specializing in the fields of the Gilded Age and Progressive Era, American Democracy and Social movements, the Scholarship of Teaching and Learning History (SoTL), and the History of Education. She is the author A Fatal Drifting Apart: Democratic Social Knowledge and Chicago Reform (2007) which examines social constructions of knowledge and the emerging tensions between democracy and expertise in modern America. She holds a Ph.D. in U.S. History from Washington University in St. Louis and has published in the Journal of American History, The History Teacher, Women's History Review, and the History of Education Quarterly. She is currently working on a book on mid-twentieth century democratic practices, titled Educating for Activism. As a joint appointment in the College of Education at UMSL, Dr. Westhoff is committed to advancing history education at all levels. She is active in social studies certification at UMSL, has served as co-author and lead historian on several Teaching American History grants, and is a frequent presenter at history workshops and colloquia for teachers. She serves as the faculty coordinator for the Student Teaching in China program.

UNIVERSITY OF MISSOURI-KANSAS CITY

Kathy Krause

Most Helpful to Students, Liberal Arts, Social Sciences, and Humanities

"Nice teacher!"

Kathy M. Krause, Professor of French, (B.A., Dartmouth College, 1984; M.A., University of Pennsylvania, 19; Ph.D., University of Pennsylvania, 1989). Chair, Department of Foreign Languages & Literatures. Medieval French literature, roles of women in medieval French literature and literary production, manuscript studies, female inheritance in 13th century France and Flanders. Author of "Generic Space-Off and the Construction of the Female Protagonist: The Chanson de Florence de Rome," Exemplaria 18.1 (Winter, 2006), 93–136. This article won Honorable Mention, Best Article Prize, Society for Medieval Feminist Scholarship, 2007.

Author of "The Falsely Accused Heroine in the Miracles de Notre Daqme par personnages and in Late Medieval Prose," in Don Maddox and Sara Sturm-Maddox, eds., Parisian Confraternity Drama of the Fourteenth Century (Turnhout, Belgium: Brepols Publishers, 2008), 113–34. Co-editor, with Alison Stones, of Gautier de Coinci: Miracles, Music and Manuscripts (Turnhout, Belgium: Brepols Publishers, 2007).

My research explores a series of interconnected issues related to women's roles and representation in the French Middle Ages. Building on my initial study of the intertextual representations of women in Old French romance, and in particular those of female protagonists, I am currently examining the ways in which female inheritance intersected with literary production in northern France and Flanders. Combining literary analysis, codicology and historical research, this work aims to challenge traditional male-centered socio-historical analyses of medieval French romance by viewing these texts within the lineage-centered politics of the 13th and 14th centuries. I am also continuing to work on the oeuvre of Gautier de Coinci, one of the major vernacular authors of Marian literature of the Middle Ages. Our edited volume on Gautier has just been published by Brepols, and I am presently developing a database of the illuminated manuscripts of his Miracles de Nostre Dame. I teach courses on French medieval and Renaissance literature, on French language and culture, and on medieval courtly history and culture.

Steven Melling

Best Teacher, Communication

Dr. Steven Melling has a Ph.D. and M.A. degrees in Communication Studies from the University of Kansas. His research focuses on the intersection of rhetoric, culture and policy. This is evident in his dissertation, which examines the role rhetoric played in the evolution of Kansas City's Crossroads Arts District. Steven currently teaches Fundamentals of Effective Speaking and Listening at UMKC. He has previously taught communication courses at the University of Kansas, Metropolitan Community College and William Jewell College.

Lorena Summerville

Best Teacher, Languages
Best Overall Faculty Member, Liberal Arts, Social Sciences, and Humanities

"She works hard to help everyone! Very smart person!! Super great helper!"

UNIVERSITY OF NEBRASKA-OMAHA

Bruce Johansen

Best Researcher/Scholar, Communication

Bruce E. Johansen is a Professor of Communication and Native American Studies, University of Nebraska at Omaha. He has been teaching and writing in the School of Communication at UNO since 1982. He had authored 39 published books as of 2014.

Johansen writes frequently about environmental subjects, including The Encyclopedia of Global Warming Science and Technology (2 vols., 2009), Global Warming in the 21st Century (3 vols., 2006), The Global Warming Desk Reference (2001), The Dirty Dozen: Toxic Chemicals and the Earth's Future (2003), and Indigenous Peoples and Environmental Issues (2004), a 200,000-word encyclopedia of indigenous peoples' struggles with corporations with a world-wide scope.

Johansen's first academic specialty was the influence of Native American political systems on United States political and legal institutions; his best-known books in this area are Forgotten Founders (1982) and Exemplar of Liberty (with Donald A. Grinde, Jr.), published in 1991. Johansen has described the present-day debate over this issue in Debating Democracy (1998), and Native American Political Systems and the Evolution of Democracy: An Annotated Bibliography (Greenwood, 1996; volume 2, 1999).

He also writes as journalist in several national forums, including the Washington Post and The Progressive, with letters to the editor in The Atlantic, New York Times, National Geographic, Wall Street Journal, et al. He was co-editor of the Encyclopedia of American Indian History, a 4-volume set (ABC-CLIO, 2007), as well as a 2-volume Praeger Handbook of Contemporary Native American Issues (2007). He recently also wrote The Global Warming Combat Manual: Serious Solutions for a Sustainable World (Praeger, 2008).

Johansen lives in Omaha, Nebraska, with his wife Pat E. Keiffer, a Spanish-English interpreter and family therapist, her son, Shannon Keiffer-Rose, and two grandchildren Samantha and Madison Keiffer-Rose.

Holly Miller

Best Teacher, Communication

Randy Rose

Best Overall Faculty Member, Communication
Most Helpful to Students, Communication

Paul Williams

Best Overall Faculty Member, Religion

"An excellent chair who puts the faculty beneath him in positions to succeed."

PROFESSIONAL BACKGROUND: Born in the Democratic Republic of Congo (DRC) and raised in both Congo and Texas, he earned a BA in Anthropology and Asian Studies (1978) from the University of Texas at Austin and a Master of Arts in Religion (1979) from the Episcopal Theological Seminary of the Southwest. At the University of Chicago, he studied history of religions, with an emphasis on African Religions, Buddhism, and the History of Christianity, earning an MA in Religious Studies (1981) and a PhD in History of Religions (2000). Based on research in the United States and the DRC, his doctoral dissertation was entitled "The Disciples of Christ Congo Mission (DCCM), 1897–1932: A Missionary Community in Colonial Central Africa." Dr. Williams is an ordained minister of the Christian Church (Disciples of Christ) and served congregations of the Disciples of Christ in Illinois from January 1993 until June 2001. Prior to coming to Nebraska in August 2001, he taught as Adjunct Faculty at St. Xavier University (Chicago, Illinois). TEACHING: Dr. Williams offers courses in African Religions, History of Christianity, Islam, New Testament, Religion and Popular Culture, Religion in America, Religious Ethics, Senior Seminar in Religion (a writing course), and World Religions, and works to develop his own pedagogy as well as the curriculum of the Religious Studies program and the College of Arts and Sciences. In 2002, Dr. Williams was awarded a fellowship to attend a National Endowment for the Humanities (NEH) Summer Institute on "Societal Transformation and the Legitimation of Power in Early Islamic States," at the University of Chicago with Dr. Fred Donner and Dr. Kenneth Hall. From that experience, he has taught courses on Islam annually since spring 2003. In addition to introducing new courses (e.g., African Religions, Religion and Popular Culture, Listening to Muslim Voices), he is the coordinator of the Religious Studies Dual Enrollment program (2007–present) and the coordinator of the Religious Studies course on "Religions in the Modern World" for the graduate Global Information Operations Certificate program. He co-developed and team-taught "Religious Diversity in the Public Schools" with Ms. Beth Katz (Director, Project Interfaith) as a graduate Teacher Education course for the College of Education (Summer 2009). During 2007–2008, after several years of teaching Islam for the Religious Studies program, he served as Director of the newly-formed program in Islamic Studies, an interdepartmental unit that includes faculty and administrators from Foreign Languages, History, International Studies, Political Science, and Religious Studies. The purpose of the Islamic studies initiative is to become a university and community resource for information about Islam and the Islamic world. In support of that mission, Dr. Williams helped to create and open a new line in the Religion Program to increase course offerings in Islamic studies, and co-wrote with Dr. Moshe Gershovich the first proposal for a Minor in Islamic Studies at UNO. Under the leadership of Dr. Bridget Blomfield, a Minor in Islamic Studies was approved in spring 2009. RESEARCH: Dr. Williams conducts research on socio-cultural dimensions of changes in religious identity from the perspective of historical anthropology in North America and Africa (African Traditional Religions, Christianity, and Islam), especially the Democratic Republic of Congo, Israel, and Morocco, as well as the rise of early Christianity in the Greco-Roman world. His publications (articles and book reviews) appear in Annales Aequatoria (Mbandaka, DRC), Discipliana (Nashville, TN), the International Third World Studies Journal and Review (ITWSJR), the Journal of Religion and Film, and the Encyclopedia of the Stone-Campbell Movement (Eerdmans, 2006). Current Research Projects: During the spring semester 2009, he conducted research on the history of Disciples of Christ in the Democratic Republic of Congo since 1960, relying primarily on semi-structured interviews with Congolese preachers, teachers, and elders. The principal purpose of the study is to employ oral history methods in a participant-observation study of a topic (Christianity in central Africa) that is long dominated by archival research from a western missionary perspective. The work to be produced includes both articles and books designed to bring Congolese voices to the fore in the study of their own history. Recently, he presented a video on liturgical innovations in recent Disciples history for the Black History Month conference at Luther College (Decorah, Iowa) on 25 Feb 2010 and the Missouri Valley History Conference (Omaha, Nebraska) on 5 Mar 2010. SERVICE: His service to the university and profession includes serving as Editor for the on-line peer-review Journal of Religion and Film (four years), Chairperson of the Department of Religious Studies (almost three years), Member of the Educational Policy Committee of the College of Arts and Sciences (three years), Member of the Search Committee for the Dean of Arts and Sciences (2007–2008), Director of Islamic Studies (2007–2008), and various other administrative responsibilities. In addition to his academic work, Dr. Williams speaks to civic organizations and religious communities about the significance of religion in the world today, the role of religious studies in higher education, and various topics like Islam, Buddhism, New Testament, Christian missions, contemporary Christian thought, inter-religious relations, and interfaith dialogue.

UNIVERSITY OF NEVADA LAS VEGAS (UNLV)

Firas Al-Douri

Best Overall Faculty Member, Architecture
Best Researcher/Scholar, Architecture
Best Teacher, Architecture

Rosa Caytas

Best Overall Faculty Member, Languages

Sean Clark

Best Teacher, Fine Arts

Sean Clark came to UNLV to create the graduate screenwriting program after more than a decade in Los Angeles as a working screenwriter and producer. In L.A. Sean wrote and/or produced for several television series including, Coach, Evening Shade, Northern Exposure, Sirens, Early Edition, and several others. The most prestigious of Clark's awards as a playwright is the Lorraine Hansberry Award, which was granted for his play, Eleven-Zulu. He attended the University of Missouri and the University of Iowa where he earned a master of fine arts degree in theatre - playwriting.

Kara Erolin

Best Researcher/Scholar, Social Science
Best Teacher, Social Science

Randy Hale

Most Helpful to Students, Architecture
Most Helpful to Students, Fine Arts

Louis Kavouras

Best Researcher/Scholar, Dance
Best Overall Faculty Member, Fine Arts

"Louis Kavouras has devised both programs and courses for the Department of Dance that have moved the program to a level of academic excellence. He teaches these academically advanced courses in a way that the Dance Majors can understand while elevating their intellectual understanding."

Louis Kavouras joined the UNLV Dance Faculty in 1992. In 1994 he was elected chair of the department. Kavouras has been a principal dancer of the Erick Hawkins Dance Company and faculty member of the Erick Hawkins School of Dance since 1996. The Erick Hawkins Dance Company is a professional, New York based, modern dance company founded by modern dance pioneer, Erick Hawkins. Kavouras has studied with modern dancers Kathryn Karipides, Kelly Holt, Erick Hawkins, Betty Jones, Lucas Hoving, Donald McKayle, Peter Pucci, Gus Solomons, Jr., Rudy Perez, Claire Porter, and Clay Taliaferro. Kavouras is accomplished as a choreographer, designer, visual and graphic artist. His dances have been showcased in Nevada, Ohio, North Carolina, Texas, Hawaii, Pennsylvania, Australia, Scotland, England, China, Korea, Japan, Denmark, Russia, Nova Scotia, and Jamaica. Kavouras believes in creating dances which are uniquely his own, and range from "hilarious" to "poignant." He comes from a total theater tradition and explores all interdisciplinary avenues between the arts.

Emily Lin

Best Teacher, Curriculum & Instruction

Emily Lin received her Ph.D. in Curriculum, Teaching and Learning from the Ontario Institute for Studies in Education-University of Toronto, Canada. Since joining UNLV, her professional and scholarly agenda has focused on teacher education reform with a concentration on international studies and comparative teaching and learning in mathematics and science education. She has been an evaluator for National Science Foundation grants and conducted analyses of cross-national international data. Her focus on teacher education reform extends beyond the United States to Canada, Australia, Asia and Europe and her work has been recognized by distinguished research organizations across the nation. Some of her articles appear in Asia-Pacific Journal of Teacher Education, Educational Researcher, Educational Research Review, Evaluation and Research in Education, The Canadian Journal of Environmental Education, International Journal of Teacher Leadership, and International Journal of Science and Mathematics Education. She served as a co-editor for the Journal of Teacher Education from 2008–11. She is currently the President of the Nevada Association of Colleges for Teacher Education and the national committee Chair of the American Education Research Association (AERA)'s Exemplary Research in Teaching and Teacher Education Award which recognizes the significant contribution to teaching and teacher education scholarship. She serves currently as the Director of Teacher Education for the College of Education at UNLV.

Collin Loader

Best Researcher/Scholar, Liberal Arts, Social Sciences, and Humanities
Best Teacher, Liberal Arts, Social Sciences, and Humanities
Most Helpful to Students, Liberal Arts, Social Sciences, and Humanities
Best Overall Faculty Member, Liberal Arts, Social Sciences, and Humanities

Colin Loader received his Ph.D. from UCLA in European intellectual history. He came to UNLV in 1986 and teaches courses in modern German history and modern European intellectual history. While he has written on subjects such as Sherlock Holmes and German silent film, his primary research interests are in the history of German sociology between 1890 and 1933 and that discipline's relationship to others such as history, economics and philosophy. Among the sociologists he has studied are Karl Mannheim, Max Weber, Alfred Weber and Werner Sombart.

Among his publications are Alfred Weber and the Crisis of Culture, 1890–1933 (New York, 2012), The Intellectual Development of Karl Mannheim (C ambridge, 1985), Karl Mannheim's Sociology as Political Education (with David Kettler, New Brunswick, 2002) and articles in The Journal of Modern History, CLIO, Sociological Theory, Film and History, The Canadian Journal of Sociology and German Studies Review.

Donald Revell

Best Teacher, English

Donald Revell is the author of twelve collections of poetry, most recently of Tantivy (2012) and The Bitter Withy (2009), both from Alice James Books. He has published five volumes of translations from the French, including Apollinaire's Alcools (Wesleyan), Rimbaud's Illuminations and A Season in Hell (both from Omnidawn) and Laforgue's Last Verses (Omnidawn). His critical writings include The Art of Attention (Graywolf) and Invisible Green: Selected Prose (Omnidawn). Winner of the PEN USA Translation Award and two-time winner of the PEN USA Award for Poetry, he has also won the Academy of American Poets' Lenore Marshall Prize and is a former Fellow of the Ingram Merrill and Guggenheim Foundations. Additionally, he has twice been awarded Fellowships from the National Endowment for the Arts. Former editor-in-chief of Denver Quarterly, he now serves as poetry editor of Colorado Review. Revell is a Professor of English & Creative Writing at the University of Nevada, Las Vegas.

Sang-Duck Seo

Best Researcher/Scholar, Fine Arts

UNIVERSITY OF NEVADA - RENO

Phill Boardman

Best Overall Faculty Member, Humanities
Best Researcher/Scholar, Humanities
Best Teacher, Humanities

Phil Boardman, who directed the Core Humanities program for fourteen years, is a professor in the English department. A specialist in medieval literature and the culture of the Middle Ages, he teaches courses in Chaucer, Arthurian literature, the Bible, and Core Humanities. His bachelor's and master's degrees are from the University of Nebraska, and he earned his Ph.D. degree at the University of Washington. With Jim Bernardi in the theatre department, he led for many years an annual London Theatre and Literary Arts tour; before that he led summer tours to England and France, and to Italy and Greece. He taught for USAC in Lueneburg, Germany. He has made numerous appearances as the Chautauqua character Geoffrey Chaucer and also as Sir Thomas Malory. Besides hiking, cross-country skiing, and kayaking, his activities include performing with musical groups, including the Nevada Opera Chorus (more than eighty productions), the vocal quartet Reno Early Music, and the men's ensemble Orfeo. He has taught at the University of Nevada since 1974 and has won teaching awards from the College of Arts and Science, the University of Nevada, and the Nevada Board of Regents, as well as the Carnegie Endowment for the Advancement of Teaching ("Nevada Professor of the Year," 2003). He was executive co-producer of the KNPB series, The Western Traditions Lectures and is editor of the reader for Core Humanities 201, Enduring Legacies: Ancient and Medieval Cultures, now in its sixth edition. His comprehensive guide to the popular traditions of King Arthur in English, The Arthurian Annals: The Tradition in English from 1250 to 2000, was published by Oxford University Press (2004), and it earned him the College of Liberal Arts Dean's Award for Research.

Dennis Cronan

Best Researcher/Scholar, Liberal Arts, Social Sciences, and Humanities
Most Helpful to Students, Liberal Arts, Social Sciences, and Humanities
Best Overall Faculty Member, Liberal Arts, Social Sciences, and Humanities

Dennis Cronan earned his PhD at the University of Minnesota in Germanic Philology, a program that enabled him to study widely in early Germanic languages and literatures. He specializes in Old and Middle English, Old Norse, Poetic Language, Poetics, and Literary History. He is currently writing a book on Beowulf.

Alison Harvey

Most Helpful to Students, Humanities
Best Teacher, Liberal Arts, Social Sciences, and Humanities

James Mardock

Best Researcher/Scholar, English
Best Teacher, English

Since joining the English faculty in 2006, James Mardock has taught undergraduate courses in Shakespeare's tragedies and comedies, the drama of Marlowe, Webster, and Jonson, early English drama, and other literatures of the sixteenth and seventeenth centuries, as well as graduate seminars in Jacobean city comedy, Spenser's epic The Faerie Queene, and the drama of Reformation. He also serves as assistant general editor for the Internet Shakespeare Editions, and worked in 2008 and 2009 as the house dramaturge for the Lake Tahoe Shakespeare Festival. He has directed productions of Shakespeare's Henry V and The Winter's Tale, and played, among other roles, Troilus, Laertes, Oliver, Antonio, and Juliet's father. In 2011, appearing with his wife and as-yet-unborn son, he played Claudius in the Nevada Rep's production of Hamlet in the original pronunciation. His early drama reading group, The Blood, Love, and Rhetoric School, stages amateur readings of lesser-known plays like 'Tis Pity She's a Whore, The Jew of Malta, and Gammer Gurton's Needle. From 2011 to 2013, Dr Mardock serves as the Joseph Crowley distinguished professor of the humanities, giving public lectures on such topics as Renaissance comedian Will Kemp's 130-mile morris dancing stunt and predestination in A Midsummer Night's Dream.

He has published articles on Dickens's debt to Shakespeare in the representation of Jews, on Ben Jonson's curious affinity for male cross-dressing, on populist poet John Taylor's politics and religion, and on what King James and bubonic plague have to do with Measure for Measure. His book Our Scene is London (2008) examines Jonson's representation of urban space as an element in his strategy of self-definition. His edition of Henry V (2014) appears in print from Broadview Press and online at the Internet Shakespeare Editions (http://internetshakespeare.uvic.ca/Library/Texts/H5/). Dr. Mardock has recently co-edited a collection of essays on drama and religion in post-Reformation England, and is at work on a book-length study of Calvin's influence on early modern drama.

Eric Rasmussen

Best Overall Faculty Member, English
Most Helpful to Students, English

My work focuses on editing Shakespeare and Early Modern drama. My first monograph, A Textual Companion to 'Doctor Faustus' (Manchester University Press, 1994), significantly changed our understanding of the early texts of Christopher Marlowe's play and paved the way for the groundbreaking editions that David Bevington and I prepared for both the Revels Plays series and for Oxford University Press. I subsequently co-edited The Norton Anthology of English Renaissance Drama, The RSC Complete Works of Shakespeare, King Henry VI Part 3 for the Arden Shakespeare series, Cynthia's Revels for the Cambridge Works of Ben Jonson, Everyman and Mankind for the Arden Early Modern Drama series, The Two Noble Kinsmen for the Malone Society, and the forthcoming New Variorum Shakespeare edition of Hamlet. In addition to producing these conventionally published works, for the last decade I have served as general textual editor of the Internet Shakespeare Editions (ISE) project, a series of critical editions specially commissioned for the electronic medium and subjected to a rigorous peer review process. These editions have become the most frequently consulted versions of Shakespeare's plays in the world; and the award-winning ISE site is now arguably the most respected Shakespearean educational resource on the Web: http://internetshakespeare.uvic.ca Since 1999, I have served as the author of Shakespeare Survey's annual omnibus review for Cambridge University Press of all new editions and textual studies. In this capacity, I have had an influence on the direction and, I like to think, the quality of work in the field. As a member of the council of the Malone Society and the editorial board of Shakespeare Quarterly, I help to maintain the high standards associated with those publications. As a member of the board of trustees of the Shakespeare Association of America and as U.S. Secretary/Treasurer for the Malone Society, I have been instrumental in steering these two preeminent professional associations. As book review editor for Medievalia et Humanistica since 1994, I have played a role in evaluating the evolving study of medieval and Renaissance culture more broadly. This year, I published The Shakespeare First Folio: A Descriptive Catalogue. This 1000-page catalogue raisonné is the result of two decades of research during which Anthony James West and I located 232 surviving copies of this immeasurably important book – a remarkable 72 more than were recorded in Sidney Lee's census over a century ago. A research team of UNR graduate students – Mark Farnsworth, Lara Hansen, Trey Jansen, and Sarah Stewart – then examined every copy in situ worldwide and prepared full bibliographic descriptions of each. I also published a trade book offering a narrative account of this field research, The Shakespeare Thefts: In Search of the First Folios, which has reached a wide (and surprisingly appreciative) audience. Here at UNR, I have received the Alan Bible Teaching Award (college-wide), the F. Donald Tibbitts Teaching Excellence Award (university-wide), and the Board of Regents Teaching Award (system-wide).

UNIVERSITY OF NEW MEXICO

Kelly Becker

Best Teacher, Philosophy

Has recently published papers in Journal of Philosophy, Erkenntinis, History of Philosophy Quarterly, American Philosophical Quarterly (July, 2002 and January, 2004), Dialectica, Metaphilosophy, Philosophical Studies, Acta Analytica, Philosophia, and Mind (book review). Becker's article, "Reliabilism," appears in The Internet Encyclopedia of Philosophy.

Prof. Becker and Tim Black are publishing a collection of new, state-of-the-art essays on the sensitivity principle in epistemology. The volume will be published with Cambridge UP in the Summer of 2012.

John Bussanich

Best Researcher/Scholar, Philosophy

Michael Chapdelaine

Best Teacher, Music

Mary Domski

Most Helpful to Students, Philosophy
Most Helpful to Students, Liberal Arts, Social Sciences, and Humanities

Mary Domski is an Associate Professor of Philosophy at the University of New Mexico. Before joining the UNM faculty in 2005, she taught at Fresno State for two years, and before that, she was a graduate student at Indiana University, where she earned a PhD in the History and Philosophy of Science. Her research focuses on the interplay of philosophy, mathematics, and science in the early modern period (and in the work of Descartes, Newton, and Kant, in particular).

She regularly teach courses in early modern philosophy and philosophy of science, and she is the recipient of four UNM teaching awards: the 2014 Presidential Teaching Fellowship, the 2011 Faculty Teaching Award from the UNM Alumni Association, the 2009 Award for Teaching Excellence from the College of Arts & Sciences, and the 2006–2007 Outstanding Teacher of the Year Award.

She is currently serving as the Chair of the UNM Philosophy Department and, in January 2015, She began a three-year term as Associate Editor of Studies in History and Philosophy of Science Part A.

Bradley Ellingboe

Best Overall Faculty Member, Liberal Arts, Social Sciences, and Humanities

Charles Kalm

Best Teacher, Liberal Arts, Social Sciences, and Humanities

Charles is a 5th generation New Mexican, born in Union County, New Mexico. A licensed New Mexico attorney, accepted into the New Mexico bar by motion of his father, Thomas L. Kalm, Esq. on September 21, 2010. In addition to his law practice, Charles Kalm is a part-time lecturer in the Department of Philosophy at UNM, where he teaches Reasoning and Critical Thinking, The Philosophy of Law and Morals, as well as Classical Social and Political Philosophy. Charles enjoys traveling, reading, conversation, coffee, philosophy, and the outdoors: running, hiking, biking, surfing, skiing, snowshoeing and snowboarding.

UNM Center For Teaching Excellence Awards Committee, Member, 2015

UNM Truman Scholarship Selection Committee, Member, 2014

UNM Lecturer of the Year, 2013–2014

Admission to the United States Federal Bar, as Attorney and Solicitor in Chancery, December 3, 2012

Admission to the Bar of the State of New Mexico, September 21, 2010

Dean's Award for Significant Contribution to the Law School Community, May 2010

International Studies Institute Graduate Assistantship, Awarded Fall 2007–Spring 2010

American Association of Justice Mock Trial Regional Champion, Spring 2009

http://kalmlawfirm.com/charles-h-kalm-esq

Ian Thomson

Best Overall Faculty Member, Philosophy
Best Researcher/Scholar, Liberal Arts, Social Sciences, and Humanities
Best Overall Faculty Member, Liberal Arts, Social Sciences, and Humanities

UNIVERSITY OF NEW ORLEANS

David Hoover

Best Researcher/Scholar, Theater
Best Teacher, Theater

In addition to being chair, Professor Hoover is the Director of the MFA Performance Program. His directorial credits at UNO include a tropical A Midsummer Night's Dream, Of Mice and Men, The Imaginary Invalid, Candida, Ghosts, Assassins, The Two Gentlemen of Verona, Educating Rita, The Merchant of Venice, and the Big Easy award winning productions of Blood Brothers and Metamorphoses. For the Tennessee Williams New Orleans Literary Festival he has directed Alec Baldwin, Elizabeth Ashley, John Goodman, Stephanie Zimbalist, Eli Wallach, and Anne Jackson.

In New Orleans his work has been seen at Le Petit Theatre du Vieux Carre, Tulane Summer Shakespeare Festival, Rivertown Rep, Le Chat Noir, Southern Rep, and Tulane Summer Lyric. He has worked at several other prestigious theatres including The Guthrie, the Dallas

Theatre Center, American Southwest Theatre Company, Shakespeare Festival of Dallas, and The Lyceum. Professor Hoover is the recipient of the Big Easy, Marquee, and Storer Boone awards, having been recognized recently for acting work in The History Boys. He has directed four original scripts for KCACTF, two of which received national recognition. He is a member of the Society of American Fight Directors, ATHE, and the Southwest Theatre Association.

Productions he has directed have been invited five times to the Regional Festival of the Kennedy Center American College Theatre Festival and his MFA Directing students have been invited three times and were all later invited to the Kennedy Center. Three of his acting students have won the Irene Ryan Acting Competition and represented Region VI at the Kennedy Center. Internationally he has been an adjudicator in London, lectured at the Universite de Sorbonne in Paris and has taught extensively in Italy and Mexico. Professor Hoover is the recipient of the Artist Fellowship award from the Louisiana Division of the Arts; the first award of its kind made to a director. He was recognized recently by the University of New Orleans Alumni Association with an award for Career Achievement for Excellence in Research.

UNIVERSITY OF NORTH CAROLINA

Harry (Michael) Coiner

Best Teacher, History

Perry Hall

Best Teacher, African-American Studies

He received a bachelor's degree in Psychology from the University of Michigan, and a doctorate in Education and Social Policy from Harvard University. In addition to teaching the survey introductory course, he has developed courses that examine various forms phases of African American cultural production, focusing especially on popular music. These include a course that surveys the historical influence of African American popular culture, and one that specifically examines the history of hip hop. Also, in addition to leading the capstone senior seminar for

Afro-American Studies majors, he has taught courses pertaining to the Black Freedom Struggle, and the development of African American social and political leadership.

Jerma Jackson

Best Teacher, Liberal Arts, Social Sciences, and Humanities

Kenneth Janken

Best Overall Faculty Member, African-American Studies

Teaching Interests: In addition to our department's survey course on the black experience, I have taught the civil rights movement, the Harlem Renaissance, black thought, and black leadership. I am on the faculty advisory board of the Office of Undergraduate Research, and I believe that all undergraduates should have a research experience at least once in their time at Carolina.

Current Research: My research focuses on 20th century African American history. I am the author of two biographies: Rayford W. Logan and the Dilemma of the African-American Intellectual (1993) and White: The Biography of Walter White, Mr. NAACP (2003), which won honorable mention in the Outstanding Book Awards from the Gustavus Myers Center for the Study of Bigotry and Human Rights in North America. I have also published several academic articles on topics such as the Harlem Renaissance, the civil rights movement in the 1940s, and African Americans and world affairs.

I am currently working on a history of the Wilmington Ten. They were civil rights activists who were falsely convicted of arson and firing upon firefighters and police officers during four days of rioting in February 1971 in Wilmington, North Carolina. The violence that rocked Wilmington occurred in the context of tensions related to the desegregation of the city's public schools. Their convictions, which were overturned by a federal appeals court in 1980, hinged on perjured testimony knowingly solicited by the state's attorney. The project will analyze the circumstances surrounding the racial violence and the state's misconduct and evaluate the nationwide "Free the Wilmington Ten" political mobilization. It will analyze Black Power forms of organizing that typified African American politics in the 1970s and their relationships to traditional forms of political action.

Lisa Lindsay

Best Researcher/Scholar, Liberal Arts, Social Sciences, and Humanities

Pat Parker

Best Overall Faculty Member, Communication Studies

"Dr. Patricia Parker has been immensely supportive to me as a PhD candidate. I really struggled through the program, but she was always patient and available for feedback and support. I definitely wouldn't have made it through this program without her. Also, her research and advocacy work has proved to be an inspiration for my own scholarly and advocacy work."

Valerie Pruvost

Best Teacher, Languages

Charlene Regester

Best Researcher/Scholar, African-American Studies

Eunice Sahle

Most Helpful to Students, African-American Studies

Frederick Vogler

Most Helpful to Students, Languages

Heather Williams

Most Helpful to Students, Liberal Arts, Social Sciences, and Humanities

Joel Williamson

Best Overall Faculty Member, Liberal Arts, Social Sciences, and Humanities
Best Researcher/Scholar, Liberal Arts, Social Sciences, and Humanities

UNIVERSITY OF NORTH CAROLINA AT CHARLOTTE

Michele Bissiere

Best Researcher/Scholar, Languages

Lil Brannon

Best Researcher/Scholar, Liberal Arts, Social Sciences, and Humanities
Best Teacher, Liberal Arts, Social Sciences, and Humanities

Education

Ed.D. Texas A&M University, Commerce, 1977 M.A. Sam Houston State University, 1973 B.A. Converse College, 1971 Areas of Interest

Composition Theory and the Teaching of Writing English Education Critical Pedagogy Selected Publications and Presentations

Books:

Michelle Comstock, Mary Ann Cain, and Lil Brannon. Composing Public Space: Teaching Writing in the Face of Private Interests. Portsmouth, N.H.: Boynton/Cook Publishers, 2010. Lil Brannon, Sally Griffin, Karen Haag, Tony Iannone, Shana Woodward, and Cindy Urbanski.

Thinking Out Loud on Paper. Portsmouth, N.H.: Heinemann, 2008. Lil Brannon and Brenda Greene, Eds. Rethinking American Literature, Urbana: NCTE, 1997. C.H. Knoblauch and Lil Brannon. Critical Teaching and the Idea of Literacy. Portsmouth, N.H.: Boynton/Cook Publishers, 1993. C.H. Knoblauch and Lil Brannon. Rhetorical Traditions and the Teaching of Writing. Portsmouth, N.H.: Boynton/Cook Publishers, 1984 Lil Brannon, V. Neverow-Turk and Melinda Knight. Writers Writing. Portsmouth, N.H.: Boynton/Cook Publishers, 1982. Recent Articles and Book Chapters:

Tony Scott and Lil Brannon, "Valuation, Assessment and the Political Economy of First-Year Writing" in On Language and Value: Political Economies of Rhetoric and Composition. Eds. Wendy Olson, Siskanna Naynhaha and Victor Villanueva. Logan, UT: Utah State University Press (in press). Lil Brannon and Cy Knoblauch, "Robert Boynton and the Making of Composition: English Educator, Editor, Publisher" in Composition's Roots in English Education," Ed. Patricia Lambert Stock. Portsmouth: Heinemann/Boynton Cook, 2012. (124–134). Lil Brannon, C. Urbanski, L. Manship, L. Arnold, and T. Iannone, "The Ebayification of Education: Critical Literacy in a Consumerocracy," in English Journal, 99 (Jan. 2010), 16–21. Lil Brannon, C. Urbanski, J Courtney, S. Woodward, J. Reynolds, K. Haag, K. Mach, M. Kendrick, T. Iannone, and L. Manship. "The Five Paragraph Essay and the Deficit Model of Education, in English Journal, 98, No. 2 (Nov. 2008), 16–21. C.H. Knoblauch and Lil Brannon, "The Emperor (Still) Has No Clothes: Revisiting the Myth of Improvement." Key Works on Teacher Response: An Anthology. Ed. Rick Straub, Portsmouth, NH: Heinemann, Boynton/Cook, 2006. 1–16. C.H. Knoblauch and Lil Brannon, "Pedagogy for the Bamboozled" in Writing with Elbow, Eds. P. Belanoff, et al. Logan: Utah Universities Press, 2002, pp. 65–83 Lil Brannon and Stephen M. North. "The Use of the Margins." The Writing Center Journal, Fall 2000. Lil Brannon. "(Dis)Missing Compulsary First-Year Composition" in Reconceiving Writing, edited by Joseph Petraglia, Mahwah, NJ: Lawrence Earlbaum, 1997. Lil Brannon. "The Problem with National Standards," College Composition and Communication. October 1995. Lil Brannon. "M(other): Lives on the Outside" Written Communication, 10, No.3 (July 1993), 457–465. Lil Brannon. "To Sir, with Love: Gender and the Imperatives of Liberatory Teaching," in Theoretical and Critical Perspectives on Teacher Change, Norwood, N.J.: Ablex, 1993. 151–54. Selected GrantS:

National Writing Project Grant, UNC Charlotte Writing Project, $60,000 2012–13 National Writing Project Grant, UNC Charlotte Writing Project, $52,000, 2011–12 National Writing Project Grant, UNC Charlotte Writing Project, $52,000, 2010–11 State of New York Initiative for Technology, $300,000 (1996–97) William and Flora Hewlett Foundation, $150,000, Project Renaissance, 1996–98. Courses Taught Undergraduate

English 2100 English 3200 Approaches to Literature English 3104 Adolescent Literature English 4254 Teaching English and Communication Skills to Middle and Secondary School Learners Graduate

EDCI 8020 Research in English Education EDCI 8020 Readings in English Education ENGL 6062 Invitational Summer Institute, UNC Charlotte Writing Project ENGL 6274 Contexts and Issues in the Teaching of English ENGL 6674 Applied Research Methods in the Teaching of English Professional Appointments Faculty Appointments

UNC Charlotte since 1998 University of Albany, SUNY, 1986–1998 New York University, 1980–1986 UNC Wilmington, 1978–1980 Texas A & M, Commerce, 1977–78 Administrative Appointments

Associate Dean, College of Arts and Sciences, 2006–present Director, UNC Charlotte Writing Project, 2003–present Coordinator, M.A. in English Education, 1998–present Director, Center for Excellence in Teaching and Learning, Univ. at Albany, 1994–1998 Director, The Writing Center, Univ. at Albany, 1988–1994 Community Service

P.E.O. Awards

W. Ross Winterowd Award for Critical Teaching and the Idea of Literacy, 1984 University at Albany, President's Award for Excellence in Academic Service, 1996 Distinguished Alumni Award, Texas A&M Commerce, 1992

Paula Eckard

Best Overall Faculty Member, Liberal Arts, Social Sciences, and Humanities

Min Jiang

Best Researcher/Scholar, Communication Studies

Dr. Min Jiang is Associate Professor of Communication Studies, Affiliate Faculty of International Studies at UNC Charlotte, and Research Affiliate at the Center for Global Communication Studies, University of Pennsylvania. She is an Advisory Board Member of Project Mosaic, a university social science research initiative and co-coordinator of Digital Arts, Sciences & Technologies (DAST), a College of Liberal Arts & Sciences initiative in digital humanities. She teaches graduate and undergraduate courses in new media technology, global media and research methods.

Highly interdisciplinary, her work blends new media studies, political communication, international communication, legal studies, and information science. She's written on Chinese digital technologies (search engine, social media), business, policies, social activism, and digital diplomacy. Her work has appeared in Journal of Communication, New Media & Society, Social Science Computer Review, Electronic Journal of Communication, Policy & Internet among other academic journals. She is writing a book tentatively titled China vs. Information.

Sam Shapiro

Most Helpful to Students, American Studies

UNIVERSITY OF NORTH CAROLINA AT GREENSBORO

Naurice Woods

Best Teacher, Ethnic Studies

"He has consistently high student evaluations and many students declare the major because they enjoyed one of his courses. He is also frequently cited by alum as a teacher they remember."

Dr. Naurice Frank Woods, Jr. currently serves as a Visiting Assistant Professor in the African American Studies Program at the University of North Carolina at Greensboro. He holds a BFA from The University of North Carolina at Chapel Hill, a MFA from UNCG, and a Ph.D. in art history from The Union Institute and University. He teaches courses in African American history, art history, music, film, and popular culture and publishes regularly in these areas. He is also a productive visual artist. He is the former director of the African American Studies Program at UNCG. His latest book is entitled African American Pioneers in Art, Film, and Music.

UNIVERSITY OF NORTH CAROLINA WILMINGTON

Rick Olsen

Best Overall Faculty Member, Communication

Dr. Rick Olsen is currently serving as department chair. He has taught many classes for the department including COM 200, COM 257, COM 258, COM 301 and COM 305. He has also participated in the Honors Scholar program by teaching several HON 120 courses on communication related topics such as humor and protest music.

His research projects are typically rhetorical/critical analyses of popular culture artifacts. He has authored entries in an encyclopedia of popular culture and sport, published research on MTV, the NBA Draft, the TV show Desperate Housewives and the popularity of SUVs among others.

Dr Olsen also actually enjoys productive committee work and serves on several committees within his department and across campus including the Osher Lifelong Learning Institute. He has also facilitated communication training and been a keynote speaker for numerous off campus groups.

When he's not working, he enjoys making music with his band mates, playing basketball, and most importantly, trying to be a husband and dad that is worthy of the family he's been blessed with.

Teresita Parra

Most Helpful to Students, Languages

Valerie Rider

Best Overall Faculty Member, Languages
Best Teacher, Languages

It has been a productive–and global!—year for Valerie Rider and her students. Rider teaches beginning and intermediate level Spanish, and is also resident director of the summer study abroad program in Valparaíso, Chile. After finishing the UNCW 2012 spring semester, Rider began the first of three summer programs, walking with an Honors group across northern Spain on the ancient pilgrimage route, the Camino de Santiago. What an experience for all! Immediately after completing the Camino, Rider led 26 students to Chile for Study Abroad Chile, an exciting program now in its eleventh year. As recipient of an Office of International Programs Global Citizenship Grant, Rider left Chile for Peru to research an Honors program she will teach and lead in 2014. She and another intrepid group visited vibrant Lima, ancient Cuzco, and hiked through remote Quechua villages on the Lares Trek to Machu Picchu. In Nov. 2012, Rider and Honors colleague Bill Atwill presented at the National Collegiate Honors Council in Boston, MA, on the Camino course they created, ¡Ultreia! (ever westward). She has also participated in two UNCW College Days, one focusing on the history, language, and culture of the Camino de Santiago, and presented at the Cape Fear Foreign Language Collaborative on applied learning (in study abroad programs). Rider collected the Spanish section's entries for the FLL website study abroad section and, along with colleagues Marianella Jara and Gina Kelley, edited last year's and the present edition of the FLL newsletter. All of these activities are extremely rewarding, but the one closest to her heart is fostering in her students an appreciation of Spanish-speaking people and places through teaching her classes and directing study abroad!

UNIVERSITY OF NORTH DAKOTA

Michael Beltz

Best Teacher, Liberal Arts, Social Sciences, and Humanities

Hans Broedel

Best Teacher, History
Most Helpful to Students, History

Hans Peter Broedel received his Ph.D. from the University of Washington, and is an Associate Professor of History at the University of North Dakota, where he teaches courses on medieval and early modern European history, witchcraft, and the history of science. He is the author of The Malleus Maleficarum and the Construction of Witchcraft, and has published articles on witches and apparitions in the late middle ages, and on early modern natural history; his current research investigates the place of fabulous animals in Early Modern thought.

William Caraher

Best Overall Faculty Member, History
Best Researcher/Scholar, History
Best Overall Faculty Member, Liberal Arts, Social Sciences, and Humanities

Robin David

Most Helpful to Students, Liberal Arts, Social Sciences, and Humanities

My primary interest areas include refugee/immigrant integration, socioeconomic class, war, politics, and consumption. I teach social science and humanities courses on these subjects, and frequently use service-learning in doing so. In addition to teaching, I coordinate our service-learning program, manage the program's tuition waivers, oversee the Sophomore Honors Portfolio process, and am responsible for general administration.

Daphine Stevens

Best Researcher/Scholar, Liberal Arts, Social Sciences, and Humanities

UNIVERSITY OF NORTH FLORIDA

Mary Baron

Best Overall Faculty Member, English
Best Researcher/Scholar, English
Most Helpful to Students, English

Alex Menocal

Best Overall Faculty Member, Liberal Arts, Social Sciences, and Humanities

UNIVERSITY OF NORTHERN COLORADO

Travis Boyce

Most Helpful to Students, Ethnic Studies

George Junne

Best Overall Faculty Member, Ethnic Studies
Best Researcher/Scholar, Ethnic Studies
Best Teacher, Ethnic Studies

Dr. George Junne, Jr. is a professor in Africana Studies. He has been a UNC faculty member since 1994. He is also treasurer of the National Association for Ethnic Studies (NAES) and a former editor for the Western Social Science Association (WSSA) journal. Dr. Junne's most recent publications are Blacks in the American West and Beyond—America, Canada, and Mexico: A Selectively Annotated Bibliography (2000) and The History of Blacks in Canada: A Selectively Annotated Bibliography (2003). During the summers of 2000, 2002, 2004 and 2006, he was visiting professor of African American Literature and Film at Bogazici University in Istanbul. For twenty-six years, each summer Dr. Junne has been a fossil collector and director of operations for paleontology field camps for the University of Michigan and Albion College.

Maria Lopez

Best Overall Faculty Member, Languages
Most Helpful to Students, Languages
Best Researcher/Scholar, Liberal Arts, Social Sciences, and Humanities

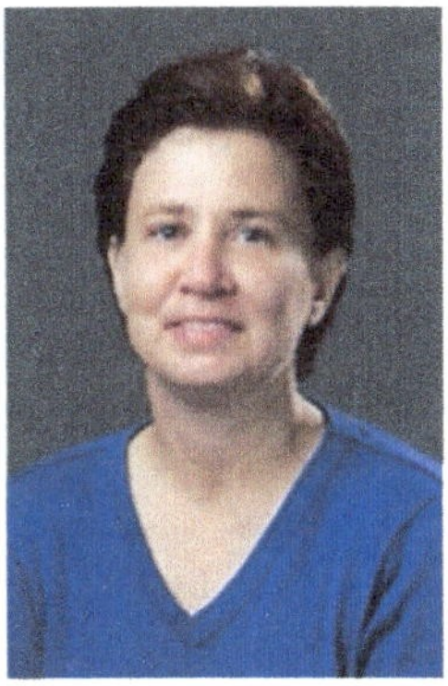

Melanie Moore

Best Overall Faculty Member, Liberal Arts, Social Sciences, and Humanities

After finishing her Ph.D. at the University of Washington, Prof. Moore did post-doctoral work at Indiana University. Her interests include Family, Social Psychology and Quantitative Methodology.

Her numerous articles and presentations cover a variety of subjects, including juvenile delinquency, gender discrimination, teaching practices, applied research, and parenting perspectives. Her current research addresses creating communities and an ethnographic analysis of community characters.

Thomas Smith

Most Helpful to Students, Liberal Arts, Social Sciences, and Humanities
Best Researcher/Scholar, Liberal Arts, Social Sciences, and Humanities

Tom Smith is Assistant Vice President for Undergraduate Studies and Dean of University College at the University of Northern Colorado. Prior to his arrival at UNC in 2006, he spent 18 years at Loyola University New Orleans, where he served as a faculty member, chair of the Department of Religious Studies, associate dean of the College of Arts and Sciences, and interim Vice President for Student Affairs/Associate Provost. He was a three-time recipient of Loyola's Student-Alumni Association Citation for Excellence in Teaching.

Dr. Smith received his bachelor's degree in History from the University of Washington, diploma and master's degree from Regent College (University of British Columbia), and master's and doctoral degrees from the University of Notre Dame. He is the author of a book and numerous book chapters and articles on topics in the early history of Christianity. In addition to his duties as University College Dean, he oversees undergraduate curriculum, supervises the Army and Air Force ROTC detachments, and teaches occasionally in UNC's Philosophy program, where he holds the rank of Professor.

Robert Weis

Best Teacher, Liberal Arts, Social Sciences, and Humanities

Dr. Weis teaches courses on Latin American history. He received his MA in Latin American Studies from the Universidad Nacional Autónoma de México and his PhD in History from the University of California, Davis. His book, Bakers and Basques: A Social

History of Bread in Mexico (University of New Mexico Press, 2012), is a history of food, workers, and politics in revolutionary Mexico. He is currently working on a new project, tentatively called God Restrains the Devil But Not the President, a cultural history of Catholic activists in Mexico City during the 1920s.

Michael Welsh

Best Teacher, Liberal Arts, Social Sciences, and Humanities

Michael Welsh received his Ph.D. degree in United States history in 1983 from the University of New Mexico. Among his fields of research interest are the history of the American West and Southwest, American Indian history, the environment, ethnicity in America, the history of American education, and the U.S. Constitution. He has taught at the University of Dayton, the University of New Mexico, Oregon State University, Cameron University (OK), St. John's College (Santa Fe, NM), and has taught since 1990 at the University of Northern Colorado, where he holds the rank of professor.

UNIVERSITY OF NORTHERN IOWA

Ken Baughman

Most Helpful to Students, English

Pierre-Damien Mvuyekure

Best Researcher/Scholar, English

Dr. Pierre-Damien Mvuyekure is Professor of English and African American literature in the Department of Languages and Literatures, the 2007–2008 Regents Awards for Faculty Excellence and the 2005 Philip G. Hubbard Outstanding Educator at the University of Northern Iowa. A Fulbright alumnus, Dr. Mvuyekure holds a B.A. in Letters and Licence (M.A.) ès Letters in English from the National University of Rwanda, Rwanda, and an M.A. and Ph.D. in English from SUNY at Buffalo. His areas of specialization include African American Literature, African and African Diaspora Literatures, African American Literary Criticism, Multicultural Literature, Post-Colonial Literatures and Theory, Nineteenth-Century American Literature, Twentieth-century American Literature, Popular Literature and Critical and Cultural Theory. Since joining UNI in 1995, Dr. Mvuyekure has been teaching a number of courses, including College Reading and Writing, Personal Essay, Introduction to Literature, American Renaissance, Multicultural Literature, Post-Colonial Literatures and Theory, African Dance in Literature, African Literature and Dance, African American Literature, African American Women Writers, The Harlem Renaissance and Blues and Jazz in African American Literature and Film. He has published four books: The "Dark Heathenism" of the American Novelist Ishmael Reed: African Voodoo as American Literary HooDoo (The Edwin Mellen Press (2007), Lamentations on the Rwandan Genocide (Final Thursday Press 2006), World Eras Volume 10: West African Kingdoms, 500–1590 (Gale/Thompson 2004), A Casebook Study of Ishmael Reed's Yellow Back Radio Broke-Down (Dalkey Archive Press 2003—www.dalkeyarchivepress.com). Among his other publications are several book chapters and articles on the Rwandan Genocide, Alice Walker, Ishmael Reed, Jewell Parker Rhodes, Gloria Naylor, Walter Mosley, Paul Laurence Dunbar, Melvin B. Tolson, Patricia Grace, Velma Pollard and Tupac. His most recent publications include: "American Nero-HooDooism: The Novels of Ishmael Reed" in The Cambridge Companion to the African American Novel (2004), "Paul Rusesabagina's Oasis of Ubu-Muntu in Hotel Rwanda: A Review Essay in Konch (Spring 2005) and "Ishmael Reed Repairs "The [African] Diaspora's Direct Line to Olódùmarè": Yoruba Language and Mythology in Japanese By Spring" in Black Humor, Irony, and Satire: Ishmael Reed, Satirically Speaking, ed. Dana A. Williams (Cambridge Scholars Press, 2007). Dr. Mvuyekure is currently completing The Cambridge Introduction to the African American Novel, to be published in 2008–2009 by Cambridge University Press. He is also a poet and fiction writer. In his teaching, scholarship, and creative activity, Dr. Mvuyekure emphasizes the connection among the African Diaspora. In this perspective, he is the producer and host of The Talking Drum on the Afro-American Community Broadcasting Network Inc. on KBBG FM88.1 every Saturday at three, an African music show designed to reconnect African-American to Africa. The Talking Drum is his way of linking the University of Northern Iowa to the Cedar Valley community and the state of Iowa and the world—the show can be streamed on the internet worldwide at www.kbbgfm.org. He is also a good dancer (Rwandan traditional dance and Congolese Rumba) who includes music in his teaching. Return to Directory

UNIVERSITY OF OKLAHOMA

Jeffrey Maiden

Best Teacher, Liberal Arts, Social Sciences, and Humanities

Dr. Maiden's research interests are focused on issues related to education funding, including intergovernmental fiscal relationships, the fiscal effects of educational reform, educational capital outlay and debt service, and education finance litigation. The results of his research have been published in such journals as The Journal of Education Finance and the Yearbook of the American Education Finance Association. Dr. Maiden teaches graduate courses in education finance, educational technology leadership, and quantitative research methods.

UNIVERSITY OF OREGON

Claudia Holguín

Best Researcher/Scholar, Romance Languages

Claudia Holguín Mendoza is Assistant Professor of Spanish Linguistics and Director of the Spanish Heritage Language program in the Department of Romance Languages. She teaches courses in Spanish Sociolinguistics and conducts research on language variation and change in specific socio-economic and political contexts.

Jocelyn Hollander

Best Teacher, Sociology

Professor Hollander received her B.A. in Linguistics from Stanford in 1987. She received an M.A. in 1991 and a Ph.D. in 1997 from the University of Washington. She joined the University of Oregon faculty in 1997. Professor Hollander is currently engaged in research on self-defense training and the prevention of violence against women.

Dean Kramer

Best Teacher, Music

"Kramer has a superb record of outstanding teaching."

Lori Kruckenberg

Best Overall Faculty Member, Music
Best Researcher/Scholar, Music

Lori Kruckenberg is associate professor of musicology, and has been on the University of Oregon music faculty since 2001. She teaches the undergraduate and graduate music history surveys of the medieval and Renaissance periods as well as speciality courses in the history of music theory and early music notations. Topics in graduate seminars have included sequences, tropes, New Song (cantica nova), women musicians in medieval German, and early music in England. She was a Fulbright scholar to Germany (1992–94) and she researched at the Universität Erlangen-Nürnberg until 1996. Prior to her arrival at the UO, she held positions at the University of Iowa and University of Louisville. Most recently, she was visiting professor at the Universität Basel (2008–09).

Dr. Kruckenberg has published articles in several international journals, conference proceedings, and lexicons, including Revue bénédictine, Tijdschrift van de Koninklijke Vereniging voor Nederlandse Muziekgeschiedenis, Cantus Planus, Berichte des Internationalen Kongresses der Gesellschaft für Musikforschung, and Die Musik in Geschichte und Gegenwart (MGG). A chapter on the liturgical sequence will appear in The Cambridge History of Medieval Music (forthcoming). Other notable articles include "Neumatizing the Sequence: Special Performances of Sequences in the Central Middle Ages" in the Journal of American Musicology (2006) and "Music for John the Evangelist: Virtue and Virtuosity at Paradies," in Harvard's Houghton Library Studies (2008). She co-edited and contributed chapters to The Sequences of Nidaros: A Nordic Repertory and Its European Context (Tapir Academic Press, 2006). She is currently preparing on the monograph A New Kind of Song: The Sequentiae novae, ca. 1050–ca. 1200 as well as articles on the traditions of the medieval German cantrix and the role of Messine chant in the propagation of cultural identity.

In 2008–09 Dr. Kruckenberg was a senior Fulbright scholar at the Bruno-Stäblein-Archiv at the Julius-Maximilians-Universität in Würzburg, Germany. Other recent awards include the 2012 Noah Greenberg Prize from the American Musicological Society, an Oregon Humanities Grant, and the Oregon Community Credit Union Research Fellowship, a partial subvention for a collaborative project with Professor Eric Mentzel and members of the Schola Cantorum Basiliensis. The recording project dealing with textless singing and reconstruction of medieval performance practices is forthcoming.

Dr. Kruckenberg has lectured and presented at several universities (Cambridge, Oxford, Southampton, Basel, Erlangen-Nuremberg, Boulder, and various conferences and symposia (Harvard, Yale, Notre Dame, Bangor [UK], Basel, Lillafüred, Munich, Niederaltaich, Paderborn, Salzburg, Trondheim, Utrecht, Venice, Weimar). She regularly presents at meetings of the American Musicological Society, Cantus Planus, International Med-Ren Music Conference, the International Congress on Medieval Studies at Kalamazoo, and International Medieval Congress at Leeds. She currently serves on the editorial board of Plainsong and Medieval Music (Cambridge University Press) and Corpus monodicum, and sits on the Council of the American Musicological Society and on the Board of Academic Advisors of the Center for Volga German Studies at Concordia in Portland.

Doralba Perez

Best Teacher, Fine Arts

"Great teacher!"

Regina Psaki

Best Teacher, Fine Arts
Most Helpful to Students, Fine Arts
Best Overall Faculty Member, Fine Arts

Melissa Redford

Best Overall Faculty Member, Linguistics
Best Researcher/Scholar, Linguistics
Most Helpful to Students, Linguistics

Luz Romero Montano

Best Researcher/Scholar, Romance Languages
Best Teacher, Romance Languages

Analisa Taylor

Best Overall Faculty Member, Romance Languages
Best Teacher, Romance Languages
Most Helpful to Students, Romance Languages

"Professor Taylor challenges her students to think critically from the other side of privilege and power, and to make important connections many people in academia avoid or deny."

In my research and teaching I seek to identify how discourses and practices that cement cultural norms, hierarchies and exclusions feed upon and feed into social inequalities and injustices and to understand how subjects marginalized by these discourses and practices reclaim epistemic authority as they occupy the art, literature, film and other types of cultural production that document their stories and strategies of cultural survival. I published Indigeneity in the Mexican Cultural Imagination: Thresholds of Belonging with the University of Arizona Press in 2009 as well as articles in Signs: Journal of Women in Culture and Society, Latin American Literary Review, Journal of Latin American Cultural Studies. I am working on two book projects, Mesoamerican Diaspora: Strategies and Stories of Cultural Survival amid Agro-industrial Crisis and Between Longing and Memory in the Lacandon Rain Forest: Maya Women Speak.

David Wacks

Best Overall Faculty Member, Romance Languages

Carl Woideck

Best Teacher, Music

Carl Woideck is an instructor of jazz, rock, and blues music histories at the University of Oregon, where he began teaching in 1982. Woideck is also a professional saxophonist and longtime jazz radio broadcaster.

Woideck is author of Charlie Parker: His Music and Life (University of Michigan Press); the book has been translated into Japanese under the title Charlie Parker: The Man Who Created Modern Jazz. Woideck is also author and editor of The John Coltrane Companion and The Charlie Parker Companion (both Schirmer Books). He has also written compact disc liner notes extensively for the Verve, Blue Note, Mosaic, and Prestige labels.

Woideck's research awards include grants from the University of Oregon, National Endowment for the Humanities, and the Institute of Jazz Studies. He has presented research papers at national meetings of the American Musicological Society and the Society of American Music.

UNIVERSITY OF PITTSBURGH

Ljiljana Duraskovic

Best Teacher, Russian

UNIVERSITY OF REDLANDS

Sara Andon

Best Teacher, Music

Degrees: Artist Diploma, Yale School of Music; M.M., University of Southern California; B.A., California State University, San Bernardino Web: http://www.SaraAndon.com/

Academic Interests and Areas of Expertise Flute Performance and Music Education Concerto Soloist Recitalist Studio Recording Artist Chamber Music Symphonic Orchestra Player Opera, Ballet and Broadway Orchestral Player Jazz Ethnic Flutes Piano and Vocal Background

Kathy Hickey

Best Researcher/Scholar, Music
Most Helpful to Students, Music

Anthony Suter

Best Overall Faculty Member, Music

Anthony Suter's music has been heard in nearly every major U.S. city, as well as a growing list of international venues. Professor Suter has a deep interest in music for younger musicians, and is a vocal advocate for public school music programs, often serving as a guest conductor and clinician for K–12 music programs.

UNIVERSITY OF RHODE ISLAND

Mark Conley

Best Teacher, Music

MARK CONLEY, a conductor, composer and singer, has conducted numerous choral ensembles from large high school and middle school festivals to small professional opera groups. He made his Carnegie Hall debut in the spring of 2009 conducting a performance of Haydn's Te Deum for the Empress Marie Therese. At the University of Rhode Island, he supervises the choral program and directs the auditioned Concert Choir and the select chamber ensemble Lively Experiment.

Winner of the Sadie Rafferty Choral Competition, his choral and instrumental works have been performed throughout the United States. He has also served as music director for numerous theater companies, as well as performing as an opera and oratorio soloist. He has directed or performed with such companies as the Aspen Opera Theater Center, Intermountain Opera, Ocean State Lyric Opera, Rhode Island Philharmonic, Papagena Opera, Ocean State Chamber Orchestra and the Bozeman, Helena and Butte Symphony Orchestras.

An active supporter of young musical talent, Mr. Conley is a frequent clinician and adjudicator. He also serves as faculty founder and sponsor of the URI student chapter of the American Choral Directors Association. With the addition of the graduate degree in Conducting Performance, he now serves as mentor to a new generation of conductors, some of whom have already established active professional careers.

Mr. Conley's music has been published by Alliance Music and earthsongs.

UNIVERSITY OF RIO GRANDE

Christopher Pines

Best Researcher/Scholar, Liberal Arts, Social Sciences, and Humanities

UNIVERSITY OF SAN DIEGO

Maria Pascuzzi

Best Teacher, Theology

Susannah Stern

Best Researcher/Scholar, Communication

Susannah Stern PhD, joined the Department of Communication Studies in 2004, after teaching at Boston College for four years. Stern offers courses that investigate the role of media in contemporary life, particularly as they involve children, adolescents, and women, as well as courses on research methods. Stern's research focuses on electronic media and youth culture, and she has conducted extensive research on the Internet as a site for cultural consumption and self-expression.

Yi Sun

Best Teacher, History
Most Helpful to Students, History

Yi Sun, PhD, has been a member of the History Department at USD since fall 1997. She teaches a number of undergraduate courses on East Asian history and U.S.-East Asia Relations. Currently she also serves as the coordinator for the Asian Studies Minor program. Her research interests include Chinese women and modernization, Sino-American relations, and globalization. She has served on the executive board of several academic organizations, including the AsiaNetwork, Chinese Historians in the United States and the Association of Third World Studies, and presently is the associate editor of the Asian section for the Journal of Third World Studies.

UNIVERSITY OF SOUTH DAKOTA

Scott Mollman

Most Helpful to Students, Theater

Scott has been Technical Director and/or Production Manager at the USD Playhouse, Omaha Theater Company for Young People, the Utah Musical Theatre Company in Ogden, Utah, Hilberry Repertory Company in Detroit, MI and the Black Hills Playhouse in Custer, South Dakota

UNIVERSITY OF SOUTHERN CALIFORNIA

Emily Anderson

Best Overall Faculty Member, Liberal Arts, Social Sciences, and Humanities

I am Associate Professor in the English Department at USC, where I have been very happily employed since earning my Ph.D. in English from Yale University in 2004. My area of specialty is eighteenth-century British literature and culture, and within this rubric I work on a range of topics: the eighteenth-century novel, drama, and the connections between them; actors, actresses, playwrights, and the emergent celebrity culture of eighteenth-century theater; and the impact of theatrical performance on debates about personal identity, literary and cultural memory, and the fact/fiction divide. I teach courses for the Thematic Option Program and within the English department on topics including the rise of the novel, eighteenth-century drama, Jane Austen and eighteenth-century women writers, recognition scenes in literature, and Shakespearean forgeries and adaptations.

Gail Borden

Best Overall Faculty Member, Architecture
Best Researcher/Scholar, Architecture
Best Teacher, Architecture
Most Helpful to Students, Architecture
Best Researcher/Scholar, Fine Arts
Best Teacher, Fine Arts
Most Helpful to Students, Fine Arts
Best Overall Faculty Member, Fine Arts

Gail Peter Borden, FAIA, NCARB Associate Dean for Faculty Affairs Architecture Discipline Head Director of Graduate Architecture Programs Associate Professor of Architecture University of Southern California

Faculty Master Parkside Arts and Humanities Residential College

Principal Borden Partnership www.bordenpartnership.com

Gail Peter Borden attended Rice University, simultaneously receiving Bachelor of Arts degrees (all cum laude) in fine arts, art history, and architecture as well as his BARCH, also cum laude. He went on to Harvard University's GSD to complete a post-professional Masters of Architecture with distinction.

In addition to holding a tenured position as associate professor at the USC, Borden is the Associate Dean, Architecture Discipline Head and Director of the Graduate Architecture Programs. As principal of Borden Partnership, his design work has won numerous recognitions including the: Architectural League Prize; AIA Young Architect Award; BD+C "40 Under 40" award; and numerous AIA, ACSA, and RADA awards. Borden received artist-in-residence awards from the: Chinati Foundation; Atlantic Center for the Arts; Borchard Fellowship; and MacDowell Colony. His teaching has been nationally recognized with an ACSA/AIAS New Faculty Teaching Award. His books: Material Precedent: The Typology of Modern Tectonics, 2010 (Wiley); Matter: Material Processes in Architectural Production, 2011 (Routledge); Principia: Architectural Principles of Material Form, 2013 (Pearson); and Process: Material and Representation in Architecture 2014 (Routledge) all focus on materiality.

As an artist, theoretician, and practitioner, Professor Borden's research and practice focuses on the role of materiality and architecture in contemporary culture.

Stephanie Bower

Best Teacher, Liberal Arts, Social Sciences, and Humanities

Joseph Dane

Most Helpful to Students, Liberal Arts, Social Sciences, and Humanities

Roberto Diaz

Best Teacher, Spanish & Portuguese

Roberto Ignacio Díaz writes about Latin American literary and cultural history with a focus on transatlantic relations. He is the author of Unhomely Rooms: Foreign Tongues and Spanish American Literature (Bucknell UP, 2002), and is presently writing a book on Latin America and opera. He has co-directed the Buenos Aires and Madrid summer programs and a comparative literature summer course in Hong Kong and Macau.

Bill Feur

Best Researcher/Scholar, Liberal Arts, Social Sciences, and Humanities

Yehuda Gilad

Best Teacher, Music

Yehuda Gilad Teacher/Clarinetist/Conductor

A teacher, instrumentalist and conductor, Yehuda Gilad strives for "total musicianship," and as a result, he has won the acclaim of both critics and audiences alike. As one of the founding members of the Colburn School Conservatory of Music, Mr. Gilad serves as Music Director of the Colburn Orchestra and Professor of Clarinet. In addition, he is Professor of Clarinet at the University of Southern California's Thornton School of Music, and across these two prestigious institutions, he has developed one of the most sought-after clarinet studios in the world. His students have achieved top honors in every premier international clarinet competition, and graduates of his studio hold positions in over 100 major orchestras worldwide, including the Cincinnati and Minnesota Orchestras; the New York, Los Angeles, Stockholm, Sweden, Hong Kong, China, and Seoul Philharmonics; San Francisco Opera; the Concertgebauw; and the Orquesta Nacional de España, to name a few. In 2012, he served as president of the jury for the Carl Nielsen 5th International Clarinet Competition in Odense, Denmark. Mr. Gilad is a recipient of numerous honors and awards, including the Distinguished Teacher Award from the White House Commission on Presidential Scholars, the Ramo Award, and USC's highest honor for excellence in teaching, and is regularly invited to present master classes and performances at music festivals and nearly every major conservatory around the globe.

An accomplished clarinetist, Mr. Gilad has performed at top music festivals worldwide, including the Marlboro Music Festival, the Bowdoin International Music Festival, the Crusell Music Festival, the San Francisco Chamber Music Festival, and Santa Barbara's Music Academy of the West. He also founded the Yoav Chamber Ensemble, which performed extensively throughout the United States and recorded for the Orion label, and the Colburn Woodwind Chamber Players, which toured Germany and China. Mr. Gilad has collaborated as performer and conductor with many of today's most prominent artists.

Mr. Gilad's success as a conductor equals that of his renown as a teacher and performer. Since its inception in 2003, Yehuda Gilad has served as Music Director and Conductor of the

Colburn Orchestra and has been instrumental in the ensemble's development from a small string orchestra to a full symphony orchestra. Maestro Gilad's additional music directorships have included the Malibu Strawberry Creek Music Festival (recognized by the County of Los Angeles for its artistic contribution to the community); the 20th Century Unlimited concert series in Santa Fe, New Mexico; the Thornton Chamber Orchestra; the Santa Monica Symphony Orchestra; and the Colonial Symphony of New Jersey—The Star-Ledger: "Gilad transformed the Colonial Symphony into one of [New Jersey's] artistic trendsetters." Under his leadership, the Colonial Symphony became the only orchestra in the area to receive an NEA grant for seven consecutive years, and in 1995 and 1996, the orchestra received the Distinguished Arts Organization Award from the New Jersey State Council on the Arts for its artistry and programming merits. Mr. Gilad received acclaim from The New York Times for his "strong imaginative programming" and championing of American and contemporary music. Also an active guest conductor, his appearances have garnered critical acclaim in many countries on three continents. In 1987 he became the first Israeli-born conductor to perform in China and has since conducted numerous engagements in Beijing and Shanghai. Mr. Gilad's repertoire encompasses a broad range of musical styles, and he has enjoyed collaborating with contemporary composers such as John Corigliano and Iannis Xenakis, among others.

Dedicated to fostering increased public appreciation and support of classical music in the United States and abroad, Mr. Gilad has appeared on various broadcasts including NBC, CBS, WQXR, WMTR, KUSC, NJN, Cablevision, and the national television of mainland China. In 2012 an 2013, the Colburn Orchestra concerts were recorded in HD and broadcast on the arts series Open Call on Southern California's independent television station, KCET. Mr. Gilad also led the Colburn Orchestra in recordings for Live Classics, Bridge Records, and Yarlung Records, collaborating with composer Menachem Wiesenberg as well as cellist Ron Leonard, violist Paul Coletti, and mezzo-soprano Sasha Cooke.

Born and raised on a kibbutz in Israel, Gilad began his musical education at the age of 16. Following his conservatory studies, he emigrated to the United States in 1975 to further his education. His former teachers include Giora Feidman, Mitchell Lurie, and Dr. Herbert Zipper. Additionally, he was an active participant in numerous intensive master classes with the late Sergiu Celibidache and Leonard Bernstein.

Sarah Gualtieri

Most Helpful to Students, History

Biographical Sketch Sarah Gualtieri is Associate Professor in the Departments of History and American Studies and Ethnicity and former Director of the Middle East Studies Program. Trained in Middle East Studies at McGill University and at the University of Chicago, her research focuses on questions of race, gender, and migration. Specifically, she explores the movement of peoples and ideas from greater Syria to the Americas and has written on the theoretical, social and empirical implications of this movement. Conducting research in Arabic, French, Spanish and English, she has travelled extensively in the Middle East and lived for two years in Damascus, Syria. Her research bridges several emerging fields, notably Middle Eastern Migration Studies, Arab American Studies, and Transnational American Studies. Gualtieri's publications include a number of articles as well as her book Between Arab and White: Race and Ethnicity in the

Early Syrian American Diaspora (University of California Press, 2009). The book examines the history of Arab racial formation in the United States with a particular focus on the problematic of "whiteness." It traces how Arabs came to be officially classified as white by the U.S. government, and how different Arab groups interpreted, accepted, or contested this racial classification over the course of the 20th century. Gualtieri is now working on a project entitled "The Lebanese in Los Angeles: Migration and Transnationalism in a Multi-racial Landscape."

At USC, Prof. Gualtieri teaches undergraduate courses on Women and Revolution in the Middle East, the Modern Middle East, Race and Class in Los Angeles, Arabs in America, and a graduate seminar in Critical Studies of Whiteness. She is currently supervising several Ph.D. theses on topics ranging from U.S. imperial interests in the Middle East, to Afro-Arab encounters, and Palestinian cinema; and is a member of several other dissertation committees. She is a recipient of two undergraduate teaching awards.

Prof. Gualtieri's fellowships include the American Council of Learned Societies (ACLS); Social Science Research Council (SSRC); and USC's Advancing Scholarship in the Humanities and Social Sciences Fellowship; and the Zumberge Research and Innovation Individual Faculty Grant.

Prof. Gualtieri is also Regional Editor for the Americas of the Encyclopedia of Women and Islamic Cultures.

Charles Paus

Best Teacher, Languages

Eric Rawson

Best Researcher/Scholar, Writing

"Organized, smart and well-prepared for every class, I recommend his class to all my 120 students."

Ramzi Rouighi

Best Researcher/Scholar, History

Terry Seip

Best Teacher, History

Carole Shammas

Best Overall Faculty Member, History

Bryan Simms

Best Researcher/Scholar, Music

Bryan R. Simms is Professor Emeritus of Musicology at the University of Southern California. He was appointed to its faculty in 1976 and served there for thirty-eight years prior to his retirement in June 2014. Before coming to USC, he taught at the University of Denver and at Yale University, where in 1971 he received a Ph.D. in musicology. He was formerly editor of the Journal of Music Theory and Music Theory Spectrum, and he has served on the Council of the American Musicological Society and Executive Board of the Society for Music Theory. He specializes in music and musical theory of the 20th century, and his books include most recently The Atonal Music of Arnold Schoenberg: 1908–1923 (Oxford University Press), Music in Western Civilization (with Craig Wright, Thomson-Schirmer Books), and Alban Berg: A Research and Information Guide, second edition (Routledge), and Pro Mundo – Pro Domo: The Writings of Alban Berg (Oxford University Press, 2014), which is an English edition of the complete writings of this composer.

David Wilkinson

Most Helpful to Students, Music

David Wilkinson is recognized as one of Southern California's leading vocal accompanists and coaches. His performances as a collaborative pianist have taken place in major concert halls and universities throughout the U.S. and Europe, including The White House, the Kennedy Center, Carnegie Hall, Wigmore Hall, the Dorothy Chandler Pavilion, Yale University, the Verdi Conservatory and Oberlin Conservatory. Wilkinson has collaborated with such distinguished artists as Frederica Von Stade, Jane Eaglen, Markella Hatziano, Julia Migenes, Adrian Gans, Gloria Parker, and Laurie Rubin. While living in Italy, he was an accompanist for the studios and masterclasses of operatic legends Tito Gobbi, Giuseppe Di Stefano, and Magda Olivero.

Beverly Younger

Most Helpful to Students, Social Work

UNIVERSITY OF SOUTHERN INDIANA

Leigh Anne Howard

Best Researcher/Scholar, Communication

Robert Jeffers

Most Helpful to Students, Communication

Renee Rowland

Most Helpful to Students, Liberal Arts, Social Sciences, and Humanities

UNIVERSITY OF ST. THOMAS

Carol Bruess

Best Teacher, Liberal Arts, Social Sciences, and Humanities

Corrine Carvalho

Best Overall Faculty Member, Theology
Best Researcher/Scholar, Theology

Massimo Faggioli

Best Researcher/Scholar, Theology

Shirley Jordon

Best Teacher, Theology

Meg Karraker

Best Researcher/Scholar, Sociology
Most Helpful to Students, Liberal Arts, Social Sciences, and Humanities

Debra Petersen

Best Overall Faculty Member, Liberal Arts, Social Sciences, and Humanities

Kevin Sauter

Best Researcher/Scholar, Liberal Arts, Social Sciences, and Humanities
Most Helpful to Students, Liberal Arts, Social Sciences, and Humanities

Susan Smith-Cunnien

Best Overall Faculty Member, Sociology
Best Teacher, Sociology
Most Helpful to Students, Sociology

Kelly Wilson

Most Helpful to Students, Theology

UNIVERSITY OF TENNESSEE - CHATTANOOGA

Rowan Johnson

Best Teacher, Liberal Arts, Social Sciences, and Humanities

Originally from South Africa, Rowan Johnson holds an Ed.D from the University of Tennessee. He also earned his MA in Applied Linguistics and English Language Teaching from the University of Nottingham, England. He has produced various forms of writing, including poetry and numerous magazine articles.

UNIVERSITY OF TENNESSEE - KNOXVILLE

Renee D'Elia-Zunino

Most Helpful to Students, Languages

Sal Dimaria

Best Overall Faculty Member, Languages

UNIVERSITY OF TEXAS

Sheldon Ekland-Olson

Best Teacher, Liberal Arts, Social Sciences, and Humanities

Julie Hardwick

Best Researcher/Scholar, History

Jacqueline Jones

Best Overall Faculty Member, History

Alberto Martinez

Best Teacher, History

Alberto Martinez is originally from San Juan, Puerto Rico. Most recently, he is the author of The Cult of Pythagoras (University of Pittsburgh Press, 2012), on the evolution of myths in the history of mathematics. He is also the author of Science Secrets: The Truth About Darwin's Finches, Einstein's Wife, and Other Myths (University of Pittsburgh Press, 2011). And previously, he published Kinematics: The Lost Origins of Einstein's Relativity (Johns Hopkins University Press, 2009), and Negative Math (Princeton University Press, 2005). He has also published articles in several journals and periodicals including The American Journal of Physics, Archive for History of Exact Sciences, Physics in Perspective, The American Mathematical Monthly, School Science Review, Physics World, and Studies in History and Philosophy of Modern Physics.

Lisa Moore

Best Overall Faculty Member, Liberal Arts, Social Sciences, and Humanities

Lisa Moore is Professor of English and Interim Director of the Center for Women's and Gender Studies. She is the author of Sister Arts: The Erotics of Lesbian Landscapes (Minnesota, 2011), which won the Lambda Literary Award and was a finalist for the Publishers' Triangle Award. She also wrote Dangerous Intimacies: Toward a Sapphic History of the British Novel (Duke, 1997), and is the is the editor, with Omi Osun Jones and Sharon Bridgforth, of Experiments in a Jazz Aesthetic: Art, Activism, Academia, and the Austin Project (Texas, 2010) and, with Joanna Brooks and Caroline Wigginton, of Transatlantic Feminisms in the Age of Revolution (Oxford, 2011). She has published articles and reviews in journals including GLQ, Eighteenth-Century Studies, Cultural Critique, Textual Practice, Signs, Albion, and Modern Philology. Her poems have appeared in Broadsided, Sinister Wisdom, Lavender Review and other journals, and she won the Museum of Fine Arts-Houston Art/Lines Juried Poetry Competition in 2012. In 2002 she was a Visiting Professor at Université de Paris X (Nanterre).

Leslie Mutchler

Best Overall Faculty Member, Fine Arts
Best Teacher, Fine Arts
Most Helpful to Students, Fine Arts

Leslie Mutchler is the Director of the Foundations Program for the Department of Art and Art History at the University of Texas at Austin. She received her BFA from Kent State University and MFA from the Tyler School of Art. Her multi-disciplinary installations are experiments in faux-functionality and imperfect design. Utilizing the material language of our everyday, Mutchler engages with issues of hand(craft), labor, sustainability, and pedagogy.

Mutchler's most recent work is WorkLAB Satellites, a commission by the Blanton Museum of Art, Austin, TX (2014). WorkLAB Satellites is an artwork, an environment, and an activity. It is an ongoing installation intended to augment the Blanton Museum of Art's educational programming objectives of participation, reflection, and intentionality.

Recent solo shows include TrendFACTORY, Kent State University Gallery, OH, Elon University Gallery, NC, Indianapolis Art Center, IN, ARPRIM Gallery, Montreal, Canada, and Lawndale Art Center, Houston, TX (2014–2012), TumblrEAL, Gensler Global Architecture, Austin, TX (2012), ASTRAL pop-up, VA Gallery, University of Wyoming, WY (2012), Geometry of Labor, Poliglota Gallery, Buenos Aires, Argentina (2011), and (F)utility Kitchen, Snowflake, St. Louis, MO (2011).

Mutchler is a 2014 recipient of The Regents' Outstanding Teaching Award.

Penne Restad

Most Helpful to Students, History

Biography: Areas: American social and cultural history, with particular interest in the history of American holidays, public ritual, consumer culture, and national identity. Research interests: Dr. Restad's scholarship explores the formation of American cultural identities and behaviors, the history of consumer culture, and the scholarship of teaching and learning in higher education. Dr. Restad regularly teaches U.S. history surveys, courses in Liberal Arts Honors, and seminars related to her research interests.

UNIVERSITY OF TEXAS - PAN AMERICAN

Tamer Balci

Best Teacher, Liberal Arts, Social Sciences, and Humanities

Brent Campney

Best Researcher/Scholar, Liberal Arts, Social Sciences, and Humanities

Colin Charlton

Best Overall Faculty Member, Liberal Arts, Social Sciences, and Humanities

Debbie Cole

Best Overall Faculty Member, English

"Dr Cole's specialty is in linguistics, but she actually collaborates with researchers in various fields and also publishes with faculty specialized in literature, etc."

Greg Gillson

Most Helpful to Students, Philosophy

Robert Johnson

Best Overall Faculty Member, English

Paul Saka

Best Researcher/Scholar, Philosophy

Alexander Stehn

Best Overall Faculty Member, Philosophy

Dr. Alexander Stehn as an assistant professor of philosophy at The University of Texas-Pan American, Stehn has been making an impact in students' lives since 2010. This year, his teaching skills have earned him The University of Texas System's 2014 Regents' Outstanding Teaching Award, considered one of the largest and most prestigious in the nation for rewarding faculty performance. Stehn was one of seven faculty members from UTPA to receive the award.

Originally from Aransas Pass, Texas, Stehn arrived at UTPA after earning a Ph.D. from Penn State University, where he was one of only 10 graduate students endowed with a university-wide teaching award in 2008. Stehn specializes in American and Latin American philosophy and is also highly qualified to teach a wide range of required and upper division philosophy courses at UTPA.

Cory Wimberly

Best Teacher, Philosophy

My research is focused on social political philosophy and post-structuralism. More specifically, I am currently thinking through the notion of democracy in post-foundational terms. This line of thought is indebted to liberalism, Marxism, existentialism and current thinkers like Foucault, Deleuze and Guattari, Wendy Brown, and William Connolly. In the broadest terms, the aim is to explore opportunities for the reformulation of democracy in post-foundational terms and its application to other areas of political life outside of its current use in politics today. I am also seeking to create a forum for the discussion of these issues and the development of a philosophical community devoted to political issues as a member of the executive board of the Society for Social and Political Philosophy.

UNIVERSITY OF TEXAS AT ARLINGTON

Tatiana Baeva

Best Overall Faculty Member, Languages

Blake Carpenter

Best Teacher, Languages

"Great guy."

My name is Blake Carpenter and I am originally from North Carolina. This fall, I begin my twenty-ninth year as a teacher. I have been at UT Arlington for 17 years and teach classes in French for the Department of Modern Languages.

Before moving to the Metroplex, I taught French for 8 years in the North Carolina Public School system at the high school level. I have also taught at the University of North Carolina at Chapel Hill and at the Université de Nice in France. My junior year was spent studying at the Université de Montpellier in southern France. I earned a Master of Divinity Degree with Biblical Languages from Southwestern Baptist Theological Seminary in May, 2001.

Currently, I am employed as the Manager of Partnerships for the Division of Digital Teaching and Learning. I oversee the new eTutroing initiative here at UT Arlington as well as coordinate an online bachelor's completion program known as the Finish@UT Program.

My interests include French grammar, pedagogy, online education, and working with students to help them achieve their goals in obtaining a degree from UT Arlington. It is indeed a privilege and an honor for me to work with such talented people both in and out of the classroom. I am looking forward to another wonderful semester!!!

Klaus Driessen

Best Overall Faculty Member, Languages

"He is devoted to teaching and makes an extra effort by working with technology to assist students in their learning. I think he truly enjoys teaching German. My vote goes to Klaus Driessen."

A. Raymond Elliott

Best Researcher/Scholar, Languages

Elliott, A. Raymond. Associate Professor of Spanish Linguistics and Former Chair, Department of Modern Languages Professor Elliott served as the Chair of the Department of Modern Languages for a ten year period from 2001–2011. Elliott's areas of specialization are Spanish linguistics, Romance linguistics, Second Language Acquisition, the acquisition of second language phonology, and the documentation and conservation of indigenous languages in Mexico. Elliott has published articles, book chapters, reviews in The Modern Language Journal, Hispania, Language Documentation & Conservation and with Georgetown University Press. He served as a panelist in the McGraw-Hill Annual Teleconference on Authentic Materials, and as a member of the Academic Advisory Board for the package to accompany Nuevos Destinos. He is the author of several grammar texts: Nuevos Destinos: Español para hispanohablantes, including the workbook and accompanying lab and instructor's manualand co-author of Puntos de partida: An Introduction to Spanish (6th, 7th, 8th and 9th editions), Puntos en Breve and ¡Qué tal! Professor Elliott's research interests have recently expanded to include the documentation of endangered and minority indigenous languages of Mexico. He has experience in documenting languages through recent fieldwork on Kak'chiquel, an indigenous language of Guatemala and Chicahuaxtla Triqui, an Otomanguean tonal language spoken in Oaxaca, Mexico. Elliott directs UT-Arlington's Study Abroad Program in Cuernavaca, México. He is currently working on a text titled, "Gramática popular del Triqui de Chicahuaxtla" and articles focusing on tone-laryngeal morphology and illustrations of the IPA for Chicahuaxtla Triqui.

Timothy Foxsmith

Most Helpful to Students, Languages

"I think Tim is a dedicated teacher to his students and to his institution."

Lonny Harrison

Most Helpful to Students, Languages

Graham Hunt

Best Researcher/Scholar, Music

Graham Hunt is Associate Professor of Musicology and Music Theory at the University of Texas at Arlington. He received his Ph.D. in Musicology from Duke University. He has presented his research at numerous venues, including the national meeting of the Society of Music Theory, Texas Society of Music Theory, and the New England Conference of Music Theorists, and is currently the President of the Texas Society Music Theory, having been elected to this office in 2011. He has also given the keynote speech for the Oklahoma City University Theory Conference and has been a guest speaker at the University of Massachussetts' Music Theory Lecture Series. He has published on subjects such as Wagner, Neo-Riemannian analysis, and the Three-Key Exposition in Sonata forms, in journals such as Journal for Schenkerian Studies, Integral, Theory and Practice, 19th-Century Music Review, and Music Theory Spectrum. He was also selected to be the Grace and Joseph Valentine Visiting Professor at Amherst College (Massachussetts) in 2009.

UNIVERSITY OF TEXAS AT EL PASO

Nikki A. Agee

Best Researcher/Scholar, English
Most Helpful to Students, Liberal Arts, Social Sciences, and Humanities

My research interests include composing and teaching with new media technologies. My dissertation, Animating Composition, examines the ways writing instructors can integrate 3D computer-generated imaging and animation in Technical Writing courses to foster rhetorical, scientific, visual, and multimodal literacies. The final chapter of my dissertation investigates student attitudes toward 3D CGI and the general and specific learning capacities they developed by composing with it.

Before pursuing an M.A. in Rhetoric from UTEP, I earned a B.A. in English from UT Austin and worked as an Army Reserve light-wheel vehicle mechanic; an Army Reserve public affairs specialist/photojournalist; a middle school and high-school English teacher; a proofreader; and an editor.

Yolanda Ainsworth

Best Teacher, English
Best Overall Faculty Member, Liberal Arts, Social Sciences, and Humanities

Lucia Dura

Best Overall Faculty Member, English
Most Helpful to Students, English
Best Researcher/Scholar, Liberal Arts, Social Sciences, and Humanities

Lynne Gilden

Best Teacher, Liberal Arts, Social Sciences, and Humanities

Kerri Harrison

Most Helpful to Students, Communication

Born and raised in northern California, Kerri Harrison has been an instructor at UTEP for 28 years. She has taught the basic speech classes for many years and has taught Small Group Communication and Interpersonal Communication during her tenure at UTEP. She has recently taught Women in Film and developed a film class on the subject of censorship. Kerri is also the Coordinator of the core course for the Communication Department and is a faculty advisor. She organizes and directs the high school debate camp and hosts a high school speech tournament every year. While teaching in the Communication Department, she was, for 20 years, the costume designer for the UTEP Union Dinner Theatre.

Richard Pineda

Best Researcher/Scholar, Communication

Dr. Richard D. Pineda is an associate professor in the Department of Communication and currently serves as director of the Sam Donaldson Center for Communication Studies. Dr. Pineda's research and teaching interests focus on politics, leadership, media, popular culture and Latinidad. Dr. Pineda holds a BA in Political Science from Baylor University, an MA in Communication from The University of El Paso at Texas, and a PhD from Wayne State University in Detroit, Michigan. His research has appeared in a number of communication journals including Text and Performance Quarterly, Argumentation and Advocacy, and The Free Speech Yearbook. Dr. Pineda serves on the editorial boards for the journals Argumentation and Advocacy and Contemporary Argumentation and Debate, the two preeminent argumentation journals in the country. Dr. Pineda has served the El Paso community as a member of the City of El Paso Ethics Review Board and the Capital Improvement Advisory Committee. He is also a member of the Director's Circle for the Stanlee and Gerald Rubin Center for the Visual Arts at UTEP. Dr. Pineda grew up in El Paso, Texas and is a proud graduate of Eastwood High School.

Thomas Ruggiero

Best Teacher, Communication

Dr. Thomas E. Ruggiero is currently an Associate Professor and the designer and coordinator of the Multimedia Journalism Program at UTEP. He recently served as a Senior Lecturer in Journalism for one year in the Department of Culture, Media & Drama at the University of the West of England in Bristol, UK., where he established undergraduate and graduate courses in multimedia journalism. As a journalism professor for 11 years, he teaches both practical and theoretical journalism courses including Public Affairs Reporting, Feature Writing, Journalism Ethics and Multimedia Journalism. He earned his Ph.D. from Bowling Green State University, his M.S. from Shippensburg University, and his B.A. from the University of Oregon School of Journalism. He has professional experience as a daily newspaper reporter with The San Jose Mercury News, The Herald Mail, Hagerstown, Md., The Record Herald, Waynesboro, Pa., The Press-Enterprise, Bloomsburg, Pa., and The Medford Mail Tribune, Medford, Or. As a researcher, his interests include: New Media Communication, Media Ethics, and Intercultural Media Effects. He has refereed articles published in many communication journals and books, including Communication Quarterly, Journal of Computer-Mediated Communication, Mass Communication & Society, and Journal of Mass Media Ethics; and regularly presents research papers at national and regional communication conferences. He co-created the borderzine.com e-magazine, serves as co-director of the Dow Jones Summer Minority High School Journalism Workshop, and is the official monitor for the Dow Jones Newspaper Fund Editing Intern Program.

Stacey Sowards

Best Overall Faculty Member, Communication

Dr. Sowards' research interests include the intersections of: rhetorical theory and criticism, feminist theories of communication, environmental communication and rhetoric, and intercultural communication and rhetoric. She is especially interested in these areas of inquiries in Asia and Latin America. She has completed research projects in and about Indonesia, Mexico, and Costa Rica. Her publications appear in Communication Theory, Philosophy and Rhetoric, Communication Studies, The Howard Journal of Communications, Hypatia, Argumentation and Advocacy, Ethics and the Environment, The Globalization of Corporate Media Hegemony, and other outlets. She also received a William J. Fulbright research grant for her dissertation research in 2000–2001 and a Fulbright-Hays grant in 2005. While at UTEP, she has led five study abroad programs/research trips in Mexico, Costa Rica, Indonesia, and Germany. Dr. Sowards also works with UTEP's new master's program with the non-governmental organization, "Rare." The "Rare" Program is based in Washington, D.C., and works to protect endangered species and their habitats in more than 40 countries. Dr. Sowards works with the Rare-UTEP training program for conservation campaigns at the

Bogor Agricultural Institute in Indonesia with more than MA students/campaign managers who earn their M.A. degrees from the Department of Communication.

UNIVERSITY OF TEXAS AT SAN ANTONIO

Joshua Beach

Best Researcher/Scholar, Writing Best Teacher, Writing

Timothy Braun

Most Helpful to Students, Writing

Timothy Braun is a writer living in Austin, TX. His essays have been published by The New York Times, Huffingtonpost, Culturebot, The Good Men Project, the Austin Chronicle, and The Weeklings, among others. He has been an artists-in-residence at the MacDowell Colony, Djerassi, Santa Fe Art Institute, Edward F. Albee Foundation, Ucross, Blue Mountain Center for the Arts and Humanities, Anderson Center for the Interdisciplinary Studies, Prairie Center For The Arts, Madroño Ranch: A Center for Writing, Art, and the Environment, the Osage Arts Community, HERE Arts Center, and was a Dorothy Norton Clay fellowship at the Mary Anderson Center for the Arts. The Ludwig Vogelstein Foundation, Hatch, John Anson Kittredge Fund, and Puffin Foundation have funded his projects. His plays and operas have been performed in theatres, site-specific spaces, and festivals across the United States and Europe. Playscripts publishes some of his work. He is the co-author of The Hand Book of Creative Writing (Edinburgh University Press) and The Half Life of 64 (Blurb). can be Education includes a BS from Ball State University in History and Theater, an MA from the University of New Mexico in Theater and Dance, an Exile fellowship at the University of Ulster in conjunction with the University of California San Diego and Yale University, and an MFA from Columbia University. He teaches at St. Edward's University, the University of Texas at San Antonio, is on the board of directors of Salvage Vanguard Theatre, and is the Editor-In-Chief of New and Social Media for Fusebox.

Renee Cowan

Best Overall Faculty Member, Communication

Dr. Renee L. Cowan received her B.S. in Communication Studies from the University of Texas at Austin and her M.A. from Texas State University — San Marcos. She received her Ph.D. from Texas A&M University with an emphasis in Organizational Communication. Dr. Cowan employs qualitative, rhetorical, and quantitative methods to investigate contemporary organizational issues including; work/life issues, workplace bullying, and issues associated with the use of communication technologies in organizations (electronic mail and weblogs). Her recent research focuses on workplace bullying and the human resource professional and work/life issues of blue collar employees. She has presented her findings at numerous international, national, and regional conferences. Her research appears in such peer-reviewed journals as Journal of Computer-Mediated Communication, Communication Education, Communication Studies, Women & Language, Communication Quarterly, Qualitative Research Reports in Communication, Communication Research Reports, and Human Communication. She has developed and published numerous instructional communication activities for the edited book, Teaching Ideas for the Basic Communication Course. She has also co-authored a chapter for an edited book, Contemplating Maternity in an Era of Choice: Explorations into Discourses of Reproduction.

Dale Davis

Best Teacher, Communication

Dale is originally from DeFuniak Spring, Florida. He enlisted in the US Air Force in 1958. He received his B.A. degree in 1972 in Police Science and Administration from California State University at Sacramento. After receiving his degree he was commissioned a Second Lieutenant. Dale received his M.A. in Communication with emphasis on Human Relations in 1979 from The University of Northern Colorado. In 1982 he retired from the Air Force. He worked for a San Antonio homebuilder until Aug., 1985 when he started teaching at UTSA. Dale is married to Bonnie Hollada Davis. They have two grown children. Their son, Leland, is a fireman with the Kern County Fire Department in Bakersfield, CA. Leland's wife, Pam, is an Associate Professor of American Sign Language at Bakersfield College. Their daughter, Cami, is a graduate of UTSA. She received a Communication Degree with a concentration in Public Relations and Speech Communication. Her husband, Robert, and his brother, own a business in San Antonio. Dale and Bonnie enjoy being grandparents, to grandkids in Texas and California, and watching 80,000 pine trees grown on their tree farm in Florida.

Curtis Fish

Best Overall Faculty Member, Architecture

Curtis Fish is an associate AIA member and a LEED Accredited Professional, who is focused on developing innovative project delivery methods and the development and application of sustainable design solutions to new projects in the region.

In 2005 he received a Bachelor of Science in Architecture and a Bachelor of Arts in Anthropology from UTSA's School of Architecture and COLFA. He had the honor to produce work during this period that received accolades, publications, and an international design award. He earned a Master of Architecture degree from UTSA's College of Architecture in 2007.

At Kell Munoz Architects, Curtis contributed to a number of projects including; the renovation of a university's student center in New Orleans which sought to revitalize its campus following the regional devastation from Hurricanes Katrina and Rita, development of a schematic design proposal for a new Smithsonian museum in Washington D.C., and numerous public and educational buildings in the region. He also had the honor to participate in multiple installation projects at the Museo Alameda del Smithsonian including the design of an installation for its grand opening.

Curtis has been affiliated with the College of Architecture since 2005 and is currently an adjunct faculty member teaching in the foundation studios. Beginning in 2008 he has also partnered with several local contractors on numerous residential and small commercial projects to promote innovative project delivery methods such as design-build and to apply sustainable and cost effective design solutions.

Christopher Hajek

Best Researcher/Scholar, Communication

Dr. Christopher Hajek completed his B.A. in Communication at Loyola University of Chicago, his M.A. in Communication at the University of Hawaii, and his Ph.D. in Communication at the University of California, Santa Barbara. His research is grounded in intergroup communication, with foci at the intersections of health, aging, and social stigma. His work has also explored relationships between group identity and police/community interaction outcomes. His teaching interests in intergroup and intercultural communication developed during his work as an intern for CNN and The Associated Press in Rome, and his Peace Corps experience in Rwanda preceding that country's genocide. His teaching has also been influenced by his experience as a trained mediator in community-based alternative dispute resolution. Dr. Hajek has published peer-reviewed articles in several journals, including the Journal of Communication, the Journal of Intercultural Communication Research, Language and Communication, Communication Research Reports, Psychological Reports, Journal of Russian Communication, and the Journal of Language and Social Psychology. Dr. Hajek has also co-authored several invited book chapters. Prior to joining the faculty at UTSA, Dr. Hajek taught courses at The University of

California, Los Angeles; the University of California, Santa Barbara; The University of Hawaii; and the Keller Graduate School of Management in Chicago.

Laura Oliver

Most Helpful to Students, Communication

Lindsay Ratcliffe

Best Teacher, Writing

Lindsay G. Ratcliffe earned her B.A. in English from Grove City College (PA) in 1999. She worked in Pittsburgh, PA for four years and moved to Texas with her husband in 2003.

In 2005, Lindsay earned her M.A. in English from UTSA, where she had been teaching Freshman Composition as a TA since 2004. She joined the Writing Program faculty in Fall 2006.

When she is not reading her students' essays, Lindsay is enjoying the outdoors. She is an avid hiker and certified Texas Master Naturalist who volunteers regularly in San Antonio's Natural Areas.

UNIVERSITY OF THE ARTS

Mark Campbell

Best Researcher/Scholar, Fine Arts

Larry Donahue

Most Helpful to Students, Ceramics

Laura Frazure

Best Teacher, Fine Arts

Michael Grothusen

Best Overall Faculty Member, Fine Arts

Michael Grothusen is an associate professor in the College of Art, Media and Design at the University of the Arts. He is a mixed-media sculptor and installation artist who has created site-specific works for numerous venues including the Philadelphia International Airport, Michener Art Museum, Abington Art Center, Crane Arts Building, Eastern State Penitentiary, Drexel University, Philadelphia Museum of Art, Carnegie Mellon University, Gallery Joe and the Philadelphia Art Alliance.

Michael is the recipient of numerous grants and awards, including three Individual Artists Fellowships from the Pennsylvania State Council on the Arts and the Pollock-Krasner Foundation. In 1998, he was awarded a Pew Fellowship for sculpture. He received a BFA from the University of Kansas in 1988 and an MFA from Tyler School of Art in 1991.

His current studio projects include a commission for a permanent public sculpture from the Redevelopment Authority of the City of Philadelphia and a kinetic piece that incorporates a solid bronze sphere that has been fabricated to correspond with the exact weight of his body.

John Woodin

Best Teacher, Photography
Most Helpful to Students, Photography
Most Helpful to Students, Fine Arts

John Woodin, adjunct associate professor of Photography at the University of the Arts, is a Philadelphia-based artist whose landscape and still-life photographs explore the complex relationships between nature and culture, chaos and order.

Woodin received a BA in Communications from the University of New Orleans and an MFA in Photography from Temple University's Tyler School of Art. His fine art photographs have been exhibited internationally and are held in private and public collections. A monograph of his work titled City of Memory, New Orleans Before and After Katrina was published in January 2010 by the Center for American Places at Columbia College Chicago.

UNIVERSITY OF THE DISTRICT OF COLUMBIA

Laurence Covington

Best Teacher, English

Alexander Howe

Best Overall Faculty Member, English
Best Researcher/Scholar, English

Amanda Huron

Best Researcher/Scholar, Liberal Arts, Social Sciences, and Humanities

Matthew Petti

Most Helpful to Students, English

UNIVERSITY OF THE INCARNATE WORD

Amalia Mondriguez

Best Overall Faculty Member, Languages
Best Teacher, Languages
Most Helpful to Students, Languages
Best Researcher/Scholar, Liberal Arts, Social Sciences, and Humanities
Best Teacher, Liberal Arts, Social Sciences, and Humanities
Most Helpful to Students, Liberal Arts, Social Sciences, and Humanities
Best Overall Faculty Member, Liberal Arts, Social Sciences, and Humanities

"Dr. Mondriguez has been the best resource for job opportunities. She always keeps former students aware of the job market and all the options available to them. She is a wonderful professor!"

Michael Tallon

Best Researcher/Scholar, Languages

Specialist in language teaching and second language acquisition. Research interests include foreign language anxiety, heritage language students, and the Spanish of San Antonio.

UNIVERSITY OF TOLEDO

David Conn

Best Overall Faculty Member, History
Best Researcher/Scholar, History
Best Teacher, History

UNIVERSITY OF TULSA

Karl Pollin

Most Helpful to Students, Languages

"For the past 8 years, no one has had more impact on the teaching mission of the French program than Professor Pollin. His leadership has assumed numerous forms: 1) he has shown great curricular innovations in teaching literature and philosophy, 2) he has been an inspiring professor to students, and 3) he has invested great energy in advising and mentoring students.

This semester, I have had the personal pleasure to co-teach a course on Exile with Dr. Pollin. Because of the emphasis he puts on the reader's personal relationship with texts, Dr. Pollin knows how to give recognition to his students' personal voice. Most importantly, Dr. Pollin's careful scaffolding of open questions leads students to explore avenues consistent with principles of critical thinking: he allows them to start with the most probable interpretation of a text, then slowly leads them through a series of investigative steps shaped by questions tied to the student's contributions to novel observations and perspectives. As the students come up with increasingly erudite observations, the relevant elements of a text are progressively highlighted, elucidated and agreed upon, thus easing students, quite comfortably, into novel assessments and interpretations of French literary texts. Most importantly, Dr. Pollin's approach, very comparable to that of solving a mystery, guides students into a clearer sense of what it means to open one's mind to multi-dimensional perspectives. His pedagogy in teaching French literature is undeniably a demonstration of excellence in reasoning through meticulous standards such as text questioning, observations, precisions, relevance, associations, insights, analyses, dialogues, depth and breadth. Dr. Pollin has a remarkable talent in creating intellectual conditions for students to both search for meaning and experience meaning, the kind of teaching that affects students deeply, a third dimension in pedagogy that goes beyond the mere acquisition of knowledge and analytical skills as both end up converging into a powerful epistemological tool. Also worthy of consideration is Dr. Pollin's devotion to all his students, which is evidenced by the remarkable amount of time he spends with each of them as he provides ongoing constructive guidelines to prepare them as they approach writing their first draft papers, and to guide them in the task of re-writing."

UNIVERSITY OF UTAH

Lindsey Adams

Best Researcher/Scholar, History

Katharina Gerstenberger

Best Overall Faculty Member, Languages

"She has done a great job under a difficult situation. I think she deserves to be acknowledged for that."

Katharina Gerstenberger is professor of German and chair of Languages and Literature at the University of Utah. She is the author of Truth to Tell: German Women's Autobiographies and Turn-of-the-Century Culture (2000) and Writing the New Berlin: The German Capital in Post-Wall Literature (2008). She co-edited German Literature in a New Century: Trends, Traditions, Transformations, Transitions (2008); and After the Berlin Wall: Germany and Beyond (2011). Her articles on topics of 20th and 21st German literary culture have appeared in Ecozona, Gegenwartsliteratur, Monatshefte, Women in German Yearbook, German Politics and Society,

German Quarterly and in several anthologies, including German Literature in the Age of Globalization (2004), Spatial Turns: Space, Place, and Mobility in German Literary and Visual Culture (2010), and Generational Shifts in Contemporary German Culture (2010). From 2007–2010 she was co-editor of Women in German Yearbook. Her current project is titled "Disaster Narratives: History and Catastrophe in the German Imagination." She is a member of the Transatlantic Research Network in Environmental Humanities.

Robert Goldberg

Best Overall Faculty Member, History
Best Overall Faculty Member, Liberal Arts, Social Sciences, and Humanities

Bob Goldberg is Professor of History, Director of the Tanner Humanities Center, at the University of Utah. He is the author of eight books with his last two, Barry Goldwater and Enemies Within: The Culture of Conspiracy in Modern America published by Yale University Press. He has won twelve teaching honors including the Graduate Student and Postdoctoral Scholar Distinguished Mentor Award, Presidential Teaching Scholar Award, and University of Utah Distinguished Teaching Award. In 2003, he held the Fulbright Distinguished Chair in American Studies, at the Swedish Institute for North American Studies, Uppsala University. He was awarded the Rosenblatt Prize for Excellence in 2008.

Lauren Liang

Best Teacher, Literature

"Dr. Liang is an excellent teacher and leader in the field of international literature."

Mark Matheson

Best Teacher, Liberal Arts, Social Sciences, and Humanities

Mark has been inspired by University of Utah students throughout his twenty years of teaching in the English Department and the Honors College. A native Utahn, he received his M.A. from the U and his Doctorate from the University of Oxford. Mark is avid about traveling with U students to London, where he teaches seminars on British Literature. Mark also serves as the Director of the University of Utah-based Tanner Lectures on Human Values. He continues to teach and greatly enjoys advocating for students' educational interests through The MUSE Project.

Maureen Mathison

Best Overall Faculty Member, Liberal Arts, Social Sciences, and Humanities

"Maureen has been a tireless advocate of education, not only at the University, but in K–12, too. She serves on the Gov.'s education task force (common core) and volunteers at local schools (elementary and HS). She has also shepherded writing at the University through one of its most remarkable periods of growth–she sits on close to a dozen different Ph.D/MA committees, she is a wonderfully supportive colleague–and she continues to be a terrific leader for the fledgling department."

Maureen O'Hara Ure

Best Teacher, Art & Art History

Assistant Professor (Lecturer) Maureen O'Hara Ure teaches studio courses in Drawing, Senior Seminar, and an introductory lecture course on contemporary trends. As faculty affiliated with the Honors College, she also conducts a seminar on Creative Process. She has had many one-person exhibitions, has participated in Sundance Institute Playwrights Lab, and has received several significant grants in support of her work. In addition to shows in Los Angeles (SPACE), Chicago (ARC/Raw Space), and Sacramento (Matrix Gallery), O'Hara Ure exhibits regularly in Salt Lake City in assorted non-profit spaces and at Phillips Gallery. Her 2012 Finch Lane Gallery exhibition, Here Be Monsters, featured animals in imaginary landscapes, work inspired by the historical forms encountered on her travels, in medieval stonework, manuscripts, and other museum artifacts. Poems by her long-term collaborator and fellow professor, the writer Katharine Coles, were interspersed with the images on panel. They are currently developing a bestiary of image and text, a letterpress artist's book.

John Reed

Best Teacher, History
Most Helpful to Students, History

Alf Seegert

Best Researcher/Scholar, Liberal Arts, Social Sciences, and Humanities

Biography Education
B.S. 1996, Philosophy, University of Utah. Project: Senior Project: Heidegger and Questions Concerning Technology and the Scientific World View M.S. 1998, Philosophy, University of Utah. Project: Ontology Recapitulates Ecology: The Relational Real in Evolution and Ecophilosophy M.A. 2004, English, University of Utah PhD 2010, English, University of Utah. Project: Cybercultural Ecologies: Interfacing Nature, Virtuality, and Narrative

Honors & Awards

Early Career Teaching Award – Annual University-wide award to faculty rewarding distinction in teaching, demonstrated by activities that result in increased learning by students, such as the development of new methods or other curricular innovation. University of Utah, 03/2014 College of Humanities Graduate Research Award. College of Humanities, 2010 Tanner Doctoral Research Fellowship. Tanner Humanities Center, 2009 Steffensen Cannon Scholarship. Graduate School, 2008 Ramona W. Cannon Award for Graduate Student Teaching Excellence. Graduate School, 2007 Vice-Presidential Fellowship. President's Office, 2004 Tanner Graduate Research Fellowship. Tanner Humanities Center, 1997

Biography

Alf Seegert's research on "cybercultural ecologies" examines the interrelations of nature, virtuality, and narrative as represented in literature, film, and new media. His current research focuses specifically on the role of the Fantastic in relation to virtuality and nature.

Ginger Smoak

Most Helpful to Students, Liberal Arts, Social Sciences, and Humanities

Margaret Toscano

Best Overall Faculty Member, Languages

Johanna Watzinger-Tharp

Best Overall Faculty Member, Languages
Best Researcher/Scholar, Languages
Best Teacher, Languages

UNIVERSITY OF VERMONT

Sarah Carleton

Best Overall Faculty Member, Theater

Sarah E. Carleton has worked in the field of professional theater, as actor, teacher, and director, since receiving her M.F.A. in Acting from Catholic University of America in 1985. She has been a member of Actor's Equity since 1983 with professional credits including lead roles in New York, Regional, and National Tours, with favorites being Dinner with Friends, The Importance of Being Ernest, Much Ado About Nothing, Blithe Spirit, Always . . . Patsy Cline, Lend Me a Tenor, Crimes of the Heart, A Phoenix Too Frequent, The Belle of Amherst, The Dining Room, A Christmas Carol, Taking Steps, The Mousetrap, Echoes for a Wooden O, Prelude to a Kiss, The Bald Soprano, A Balancing Act, and The Taming of the Shrew (directed by Maurice Daniels of the Royal Shakespeare Company). Sarah has had the distinguished opportunity to study physical theater at the Celebration Barn Theater with Tony Montanaro, a former student of Marcel Marceau, Davis Robinson, and Mark Olsen. She has also studied at the DellArte' School of Physical theater in Blue Lake California.

Sarah is a certified actor combatant with the Society of American Fight Directors and has served as fight director/movement coach on productions at UVM, St. Michael's Playhouse, and Vermont Stage Company. She has directed productions at St. Michael's Playhouse, The Academy of Performing Arts, Adam Mickewicz University in Poznan, Poland, Seton Hall University, and a special theatrical presentation for the United States Information Agency in Washington, D.C. Teaching credits include Kean College, Adam Mickewicz University, and Seton Hall University. While earning her M.F.A. at Catholic University, Sarah as recipient of the G. V. Hartke Award for Best Actress and was additionally awarded a Young Artist Grant for outstanding service to the performing arts. She has also done television, film, commercial, and voice–overs. Sarah is the co-founder of Foxhill Stage, a theatrical production partnership, and a member of the Association of Theater Movement Educators, and the American Federation of Television and Radio Artists.

David Feurzeig

Best Teacher, Music Most Helpful to Students, Music

The music of David Feurzeig has been performed throughout the United States as well as in New Zealand, Asia, and Europe. He was twice a featured guest (2005–2006) at the International Composer's Festival in Bangkok. His Songs of Love and Protest were selected by the Dresden Chamber Chorus for the city's 800th jubilee in 2006 and premiered in the fabled Semper Opera House. His work has been awarded the Silver Medal of the Royal Academy of Arts (London), the Hugh MacColl and Thomas Hoopes prizes from Harvard University, a Beebe fellowship, and a John James Blackmore prize.

A pianist as well, he specializes in eclectic lecture-recitals featuring music from the classics to stride piano, jazz, and new music. He is the silver medalist of the 2001 World Championship Old-Time Piano Playing Contest and recipient of the 2003 "Best New Rag" award of the Old-Time Music Preservation Association (OMPA). His research interests range from the playing of James P. Johnson and Thelonious Monk to the music of György Ligeti.

At the University of Vermont, Feurzeig teaches composition, orchestration, and theory. Before coming to UVM he held teaching positions at Illinois State University and Centre College. His students have received numerous accolades for their music, including the Toru Takemitsu and the Alexander Zemlinsky Prizes, awards from ASCAP and from the Society for Electro-Acoustic Music (SEAMUS).

John Gennari

Best Researcher/Scholar, Liberal Arts, Social Sciences, and Humanities

John Gennari is an American Studies-trained U.S. cultural historian and nonfiction writer with specializations in jazz and popular music studies, Italian American cultural studies, food studies, race and ethnic studies, and cultural criticism. He is the author of "Blowin" Hot and Cool: Jazz and Its Critics (University of Chicago Press, 2006), which won an ASCAP-Deems Taylor Award for Excellence in Music Criticism and the John Cawelti Award for the Best Book in American Culture. He is currently completing a book examining how practices of expressive ethnicity in music, film, sports, cooking, and eating reconfigure our understanding of Italian American culture. He has held fellowships from the National Endowment for the Humanities, the W.E.B. Du Bois Institute at Harvard University, and the Carter G. Woodson Institute at the University of Virginia. An active member of the American Studies Association since 1993, he chaired the association's Gabriel Dissertation Prize committee in 2008, and served on the Romero Book Prize committee in 2010.

Sylvia Parker

Best Researcher/Scholar, Music

Sylvia Parker is Senior Lecturer of Music at the University of Vermont, where she teaches piano and music theory. Parker performs frequently as piano soloist and ensemble player in Vermont, nationally and internationally. Her new CD entitled "Béla Bartók: Peasant Jewels" is published by Centaur Records (2014) and includes his complete piano settings of authentic peasant melodies. Her earlier CD of solo piano music by Bartok, Griffes, Mozart and Scarlatti is also published by Centaur Records (2006). Her recent journal publications include "Béla Bartók's Other Microcosmos" in MTNA e-Journal (2014), "Claude Debussy's Gamelan" in College Music Symposium (2013), "A Riverton Retreat: Royal Charter to State Forest" in Vermont History (Winter/Spring 2010), "Béla Bartók's Arab Music Research and Composition" in Studia Musicologica (2008), "Understanding Sonata Form through Model Composition" in Journal of Music Theory Pedagogy (2006), and "Bartók at the Crossroads: A Classical Sonatina from Five Rumanian Folk Dances" in College Music Symposium (2003).

Carmen Pont

Best Teacher, Liberal Arts, Social Sciences, and Humanities
Most Helpful to Students, Liberal Arts, Social Sciences, and Humanities
Best Overall Faculty Member, Liberal Arts, Social Sciences, and Humanities

Carmen Ana Pont has taught language and literature in universities on both sides of the Atlantic, is the author of two monographs (Yeux ouverts, yeux fermés: la poétique du rêve dans l'œuvre de Marguerite Yourcenar, Amsterdam, Rodopi, 1994 and L'autobiographie au XXe siècle à Porto Rico, Paris, L'Harmattan, 2008). She is also the author of a book of poetry (Vitral, Madrid, Ediciones Torremozas, 2007). Her scholarly articles and her poems have been published in scholarly journals and anthologies in Europe and the United States. Her most recent work is a poetic preface to the paintings of Puerto Rican artist Rafael Trelles (Metamorfosis, Puerto Rico, Terranova Editores, 2009). She currently teaches French and Spanish at the University of Vermont and works on several research and poetry projects.

Tom Toner

Best Overall Faculty Member, Music

D. Thomas Toner is a Professor and Chair of the Department of Music and Dance at the University of Vermont, where he conducts the Concert Band and teaches percussion. Dr. Toner was previously on the faculties of SUNY Fredonia, SUNY Plattsburgh, and the Summer Session of the Eastman School of Music. Dr. Toner was awarded both the Doctor of Musical Arts degree and the Performer's Certificate from the Eastman School of Music, a Master of Music degree and Artist Diploma from the Yale School of Music, and a Bachelor of Music degree from the University of Massachusetts/Amherst. He has studied percussion with John Beck (Eastman School of Music), Gordon Gottlieb (Yale School of Music), Peter Tanner (University of Massachusetts), Arthur Press (Boston Symphony), Glen Velez (Steve Reich Ensemble), Leigh Howard Stevens (marimba virtuoso), Doug Howard (Dallas Symphony), and Pascal Younge (Ghana, West Africa).

As a percussionist, Dr. Toner's eclectic career has included performances with the Clifford Ball Orchestra (for the rock band

Phish), before Princess Grace and Princess Caroline of Monaco, at a world music festival in Bali, Indonesia, and with a native troupe in a performance in a torrential lightening storm in Ghana, West Africa. Dr. Toner joined the Vermont Symphony Orchestra at the age of 18, and became Principal Percussionist a few years later. He has performed a wide variety of music with the VSO, including most of the standard orchestral repertoire, unusual 20th Century pieces (such as the solo snare drum part in Carl Nielsen's Clarinet Concerto with soloist Richard Stoltzman), and "Pops" music with the likes of Peter, Paul, and Mary, and Tish Hinojosa (where he played drum set, congas, and hand drums). Dr. Toner was the featured soloist of the VSO's fall 2002 tour of Vermont in David Gunn's "A Tangoed Web" (commissioned by the orchestra specifically for that occasion) and was the soloist in Shane Shanahan's "Ahshutmas" with the Eastman Percussion Ensemble in April 2003. A frequent substitute percussionist with the Rochester Philharmonic Orchestra from 1993–1995, he has also appeared with the Moscow Chamber Orchestra (on their 1998 tour of New England), the Massachusetts Chamber Players, and many other ensembles in the northeast. Dr. Toner has also been a featured artist on faculty recitals at the University of Massachusetts and Dartmouth, Williams and Smith Colleges, as well as at the Music Teachers National Association (MTNA) National Convention and the Northeast Conference of the College Music Society. In 2011 and 2012 he presented recitals and masterclasses at the Inner Mongolian Arts College in Hohhot, China. Dr. Toner appears on recordings with the Vermont Symphony and the Robert DeCormier Ensemble, issued by Arabesque Records, and on two solo recordings of Trey Anastasio, issued on the Elektra label. In addition, he is a featured percussionist on numerous audio and video recordings by the vocal group Counterpoint, conducted by Robert DeCormier, and has been heard with that group on National Public Radio's "Performance Today". Dr. Toner is an endorser of Grover Pro Percussion.

An active guest conductor, Dr. Toner has conducted wind groups at the District level throughout Vermont, New Hampshire, and Massachusetts, was the conductor of the 1993 Vermont All-State Orchestra, and has conducted the New Hampshire All-State Percussion Ensemble three times. He has also appeared as guest conductor of the Yale Concert Band, the SUNY-Fredonia Percussion Ensemble, the Green Mountain Mahler Festival Orchestra, and has been a co-presenter at the Eastern Conference of the College Band Directors National Association. From 2002–2010, he was Director of the Northeast Instrumental Music Festival, in Lake Placid, New York. Dr. Toner is a member of the Percussive Arts Society, the National Association of College Wind and Percussion Instructors, the College Band Director's National Association, and the American Federation of Musicians, all of which he has served as an officer at the state level.

UNIVERSITY OF VIRGINIA

Jeffrey Grossman

Best Researcher/Scholar, Literature
Best Teacher, Literature

UNIVERSITY OF WASHINGTON

Ia Dubois

Best Teacher, Scandinavian
Most Helpful to Students, Scandinavian

Ia Dübois' research initially focused on issues of ethnic identity in Sweden and in Swedish literature. Inspired by her interest in interdisciplinary approaches to her research, and by a course she created a few years ago, Dr. Dübois turned her research from Scandinavian literature for a while to focus on a comparative study of sexuality in Scandinavia with the intent to find out if, and how, the Scandinavian countries differ in their perceptions of sexuality in society. More specifically, what made Sweden so progressive and liberal after the welfare state was established in 1933?

As Dübois soon found more differences between the Nordic countires than she expected, the initial phase of her research focused on socio-political issues, particularly on legislative actions with regards to human sexuality. While issues like homosexuality, prostitution and trafficking of women and children are debated globally today, each Scandinavian country has its own interpretation of how their society should handle these issues.

Dr. Dübois spent the Fall Quarter 2006 at Linköping University in Sweden, where she was a Jonsson Family Scholar.

Richard Salomon

Best Researcher/Scholar, Liberal Arts, Social Sciences, and Humanities

"Prof. Salomon is a first-class scholar, who has contributed by publishing text books as well as highly specialized editions of the earliest Buddhist texts, which has revolutionized his field. He is a very inspiring teacher, inspiring students to live up to their fullest potential and an encouraging colleague, always helpful with valuable advice."

Director of British Library/University of Washington Early Buddhist Manuscripts Project and general editor of Gandharan Buddhist Texts series. Edition of two Gandhari versions of the "Songs of Lake Anavatapta." Editions of various Gandhari manuscripts in the Schøyen and Senior collections.

Michael Shapiro

Best Teacher, Liberal Arts, Social Sciences, and Humanities

"Professor Shapiro is an amazing teacher and a great mentor to graduate students and colleagues alike. He has an open door policy and is always ready to listen to his students and colleagues and has an incredible sense of how to help people along."

Michael C.Shapiro Professor, Divisional Dean of Humanities Ph.D. Linguistics, University of Chicago, 1974 M.A. Linguistics, University of Chicago, 1970 B.A. Linguistics, Queens College (CUNY), 1967. FIELDS OF INTEREST South AsianHindiLinguistics Indo-Aryan languages and linguistics.

Current Research Projects: Book-length study on the structure and history of the Hindi language. Studies on the linguistic structure and rhetorical structures of early New Indo-Aryan texts. Work on aspects of early Sikh scripture.

Jan Sjavik

Best Overall Faculty Member, Scandinavian

A specialist in Norwegian literature and literary theory, Sjåvik helped introduce deconstruction criticism to Norway with a ground-breaking article published in Norway's premier literary publication, Edda, in 1985. Additionally, has published extensively in North America where he was fascinated by the authorial rhetorical techniques of such authors as Knut Hamsun, Arne Garborg, Ole Rølvaag, and Henrik Ibsen, among others. His second book, Reading for the Truth: Rhetorical Constructions in Norwegian Fiction, focuses on the role of rhetoric and irony in achieving meaning through an intersubjective approach to literature.

A productive and imaginative scholar, Sjåvik was also a challenging and supportive teacher and mentor. He served as dissertation advisor for ten (10) students, seven of whom would go on to faculty appointments where they have built their own successful scholarly careers.

His colleagues and students in the Department of Scandinavian Studies congratulate Professor Sjåvik for his exceptional contributions to the field of Norwegian and Scandinavian studies and thank him for his thirty-six years of service to the University of Washington. Our sincere best wishes as he begins a new phase of his life under the guise of retirement.

Marianne Stecher-Hansen

Best Researcher/Scholar, Scandinavian

Marianne Stecher investigates Scandinavian and Danish literature in various critical, historical, and literary contexts of the 19th and 20th centuries. She employs recent criticism and theory to write about the works of Hans Christian Andersen, Karen Blixen (Isak Dinesen), Thorkild Hansen, and other writers. Professor Stecher has served as the editor for two reference volumes on Danish writers and her articles appear in Scandinavian Studies, Edda, and other journals. Marianne Stecher's most recent book, The Creative Dialectic on Karen Blixen's Essays (2014) is published by Museum Tusculanum Press in Copenhagen and distributed by the University of Chicago Press.

UNIVERSITY OF WEST FLORIDA

William Belko

Best Researcher/Scholar, History
Best Researcher/Scholar, Liberal Arts, Social Sciences, and Humanities

Melissa Gonzalez

Most Helpful to Students, History

Daniel Miller

Best Overall Faculty Member, History

Derek Zumbro

Best Teacher, History

UNIVERSITY OF WISCONSIN - EAU CLAIRE

Robert Baca

Best Teacher, Music

Phillip Ostrander

Best Overall Faculty Member, Music

Phil Ostrander is Associate Professor of Trombone at the University of Wisconsin-Eau Claire where he conducts the Symphony Band and teaches private trombone and trombone ensemble. Dr. Ostrander completed his doctoral studies at the Eastman School of Music. Dr. Ostrander received master's degrees in both trombone and wind conducting from the New England Conservatory, as well as a bachelor's and Performer's Certificate from Eastman. He has performed with the Minnesota Orchestra, the St. Paul Chamber Orchestra, the Boston Pops Esplanade Orchestra, the Rochester Philharmonic, the Kansas City Symphony, the Buffalo Philharmonic and the popular chamber group Burning River Brass Ensemble. Currently, he is a member of the IRIS Chamber Orchestra in Memphis, Tennessee under Michael Stern and also Principal Trombone of the Minnesota Opera Orchestra. An accomplished jazz trombonist, Dr. Ostrander has collaborated with jazz artists Maria Schneider, Jim McNeely, Jimmy Heath, Claudio Roditi and Rich Beirach. He teaches jazz at the Shell Lake Arts Camp in northern Wisconsin. He has recorded on Sony Classical with the Eastman Wind Ensemble and Naxos with the IRIS Orchestra. Dr. Ostrander is a clinician for the Conn-Selmer Musical Instrument Company.

Gretchen Peters

Best Researcher/Scholar, Music

Gretchen Peters is a professor of music at UW-Eau Claire with teaching responsibilities in music history and world music. She holds a Ph.D. in musicology from the University of Illinois, masters degree from the University of Illinois in musicology, baccalaureate degree from the University of Wisconsin-Madison in applied cello, and has studied historical performance at Indiana University's Early Music Institute.

Dr. Peters' research interests are wide-ranging. She is an expert in urban musical culture in southern France in the late middle ages and has authored articles and presented numerous presentations on the subject. Some of her articles include "Urban Musical Culture in Late Medieval Southern France: Evidence from Private Material Contracts" in Early Music (1997); "Urban Minstrels in Late Medieval Southern France: Opportunities, Status and Professional Relationships" in Early Music History (2000); "Civic Subsidy and Musicians in Southern France During the Fourteenth and Fifteenth Centuries: A Comparison of Montpellier, Toulouse, and Avignon" in Music and Musicians in Renaissance Cities and Towns (Cambridge, 2001); and "Musical Culture in Medieval Avignon: An Opportunity for Exchange" in The Cultures of Papal Avignon (2007). Recent research interests also include the traditional music of the Laotian Hmong culture and the music of the Ojibwe Native American culture. In these areas she and her students have received awards for their collaborative work.

Alan Rieck

Most Helpful to Students, Music

Alan J. Rieck, Professor of Music, conducts the Women's Concert Chorale and Symphonic Choir, and teaches courses in music education. Prior to his current appointment, Dr. Rieck served as Assistant Director of Choral Studies and professor of music education at Northern Arizona University in Flagstaff, Arizona where he taught courses in conducting, current issues in music education, and music education teaching methods; and conducted the Men's Chorale and Women's Chorale. While in Flagstaff, he also served as assistant director of the Master Chorale of Flagstaff, was the founding director of the Children's Chorale of Flagstaff, and during the 2000–2001 school year served as the NAU Director of Choral Studies. Prior to moving to Arizona, Dr. Rieck taught and conducted at the University of Wisconsin-Madison, conducted the top performing groups of the Madison Children's Choir, and served as chorus master for the Madison Symphony Orchestra Chorus. Dr. Rieck taught for six years in the public schools in West Allis, Wisconsin and is an active member of the American Choral Directors Association and Music Educators National Conference. He holds a B.M. degree in Music Education, M.M. degree in Music Education, and Ph.D. in Curriculum and Instruction, all from the University of Wisconsin-Madison.

UNIVERSITY OF WISCONSIN - MADISON

Stacy Bryant

Best Overall Faculty Member, Languages

Currently a teaching assistant in the Spanish Dept. and a Ph.D. student in Spanish linguistics. Experienced teacher, tutor and translator.

UNIVERSITY OF WISCONSIN - MILWAUKEE

Carl Bogner

Best Overall Faculty Member, Film
Best Teacher, Film

I teach a two-semester class on experimental media as part of our Department's foundational curriculum. I also work with graduate students, conducting, upon occasion, the Graduate Seminar as well as offering other graduate classes (directed readings/independent studies; a course on experimental writing/criticism). In addition to contributing programming and inviting guests to the Department's ongoing occasional screening/visiting artists series, I also direct the Milwaukee LGBT Film/Video Festival, which is hosted by the Film Department. Previous experience includes bookselling, various positions as a film curator. Book I am currently (Spring/Fall 2012) recommending: Michael Taussig's I Swear I Saw This: Drawings in Fieldwork Notebooks, Namely My Own (University of Chicago Press, 2011)

Nancy Burrell

Best Researcher/Scholar, Communication

Vicki Callahan

Most Helpful to Students, Film

Professor Callahan's work is focused on the intersection between theory and practice as both historical/cultural phenomena and practical artistic strategy. Her research interests include feminist theory, silent film history, digital media, and intermedia. She is particularly engaged with work that explores new ways of writing history and theory, for example, database narratives, participatory archives, web-based interactive essays, and other parafilmic or multimodal forms of critical thought. She is the editor of the collection Reclaiming the Archive: Feminism and Film History (Wayne State University Press, 2010) and author of Zones of Anxiety: Movement, Musidora, and the Crime Serials of Louis Feuillade, (WSUP, 2004). Other publications include; "'Gravity and grace ... turn a camera movement into a prayer': on the "Sacred" and Cinematic Vision in the Films of Jean-Luc Godard," Forever Godard, ed. James Williams, Michael Temple, Michael Witt, London: Black Dog Press, 2004; "Representations 1895–1930," The French Cinema Book, eds. Michael Temple and Michael Witt, London: BFI, 2004; "Screening Musidora: Inscribing Indeterminacy in Film History" Camera Obscura 48 February 2002; and she was a respondent to Eric Faden's essay "Tracking Theory" in Vectors issue Perception (Fall 2006). Currently, she is working on a study of the silent film actress/director Mabel Normand.

She has taught courses on Multicultural America, Interactivities, Media Archaeology, and Animation Aesthetics, as well as graduate seminars on Feminist Film Theory and the work of Godard/Mieville. She organized the Colloquia in Conceptual Studies, a speaker series, which explored contemporary topics in media arts such as "The Postmedium Condition" and "Interactivities" (and included presentations by Trinh T. Minh-ha, Victor Burgin, Norman Klein, and artist/theorist collaborations such as Perry Hoberman/Anne Friedberg and Mary Lucier/Melinda Barlow).

Professor Callahan always inspired by Sun Ra and Donna Haraway's understanding of extended being, includes the photo of her close friend, M. Charles, for your visual pleasure.

Cecelia Condit

Best Researcher/Scholar, Film

Cecelia Condit is an artist/storyteller who creates contemporary fairy tales in video, installation and photography. She has received grants from the John Simon Guggenheim Foundation, the American Film Institute and the National Endowment for the Arts, and the Greater Milwaukee Foundation's Mary L. Nohl Fund. Condit's work is in many prestigious collections including the Museum of Modern Art in New York City and the Centre Georges Pompidou Musee National d'Art Moderne Paris, France. Her work has focused on the relationship between women and their lovers, mothers, families and friends. In 2012, she completed a video and photographic installation, "Within a Stone's Throw," in which the artist appears walking across rocks, climbing berms, and throwing stones. Within a Stone's Throw plays with the unfolding of human and geologic time, and probes the connections and displacements that exist between ourselves and the natural world. Condit is also the director of the graduate program in film and teaches introduction to video editing, an elective class called "Dreams," "From Script to Screen" and "Performance and Movement Collaborations" which she co-teaches with a faculty from the Dance Department.

Jennifer Johung

Best Researcher/Scholar, Fine Arts

Arijit Sen

Best Teacher, Fine Arts

Arijit Sen, Assistant Professor of Architecture, teaches architectural design urbanism and cultural landscapes at the University of Wisconsin - Milwaukee. He is the co-coordinator of the Buildings Cultures Landscapes doctoral program initiative between the University of Wisconsin-Madison and Milwaukee. His research interests include physical and cultural landscapes of immigration in the United States. He is currently completing his book "Creative Dissonance: The Politics of Immigrant World Making" and a co-edited monograph "Devon Street, Chicago: Interpreting Landscapes of Transnationalism." Sen received his PhD from the University of California, Berkeley and served as a Center for 21st Century Studies fellow at the University of Wisconsin Milwaukee and a Quadrant Fellow at the Institute of Advanced Study at the University of Minnesota, Minneapolis.

UNIVERSITY OF WISCONSIN - RIVER FALLS

Kathleen Hunzer

Best Overall Faculty Member, English

Kathleen is a Composition and Rhetoric specialist who currently teaches both undergraduate and graduate courses including ENGL 100: Academic Reading and Writing; ENGL 200: Investigating Ideas; ENGL 251: Advanced Composition; ENGL/FILM/WGST 300/500: Women, Film, and Society; ENGL 361/561: Composition Theory; and ENGL 362: Introduction to Rhetorical Studies. She has also taught honors seminars on "Interrogating Literary Utopias" and "Freedom of Expression in America" as well as ENGL 250: Literature for Children (online), ENGL/FILM 307: Ethnic Film, Literature, and Culture, ENGL 355: Literature for Adolescents, FILM 389: The Films of Alfred Hitchcock, and WGST 200: Introduction to Women's Studies. Kathleen embraces the interdisciplinary nature of all of her classes and hopes that she can help her students appreciate the power of language in our culture and our lives.

UNIVERSITY OF WISCONSIN - WHITEWATER

Margarita Alario

Best Researcher/Scholar, Sociology

Alicia De Gregorio

Best Overall Faculty Member, Languages

"Dr. Alicia de Gregorio is an amazing person who is always ready to help peers in anything they need in despite the work load that she might have."

Peter Hoff

Most Helpful to Students, Languages

"Dr. Peter Hoff is an exceptional Languages Coordinator & teacher who goes beyond limitations to help students succeed."

James Kates

Best Teacher, Journalism
Most Helpful to Students, Journalism

James Kates is an Associate Professor in the Department of Communication at UW-Whitewater. He teaches courses in media history and law, digital journalism, writing, reporting and editing. He earned a Ph.D. in mass communication from the University of Wisconsin-Madison in 1997 and joined the UWW faculty in 2007. Trained as a historian, his research specialty is the role of writers and editors in social movements, particularly natural-resources conservation in the early 20th century. His book, "Planning a Wilderness," was published by the University of Minnesota Press in 2001. Dr. Kates has 25 years of experience as an editor at metropolitan newspapers, including The Philadelphia Inquirer and the Milwaukee Journal Sentinel.

Leda Nath

Best Overall Faculty Member, Sociology

Leda Nath holds a Ph.D. (2002) from the University of Iowa in Sociology, emphasis in Complex Organizations. She arrived at UW-Whitewater in Fall 2002. Her research interests span across various areas: social psychology; unpaid work, paid work, and leisure; gender; decision-making; and teaching & learning. She teaches face-to-face and online social research methods, basic social statistics, social psychology, sociology of family & work, organizations & society, plus general education courses. She enjoys mentoring undergraduates on research and advising students, She is also the Interim Director of the Family, Health & Disabilities Minor, and Department Chair. She also holds the President role for the Wisconsin Sociological Association.

UTAH STATE UNIVERSITY

John Barton

Most Helpful to Students, History

John has been teaching history with an emphasis in Western, Utah, and Native American history at USU Uintah Basin since 1988. He received his master's degree in history from Brigham Young University. John has worked to build the history program at the Uintah Basin Campus and has been integral in expanding this program statewide via interactive video conferencing, online courses, and face-to-face classroom settings.

Troy Beckert

Best Teacher, Family Consumer & Human Dev.

My interests in adolescents and their families shape a quantitative research focus on intergenerational learning and cognitive development. Currently, I have three main directions in my research. We are assessing cognitive autonomy in adolescents of both at-risk and culturally diverse populations. We are also looking at the influence of technology on adolescent psychosocial development. Additionally, I am involved in a large-scale project that evaluates the educational benefits of polling in secondary education students. Other areas of research include multi-generational assessments of mothers, fathers, and grandparents of early adolescent children.

Jeffery Dew

Best Researcher/Scholar, Family Consumer & Human Dev.

Elizabeth Fauth

Best Teacher, Family Consumer & Human Dev.

My research focuses on human development and family studies in adulthood and late life, specifically 1) family caregiving for persons with dementia and dementia care (with a focus on services and program evaluation), 2) well-being and physical functioning of the oldest-old, and 3) intergenerational interactions across the lifespan.

I collaborate with the Alzheimer's Association Utah Chapter and the State of Utah Division of Aging Services on several program evaluation projects assessing the extent to which the educational, psychological, and family-focused counseling sessions improve outcomes for family caregivers of individuals with dementia. I will also evaluate staff interactions and recreational activities in memory care units, to determine high quality care and recreation based on client behavior and affect. As an investigator on the Cache County Dementia Progression Study, I am interested in uncovering how the caregiving environment affects the progression of dementia, as well as the financial cost of care. I collaborate with the Institute for Gerontology at Jönköping University in Sweden, utilizing multiple longitudinal studies of late life. I examine patterns of health and functional decline and the role of psychosocial variables such as depressive symptoms and social support as buffers in the disablement process. I received NIH funding to combine Swedish data into a large longitudinal sample of older adults, studying interrelationships between disease, cognition, physical functioning, social support and well-being over time. I also currently collaborate with Sunshine Terrace and collect data on functional performance in clients across multiple care settings, to determine interdepartmental trends, as well as intra-departmental changes over time in client outcomes. I co-founded the Intergenerational Lifespan Lab, a large-scale project evaluating the benefits of intergenerational interactions across the lifespan.

Evelyn Funda

Best Researcher/Scholar, English

Brian McCusky

Best Teacher, English

"Holds students to a higher level of expression in language and writing. Very witty. Creative with subject matter."

Kristine Miller

Best Researcher/Scholar, English

"She wrote a significant book."

Robert Mueller

Best Overall Faculty Member, History
Best Teacher, History

Michael Nichols

Best Researcher/Scholar, History

Steve Shively

Most Helpful to Students, English

"Always working with students. Very caring"

UTAH VALLEY UNIVERSITY

Kathren Brown

Best Researcher/Scholar, Liberal Arts, Social Sciences, and Humanities

Alex Caldiero

Best Overall Faculty Member, Humanities
Best Researcher/Scholar, Humanities
Best Teacher, Humanities
Most Helpful to Students, Humanities

Chris Clark

Most Helpful to Students, Theater

Robert Cousins

Most Helpful to Students, English

Robert joined the English and Literature Department in 1998. His academic background is in American Studies, the interdisciplinary examination of American literature, history, and culture. His specialties include early American literature and multi-ethnic literature.

Courtney Davis

Best Researcher/Scholar, Art

"Outstanding research for art history classes"

Courtney Davis has taught art history at Utah Valley University since 2007, as both an assistant professor and a lecturer. She has twice been recognized as the UVU Alumni Outstanding Educator of the Year for the School of the Arts, and has also received a Faculty Senate Award of Excellence. Courtney is the founder and director of the AVC Department's annual art history symposium, a tradition that began in 2010. Courtney has led three study abroad programs to Europe, garnering national attention in 2012 with an award from Abroad101 for short-term travel study programs. Courtney holds a Masters of Art in Art History & Curatorial Studies and a Juris Doctor, both from Brigham Young University. Her research areas include early eighteenth-century European painting, the American Gilded Age, Modernism, and copyright law.

Catherine Downing

Most Helpful to Students, Fine Arts

Catherine is a member of the Art and Visual Communication Department at UVU. She started at UVU in 1996. She specializes in teaching in the areas of Drawing, Oil Painting, and Landscape Painting. Her web page is catherinedowning.com

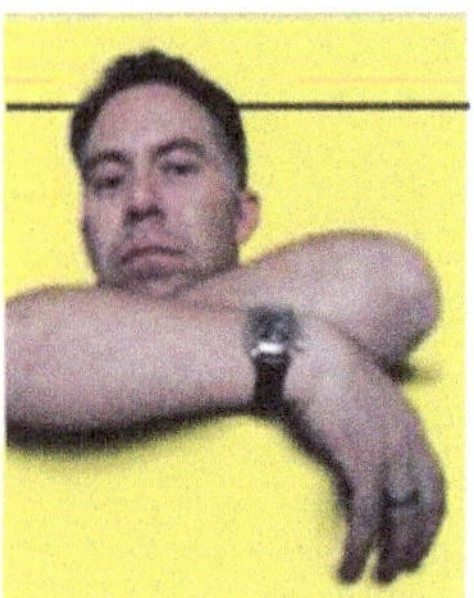

Howard Fullmer

Best Researcher/Scholar, Liberal Arts, Social Sciences, and Humanities

Lisa Hall

Best Overall Faculty Member, Theater
Best Researcher/Scholar, Theater
Best Teacher, Theater
Best Researcher/Scholar, Liberal Arts, Social Sciences, and Humanities
Best Teacher, Liberal Arts, Social Sciences, and Humanities
Most Helpful to Students, Liberal Arts, Social Sciences, and Humanities
Best Overall Faculty Member, Liberal Arts, Social Sciences, and Humanities

I currently serve as an Assistant Professor of Theatre History and Dramaturgy, and also teach with the Honors Program. I am the Dramaturgy Advisor and have served in the past as Coordinator of the Gender Studies minor. I am also proud to coordinate the Theatre Department's yearly service project, which has benefited the Center for Women and Children in Crisis, as well as the Newborn Intensive Care Unit at Utah Valley Regional Medical Center.

Mike Harper

Best Overall Faculty Member, Multimedia
Best Researcher/Scholar, Multimedia
Most Helpful to Students, Multimedia

Michael Harper received his Undergraduate and Master's Degree's from Utah State University in Geography Education with a specific interest in building multimedia-based curricula for secondary education. After teaching at USU for a one-year appointment after graduation, he then moved his family to Phoenix to participate in what would later become known as the Dotcom boom and bust! After several years as a Web Content Manager and Online Media Director at a U.S. Latino focused startup he looked for new adventures in a rapidly expanding industry. Post Dotcom adventures included stints at a video production company, a unified school district integrating media technology in the classroom, and a healthy dose of media consulting for Web and optical media development. Michael then chose to follow his original goal of teaching at the college level and came to UVU with industry insights that have served students well. Since joining Digital Media, Michael has been an advocate for infusing emerging technologies in the classroom, a staunch adherent for improved standards in the mobile media markets, and heavily involved in cultivating International Media engaged learning opportunities for DGM students.

Marilyn Jenkins

Best Teacher, Liberal Arts, Social Sciences, and Humanities
Most Helpful to Students, Liberal Arts, Social Sciences, and Humanities
Best Overall Faculty Member, Liberal Arts, Social Sciences, and Humanities

Pierre Lamarche

Best Overall Faculty Member, Philosophy
Best Researcher/Scholar, Philosophy
Best Teacher, Philosophy
Most Helpful to Students, Philosophy

I work primarily in 20th Century Continental Philosophy, in particular Existential Phenomenology, Critical Theory, and Post-structuralism and Post-Marxism. I also work in Marxism, Aesthetics, Scepticism, Philosophy of Literature, and Hellenistic Philosophy.

Rick McDonald

Best Overall Faculty Member, English
Most Helpful to Students, English

Jana Parkin

Most Helpful to Students, Art

"Gets on the students level and teaches them well."

Jana Winters Parkin is a passionate, hands-on instructor with 12 years experience teaching on the college level, first at Glendale Community College, and now at Utah Valley University. She inspires students to work outside their comfort zone, experiment with new color combinations and techniques, and discover new levels of excellence in their work. She regularly takes a group of students to paint on location for a week at Capitol Reef—an experience which has been described as "one of the most valuable educational experiences . . . ever."

Donald Seegmiller

Best Overall Faculty Member, Art

"Works very hard for students success"

Don Seegmiller is a successful artist, author, and instructor. He has a Bachelor of Fine Arts degree from Brigham Young University in Provo, UT. Don has been teaching at the university level for 17 years. He is faculty and coordinates the Illustration Area of the Art and Visual Communication Department at Utah Valley University in Orem, UT. Don teaches both traditional and digital painting and conducts classes for the Academy of Art University online. His traditional oil painting work is represented in major art galleries in the United States. Don worked as the Art Director for Saffire Corporation, Art Director at Imagine Learning, has been a regular speaker at the annual Game Developers Conference, and travels regularly giving lectures and workshops at institutions as diverse as the American Medical Illustrators Conference and Disney in Orlando Florida. He is author of Digital Character Design and Painting: Digital Character Design and Painting; the Photoshop CS Edition: Digital Character Painting Using Photoshop CS3: Advanced Painter Techniques: and Coauthored Ballistic Publishing's Digital Painting 2. Don's work has been featured in the Painter Wow series of books from version 6 to the current edition, Electronic Step by Step Design, Art International magazine, Spectrum 7, and Spectrum 8. He was one of the Judges for Ballistic Publishing's Painter book. He demonstrated Painter at Siggraph a number of years for Corel and did demonstrations on digital painting in the Guerilla Studio. Don conducts regular sold out workshops for the CG Society. He is on the Corel Painter Advisory committee. He has used Painter from its earliest days as Fractal Design Painter® 1 and continues to use it as his main digital tool. He consults with Adobe on an occasional basis about Photoshop which he also uses regularly. When he has free time, he enjoys riding his motorcycle and fly fishing with his son.

Perry Stewart

Best Teacher, Art

"Excellent teacher"

Sandy Vogel

Best Teacher, English

Mike Wisland

Best Teacher, Multimedia

Mike Wisland graduated from the Missouri Institute of Technology with a degree in Electrical Engineering and a masters in Digital Signal Processing, and immediately went to work on fullsize military flight simulator domes (really big video games) at McDonnell Douglas Corp. Mike moved to Utah in '91 to work with a government contractor as a Digital Image and mapping scientist. In 1997, he helped co-found another government contracting company working in Infrared imaging and advanced Thin Films where he currently contracts. Mike built one of the first all-digital recording studios in the state, and formed Wiz Productions in 1992, and has recorded local bands, radio commercials and video projects ever since. He also co-hosts a local radio show, "The Vintage Vinyl Radio Show", and is a DJ on Park City's KPCW radio station on Friday nights. He is an avid collector of all things audio, from Edison cylinders to Blu-ray Audio disks. Don't get him started . . . Mike taught first in the Music Dept, for 2 years, and came to the Digital Media Dept in 2003, designing most of the audio courses. He recently built a new BBR (Bed-Breakfast-Record) studio in PG to service even more clients in-state and out-of-state.

VALDOSTA STATE UNIVERSITY

Byron Brown

Most Helpful to Students, English

Donna Sewell

Best Teacher, English

VALENCIA COLLEGE

Mary Allen

Best Teacher, Liberal Arts, Social Sciences, and Humanities

"most helpful for students"

Rachel Allen

Best Overall Faculty Member, Liberal Arts, Social Sciences, and Humanities

"Rachel is an amazing colleague. She goes beyond the call of duty. As the coordinator of Valencia's Peace and Justice Initiative, Rachel is a true inspiration. She is always looking for new ways to create, transform and integrate the culture of peace at not only our campus and college wide but also throughout our

community and abroad. Through her leadership, the Peace and Justice Initiative is partnering with many community organizations and leaders in Central Florida to work together in bringing peaceful and practical solutions to many difficult problems we face such as homelessness, bullying, understanding diversity and other issues. Her ability, enthusiasm and passion has also opened doors to serve abroad partnering with a local church to take "Peace and Conflict Resolution" skills to Haiti. Ms Allen is a true Gem to Valencia College."

Irma Berner

Best Teacher, Fine Arts

Guy Calabrese

Best Teacher, Humanities

"Dr. Guy Calabrese is one of the most dedicated and hardworking professors with whom I have ever met. Even though he is only part-time and have been overlooked on several occasions for a full-time position, he comes through with the students, no matter how many part-time classes he is teaching. He deserves some recognition."

Rose Casterline

Best Teacher, Fine Arts

Jennifer Copp

Best Teacher, Art

Jennifer Bronwynn Copp Photography, Video & Multi-Media Installation

Born 1972 - Syracuse, New York BFA Photography - Savannah College of Art & Design, Savannah, GA, 1995 MFA Photography and Related Media, Video Concentration - School of Visual Arts, New York, NY, 2004

Most of my work deals with a sense of place, and how a particular location can help shape one's state of mind. The second theme in my work involves an attempt to figure out how one fits into the world, the idea of one's placement in society. My newest project, a feature length documentary, uses adoption as subject to research who we are and why we feel a connection to the people that we are biologically related to.

As the Child of theatre academics I got to experience the arts early. Both my parents worked in community theatre as a director (Mom) and additionally (Dad) as a playwrite, set, lighting and custom designer. The focus of many family discussions was about the setting of events, both staged and real. The dinner table talk had to do with where, when and how life situations took place. As a photographer, I digested this view of life and began to see staged tableaus in every place that I encountered. I could see in my mind the moment before and after the event that I was capturing with my camera.

In rebellion to the "in the moment" focus of a live theatrical event, I worked instead to try to collect and obsessively control moments in time. There was a prevalent feeling for me that I wanted to save moments, more than I wanted them to happen briefly and then vanish and be gone. By recording moments in time, I could go back and look at them again and again.

One of my projects, Alone in New York (B&W photography, 2004–ongoing) is a series of images of a New York that is uncharacteristically quiet. Where people are held in solitary moments in the midst of the movement and energy of a huge city. Where did these people come from, and where are they going? My images talk about the architecture and visual uniqueness of a city caught at rest. Sometimes we feel alone and sometimes we feel content, and at certain moments there is time to think. I have been photographing in New York for several years now. I am captivated with the light and the small moments in time that occur with every neighborhood that I find.

For me it is important to bring the ideas of land use, public space and how we use that space into the public vocabulary. Of equal importance is to bring to light the ideas of the fractured identity through adoption, divorce, family and the public view of what is family. I am intrigued with the power that the arts have to enact change. I use photography and video as a mirror and spotlight. I hope to encourage people to see their common world in a new way, which might instigate internal and external change in their lives.

Theatre production has never left my creative impulses, as I continue to use still photography. I am also in the process of producing an installation, Dis/Connected, which is a multi-media project based on my documentary about adoption, Date of Birth. Dis/Connected is also currently a touring workshop and installation addressing the emotional impact of adoption and identity.

My hope is to continue to bring my projects, both still photography and multi-media to galleries as well as un-traditional locations. I want to bring the idea of questioning identity and our place in our contemporary world into the public consciousness through my ongoing projects.

Karen Cowden

Best Overall Faculty Member, Reading

Karen is an innovator in developing reading activities that creatively connect students' learning styles and integrate curricular pathways. She also is skilled at building collegiality through collaborative college wide initiatives. As a leader, and the calm in the center of a storm, she is thoughtful by nature, provides an atmosphere of open, honest feedback, and lives by mutual respect. She supports college-preparatory students in the tricky balance of adoption into the college environment. Her lessons build students' skills with encouragement and a lot of fun. Students are regularly offered opportunities for interactive peer-pairing learning experiences, collaborative activities, and opportunities to repair work that was not their desired result on their first attempt. For her, the foundation of her practice is to help students develop a different way of thinking, behaving, and living to contribute as citizens who display academic demeanor. All of her work is built on the joy of being allowed to stand alongside her students as a mentor.

Carl Creasman

Best Teacher, History

"Highly intelligent, knows the subject matter well and understands the concept of effective teaching."

Professor Carl E. Creasman Jr. has been teaching at Valencia since early 2001. His first classes at Valencia were in our vaunted Dual Enrollment Program; Carl taught at Jones High School. In the Fall, as an adjunct, Professor Creasman taught classes both on campus at East as well as continuing the Dual Enrollment area. In 2004, Professor Creasman accepted a four month position remaining at East. He teaches all 4 of the History classes (AMH 2010, AMH 2020, EUH 2000, EUH 2001) as well as offering some special topics courses such as The History of World War 2. Previous to Valencia, Professor Creasman has held a wide and varied career working with children, teens and adults in various areas of education, youth sports and Christian education. Creasman has also travelled extensively within the United States as a communicator and speaker.

Elizabeth Eschbach

Best Overall Faculty Member, Humanities
Best Researcher/Scholar, Humanities
Most Helpful to Students, Humanities

"Elizabeth Eschbach was my mentor back in 1992. She inspired me to be the professor that I am and deserves some recognition.

I had Mrs. Eschbach when I first returned to college in 1982. She has inspired me to go on with my education, as well as, many other students over the years. She had my vote!

Mrs. Eschbach is a true teacher in every sense. She is caring about the students, dedicated to Valencia College, and is knowledgeable about every aspect of the Humanities she teaches. A true inspiration for educators and students."

It is a joy and privilege to be part of one of the richest and most dynamic humanities programs in the country. I am thrilled to work and interact with Valencia students who bring energy, vitality, curiosity, and diverse backgrounds to our humanities classrooms and to the campus as a whole. Thanks to this extraordinary student body, every class experience brings fresh perspectives and new opportunities to deepen our pursuit of beauty, truth, and knowledge. I am very grateful for these opportunities.

My own academic background has been broad and varied, from the small liberal arts college where I began my higher education, Agnes Scott College, to the second largest university in the country, the University of Central Florida, where I completed my Master's degree. I have also studied French language and culture at the Université Paul Cezanne in Aix-en-Provence, France, as well as medieval history at Cambridge University. My studies are often combined with travel experiences, from the high Andes Mountains of Peru to the Himalayan foothills in India. Every year I travel overseas to nourish my love of history, art, music, and culture. My continuous studies and love of travel enable me to endlessly renew and enrich my passion for the humanities.

My classes are interactive with an emphasis on discussion and student engagement. I believe firmly in learning beyond the walls of the classrooms, so part of my curriculum involves participation in our fine and performing arts program, as well as the cultural offerings of Central Florida. I have been involved in the Humanities Speaker Series, that brings internationally recognized authors and scholars to our campus, as well as the book circles and reading projects that complement our series.

Every semester brings endless possibilities. Welcome to the journey.

Edie Gaythwaite

Best Researcher/Scholar, Communication

Bill Gombash

Best Teacher, Communication

Clay Holliday

Best Overall Faculty Member, English

Mayra Holzer

Best Overall Faculty Member, Communication

"Mayra Holzer gives 110% to her students and her professionalism and dedication go above and beyond the call to positively affect both students and colleagues."

Lula Keyes

Best Researcher/Scholar, Languages

Mandy Mahaffey

Best Researcher/Scholar, Liberal Arts, Social Sciences, and Humanities
Best Teacher, Liberal Arts, Social Sciences, and Humanities
Most Helpful to Students, Liberal Arts, Social Sciences, and Humanities
Best Overall Faculty Member, Liberal Arts, Social Sciences, and Humanities

Kristy Pennino

Best Overall Faculty Member, Graphic Arts

"Amazing individual that could easily earn my vote for any category. Valencia and it's students are blessed to have Kristy."

Maria Perez-Boudet

Best Teacher, Languages

I was born in Havana, Cuba, an amazing city filled with art, music, and gorgeous architecture that ranged from the 1500s to modern times. People from all over the world came to visit the city. Havana was a vibrant, exciting place even as a child. When I was 13, my mother sent me to school in New York City (in addition to my year of school in Havana) to learn English. I was excited to go to New York, but hated it when I could not speak or understand one word of the language. So I know what it is like to learn a language and to feel frustrated when you can't find the words. But I learned and you can too.

Today, it is even more important to learn another language because technology connects us with other people, cultures, and countries instantly. As a global community, learning another language can help you in work and in life. What I hated at first turned out to be one of the best gifts of my life. I did not have to speak my new language perfectly. People understood me and they were happy that I respected them enough to try to learn their language and culture. They thought my many mistakes were 'cute'.

So make a consistent effort in class and keep trying when you make mistakes. You don't have to speak perfectly to discover new opportunities. At the very least, you get the 4 credits you need in a foreign language! :) (But I hope it becomes more than that!)

Deborah Poulalion

Best Overall Faculty Member, Communication
Best Researcher/Scholar, Communication
Best Teacher, Communication
Most Helpful to Students, Communication

My expertise in communication comes from twenty years of work as a professional book editor and writer. My career started as the editorial manager for a local book publishing company. Since then I have spearheaded numerous book, magazine, and curriculum projects. I've also ghostwritten six books with an international bestselling author. I graduated magna cum laude from the University of Florida with a Master of Arts in Mass Communication in 1991.

Liza Schellpfeffer

Most Helpful to Students, Communication

Karen Styles

Best Researcher/Scholar, Humanities

VANDERBILT UNIVERSITY

Marshall C. Eakin

Best Overall Faculty Member, History
Most Helpful to Students, History

Marshall Eakin is a historian of Latin America specializing in the history of Brazil. Although his work spans all of Brazilian history, his major publications have concentrated on the processes of nationalism and nation-building, economic and business history, and industrialization—primarily in the twentieth century. His first book, British Enterprise in Brazil: The St. John d'el Rey Mining Company and the Morro Velho Gold Mine, 1830–1960 (Duke, 1989), traces the history of the most successful foreign enterprise in 19th- and 20th-century Brazil. Tropical Capitalism: The Industrialization of Belo Horizonte, Brazil (Palgrave, 2001) examines the industrialization of the second-largest industrial center in Brazil. Much of his work addresses audiences beyond the academy. This work includes Brazil: The Once and Future Country (St. Martin's, 1997), a one-volume introduction to Brazil for beginners and two video courses with the Great Courses, The Conquest of the Americas and The Americas in a Revolutionary Era. His more recent book is The History of Latin America: Collision of Cultures (Palgrave Macmillan, 2007). Eakin's latest book project is "Shadows in the Soul: Brazilian Identity in the Twentieth Century."

Jonathan Lamb

Most Helpful to Students, English

In the eighteenth century a new kind of prose fiction—first person narratives of things such as coins, coaches, clothes, animals and insects—emerged alongside the novel, the oriental tale, spy narratives and pornography. The experiences of these things for the most part are not happy and, like the king of Brobdingnag when he hears Gulliver boasting of his knowledge of gunpower, they accuse human beings of inhumanity. What standard of humanity can be invoked by a thing? The answer, on the basis of their testimony, is a value outside the rights and practices of civil society, beyond the reach of institutionalised self-interest. They inhabit a state of nature, either privative, like Hobbes's, or full, like Rousseau's or Locke's. The aspiration of a thing to be an author, an agent capable of speech and self-constitution outside an established community, endowed with unconditional right of uttering sentiments that give it form of life entirely original and its very own, echoes that of the human authors who stand behind them: impoverished professionals who are obliged to part with whatever they invent so that they can have shelter and food. That is to say, things find common ground with humans at the rim of the commercial world where the circulation of commodities (including print) and the ownership of property has few consolations left for them. Neither the human nor the thing is an apologist for a system of global exchange that leaves them in pain and beggary; nor are they willing to take on the duties and obligations that herd under the name of the person, the artificial representative within the state who bears the persons of other persons. Instead they occupy the Hobbesian position of author, which he also called the natural person: a pre-civil solitaire with rights as yet unmortgaged to a social contract. Under certain emergencies, some pleasurable and others painful, humans and things are driven towards this edge, where the hierarchies not only of state but also of matter, imagination and divinity, get upset and rearranged. The benefits of this disorder are freedom that may be expressed in a variety of ways: as idolatry or godhead; as the overload of the nerves Defoe's Roxana and Charles Gildon's talking coins call 'the Sence of things;' as 'radical first personality' that is primordial and original; as a reflexive sensibility that transports matter and flesh into sympathies indistinguishable from metempsychosis; as the conduct of a narrative where the aim is not to represent a sequence of events but to cause them to happen, as if reading and writing themselves were events; but most of all, as the handling of pen and ink, the first and last badges of true authorship.

Thomas Schwartz

Most Helpful to Students, History

Thomas Alan Schwartz is a historian of the foreign relations of the United States, with related interests in Modern European history and the history of international relations. He is the author of America's Germany: John J. McCloy and the Federal Republic of Germany (Harvard, 1991), which was translated into German, Die Atlantik Brücke (Ullstein, 1992). The book examined the "dual containment" policy of the United States in Germany, a policy which sought to integrate Germany into the West while using her resources and strength to contain the Soviet Union. This book received the Stuart Bernath Book Prize of the Society of American Foreign Relations, and the Harry S. Truman Book Award, given by the Truman Presidential Library. He is also the author of Lyndon Johnson and Europe: In the Shadow of Vietnam (Harvard, 2003), which examined the Johnson Administration's policy toward Europe and assessed the impact of the war in Vietnam on its other foreign policy objectives. He is the co-editor with Matthias Schulz of The Strained Alliance: U.S.-European Relations from Nixon to Carter, (Cambridge University Press, 2009). He is currently working on two books: a biography of former Secretary of State Henry Kissinger, tentatively entitled, "Henry Kissinger and the Dilemmas of American Power," and "The Long Twilight Struggle: A Concise History of the Cold War."

Professor Schwartz has held fellowships from the Social Science Research Council, the German Historical Society, the Norwegian Nobel Institute, the Woodrow Wilson International Center for Scholars, and the Center for the Study of European Integration. He has served as President of the Society of Historians of American Foreign Relations. He served on the United States Department of State's Historical Advisory Committee as the representative of the Organization of American Historians from 2005–2008. Professor Schwartz received The Madison Sarratt Prize for Excellence in Undergraduate Teaching on April 3, 2013 at the Spring Faculty Assembly, Vanderbilt University. In 2008 Professor Schwartz received the Annual Alumni Education Award from the Vanderbilt Alumni Association. Schwartz is the recipient of the 2008 Book Award by Chi Chapter of the Kappa Alpha Order. This award is given to a faculty member who has been particularly influential in the lives and education of members of KAO. Professor Schwartz presented The 2010 Herbert S. Schell Annual Lecture in American History, "Henry Kissinger, Vietnam, and Iraq: The Problem of Realism in American Foreign Policy," on October 18, 2010. Professor Schwartz presented, "The Arab Spring: Revolution in the Middle East," on April 19, 2011, as part of the Samuel L. Shannon distinguished Lecture Series at Tennessee State University. Professor Schwartz has presented lectures for the OAH Distinguished Lecturers Program, please click here to see the list of lectures.

Helmut Smith

Best Researcher/Scholar, History

Helmut Walser Smith is a historian of modern Germany, with particular interests in the history of nation-building and nationalism, religious history, and the history of anti-Semitism. He is the author of German Nationalism and Religious Conflict, 1870–1914 (Princeton, 1995), and a number of edited collections, including The Oxford Handbook of Modern German History (Oxford, 2011), Protestants, Catholics and Jews in Germany, 1800–1914 (Oxford, 2001), The Holocaust and other Genocides: History, Representation, Ethics (Nashville, 2002), and, with Werner Bergmann and Christhard Hoffmann, Exclusionary Violence: Antisemitic Riots in Modern German History (Ann Arbor, 2002). His book, The Butcher's Tale: Murder and Anti-Semitism in a German Town (New York, 2002), received the Fraenkel Prize in Contemporary History and was an L.A. Times Non-Fiction Book of the Year. It has also been translated into French, Dutch, Polish, and German, where it received an accolade as one of the three most innovative works of history published in 2002. Smith has also authored The Continuities of German History: Nation, Religion, and Race across the Long Nineteenth Century(Cambridge University Press, 2008), and is presently working on a book on German conceptions of nation before, during, and after nationalism. His research has been funded by the NEH, the German Academic Exchange Service, the Volkswagen Foundation, and the Alexander von Humboldt Foundation. At Vanderbilt, he has served as Director of the Robert Penn Warren Center for the Humanities and the Max Kade Center for European and German Studies. He teaches a wide variety of courses in European history and in historical methodology. In 1997, he received the Jeffrey Nordhaus Award for Excellence in Undergraduate Teaching.

Hortense Spillers

Most Helpful to Students, Liberal Arts, Social Sciences, and Humanities

Having taught in the U.S. academy for three decades now (and counting), I am reluctant to look back for all those mythical reasons that warn against the backward glance. (I never figured out why doing so might turn one into a pillar of salt, but it is alleged from quite reputable sources to have happened to at least one person!) In any case, when I conjure up 1974, when I started post-doc teaching at Wellesley College, I always find something to cut the memory short, since remembering is to suggest that you have more past than future, but it doesn't feel so to me at all. Actually, I feel as though I'm just getting started good! Though I think I wouldn't have been half bad at either, I am nevertheless grateful to myself that I didn't pursue a career in the practice of law, or tv/radio broadcasting, having spent my last two years in undergraduate school at the University of Memphis as a disc jockey at WDIA radio in Memphis. This historic organization—among the first, if not the dead absolute first, all-black radio station in the United States—might have been my launching pad, I'd hoped, to a career in national news; as I recall,

I was preparing to take the broadcasters' examination, administered by the Feds (and the equivalent of our SATs, or in those days, CEEBs) and about to cut a tape, at their request, to post to the executives who ran WHER in Memphis—the first all-woman radio station in the country, I think. But after all that, William Blake's prophetic books won the charm offensive! Is that not a surprise, or what! Not many things were more interesting to me then than Walter Cronkite, Pauline Fredrick, and Edward R. Morrow, unless it was "Vala, or the Four Zoas"! And one thing led to another and another and finally a career of literary and cultural interrogation that has taken me literally from my birthplace on the southern tier to the East and Mid-West of the country and several decades later, back again. It would be an understatement to assert that it is not today the same South from which I departed my parents' driveway in my little Buick Skylark, three months after MLK's assassination, enroute to Boston and Brandeis. The changes have been momentous for everyone and precisely frame my own professional development.
Try to imagine this: I hired someone to type my doctoral dissertation, though I was competent enough to have done it myself. But one hired out the work before computers because the professional typist was expected to be very fast, very capable, and expert at the proper formatting. You did everything else. The distance that separates the mid-70s from the turn-of-the-century world is a matter of light years, but I wonder how we are doing today with an old-fashioned aim in mind, and that is to say, teaching reading and writing in the age of twitter, although we apply far fancier names to what we do. It is likely that I wrote my dissertation on the rhetoric of black sermons by hand first, then made a rough copy of it on my Olivetti, then gave the secretary the rough draft from which to make the perfect draft. I think I paid the lady $200.00 and change, as the first "real" painting I bought from the same era—a striking head of Miles Davis on a black ground–cost five hundred. Living in Haverford rather than Philly, Ithaca rather than the Big Apple, not taking a job in Chicago, but staying in central New York, I survived the 80s, 90s, and the new millennium; what bothers me now is that we haven't figured out yet the implications of inflated costs, e.g;, that of higher education and the speeds that are supposed to match the global flows of capital. I think we need to spend a little time trying to imagine what all the latter means to and for the tasks of higher education.

Frank Wcislo

Best Teacher, History

Frank Wcislo is a historian of modern Russia, Eurasia, and Europe. He came to Vanderbilt in 1984 from Columbia University and has taught as Visiting Professor in the School of History, University of Leeds (1994–95). His research encompasses the politics, society, economy, and culture of the pre-revolutionary Russian Empire. He is the author of Reforming Rural Russia: State, Local Society, and National Politics, 1855–1914 (Princeton, 1990), a study of imperial political culture and bureaucratic attempts to transform it during the last half century of tsarist rule, and the implications of these efforts for the multiple crises that weakened the old regime before 1917. He was a collaborator and member of the editorial board in the publication project of the St. Petersburg Institute of History [Russian Academy of Sciences] and The Bakhmeteff Archive of Russian and East European History and Culture [Columbia University], Iz arkhiva S. Iu. Vitte. Vospominaniia. [From the Archive of

S. Iu. Witte. Memoirs], 3 vols. [St. Petersburg, Dmitrii Bulanin, 2003]. His Tales of Imperial Russia: The Life and Times of Sergei Witte, 1849–1915 (Oxford University Press, 2011), is a history and biography that reconsiders these same last decades from the perspective of Russian elite identity, culture, and experience in the Victorian Age of Empire. His current book project, "The Death of Autocracy, 1914–1920. A Study of Organizational Failure," is a collection of chapters that examines, from the perspective of organizational and managerial theory, the political elites of the Russian Empire and the behaviors during the era of World War I that led to that polity's loss of legitimacy, authority, and power. Wcislo's research has been supported by the International Research and Exchanges Board, the Kennan Institute (Smithsonian), Vanderbilt University and its Robert Penn Warren Center for the Humanities, and The Harriman Institute of Columbia University. Wcislo is a prize-winning undergraduate teacher and adviser. He is a recipient of the Jeffrey Nordhaus Prize for Excellence in Undergraduate Teaching in the College of Arts and Science (1993), the Alumni Education Award of the Vanderbilt Alumni Association (1998), The Madison Sarratt University Prize for Excellence in Undergraduate Education (2003), and the Alumni Freshman Adviser Award of the College of Arts and Science (2006). He has taught and team-taught a range of courses to undergraduates, graduate students, and continuing adult education: the introductory Columbia University Contemporary Civilization core curriculum; survey lecture courses on medieval and early modern Russia, pre-revolutionary imperial Russia, soviet and post-soviet history, European civilization since 1700, comparative communisms in Russia and China, and concepts of liberty; seminars on the end of the Soviet empire, the Russian revolution, late imperial Russian intellectual history, agrarian Russia from serfdom to collectivization, gender in modern Russian history and literature (seminar), and historical methods and research. He established and has frequently served as the Director of the Honors Program in History. Since July 2006, Wcislo also has served as the inaugural Dean of The Martha Rivers Ingram Commons, Vanderbilt's residential college and living-learning community for all first year students. The Commons, which combines ten residential Faculty Heads of House, non-residential faculty participants, student affairs professionals, undergraduate peer mentors, and 1600 first year undergraduates, is a signature experience for Vanderbilt undergraduates from all four colleges and schools of the university (Arts and Science, Blair School of Music, Engineering, and Peabody College of Education and Human Development). He lives with his family on The Commons campus in the Dean's Residence. He invites faculty, professional colleagues, graduate students, and undergraduates to come to The Ingram Commons and participate in Vanderbilt University's ongoing creation of a residential undergraduate education for the 21st century.

VENTURA COLLEGE

Elizabeth Helms

Best Teacher, Music

Robert Lawson

Best Overall Faculty Member, Music
Most Helpful to Students, Music

Robert E. Lawson, Music Director/Conductor of the Ventura College Symphony Orchestra and the Santa Clarita Symphony has been involved in many areas of music. As a composer, Mr. Lawson's compositions have been performed throughout the United States. His Harp Sonata has received premieres at Florida State University, the XII Foro Internacional de Musica Nueva in Mexico City, and in Paris at the International Harp Festival. His String Trio was selected as one of the most innovative compositions for the Ernest Bloch Music Festival. In 2003, he was commissioned to compose "Duets for Orchestra" which had its world premier by the Ventura Symphony on December 1, 2004. In April 2010, a new work for orchestra "Tumaini" was premiered both with the Santa Clarita Valley Youth Orchestra and the Ventura College Symphony Orchestra. Mr. Lawson served as Music Director/founder of the Ventura City Hall Concert Series for eleven seasons and was the founding Music Director/Conductor of the Channel Islands Chamber Orchestra for twelve seasons. He was Music Director/Conductor of the Santa Clarita Valley Youth Orchestra for eighteen years until he turned over the orchestra in 2008 to its current director. From 1981 through 1988, he served as associate conductor of the Ventura County Symphony and Ventura Civic/Youth Orchestra. This will be the fifth season he has been music director with the Ventura County Ballet Company with it's performance of Tchaikovsky's Nutcracker Ballet. Mr. Lawson has been guest conductor of numerous orchestras both here in the United States and Europe. Mr. Lawson has been Department Chair of Music at Ventura College since 2000. In 2012, he was elected to Chair the Performing Arts, where he continues to develop and expand the department's current student population of over 1400 students and 28 faculty positions. Mr. Lawson received his M.F.A. and B.F.A. Degrees from California Institute of the Arts. His principal teachers have been Pulitzer Price winner Mel Powell, Dean of Music at Yale University and R.O. Disney Chair of Cal Arts. Electronic music innovator Mort Subotnick, developer of Apples music software and the first composer to produce electronic music on long playing Albums, conductor/composer Stephen Mosko, director of the 21st century players, and Arnold Schoenberg institute director/professor Leonard Stein. While at Ventura College early in his musical education he studied with Dr. Burns Taft and Ventura County Symphony founder and Music Director Frank Salazar.

Ola Washington

Best Overall Faculty Member, History

VICTOR VALLEY COLLEGE

Nancy Christian

Best Researcher/Scholar, Theater

"Uber smart and regurgitates this information to the students in class as well as out of class. There is no time that she is not focused on Theatre and the elements that make it unique and rewarding."

VILLANOVA UNIVERSITY

Carol Anthony

Best Teacher, Humanities

Thomas Busch

Best Researcher/Scholar, Philosophy

Hugh Ormsby-Lennon

Best Researcher/Scholar, English

VIRGINIA COMMONWEALTH UNIVERSITY

Liz Canfield

Most Helpful to Students, English

I teach the department's Introductory course as well as upper-level courses on feminist/queer theory, literature, and film. I am also the academic advisor for all of our undergraduate Women's Studies majors and minors.

Tom Dehaven

Best Overall Faculty Member, English
Best Teacher, English

Humberto Fabelo

Most Helpful to Students, Social Work

Nicole Lee

Best Overall Faculty Member, Social Work

Ellen Netting

Best Researcher/Scholar, Social Work

Nick Sharp

Best Overall Faculty Member, English

Judy Thomas

Best Teacher, Social Work

VITERBO UNIVERSITY

Jean Saladino

Best Overall Faculty Member, Music

"Jean Saladino directs the choirs at Viterbo University, as well as teaching applied voice and sight singing. She is an exceptional colleague on both intellectual and collegial levels. I admire her expertise and seek her wisdom; there is no one else who can claim the same excellence in both areas."

WAKE TECHNICAL COMMUNITY COLLEGE

Tyler Dockery

Best Teacher, Graphic Arts

Linda Hill

Best Teacher, Humanities

Jean-Pierre Kamuabo

Best Overall Faculty Member, Humanities
Most Helpful to Students, Humanities

Barry Malone

Best Teacher, History

"Barry is passionate about the subject material and challenges his students in his classes."

WALLACE STATE COMMUNITY COLLEGE AT HANCEVILLE

Lauren Cantrell

Best Teacher, Fine Arts
Best Overall Faculty Member, Fine Arts

Beth Johson

Best Researcher/Scholar, Liberal Arts, Social Sciences, and Humanities

Gayle Ledbetter

Most Helpful to Students, English

Rob Metcalf

Most Helpful to Students, Liberal Arts, Social Sciences, and Humanities

Michael Salerno

Best Overall Faculty Member, English
Best Researcher/Scholar, English
Best Teacher, English

WASHINGTON AND LEE UNIVERSITY

Deborah Miranda

Most Helpful to Students, English

Deborah A. Miranda's primary teaching and research topics include American Literature that is off the canonical map - in other words, contemporary American Lit by authors from the margins of U.S. culture. This literature typically approaches America culture from directions previously unconsidered, as we see the world through the eyes of Native Americans, Chicanos and Chicanas, African Americans, Japanese Americans, Chinese Americans, Appalachians, Southern Americans, and more. Contexts such as environmental issues, economic limitations, linguistic barriers/advantages, and historical trauma are all explored in voices typically not heard in mainstream literature. Many of Professor Miranda's courses are cross-listed with Women's Studies, Shepherd Poverty Program, Latin American & Caribbean and African American Studies programs and, with permission of each program's director and Professor Miranda, may be taken for credit in those areas.

Professor Miranda is in demand as a guest speaker on these topics, and has lectured at numerous universities, including Ohio State University, UC Santa Cruz, Bard College, University of Illinois at Springfield, University of Texas at San Antonio, Aquinas College, UC California at Los Angeles, Oregon State University, Willamette College, Seattle University, University of California at Long Beach, Colby College, Texas Women's University, Texas A & M University, Kenyon College, Syracuse University, SUNY Oneonta, and many others.

Holly Pickett

Best Overall Faculty Member, Liberal Arts, Social Sciences, and Humanities

Lesley Wheeler

Best Researcher/Scholar, English
Best Teacher, English

*"Active poet and scholar.
Dedicated teacher of writing."*

WASHINGTON STATE UNIVERSITY

Chris Arigo

Best Teacher, English

Christopher Arigo received his MFA in Poetry from Colorado State University (2000) and his PhD in English from University of Nevada, Las Vegas (2008). His first poetry collection Lit interim won the 2001–2002 Transcontinental Poetry Prize (selected by David Bromige) and was published by Pavement Saw Press (2003). His second collection In the archives (2007) was published by Omnidawn Press. He currently edits the literary journal Interim with poet Claudia Keelan.

Maria Deprano

Best Researcher/Scholar, Fine Arts

Jolanta Drzewiecka

Best Overall Faculty Member, Communication

Jolanta A. Drzewiecka (PhD, 1999, Arizona State University) is an associate professor of communication in the E. R. Murrow College of Communication. Her research interests center on construction of cultural, racial, and national differences and identities in discourse. Her two focus areas are immigrant identities and public memories. She examines how immigrant identities are negotiated and represented in personal and media narratives. Her work advances understanding of immigrant identity as racial incorporation and intercultural translation. She also examines negotiations of public memories, particularly memories of racist violence. Her current book project examines how memories of how others recuperate and purify the nation in response to ongoing and new global challenges to national purity and exclusivity.

Glenn Johnson

Best Overall Faculty Member, Communication

Louis Kirk-Macauley

Best Overall Faculty Member, English
Best Researcher/Scholar, English
Most Helpful to Students, English

Kirk McAuley received his Ph.D. in British and American literature from the State University of New York, University at Buffalo in 2006. Since then he has taught in the Expository Writing Program at the University of Oklahoma, and the Division of Humanities at New College (the honors college) of Florida. He joined the English Department at WSU in August 2008.

WASHINGTON UNIVERSITY IN ST. LOUIS

Younasse Tarbouni

Best Teacher, Languages

Colette Winn

Best Overall Faculty Member, Languages

Colette H. Winn (Ph.D University of Missouri-Columbia) is Professor of French at Washington University where she teaches French language, culture, and literature with an emphasis on the early modern period. Since first coming to Washington University she has been involved with study abroad. Eighteen years ago, she developed the popular France for the Premed program for premed students with an interest in French. Her research and teaching interests concern women and gender in sixteenth-century France, specifically women's writing. One of her long-term goals has been to make available texts by early modern women. She has published various critical editions of early modern women writers including Madeleine de L'Aubespine, Mlle de Beaulieu, Louise Boursier, Charlotte de Brachart, Marguerite de Cambis, Gabrielle de Coignard, les dames du Verger, Marie Le Gendre, Jacqueline de Miremont, Suzanne de Nervèse, etc. In the course of her career, she has coordinated and produced numerous collections of essays among which The Dialogue in Early Modern France (1993), Ronsard, figure de la variété (2002), Veufs, veuves et veuvage dans la France d'Ancien Régime (2003), Vieillir à la Renaissance (2009, with C. Yandell); as well as special issues on biblical exegesis in the XVIth century, the education of women in Old Regime France, and memory in early modern times. She has also published on early modern medicine, in particular the "books of secrets," midwifery, and pediatrics. Her critical edition of Simon de Vallambert's De la maniere de nourrir et gouverner les enfans dès leur naissance appeared in 2005. She is currently preparing a volume on Teaching French Women Writers of the Renaissance and Reformation for the MLA Teaching Option series, which follows her recently-published volume on Teaching Marguerite de Navarre's Heptameron.

WASHTENAW COMMUNITY COLLEGE

Ruth Hatcher

Best Researcher/Scholar, English

Mary Mullalond

Best Overall Faculty Member, English
Best Teacher, English

Mary Mullalond earned a Bachelors of Art degree in English from the University of Portland, in Oregon, a Masters of Arts in Gender & Cultural Studies from Simmons College, in Boston, and a Masters of Arts in English Education from the University of California, Santa Cruz. Before coming to WCC, Mary worked in alternative education at the high school level. Mary has been at WCC since 2005 teaching Composition 1, Composition II, Introduction to Technical & Workplace Writing, Technical Writing 1, and Children's Literature. She's developed two online courses for the college, and enjoys making her courses fun and engaging learning environments for students. She lives in Ann Arbor with her wife and two cats.

Dave Waskin

Most Helpful to Students, English

Dave Waskin received a bachelor of arts in English and education from the University of Michigan and a master of arts in journalism from the University of Miami. He worked as a copy editor for The Miami Herald and a freelance journalist before joining the WCC English department full-time in 2005. He teaches composition and journalism, and serves as the academic advisor to the journalism program. He remains divided as to which of his alma maters should be regarded as "the real UM."

WAYNE STATE UNIVERSITY

Robert Aguirre

Best Researcher/Scholar, English

"Despite moving into administration, Robert has kept up his research and has just had an excellent article on Anthony accepted for publication."

Ellen Barton

Best Overall Faculty Member, English

"Ellen has served as chair of the English Department for the last several years and has done a remarkably good job at it. This has not prevented her from being an outstanding teacher and researcher as well."

Norah Duncan

Most Helpful to Students, Music

Norah Duncan IV is Interim Chair and Associate Professor of Music at Wayne State University. He is also the Area Coordinator for Organ Performance. Additionally, he directs the WSU Concert Chorale and coordinates many of the choral concerts presented by the Department of Music.

Duncan received an undergraduate degree (English Literature) from the University of Detroit, a Master of Fine Arts degree (Organ Performance) from Wayne State University, and a Doctor of Musical Arts (Church Music/Organ Performance) degree from the University of Michigan.

Norah Duncan IV is a multi-faceted musician. As an organist, he has performed extensively in both the United States and Europe. Recently, he presented organ recitals in Trier, Germany and Sydney, Australia. As a liturgical musician, he has been music director for the Cathedral of the Most Blessed Sacrament for over 26 years, both planning and directing music for both major Archdiocesan liturgies and local parish celebrations. He is the second longest serving Roman Catholic cathedral musician in the country. Among his greatest achievements as a liturgical musician was as director of music for the historic visit of Pope John Paul II to Detroit in 1987. With the Benedictine Monastery in Solesmes, France, Norah established the Gregorian Institute of Detroit for the study of Gregorian chant. He has presented papers on liturgical music at international congresses, clinics and workshops in Rome, Dublin, and for the Diocese of Engu and the Archdiocese of Abuja, in Nigeria.

As a consultant, he has been on the editorial review board for Oregon Catholic Press, and presently is a member of the music review committee for the revision of "Lead Me, Guide Me," the African American Catholic Hymnal. He is also a member of the Education Board of the Detroit Symphony Orchestra. As a choral director, Norah has presented major concerts under the Cathedral Cultural Series with the Archdiocesan Chorus, and has collaborated in concerts with the Wayne State University choral ensembles, the Brazeal Dennard Chorale ensembles and the Detroit Symphony Civic Orchestra. In both 2004 and 2005, he directed the Michigan Youth Choral Ensemble in Carnegie Hall in New York City. Duncan is the recipient of the Award for Excellence for Musical Direction from the American College Theater Festival, Washington, DC, the Sr. Thea Bowman Award of the Archdiocese of Detroit, the Spirit of Detroit Award, the Mother Teresa Duchemin Award for exemplary community service, and this year was the first recipient of the "Changing Lives Through Music" Award presented to him by Anne Parsons, President of the Detroit Symphony Orchestra.

Donald Haase

Best Researcher/Scholar, Liberal Arts, Social Sciences, and Humanities

"Donald has been internationally recognized as a major contributor to folk- and fairy-tale criticism."

Ken Jackson

Best Teacher, English

"Ken Jackson consistently gets high grades from his students, and students seek him out particularly, no matter what he is teaching."

Arthur Marroti

Most Helpful to Students, English

"Although he is retired, Arthur continues to assist with the Wayne-Brazil exchange program, and he continues to mentor graduate students and assist with educational seminars on writing grant proposals."

Dennis Tini

Best Teacher, Liberal Arts, Social Sciences, and Humanities

"Dennis has been a highly successful teacher in the music department for many years."

Dennis J. Tini, WSU Distinguished Professor of Music, is Senior Associate Dean, Interim, of the College of Fine, Performing and Communication Arts, and Area Coordinator for Music Business and Music Technology at Wayne State University in Detroit. Professor Tini is a jazz pianist/composer, choral conductor and clinician, and has appeared frequently at major music festivals, universities, and conferences throughout Europe, South Africa, Canada, Jamaica, Argentina and the United States during the past forty years. Professor Tini is the recipient of numerous National, State, University and International awards for excellence in music performance, teaching, leadership and service.

Co-Founder of the WSU Jazz Studies Program in 1979, Professor Tini is founding director of the WSU Jazztet, founding conductor of the WSU Concert Chorale, and Conductor-Emeritus of the WSU Choral Union & Orchestra. Professor Tini served as Chairman of the Wayne State University Department of Music from 1992–2005, as Director of Jazz Studies from 1978–1988, Co-Director of Jazz Studies from 1988–2005, and as Director of WSU Choral Activities from 1988–2005. Tini is currently Area Coordinator for the WSU Music Business and Music Technology Programs. Tini serves as interim Senior Associate Dean for Academic Affairs & Policy, also overseeing Entrepreneurship, Mentoring, and Arts Administration for the College's departments of Art/Art History, Communication, Music and Theatre/Dance.

Dennis Tini has appeared with Buddy Rich, the Detroit Symphony Orchestra, Larry Nozero, J.C. Heard, Marvin Stamm, Chris Collins, the Dennis Tini Trio/Quartet, April Arabian-Tini, Jon Faddis, Don Ellis, Woody Shaw, Matt Michaels, and Bunky Green, among others. Tini has conducted orchestras for Hollywood movies, composed music for documentaries/commercials, and performed as musical director for national corporate musical productions. Professor Tini has conducted and appeared as piano soloist with the WSU Jazz Lab Band, Vocal Jazz Ensemble, Jazztet and Concert Chorale in numerous concert tours throughout Europe. Dennis Tini's jazz trio, duo, quintet, and quartet with strings CDs are available at: http://cdbaby.com/cd/tinitrio and http://www.cdbaby.com/cd/dennistinirussmillerjazz.

Dennis Tini has served on National Endowment for the Arts Music Review Panels, the United States Information Agency International Arts America Evaluation Committee, and was designated an External Evaluator for the National Association of Schools of Music. Tini is a Past President of the former International Association of Jazz Educators (9,000 members in 35 countries), and is founder of the IAJE International Music Relief Effort for South Africa. In September, 2013, Professor Tini was a featured performer/clinician at the South African Jazz Educators Conference in Durban, South Africa - his 10th trip in support of the Music Relief Effort for South Africa.

Dennis Tini has served on the Chamber Music Society of Detroit Board of Trustees, Detroit Symphony Orchestra Education Advisory Council, the IAJE Past-Presidents Advisory Council, and is a Past-President of the Pi Kappa Lambda National Music Honor Society-WSU Chapter. Professor Tini has served as a Judge for the American Choral Directors Association National Conducting Competition, and appeared frequently as a choral director with the Detroit Symphony Orchestra, preparing Symphony Choruses for Neeme Jarvi, Antal Dorati, Erich Kunzel, and Itzhak Perlman, among others.

Recipient of the WSU President's Award for Excellence in Teaching in 1996 and the 1999 President's Exceptional Service Award, Tini was appointed WSU Distinguished Professor of Music in 2005. In addition to his frequent jazz and choral concerts, Professor Tini's recent research involves The Brain in Music and the Martial Arts, integrating his University teaching, global performance activities, training in Karate & Filipino Martial Arts, and Black Belt in Attributes of the Modern Warrior under Guru Brad Rabe. Dennis Tini was honored to receive the 2014 Global Music Awards Gold Medal for Composition, the 2010 Honorary Lifetime SAJE Membership Award for Outstanding Service to Jazz Education in South Africa, and the 2009 Jazz Artist Dedication from the Michigan Jazz Festival.

WEBSTER UNIVERSITY

Mary Cox

Most Helpful to Students, Communication

Cox serves as advisor to the student staff at The Galaxy, Webster University's radio station. She served two terms as chairman of the Publications Board, which oversees the publication of The Journal, Webster University's nationally award-winning student newspaper and website. She was chosen as a member of the inaugural 2012 Global Leadership Academy.

Bernie Hayes

Best Teacher, Communication

Clara Nunez-Reguiero

Best Researcher/Scholar, Liberal Arts, Social Sciences, and Humanities
Best Teacher, Liberal Arts, Social Sciences, and Humanities
Most Helpful to Students, Liberal Arts, Social Sciences, and Humanities
Best Overall Faculty Member, Liberal Arts, Social Sciences, and Humanities

Leadership Experience: Highly accomplished, results-oriented decisive leader and administrator with proven success in strategic thinking and problem-solving, nationally and internationally. Excellent team-building skills and tendency to thrive in dynamic environments while remaining pragmatically focused. Designed and managed inter-institutional and cultural development projects for 4 years in St. Louis, MO. Experience in all aspects of running own consulting business. Bilingual English/Spanish. • HR Experience: Diplomatic and accustomed to handling sensitive, confidential information. Demonstrated history of producing accurate, timely reports while meeting stringent guidelines. Flexible and versatile –able to maintain a sense of humor under pressure and succeed in deadline-driven environments. Poised and competent with proven ability to easily transcend cultural differences and personal abilities. Organized Office and HR department for startup company in Rochester, NY. Recruiting and supervision as a consultant for 4 years in St. Louis, MO and Pawhuska, OK. Recruiting, training and development in a mining operation in Catamarca, Argentina for 5 years as well as HRIS. • Teaching Experience: Seventeen university courses in Missouri, Illinois, Oklahoma, and Tucuman, Argentina. Components included classroom, lab, and field instruction. Reading and writing in literacy campaign. Zulia, Venezuela. Teaching technologies and multicultural education and diversity training. • Scholarly experience: Research and publication as well as conference presentations and guest lectures.

WENTWORTH INSTITUTE OF TECHNOLOGY

Weldon Pries

Best Teacher, Architecture
Most Helpful to Students, Architecture

WEST CHESTER UNIVERSITY OF PENNSYLVANIA

Sylvia Ahramjian

Best Overall Faculty Member, Music
Best Researcher/Scholar, Music
Best Teacher, Music
Most Helpful to Students, Music

Michelle Blake

Best Researcher/Scholar, Writing

Steve Gimber

Best Teacher, History

Dr. Gimber is the director of the American Studies program at West Chester University. He has his doctorate in U.S. history from the American University in Washington, D.C. and a Master's degree in American Studies from the University of Pennsylvania. He received his B.A. in history from Glassboro State College (now Rowan University) in 1988 and taught as an adjunct professor of history at his alma mater for several years.

Mark Hickman

Most Helpful to Students, Communication

Professor Hickman is a scholar of speech, with a particular interest in public speaking and political communication. He is the department's director of Forensics. Professor Hickman received an M.A. from Miami University of Ohio, and is ABD in his Ph.D. program at the University of Maryland. He joined the faculty of West Chester University in 1997.

David Levasseur

Best Researcher/Scholar, Communication
Best Teacher, Communication

David Levasseur, Ph.D., holds the rank of Professor in the Department of Communication Studies. Dr. Levasseur specializes in classes and research in Communication Theory, Political Communication, and Public Speaking. He holds a B.A. and an M.A. from the University of Maryland-College Park, and a Ph.D. from the University of Kansas. Dr. Levasseur joined the faculty of West Chester University in 1997.

Jordan Schugar

Best Teacher, Writing
Most Helpful to Students, Writing

WEST VALLEY COLLEGE

Kelley Carey

Best Researcher/Scholar, Design

Christopher Cryer

Best Teacher, Fine Arts

WEST VIRGINIA UNIVERSITY

Amy Hirshman

Best Overall Faculty Member, Sociology

Amy Hirshman is an associate professor of Anthropology at West Virginia University. She completed her Ph.D. in Anthropology in 2003 at Michigan State University. Dr. Hirshman's research broadly encompasses social stratification, power and the political economies of emergent state societies, and positioned research results as an evaluation of the relative explanatory power of dominant theoretical models on these topics within Mesoamerica archaeology.

She studies these broader concerns through the lens of craft production, with an emphasis upon ceramics. Ceramics are easily manipulated in both form and decoration, are used by individuals from all social strata, and are crucial to tasks ranging from the mundane, such as daily food storage and preparation, to the most symbolic, including religious expression and political rituals reifying state power. Therefore, products of craft specialists (or non-specialists, as the case may be) and their final distribution within the archaeological record provides insight into cultural change, including, but not limited to, economic, political and social centralizing and decentralizing forces within a given cultural context. The emergence of the Late Postclassic Tarascan state of western Mesoamerica (emerged ca. AD 1350 in what is now Michoacán, Mexico), and the concurrent social, political, and economic changes involved in state emergence, is the particular focus of her work. However, her data crosscut multiple temporal periods (over 1000 years, from the Classic to the Postclassic periods in Mesoamerica) and place results within that larger temporal context.

Using the lens of ceramic artifacts, she currently engages in studying the organization of ceramic production and distribution prior to and concurrent with the emergence of the Tarascan state. With an emphasis on "ceramic ecology" approach to her work, which underscores people and their behavior in the past. She also utilizes multiple lines of evidence, from ceramic style and morphology to archeometry (including chemistry and petrography) to ethnographic analogy and ethnoarchaeology to interpret data and develop a model for the context of ceramic production, distribution, and consumption in the Prehistoric Lake Pátzcuaro Basin. Dr. Hirshman currently teaches an introductory physical anthropology and upper-division archaeology courses, though she has taught across the discipline of anthropology since arriving at WVU. Her courses highlight multiple lines of evidence in a comparative framework, with an emphasis on student discovery. She also is an academic adviser to Sociology and Anthropology majors at WVU.

Andrew Kohn

Best Overall Faculty Member, Music

Andrew Kohn teaches string bass and music theory at WVU. He also composes. His terminal degrees are a Ph.D. from the University of Pittsburgh and an Artist Diploma from The Peabody Conservatory. He has published in American String Teacher, Bass World, Double Reed, Fulcrum, Musik-Konzepte, Música Hodie, Perspectives of New Music, The Strad, and in the monograph, On the Music of Stefan Wolpe, and has recorded for Albany, Music Minus One, and Ravello.

James Nolan

Best Overall Faculty Member, Liberal Arts, Social Sciences, and Humanities

James Nolan is an associate professor in the Division of Sociology and Anthropology at West Virginia University where he teaches courses in the area of crime and social control. His research currently focuses on police procedures, crime measurement, and hate crimes and homicide. Dr. Nolan's professional career began as a police officer in Wilmington, Delaware. In 13 years with that department, he worked in a variety of divisions, including patrol, community policing, organized crime and vice, and planning and research. In 1993, Dr. Nolan was the project director of the community-policing component of the Weed & Seed Program in Wilmington, Delaware (one of the first 16 sites). He is a 1992 graduate of the Federal Bureau of Investigation (FBI) National Academy. Just prior to joining the faculty at West Virginia University, Dr. Nolan worked for the FBI as a unit chief in the Crime Analysis, Research and Development Unit that provided management oversight for the National Hate Crime Data Collection Program.

Dr. Nolan is co author of the book The Violence of Hate: Confronting Racism, Anti Semitism, and Other Forms of Bigotry 3rd edition and co editor of the book The Essential Hate Crime Reader. In addition his research publications have appeared in the American Behavioral Scientist, Journal of Quantitative Criminology, Journal of Contemporary Criminal Justice, The Justice Professional, Policing & Society, Criminal Justice Studies, Homicide Studies, Journal of Criminal Justice, and The American Sociologist. Dr. Nolan was the 2010 recipient of the Carnegie Foundation for the Advancement of Teaching and Council for Advancement and Support of Education (CASE) West Virginia Professor of the Year.

Jen Steele

Most Helpful to Students, Liberal Arts, Social Sciences, and Humanities

Jennifer Steele is a Clinical Associate Professor in Sociology. She received an M.S. and Ph.D. in Rural Sociology from Penn State University, specializing in natural resource sociology and rural community and economic development. Her appointment is service and teaching-oriented. She coordinates departmental advising, directly advises current and prospective students, and plays a lead role in the department's curriculum assessment efforts. Her teaching centers on core requirements for majors, including SOCA 199 (Orientation to Sociology and Anthropology) and SOCA 311 (Social Research Methods), as well as occasional electives. These include SOCA 323 (Sociology of Rural Life) and most recently, SOCA 463 (Economy and Society). While research is a smaller component of her position, Dr. Steele pursues inquiry aimed at enhancing her service and teaching and collaborates with Extension Service colleagues and others on a variety of projects related to agricultural and forest management practices and to program evaluation.

Rachel Stein

Best Teacher, Sociology
Most Helpful to Students, Sociology
Best Teacher, Liberal Arts, Social Sciences, and Humanities

Rachel E. Stein is an associate professor of Sociology at West Virginia University. She received her M.A. and Ph.D. in sociology from the University of Akron. Dr. Stein's research has focused primarily on the multilevel analysis of cross-national victimization within a routine activities/lifestyles theoretical framework. Her work has appeared in the International Criminal Justice Review and the International Journal of Criminal Justice Sciences. A second area of Dr. Stein's research explores the portrayal of violent females in popular film through content analysis. This area of interest is highlighted in her Gender, Crime, and Film capstone course. Dr. Stein also teaches courses in Criminology, Juvenile Delinquency, and Gender and Crime.

Matthew Vester

Best Researcher/Scholar, Liberal Arts, Social Sciences, and Humanities

In general, I am interested in political culture, which I understand in a very broad sense to include institutional practices, patronage systems and other informal political ties, and political thought (customary and written). My research is informed by a desire to uncover the rich variety of political interactions in the past, and therefore to avoid an anachronistic exaggeration of the role of 'the state' in early modern political life. My biography of Jacques de Savoie, Duke of Genevois-Nemours, traces the life of an illustrious renaissance prince while showing how sovereign dynasties functioned as political actors in sixteenth-century Europe, creating contradictory pressures within and between families and patronage networks. The house of Savoy responded to these pressures by creating institutional structures such as the princely apanage, which both increased dynastic prestige but also undermined the authority of the senior branch of the ruling family.

Karen Weiss

Best Researcher/Scholar, Sociology

WESTCHESTER COMMUNITY COLLEGE

Kent Trickel

Best Teacher, Languages

WESTERN CONNECTICUT STATE UNIVERSITY

Bill Petkanas

Most Helpful to Students, Communication
Most Helpful to Students, Liberal Arts, Social Sciences, and Humanities

"Dr. Bill Petkanas' outstanding support and guidance to students and faculty have helped shape the high standards of excellence in the Dept. of Communication & Media Arts at Western CT State University."

Cathy Vanaria

Best Overall Faculty Member, Fine Arts
Best Teacher, Fine Arts
Most Helpful to Students, Fine Arts
Best Teacher, Fine Arts
Most Helpful to Students, Fine Arts
Best Overall Faculty Member, Fine Arts

Catherine Vanaria is a photographer, master darkroom technician, and educator. Her work is included in the Polaroid Collection, the Rock and Roll Hall of Fame collection, as well as many private collections, and has been included in numerous group shows throughout the northeast United States including the Aldrich Museum (Ridgefield, CT).

WESTERN ILLINOIS UNIVERSITY

Lisa Kernek

Best Researcher/Scholar, Journalism
Most Helpful to Students, Journalism

Lisa Kernek joined the journalism program in 2006 and teaches news reporting and photojournalism. Before coming to WIU, she worked as a newspaper reporter for The State Journal-Register in Springfield, Ill., for 12 years, and at smaller papers before that. She is the faculty adviser for WIU's student chapter of the Society of Professional Journalists.

Christine Lapka

Most Helpful to Students, Music

Professor of Music Education. Dr. Lapka teaches elementary music methods courses, leads exceptionality in music class, coordinates the Music Education program and organizes music student teaching. Prior to her term at Western, Dr. Lapka worked in Illinois public schools in various music positions. Elementary music education made up the majority of her practical experience. During that time she found her two undergraduate degrees invaluable to her teaching (Music Education and Music Therapy, WIU). Dr. Lapka holds a Master of Arts in Music Education (WIU, 1993) and completed her doctorate in Music Education at the University of Illinois (2005). She received the Mary Hoffman Award of Excellence from the Illinois Music Educators Association (ILMEA) in 2014. As a researcher in the area of disabilities in music, she finds herself leading numerous workshops about students with different abilities. Likewise, several articles on the same subject appear in state, national, and international journals of music education. WIU Appointment 1998.

Scott Palmer

Best Researcher/Scholar, History

Dr. Scott W. Palmer (Ph.D., Univ. of Illinois, Urbana-Champaign, 1997) is a specialist in Russian history and the history of technology. He is author of Dictatorship of the Air: Aviation

Culture and the Fate of Modern Russia (Cambridge University Press, 2006) as well as numerous scholarly articles and essays. His current project, From the Icon to the iMac: A History of Russia, Technology and Culture examines the foreign origins and domestic transformation of major technological systems from the founding of the Kievan State through the opening decades of the twenty-first century.

In addition to the department's two-semester Russian history sequence covering the Imperial and Soviet eras, Prof. Palmer regularly teaches "US Military History" (HIST 304); "History of Flight Culture" (HIST 311); "Technology, Culture & Society" (HIST 312); and offerings in modern European history. He has recently debuted new classes on "The Culture and Civilization of Stalinism" and "The Great War, 1914–1918" (HIST 380).

Katharine Pawelko

Most Helpful to Students, Recreation, Parks, & Tourism

Robert Porter

Best Researcher/Scholar, Recreation, Parks, & Tourism

Paul Schlag

Best Overall Faculty Member, Recreation, Parks, & Tourism
Best Teacher, Recreation, Parks, & Tourism

Barry Witten

Best Overall Faculty Member, Curriculum & Instruction

Dr. Witten earned his Ph.D. from Kansas State University in Curriculum and Instruction, with an emphasis in social studies education in 1999. He also holds B.A. and M.A. degrees in history from Fort Hays State University. He teaches undergraduate and graduate courses in Social Studies Education at both elementary and secondary levels in the department. Dr. Witten is active at the state and national levels of the National Council for the Social Studies and is also active in the National Council for Geographic Education. Dr. Witten worked with two U.S. history grants for area middle level and high school American history teachers funded by the U.S. Department of Education. He edited seven volumes of teaching strategies as editor of publications for both grants. His research interests include the American West, historical inquiry, performance assessment, and current issues and trends in elementary and middle level education. Dr. Witten currently serves as the chair of the Graduate Committee for the Department of Curriculum and Instruction and is the coordinator for the Graduate Program in Elementary Education. He serves as an advisor for about 75 candidates in this program.

WESTERN KENTUCKY UNIVERSITY

Fabian Alverez

Best Overall Faculty Member, English

Throughout my teaching career, I've been focused upon understanding the role of language in the creation and transformation of realities. I find it my duty to help students participate in these investigations, believing that rhetorical analysis will be the most helpful skill they acquire.

Joe Glaser

Best Researcher/Scholar, English

Mary Ellen Miller

Best Teacher, English
Most Helpful to Students, English

Mary Ellen Miller has her B. A, degree in English and modern languages from Berea College; her M. A. in English from the University of Kentucky; and all course work, written exams, and language exams in French and German for her Ph. D. from

Vanderbilt University. She started her teaching career at Western Kentucky University in the fall of 1963.

WESTERN MICHIGAN UNIVERSITY

Patricia Montilla

Best Researcher/Scholar, Languages

Dr. Patricia Montilla is associate professor and graduate advisor of Spanish in the Department of Spanish at Western Michigan University.

Montilla specializes in Spanish American poetry and U.S. Latino literature and culture. She has published articles on the works of Oliverio Girondo, Matías Montes Huidobro, Judith Ortiz Cofer, Carlota Caulfield, Mariano Brull, and Eugenio Florit. She authored a book, Parody and the Poetics of Subversion in Oliverio Girondo, as well as a panoramic study on contemporary Puerto Rican literature of the United States for a companion to U.S. Latino letters. Most recently, Montilla edited a book titled Latinos and American Popular Culture for Praeger. She serves the department as faculty director of the undergraduate study abroad program in Quito, Ecuador.

Cynthia Running-Johnson

Best Overall Faculty Member, Languages

Dr. Cynthia Running-Johnson is an associate professor and advisor of French in the Department of World Languages and Literatures, with a joint appointment as associate professor of Gender and Women's Studies at Western Michigan University. Running-Johnson knew the joys of languages and art from an early age, accompanying her artist parents when they led study abroad trips in Europe. In college she learned to appreciate literature through her French classes. In graduate school, she worked in theatre production while studying French literature and culture. During her graduate studies she taught as an assistant to English teachers at a high school in Lyon, France, and earned a certificate in art at the Ecole du Louvre in Paris.

WESTERN WASHINGTON UNIVERSITY

Susan Haines

Best Researcher/Scholar, Dance
Best Teacher, Dance

Susan Haines is a dancer, choreographer and educator with performance credits with ballet and modern companies in Colorado, New York, North Carolina and Virginia, independent choreographers Gerri Houlihan, B.J. Sullivan, KT Niehoff, as well as collaborative work with Brad Parquette, and study at The Place; London Contemporary Dance Theatre, and Dock 11 in Berlin. Her dance training includes master teachers JoAnna Kneeland, Karen Clippinger, David Howard, and Frano Jelincic of English National Ballet. Susan's choreography has been presented by Danceworks in Jacksonville, FL, ACDF, the NC Dance Festival, Highlands Arts Council, Black Mountain Center for the Arts, Labor Force Dances, Bellingham Electronic Arts Festival, the Power Company in residence at Columbia College (SC), Bellingham Repertory Dance, Dance Gallery, the Center for Performance Research in Brooklyn, and Open Flight Studio in Seattle. She has worked with composer Bruce Hamilton for Do Ho Suh's sculpture installation ceremony at WWU in 2012. Her dance films have been screened at the American College Dance Festival, as part of the "Next Generation of Dance Filmmakers" at the Dance for the Camera Festival in Salt Lake City, and at Tacoma Museum of Glass Dance Film series. She has an MFA in Dance from UNC-Greensboro, and is a certified Comprehensive Pilates Instructor through Balanced Body. Susan has received two NEA grants to restage works by the Trisha Brown Dance Company and the Martha Graham Dance Company, and has presented her dance conditioning workshops around the country. She is the founding director of It Must Have Been Violet Dance Productions, a professional dance company that presents contemporary work for stage and screen nationally. BA Radford University, MFA University of North Carolina Greensboro.

Jennifer Keller

Best Overall Faculty Member, Journalism
Best Teacher, Journalism
Most Helpful to Students, Journalism

Carla Rutschman

Best Teacher, Music

"An excellent and thorough teacher who cares about her students."

Senior Instructor, Tuba, Euphonium

WESTFIELD STATE UNIVERSITY

Susan Leggett

Best Overall Faculty Member, Communication

Susan Leggett joined Westfield State University in 2004. She is currently Chair of the department and teaches Communication Research, Communication Theory, Special Topics in Environmental Communication, Sport, Media, and Culture. Her research concerns include commodification of higher education, feminist theory and practice, and the implementation of critical pedagogy in Communication.

She holds a Ph.D. from the University of Massachusetts/Amherst.

Jaime Wainright

Most Helpful to Students, Fine Arts

WICHITA STATE UNIVERSITY

Ted Adler

Most Helpful to Students, Fine Arts

Amy Baker

Most Helpful to Students, Fine Arts

Amy Baker Schwiethale, Associate Professor of Musical Theatre/Choreographer, is honored to be part of the School of Performing Arts at WSU. Professional Choreography Credits include: Music Theater of Wichita's HONK! (Mary Jane Teall Award), GYPSY(Mary Jane Teall Award), and CAMELOT. Amy's choreography has been showcased in NYC at the New York Jazz Choreography Project (Alvin Ailey Studios) and Traditions in Tap, with the Jacksonville Symphony Orchestra, the Newton Mid-Kansas Symphony Orchestra, Opera Kansas, across Northern Italy, and in numerous capacities at WSU as Director/Choreographer. Professional Performing Highlights include: Phyllis Dale on the First National Broadway Tour of 42nd STREET, performing for two years with the RADIO CITY CHRISMAS SPECTACULAR, receiving her Actor's Equity card while doing SINGIN' IN THE RAIN at the Westchester Broadway Theatre, and performing in the resident Music Theatre of Wichita company for five summers and played the role of "Anytime Annie" in their production of 42ND STREET. She has a BFA in Musical Theatre from our very own WSU and an MFA in Choreography from Jacksonville University. Amy is fiercely committed to the preservation of Original Broadway choreography and preparing WSU Performing Arts students to make a living doing what they love.

Deborah Baxter

Best Teacher, Music

Robert Bubp

Best Overall Faculty Member, Fine Arts

Robert Bubp has participated in over 50 exhbitons spanning solo, two-person, and group shows. His work has been displayed across the U.S. and Europe, including the Salina Art Center, Salina, KS; Rose Museum at Brandeis University, Waltham, Ma.; Dowd Fine Art Center, SUNY Cortland, Cortland, NY; Indianapolis Art Center, Indianapolis, IN.; Museum of Fine Arts, Florida State University, Tallahassee, FL.; Kouvola Museum of Art, Kouvola, Finland; and Kspace Contemporary, Corpus Christi, TX. Bubp's work has been noted and reviewed in numerous publications including, ArtPapers, the Atlanta Journal-Constitution, and Art In America.

Bubp's artwork is held in numerous permanent collections, including Georgia State University, Atlanta, GA.; UPS, Atlanta, GA.; Kansas Health Foundation, Wichita; Emprise Bank, Wichita; and the University of Kansas. He has been the recipient of several awards for artwork and scholarship, including the 2005 Kansas Artist Fellowship for Interdisciplinary or Performance Art from the Kansas Arts Commission, and the 2009 Excellence in Creative Activity Award at Wichita State University.

Jim Hellman

Best Teacher, Fine Arts

Jim Hellman is a Kansas native. He holds degrees in graphic design from Wichita State University and Kansas State University. Professionally, he art directed Apartment Living, Amateur Archeology, three Conoco magazines and designed numerous books for New York and Boston publishers. He has designed advertising for various local and regional banking, real estate, insurance, education, manufacturing and retail companies while working for Art for Living, McCormick-Armstrong, and Bruce Hinel Graphic Design. Associate Professor Hellman has taught at WSU in various capacities since 1975.

Victor Markovich

Best Teacher, Music

Dr. Victor Alan Markovich is a Professor of Music at the School of Music at Wichita State University (WSU) in Wichita, KS. At WSU he serves as Director of Bands, Director of Winds and Percussion Studies, Coordinatior of Recruiting for the School of Music, and Coordinator of the Undergraduate and Graduate Instrumental Music Conducting Programs.

Additonally, at WSU he teaches graduate instrumental conducting classes, conducts the WSU Symphonic Wind Ensemble and concert band, and administers and guides all aspects of the WSU band program including the athletic bands.

Under the direction of Dr. Markovich, the WSU Symphonic Wind Ensemble has performed for numerous national, regional, and state music conventions. The WSU Symphonic Wind Ensemble has been heralded as "one of the finest university bands in the nation," receiving national critical acclaim for its "musically emotional and technically detailed performances."

Bands under his baton have performed for the American Bandmasters Association National Convention; College Band Directors National Association (CBDNA) Biennial Conference; CBDNA Southwestern Division Conference (twice), CBDNA North Central Division Conference; Music Educators National Conference (MENC) Biennial National In-Service Workshop (twice); the Kansas Music Educators Association (KMEA) In-Service Workshop (seven times); Kansas Bandmasters Association (KBA) Convention (four times); National Band Association North Central Conference; the Illinois Music Educators Association (IMEA) Conference.

Dr. Markovich received the Doctor of Philosophy in Music with an emphasis in wind ensemble/band conducting degree, and Master of Music degree from The University of Michigan (Ann Arbor); and a Bachelor of Music degree from Kent (OH) State University He has done additional study at the Cleveland (OH) Institute of Music, Bowling Green (OH) State University, and Baldwin-Wallace College (Berea, OH).

Dr. Markovich has served as an adjudicator and guest conductor for all-state and honor bands in forty-two American states and seven European countries. He has nationally published articles dealing with bands and music education. He is a contributor to volumes III and IV of Teaching Music Through Performance in Band.

Dr. Markovich is a Past-President of the Southwestern Division of the College Band Directors National Association and a Past President of the Kansas Bandmasters Association.

Walter Mays

Best Teacher, Music

Royce Smith

Best Researcher/Scholar, Fine Arts

Deborah Soles

Best Overall Faculty Member, Philosophy
Best Teacher, Philosophy
Most Helpful to Students, Philosophy
Best Overall Faculty Member, Liberal Arts, Social Sciences, and Humanities

"Deborah Soles is an exceptional teacher who has received both the John R. Barrier Distinguished Teaching Award and the University's Excellence in Teaching Award. Her classes are well conceived, carefully prepared, rigorous, appropriate in terms of content and coverage and apposite. She teaches some of the most difficult and formally challenging courses in the college, as well as the most popular. Her student evaluations consistently rate her courses as excellent in all categories. For Deb teaching is more than just a class room activity; she spends countless hours working with students one-on-one or in small groups. One of the best indicators of the quality of her teaching is the success of the students who have worked closely with her; students whom she has mentored have been admitted to some of the finest Ph.D programs and law schools in the nation, including Harvard, MIT, Cornell, Berkeley, Claremont, etc.

David Soles and Deborah Soles. They are the most helpful to the students.

I nominate Deborah Hansen Soles for recognition as best overall faculty member. Deborah excels in all categories. She is a Full Professor with a national reputation in philosophy, a reputation achieved through her record of publications and presentations at professional meetings. She is an extraordinary teacher who Has been recognized by receipt of both the John R. Barrier Distinguished Teaching Award and University Award for Excellence in Teaching. She also has been selcted as an Emory Lindquist Honors Society Mentor and a Mcnair Mentor several times. Deborah's service record is nothing less than spectacular. At the university level she has served on and chaired both the LAS and the University Tenure and Promotion Committee innumerable times; served on the Faculty Senate both as a member and as President; served as a member of the LAS College Council; chairs several departmental committees; volunteers her time for the McNair Program; serves on or chairs several college committees, etc. At the professional level, she has been active in several regional and national professional societies, holding various offices, including President, of the Southwestern Philosophical Society; she is a tireless paper and book referee for various programs, journals and presses."

Professor Deborah Soles routinely teaches Introductory Logic, and is responsible for Philosophy of Language, Philosophy of Logic, Theory of Knowledge, Analytic Philosophy, and offers special courses such as seminars on Davidson and Wittgenstein. Professor Soles received her B.A. from George Washington University and her Ph.D. from the Johns Hopkins University. She has been teaching continuously at WSU since 1975. Her major research area is the philosophy of language and adjacent topics in epistemology and logic, and she has published work on both historical and contemporary treatments of these topics. Currently she working on a book on Donald Davidson's epistemology. [click here for Dr. Soles's cv.] Since coming to WSU she has taught some dozen and a half different courses. She has served several times as an Emory Lindquist Honors program mentor, as a McNair program mentor, in 1999 was awarded the college's John R. Barrier Distinguished Teaching Award, and in 2008 was awarded the University Excellence in Teaching Award. In 1998 she received the President's Award for Distinguished Service to WSU. Dr. Soles is a Past President of the Southwestern Philosophical Society, and served as President of the Faculty Senate in 2009–2010. She is the advisor for the undergraduate Philosophy Society.

Linda Starkey

Best Teacher, Liberal Arts, Social Sciences, and Humanities

Linda Starkey (Acting, Voice, Musical Theatre History), Associate Professor and Chair of the School of Performing Arts, earned the B.M.E. from the University of Kansas, an M.A. in Theatre from WSU, and an M.M. in Voice Performance from Fort Hays State University. At WSU for the past 24 years, she has taught voice, acting, and Music Theatre History. She has been the music director or stage director for over 50 shows for University Theatre, WSU Summer Theatre and WSU Opera Theatre. Active in the Wichita theatre community, she has also worked for Music Theatre of Wichita, Wichita Center for the Arts, Opera Kansas, Cabaret Oldtown, Stage One, Wichita Children's Theatre and Crown Uptown Dinner Theatre. This summer she is directing the popular Music Theatre of Wichita Target Teen Choir. Many of Linda's former students are performing professionally in regional theatres, national touring companies, Off-Broadway, and Broadway. She is a past Regional Governor and Auditions Chairman for the West Central Region of National Association of Teachers of Singing.

Aleks Sternfeld-Dunn

Best Researcher/Scholar, Music
Best Researcher/Scholar, Liberal Arts, Social Sciences, and Humanities

Aleksander Sternfeld-Dunn joined the faculty of Wichita State University in 2011. His music is performed widely throughout the U.S. and Europe by groups like Composers Inc., The Kiev Philharmonic, The Conundrum Ensemble, the Contra Costa Chamber Orchestra, and the Interlochen Chamber Players. Awards include a first place prize in the William R. Wiley Exposition in Fine Arts for his composition "Sonata for Alto Saxophone and Piano" in 2005 and an honorable mention for his string orchestra work "Evolutions" in the 2007 adagio competition and Joker's Wild in the 2009 concerto competition held by the Fauxharmonic. He has also received grants from Meet the Composer and the Visual, Performing, and Literary Arts Committee. His music is published by FJH Music, Jomar Press, Dorn Publishing and can be heard on the ERM Media and Capstone record labels.

Mr. Sternfeld-Dunn completed his doctorate in composition with a minor in music theory from the Hartt School of Music. He holds a Masters Degree in Composition from Washington State University as well as a Bachelors Degree from California State University East Bay. His primary teachers have included Robert Carl, Martin Rokeach, Stephen Gryc, Charles Argersinger, Ryan Hare, Greg Yasinitsky, and Frank LaRocca

As well as being an accomplished composer, he is also active as a conductor and string bassist. As a conductor he recently guest conducted the new music ensemble Turn on the Music for their CD "The Darwin Effect" on Capstone Records. He has conducted various orchestras, wind ensembles, and theatre productions. As a bassist, he has held principal positions with several San Francisco Bay Area Orchestras, as well as numerous jazz small groups and big bands. He is the featured bassist on Bay Area jazz pianist Marty Namaro's CD "Abstractionisms".

Levente Sulyok

Best Researcher/Scholar, Fine Arts

Levente Sulyok was born and raised in Hungary and moved to the U.S. in 1991. His interest in philosophy - particularly the relationship between aesthetics, language, and the politics of resistance - can be seen throughout his work. Sulyok's career boasts numerous solo exhibitions, as well as over 30 group exhibitions in Kansas, Texas, California, Massachusetts, New York, New Jersey and Rhode Island.

In Postcards from the Edge, art historian Dr. Rachel Epp Buller observes that his work, "... creates a multifaceted layering of geographic detachment: foreign artist as outsider citizen; the mythologized American West as both a commercial export and an outpost from urban civilization; and contemporary corporate greed as morally suspect disjuncture."

Martina Johnston, an artist-run house gallery in West Berkeley, hosted Sulyok's 2012 solo exhibition, All the Paranoid Monoliths. The review of this solo exhibition in the East Bay Express proclaims, "[s]uch voluble silences, compressions, excisions, and about-faces pervade All the Paranoid Monoliths — an aesthetically cohesive exhibition if ever there was one."

Recent projects for Sulyok include the collaborative intervention "Documenta Detour" in Kassel, Germany, and Small Pleasures (2012–2016), a large-scale interactive public installation on the facade of the Ulrich Museum of Art in Wichita, Kansas. Currently, he pursuing several collaborative projects and furthering his philosophical and aesthetic concerns.

Sabrina Vasquez

Best Teacher, Fine Arts
Best Overall Faculty Member, Fine Arts

Sabrina Vasquez began training in Tucson Az, where her early performance experience was with Arizona Dance Theatre under the artistic direction of Jean Paul Comelin as an apprentice. While at the University of Arizona, she performed the modern, ballet, jazz, and mime theatre works of Douglas Nielsen, Jory Hancock, Melissa Lowe, Susan Quinn Williams and Nick Johnson. Hubbard Street Dance Chicago hired Sabrina in 1995 as a member where she performed the works of Twyla Tharp's Sinatra Suites, Step out of Love by Margo Sappington and Daniel Ezralow's Read my Hips. During her years in Chicago, she also performed with the James Kelly Choreography Project, Chicago Tribune Charities Nutcracker choreographed by Ruth Page, under the direction of Larry Long and Ballet Theatre of Chicago. Since her arrival to Wichita, Ms. Vasquez has performed with Music Theatre Wichita in Kismet and choreographed Chicago for the Crown Uptown Theatre. While at Wichita State University, she has collaborated with WSU OperaMusical Theatre on the productions of Ruddigore, Candide. The Desert Song and Pippin, which was an alternate for the 2005 ACTF Kennedy Center Festival. In 2007 she received an Individual Artist Fellowship Award for Choreography from the Kansas Arts Commission and most recently was the recipient of the Dorothy Johansen Hauck Faculty Fellow in Dance and Musical Theatre in 2012.

Sabrina Vasquez is currently an instructor at Wichita State University, Associate Artistic Director of Wichita Contemporary Dance Theatre and Associate Artistic Director of Alithea Mime Theatre with her husband of 15 years Nick Johnson. Her dance choreography has been seen nationally and Internationally with WCDT, which toured Kansas with Wiyo's of Oz and repertory concerts to Taiwan and Italy. Alithea Mime Theatre has toured to the International Mime Art Festival, Warsaw, Poland, in May 2000 and August 2009, The First International Mime and Physical Theatre Festival in the Caribbean, San Juan, Puerto Rico, in May 2001, the 2005 Shanghai International Arts Festival and New York City at the United Nations in 2006.

Russell Widener

Best Overall Faculty Member, Music

Professor Widener has been on the faculty of Wichita State University since 1981. For the six years prior to this, he had been freelancing in the Los Angeles area. Some of his experiences in Los Angeles included performing for various recordings, teaching at the University of California at Santa Barbara and the California State University at Fullerton. He was Bass Trombonist for the Santa Barbara Symphony and Principal Trombonist with the Ventura Symphony. Mr. Widener also performed with various chamber groups in the area and performed with the Los Angeles Philharmonic Orchestra on a performance of the "Suntreader".

Before moving to Los Angeles, Professor Widener had taught at the University of Georgia and performed on trombone and euphonium as a member of the First Army Band.

Mr. Widener's education includes a Bachelor's degree from Baylor University with a major in Music Theory, a Master's degree from Catholic University of America with a major in Trombone Performance, and doctoral work at Indiana University in Brass Pedagogy. He studied with John Marcellus, Lewis Van Haney, Keith Brown, Ned Meredith, Byron Peebles, and Dee Stewart.

Since coming to WSU, Professor Widener has performed as the Trombonist with the Wichita Brass Quintet, Principal Trombonist with the Wichita Symphony, and Bass Trombonist with Music Theatre of Wichita.

WILLIAM CAREY UNIVERSITY

Dewey Douglas

Best Teacher, Theater

"Design faculty at William Carey University and mentor to students, Dewey Douglas continually goes above and beyond to see his students succeed."

WILLIAM PATERSON UNIVERSITY

Alejandro Anreus

Best Researcher/Scholar, Fine Arts
Best Researcher/Scholar, Fine Arts

Born in Havana, Cuba, Professor Anreus received his B.A. in art history from Kean College in 1984. He completed his M.A. (1995) and PhD (1997) in art history at the Graduate Center, City University of New York. Before joining the William Paterson faculty, Prof. Anreus was a curator at the Montclair Art Museum (1986–93) and the Jersey City Museum (1993–2001), as well as a critic in residence at the Latino Center for Art and Culture, Rutgers University (1999–2000). Prof. Anreus has taught art history at New Jersey City University, Seton Hall University and Kean University.

His most recent publications are Ben Shahn and The Passion of Sacco and Vanzetti (Jersey City Museum and Rutgers University Press, 2001), Orozco in Gringoland: The Years in New York (University of New Mexico Press, 2001), and The Social and The Real; Political Art of the 1930s in the Western Hemisphere (Penn State Press, 2006), which he co-edited with Diana L. Linden and Jonathan Weinberg. His articles have appeared in Art Journal, Third Text, Art Nexus and Encuentro de la Cultura Cubana.

Since 2003, Prof. Anreus has been part of the national advisory board of the "A Ver" series of monographs. This is the first series of monographs focused on living Latino artists. This multi-volume project is funded by the Getty and Rockefeller Foundations and is based at the Chicano Studies Department at UCLA.

Prof. Anreus has just completed Mexican Muralism, A Critical History (in collaboration with Leonard Folgarait of Vanderbilt and Robin Adèle Greeley of UConn), which will be published by University of California Press at the end of 2010. He is currently writing a monograph on Cuban-American painter Luis Cruz Azaceta.

Maboud Ansari

Best Overall Faculty Member, Sociology

"He is a great teacher, great scholar and a very nice colleague to work with."

Phil Ciofari

Most Helpful to Students, English

"Phil will put in as many hours as students need to help them be successful."

Zhiyuan Cong

Best Overall Faculty Member, Fine Arts
Best Overall Faculty Member, Fine Arts

Professor Zhiyuan Cong earned an MFA in printmaking from Indiana University in 1994. Prior to his arrival in the United States, he earned a BFA (1980) and MFA (1986) in Chinese Painting from the Nanjing Arts Institute, China. He has taught at William Paterson University since 1994. Prior to that, he was a professor at Nanjing Arts Institute from 1980 to 1988, a guest artist at the Chicago Art Institute in the summer of 1992, and an instructor at Indiana University from 1989 to 1994. Professor Cong's most recent solo exhibition was at Nantong Museum of China, in 2008. He has also had solo exhibitions at the Butler Institute of American Art, Youngstown, OH (2006), at the United Nations, New York City (2003), at the East-West Center, Hawaii (2002), St. John's University, New York (2001), McDonough Museum of Art, OH (2000), Bergen Museum of Art and Science, the Indiana University Art Museum, the Nanjing Arts Institute, Qinghai Provincial Museum, China, and the Elisabeth Wang Gallery in New York City. Cong's group exhibitions include the Indianapolis Museum of Art, the New Jersey State Museum, the Newark Museum, the Nanjing Museum, the Shanghai Art Museum, and the National Fine Art Museum of China. His art demonstrations and lectures include Princeton University, UC Berkeley, the University of Oregon, and the Chicago Art Institute. Cong also was a consultant and project coordinator for the Indianapolis Museum of Art for an art project funded by the National Endowment for the Humanities, and has been a director of the Summer

Art in China program at William Paterson since 2001. Currently Professor Cong is a director of Center for Chinese Art at William Paterson University. In addition to his exhibitions, his work has earned him several awards for Artist-in-residency, East-West Center, HI, Spring, 2002; The New Jersey Printmaking Fellows for 2001, Rutgers University; Best of Show: "Contemporary Prints Exhibition," NJ, 1999, and "Imaging New Jersey," 1995. He received the First-Place Award at the American State Arts Competition in Indianapolis in 1993, and the Outstanding Works Award from the National Committee of the International Youth Year in Beijing in 1985. He have received reviews of his work from New York Times, Star-Ledger, Herald news, World Journal, Chinese news Net, People's Daily, NJN and WLIW. His art works have been collected at a lot of museums, such as New Jersey State Museum, Jersey City Museum, Nanjing Museum, and National Art Museum of China.

Angela Delaura

Best Overall Faculty Member, Fine Arts

"My very favorite professor of all. She's tough of totally fair as a teacher. The best!"

Professor DeLaura enjoys a successful career as a graphic designer, medical illustrator, and fine artist. Her paintings and sculpture are in the permanent collections of major museums and collectors. She is a graphic designer and medical illustrator for major corporations in the tri-state area. She has won several fellowships including the NJ State Council on the Arts and the Stockbridge Artist in Residence program. She is a member of many professional organizations and is highly active within the community and University wide events.

Jacob Felson

Best Teacher, Sociology

Jacob Felson received his doctorate in sociology from Penn State University in December 2009. He received a master's in sociology from Penn State in 2004, and a bachelor's degree, also in sociology, from the University of Chicago in 2002. Prof. Felson's work has focused on issues related to the debate about nature versus nurture. For example, Prof. Felson published a paper in the past year which explores the extent to which sexual orientation is simply a product of biological predispositions, or whether social forces, such as religious background, also play a role ("The Effect of Religious Background on Sexual Orientation"). Prof. Felson has also written a book with Edwin Mellen Press, entitled What Can We Learn from Twin Studies? A Comprehensive Evaluation of the Equal Environments Assumption. In his spare time, Prof. Felson blogs occasionally about NYC architecture and real estate at the website www.newyorkyimby.com. The link to Prof. Felson's class is http://tinyurl.com/soc3020.

Gennifer Furst

Best Researcher/Scholar, Liberal Arts, Social Sciences, and Humanities

Gennifer Furst received her B.A. in psychology (with a sociology minor) from Connecticut College and her M.A. in psychology (with a concentration in evaluation methodology) from Claremont Graduate University. She received her doctorate in criminal justice (with a concentration in corrections) from CUNY Graduate Center. Dr. Furst is the Department of Sociology's Criminal Justice Director. Dr. Furst's research interests focus on issues of incarceration. She published the first national survey of prison-based animal programs in the US. A book based on that work was recently published. Additionally, she is interested in race and the administration of criminal justice, the death penalty, the use of animals in the criminal justice system, and the relationship between drugs and crime. Most recently, she co-authored (with Kimberly Collica) a textbook, Criminology and Society. She also has a chapter, "The Experiences of an Outsider Spending Time Inside," appearing in Experiences in Corrections, and a book review of Do Prisons Make Us Safer? The Benefits and Costs of the Prison Boom is in Criminal Justice Review. She is the co-author of another book chapter examining the inclusion of issues of race and ethnicity in college criminology/criminal justice courses. Her article about how prison-based animal programs change participants was included in a leading reader on human-animal relations. She edited a reader that serves as a supplement to published criminology texts. Her current research examines prison programs that teach inmates how to train shelter dogs who are then given to veterans with Post Traumatic Stress Disorder (PTSD). Prior to teaching full-time, Professor Furst worked at one of the country's three independent prison monitoring agencies, The Correctional Association of New York. Before coming to WPU Professor Furst taught at CUNY, John Jay College and Bronx Community College, as well as The College of New Jersey. Dr. Furst lives in Englewood, NJ with her dog and cats. She loves all things Pearl Jam, especially Eddie Vedder.

Dave Kerzner

Best Overall Faculty Member, Music

"Puts students first, always thinking forward. Stays positive but keeps the rest of the faculty on our toes."

David Kerzner, associate professor, coordinator of BA in Popular Music, coordinator of BM in Sound Engineering Arts, M.S., B.S. College of Fredonia, State University of New York.

Professor David Kerzner holds a master's degree in computer-aided composition from the State University of New York at Fredonia and a bachelor's degree in sound recording technology with a dual major in mathematics/physics. His students have been the recipients of Grammy and Emmy awards and have worked with artists at the highest levels of the music industry.

Professor David Kerzner has been responsible for the creation of 3 academic programs during his 20-year academic career. Professor Kerzner has also written music for film and for video. Clients have included: Alarm Will Sound, Clark Terry, Albany Records, Naxos Records, Crystal Records, Warren Miller Enterprises, Sherri Lewis, Independence Jazz Reunion, and

Temple Town records. In addition, Professor Kerzner continues to be active as a guitarist playing in a wide variety of jazz, rock, soul, and theatrical ensembles.

Payton MacDonald

Best Researcher/Scholar, Music

"Travels the world to learn more about his art form and share it with his students and audience, through performance, lecture, and publication."

Payton MacDonald is a composer/improviser/interpreter. He has created a unique body of work that draws upon his extensive experience with East Indian tabla drumming, American military rudimental drumming, Jazz, European classical music, and the American experimental tradition. He works across multiple musical genres, often at the same time. Several recordings of MacDonald's music exist, including the critically acclaimed Payton MacDonald: Works for Tabla on the ATMA label. MacDonald recently performed his percussion concerto in Carnegie Hall with Alarm Will Sound and again at the Walt Disney Concert Hall in L.A. with the L.A. Philharmonic with John Adams conducting as part of the Green Umbrella series. The New York Times described him as an "energetic soloist." The Los Angeles Times described him as an ". . . inventive, stylistically omnivorous composer and gifted performer . . ." His music has also been described as ". . . engaging and utterly beautiful." (Sequenza 21) MacDonald has been a featured performer of his own music on festivals in Montreal (Voyages) and Minneapolis (Electric Eyes). Numerous ensembles have performed his music around the world, including Alarm Will Sound, Los Angeles Philharmonic, So Percussion, Quintet Mont Royal, Classical Jam, and tabla soloist Shawn Mativetsky. He has received grants and awards from ASCAP, Meet the Composer, American Music Center, American Institute of Indian Studies, as well as fellowships from Yaddo and Ragdale. He has also composed music for the Silken Dance Company. As a performer, he is a founding member of Alarm Will Sound, a new-music group based in New York City. Alarm Will Sound is currently regarded as one of the foremost new music ensembles in the country and they have made five recordings, on the Nonesuch and Cantaloupe labels. The New York Times wrote that they are "the future of classical music." Alarm Will Sound has performed in the finest concert halls all over the world. MacDonald also frequently appears as a marimba soloist with Super Marimba. With Super Marimba MacDonald performs his own music using looping machines and delay pedals. MacDonald expands that project with the Super Marimba Band, which includes drums, bass, and two bari saxophones. He also performs full-length concerts of freely improvised music. As an interpreter he has commissioned works from other composers, including Charles Wuorinen, Robert Morris, Caleb Burhans, David Saperstein, and Stuart Saunders Smith. From 1994–2004 MacDonald performed with Verederos, a flute and percussion duo. Verederos recorded two CDs under the Equilibrium label. He has also appeared as a soloist in England and Croatia, performed with Present Music and the New Jersey Percussion Ensemble, and toured Japan with Keiko Abe and the Galaxy percussion group. MacDonald studied music at the University of Michigan and the Eastman School of Music. His composition teachers include Sydney Hodkinson, Robert Morris, Dave Rivello, Bright Sheng, and Augusta Read Thomas. His percussion teachers include John Beck and Michael Udow. Further studies include tabla with Bob Becker and Pandit Sharda Sahai. MacDonald is a disciple of Mr. Sahai.

Charles Magistro

Most Helpful to Students, Fine Arts

Professor Magistro has been teaching at the university level for the past thirty years. His career as teacher, artist, lecturer and administrator began at Carnegie Mellon University, where he received his BFA in 1964. He then went on to complete his MFA at Ohio State University in 1967. He has been teaching at William Paterson since 1977, and served as Chair of the Art Department from 1996 to 2000.

Vince Parrillo

Best Researcher/Scholar, Sociology
Best Researcher/Scholar, Liberal Arts, Social Sciences, and Humanities

Steve Rittler

Most Helpful to Students, Fine Arts

Steve Rittler is an independent Animator, Illustrator, Designer and Educator, who specializes in 3D character animation, visual storytelling and figurative art. He graduated with honors from the University of the Arts in Philadelphia in 1990 with a BFA in Illustration and an Animation minor. After launching a career as an illustrator in Philadelphia, Steve moved to New York City in 1991 and a year later enrolled in the MFA Computer Graphics Animation program at Pratt Institute in Brooklyn, New York. Becoming a member of the Pratt Faculty in 1993, Steve embarked on his career as a 3D character animator in NYC in 1995, moving to Los Angeles for a year in 1996 to work for the Walt Disney Company. Returning to New York in 1997, Steve resumed his faculty position at Pratt Institute and his character animation career in the city. Steve was invited to join the faculty of the SVA MFA Computer Art Department in 1999 to teach character animation, and completing his MFA thesis in 1999, received his MFA in Computer Graphics Animation With Distinction from Pratt Institute in 2000.

Today Steve has his own independent animation studio, MuseMachine Productions, is a freelance animator, designer and illustrator, and a tenured Associate Professor of Animation at the William Paterson University Department of Art, where he joined the faculty in 2001. Films featuring his work as an animator have been screened and won awards nationally and internationally, including the Annecy Animated Film Festival in France, and his own animated film, Nor Gloom of Nyght, has been screened at the International SIGGRAPH Conference, The first Beijing International New Media Arts Exhibition and Imposium and at BUDi 2005 in Seoul. Elected in 2000, 2002, 2004 and 2006, he currently serves on the board of directors of New York City ACM SIGGRAPH, local chapter of the Special-Interest Group for Graphics of the Association of Computing Machines, and has presented courses on animation production at the 2003 and 2004 International SIGGRAPH Conferences in San Diego and Los Angeles, and at the 2008 SIGGRAPH Asia Conference in Singapore. He served as chair of MetroCAF 2006, the New York Metropolitan Area College Computer Animation Festival, and was also the principal designer, illustrator and animator for the event opener, titles and promotional materials. Steve lives in New York City, plays electric bass guitar and sings in a rock band.

Lynne Rogers

Best Teacher, Music

"Very devoted to music theory pedagogy and to giving her students the best her field has to offer. And at the same time displays the highest academic and personal standards."

Lynne Rogers performs research primarily in the field of twentieth-century music. Her articles appear in Contemporary Music Review, Journal of the Royal Musical Association, Journal of Musicology, Music Theory Spectrum, and other professional journals. She is co-editor of Selected Essays from the Fourth International Schenker Symposium (Olms, 2013). She has presented numerous papers in the United States and abroad as an invited speaker and at meetings of scholarly societies. She received grants from the National Endowment for the Humanities, Paul Sacher Stiftung, Oberlin College, and the University of Texas. She served as President of the Society for Music Theory from 2010–2011. In 2010, she received the Faculty Excellence Award for Teaching from William Paterson University. Before coming to WPU, she taught at Oberlin Conservatory and the University of Texas. She received her Ph.D. from Princeton University.

Alicyn Sabol

Best Overall Faculty Member, Liberal Arts, Social Sciences, and Humanities

Arlene Scala

Best Teacher, Women's Studies

Robin Schwartz

Best Teacher, Fine Arts
Best Teacher, Fine Arts

Robin Schwartz's earned a Master of Fine Arts in Photography from Pratt Institute and her photographs are held in several museum collections, including The Metropolitan Museum of Art, The Museum of Modern Art, The Smithsonian American Art Museum, Washington, D.C. and The San Francisco Museum of Modern Art. The Aperture Foundation published a comprehensive book of the twelve-year project of Amelia photographs, Amelia and The Animals, 2014. Book quotes from Lena Dunham, Film Maker, Writer, Actress and Mia Fineman, Curator, and Elizabeth Krist, Senior Editor at National Geographic Magazine. Schwartz's two earlier books are LIKE US: Primate Portraits, W.W. Norton & Co., and Dog Watching and Amelia's World, edited by Tim Barber, Aperture, NYC. Schwartz's photographs have been published in The New York Times Magazine, The New Yorker Magazines, TIME Magazine LIGHTBOX & Print, and blogs New York Times Lens Blog, Slate Magazine, PetaPixel, CBSNEWS.com (covering The FENCE at Photoville/United Photo Industries, Brooklyn,) The Guardian UK, My Modern Met and other selected magazines; O –Oprah Interview, PDN, Fader, Juxtapoz, Stern (Germany), Liberation (France), H (Spain), Bang (Italy), Esquire (Russia),: Marie Clair Magazine, The British Journal Photography and in books as Photography Speaks: 150 Photographers on Their Art, 2004 and 1995. Schwartz was one of the ten short-listed finalists at the Hyeres 2010 Photography Festival in France. She was a guest speaker and present the Amelia Series at The National Geographic Magazine's Annual Photography Seminar in Washington D.C. and gave Master Talk at LOOK3, The Charlottesville Photography Festival, David Alan Harvey Workshop, Photo Workshop New York with Maggie Steber & Erica McDonald, The Eddie Adams Workshop and the International Center for Photography, New York. Schwartz is also on the faculty at The International Center for Photography and has worked as an editor/curator for National Geographic Magazine's Your Shot assignments including a chapter in an upcoming National Geographic Magazine Book regarding shooting animals on assignment.

Paul Whitehill

Best Overall Faculty Member, Italian

Martha Witt

Best Overall Faculty Member, English

"Prof. Witt puts in a lot of time on her course work, with her students and on the department level doing a great deal of service for the department. She is an all-around excellent colleague and professor."

Academic Interests: Creative Writing, Translation, English Education.Books: Broken As Things Are (hardback, Holt 2000; paperback, Picador 2004). National book tour.Short Stories: Published in This is Not Chick Lit: Original Stories by America's Best Women Writers (Random House 2001); One Story Magazine; The Chatahoochee Review; Boulevard Magazine; Harpur's Palate; Portland Review, and many other national literary journals.Artist Colonies: Yaddo; VCCA; Ragdale; Wassard Elea (Italy).Fellowship Recipient: Thomas J. Watson Traveling Fellowship; Spencer Grant; Walter E. Dakin Fellowship; McCracken Fellowship; New York Times Fellowship

WINSTON-SALEM STATE UNIVERSITY

Darius Cureton

Most Helpful to Students, English

Darius M. Cureton is a native of Charlotte, NC but has lived in Winston-Salem since 2001. He is a graduate of Garinger High School of Charlotte, NC. In the fall of 1996, Darius entered to learn at Winston-Salem State University and departed to serve in May of 2000. During his matriculation, he participated in the WSSU Marching Ram Band, the Ram Ambassadors, and Campus Pals. He was inducted in to Alpha Kappa Mu, Honor Society and was a founding member of the Kappa Lambda Chapter of Kappa Kappa Psi, National Honorary Band Fraternity, Inc. The following spring, upon graduation from WSSU, he began his matriculation at North Carolina Agricultural and Technical State University where he received his master's degree in English and African-American Literature in December of 2001.

During his tenure at A&T, he was exposed to the world of teaching as he was a Graduate Teaching Assistant. In the spring of 2002, he was hired as an adjunct professor and part-time assistant lab coordinator in the English and Foreign Languages Department. He held that job until he was officially hired by the University full time in the fall of 2003 as the Director of Writing/Computer Literacy Lab.

Since then, Darius has served on numerous university committees which include the University Assessment Committee, the Academic Athletic Council, and the United Way Campaign Committee. He has also served the Department of English and Foreign Languages in a variety of capacities including being the Chair of the Creative Writing Committee, being a member of the Department's Assessment, Strategic Planning & Recruitment teams, assisting with the department newsletter, and being the technical liaison for the building's classrooms and the content manager for the department's webpage.

Darius's passion is still teaching as he teaches traditional student Freshman Composition Courses each semester. His hope is that he can continue to help students gain a better understanding of English and writing as they are skills that are highly regarded in today's society; as well as having a little fun while doing it.

Dwedor Ford

Best Researcher/Scholar, English

Lavie Leasure

Best Overall Faculty Member, English

WINTHROP UNIVERSITY

Shaun Cassidy

Best Teacher, Fine Arts

British-born artist Cassidy has been making work in the U.S since 1991. He studied sculpture in England and Canada and worked as a studio assistant for Sir Anthony Caro in London. His work spans a diverse range of materials, processes, and ideas. His work has been shown extensively, including exhibitions at Socrates Sculpture Park, N.Y., Grounds for Sculpture, N.J., and at the Museum of Contemporary Art in Marseille, France. His art has been featured in Sculpture Magazine, Art in America, and the Wall Street Journal.

Cassidy has been awarded artist residencies at over ten venues including the McDowell Colony, N.H., Sculpture Space, N.Y., Bemis Center for Contemporary Art, Neb., and the Djerassi Resident Artist Program, California. From 2001–2002 he was an affiliate artist at the McColl Center for Visual Arts in Charlotte, N.C., and a CATS-sponsored artist in residence in 2006. Cassidy has completed several public art projects throughout the region and is currently working with Tom Stanley on two major works.

Cassidy serves on the board of directors of the McColl Center for Visual Art in Charlotte. He teaches sculpture at Winthrop, where he was named Kinard Teaching Professor of the Year in 2007.

Mark Hamilton

Best Researcher/Scholar, Fine Arts

Born in Brooklyn, New York in 1957, Mark Hamilton's interest in photography began with hours looking at the great picture magazines of the sixties: Bazaar, Life, Look, and Vogue. The first camera he owned was Canon rangefinder purchased with Christmas money from his parents at Camera Barn on Broadway in New York City. A family vacation to Germany in the seventies started with the purchase of an Exakta VX1000 camera in Frankfurt. The next few months were spent documenting his travels in Europe while visiting his uncle's family stationed in Germany with the Army.

While pursuing studies in architecture at Brooklyn Technical High School, an instructor suggested a career in photography stating, "that you have had more experiences in your life at sixteen than most people who are in their thirties." Maybe it was the suggestion of a life path from an authority figure, or just stubbornness that led him to join the Marine Corps, flee home, and the turmoil of parents going through a divorce. Purchasing a Asahi Spotmatic camera at a pawnshop while waiting at a bus stop in Okinawa, Japan, his off duty time was filled with taking and printing photographs at the base hobby shop. The purchase of several cameras in Japan led to the purchase of a Nikon F2AS professional camera. His next duty station was Marine Barracks in Rota, Spain where he spent the next four years teaching himself lighting and studio skills. His photos were spotted on the walls of the base dining facility where he worked as a cook. Transferred to the base photographic unit and as a self-taught photographer led a team of 22 photographers in various photographic assignments with the Naval Atlantic Fleet Audio Visual Command.

He returned to New York City after the military enrolled in the new photo program at Parsons School of Design led by Benedict J. Fernandez, the former personal photographer of Martin Luther King Jr. Mark left Parsons after one year to pursue a professional career. His photo work for the schools public relations department led to assignments with Andy Warhol's Interview Magazine. Fernandez offered a few classes for him to teach at the New School for Social Research to financially assist his pursuits. Early assignments were for Cole of California with Eleanor Lambert Public Relations and New York Apparel News. It was his work with editor Bridget Foley at Apparel News Group that caught the eye of her new employer Woman's Wear Daily. Foley's introduction to the staff at WWD opened the doors to assignment work at WWD, W & M magazines and work with Anne Klein, Donna Karan, Liz Claiborne, Cynthia Rowley, Malcon Forbes, Frank Stella, Ralph Lauren, Michael Graves, Carolina Herrera, Norma Kamali, among others.

Shooting the New York Collections with editors like Etta Froio, Bobbi Queen, Bridget Foley, Ruth LaFerla, and Karyn Monget provided an education in fashion that led to opening a commercial studio and the confidence to explore the market in Milan and published work with Italian publications Amica & Annabella. Representation followed from the Asian markets when he signed with MEGAPRESS in Tokyo. Mark was asked to teach at the School of Visual Arts by Photography Chair, Alice Beck-Odette while maintaining the studio in NYC. Hamilton left New York City after maintaining the studio for nearly a decade.

A move to North Carolina began a change to personal projects and a return to the study of fine arts at the University of North Carolina at Charlotte; pursuing primarily painting at first, along with work in Religion and Women's Studies. Mark's first fine art projects debuted at the Blue Pony Gallery in Charlotte. He received an invitation from the Union of Students Artists to speak about his work at Winthrop University. The talk that evening led to an invitation by Phil Moody to teach at Winthrop University in a part-time capacity as a lecturer. Several years later, Mark started teaching in a full-time role and is presently a tenured Professor of Fine Arts at Winthrop University. Mark pursues personal fine arts projects, collaborative projects (www.jenniferandmarkhamilton.com) with photographer Jennifer Hamilton, along with commercial projects for select clientele.

Paul Martyka

Best Teacher, Fine Arts

Martyka, a native of Detroit, Mich., earned his B.F.A. and M.F.A. degrees at Wayne State University and the University of Michigan, respectively. Before coming to Winthrop in 1979, he worked as master printer at the Michigan Workshop of Fine Prints and completed a staff assistantship at the University of Michigan. Locally, his work is on display at Gallery Up in Rock Hill and the Joie Lassiter Gallery in Charlotte, N.C. His work has garnered awards and been included in a number of local, regional, and national exhibitions. In addition, his art has been displayed in many private, institutional, and corporate art collections.

At Winthrop, Martyka is the fourth recipient of Winthrop's Elizabeth Dunlap Patrick Faculty Grant, a grant established to further new work, research or collaboration by faculty culminating in an exhibition project in the Patrick Gallery. An exhibition of his hand-printed cut paper collages, Conversations with an Echo, ran in fall 2009 in the Patrick Gallery. In the intricate details of this work, Martyka references cultural identities or art historical pieces as inspiration for color and form.

The collages have garnered regional attention and accolades. In fall 2008, Martyka's collage Totemic Talk, featured in the S.C. State Museum's 20th Anniversary Juried Exhibition, won Best in Show and the museum's Purchase Award. His collages also have been featured in an Arts Council of York County exhibition.

Stacy McConnell

Most Helpful to Students, Fine Arts

In addition to her studies at Columbia College and Florida State University, McConnell has studied at the Alvin Ailey American Dance Center's Summer Intensive and the American Dance Festival.

As a performer, McConnell has worked with choreographers such as Lynda Davis, Alan Danielson, Martha Brim, Angela Gallo, Nancy Smith Fichter, and others. She was a member of the

Power Company, a contemporary dance company in Columbia, S.C.; Dance Repertory Theater, a resident company affiliated with the FSU Dance Department; and Eight Feet Repertory Dance Company in Spartanburg, S.C. McConnell interned with Urban Bush Women, a modern dance company located in Brooklyn, N.Y., directed by Jawole Zollar. Today she continues to perform her own choreography and is a member of Saffire Moon Dance Company.

Some of McConnell's choreographic works include Untitled U, Zalophus, Feelodin, Two: A Piece for Jack, Applause, Unwanted, and Suite Dean. Some of her choreographic works have been performed at Columbia College's Faculty/Guest Artist Performance Series, Sarasota Ballet's Summer Sizzler, the Atlantic Center for the Arts' Master Artists Showcase, Moving Currents' Emerging Choreographers Series, and the South Carolina Dance Association's Dancing Festival's gala concert, among others.

While continuing to pursue a professional career in performance and choreography, McConnell served as co-artistic director alongside Kurt Heinlein of Hot Southern Nights, an outdoor festival of music, theatre, and dance, which premiered summer 2003 and again in summer of 2004. She continues to choreograph for contemporary project-based company Woodward Street Dance Project, of which she was co-founder and co-artistic director.

Marge Moody

Best Researcher/Scholar, Fine Arts

Since graduating from traditional training at art college in Scotland in 1972, Moody has continued to develop her work as a painter, first based on observation and then working non-objectively in the medium of abstract collage and painting on canvas. Subject matter for her work varies widely but could be described as being landscape-based. Moody believes that everything she experiences is potential for her work, whether at home or traveling, observed or imagined. She is actively engaged as an artist on an ongoing basis and exhibits locally, regionally, and nationally in the U.S. and formerly in Great Britain.

Moody has been the recipient of numerous grants in support of her work, notably from the South Carolina Arts Commission, the Arts & Science Council of Mecklenberg County, N.C., and from Winthrop. Meritorious achievements include one-person exhibitions in Washington, D.C., two-person shows in New York City, and selection for a summer residency at McColl Center for Visual Art in Charlotte, N.C., in 2006. Her work is found in numerous private and public collections, including the S.C. Arts Commission Permanent Collection and, more recently, the Carolina Contemporary Collection of the Medical University of South Carolina in Charleston, S.C.

Born of British parents and raised in Kenya, East Africa, Moody moved to Scotland at age 14, where she completed her formal education. She moved to the U.S. in 1982 and has lived in Rock Hill for the past 30 years with her husband, Phil Moody, photographer and Winthrop fine arts faculty member, and two (now grown) sons, James and Paul.

Phil Moody

Best Overall Faculty Member, Fine Arts
Best Overall Faculty Member, Fine Arts

Moody's work has been included in more than 80 group exhibitions in the U.S., England, Scotland, and Poland, and he has done 37 one-man exhibitions since 1985. He has received grants from the Charlotte-Mecklenburg Arts & Science Council, the Scottish Arts Council, the Hope-Scott Foundation (Scotland), Winthrop and the Arts Council of York County. Moody's artwork also appears in the collections of Bank of America, B.F. Goodrich, S.C. State Art Collection, Scottish Arts Council, Kodak Pharmaceuticals, North Philadelphia HealthCare, Midlands Technical College, Charlotte-Douglas International Airport, and Clemson University, among others.

Moody has won a number of awards in his career, including the Carnegie Foundation for the Advancement of Teaching South Carolina Professor of the Year Award (1997), the Winthrop University Kinard Award for Excellence in Teaching (1996), and the Winthrop University Outstanding Junior Professor Award (1990). He also was the recipient of an S.C. Arts Commission Fellowship in 2004.

Seymore Simmons

Most Helpful to Students, Fine Arts

Simmons, a native of Colorado, has spent the past 35 years working as an exhibiting artist and illustrator. He has taught drawing and painting at all levels, including gifted and talented middle school, college, and adult education. He came to Winthop in 1993 from Cambridge, Mass., where he taught art education and foundations at Massachusetts College of Art and did research at Harvard Project Zero with Howard Gardner.

Since coming to Winthrop, he has taught undergraduate and graduate art education courses as well as studio classes in drawing and illustration. He is the co-author of Drawing: The Creative Process (Prentice-Hall, 1976); co-editor of the Arts PROPEL Visual Arts Handbook (Harvard University, 1993); and co-editor of The Heart of Art Education: Creativity, Connections, and Transformation (National Art Education, in press). His research, publishing, and presentation interests include: art education, creativity, drawing, experiential education, and multiple intelligence theory. In addition, he and his wife, Martine, lead summer trips to France to study painting, photography, and French culture and cuisine.

WOODBURY UNIVERSITY

Vic Liptak

Most Helpful to Students, Architecture

Louis Molina

Best Teacher, Architecture

Mark Owen

Best Overall Faculty Member, Architecture
Best Researcher/Scholar, Architecture

Mark Owen was born and raised in rural western Pennsylvania, not far from Frank Lloyd Wright's "Fallingwater". It was from his very first visit to that architectural masterpiece, at age 6, that he knew he wanted to be an architect.

As with most kids growing up at that time, he also became fascinated with the "California Lifestyle" and naturally left the farm within days of graduating high school to pursue his education at a little known school in Burbank, CA by the name of Woodbury University. He received his BArch in 1994, the same year that Woodbury's Architecture Program achieved NAAB accreditation. Shortly after graduating from Woodbury, he decided to go back for a Master of Architecture degree and focused on integrating "the computer" into his design process. Since graduation from UCLA, Mark has worked primarily on large scale transportation and aviation projects as well as a series of smaller, experimental residential projects including his own house. Mark also began teaching at Woodbury in 1999 and has since taught at Art Center College of Design, UCLA Extension, and Otis College of Art and Design. In his spare time, he can be found building, racing, and riding anything with 2 wheels, as well as spending time with his family.

WRIGHT STATE UNIVERSITY

David Garrison

Best Researcher/Scholar, Liberal Arts, Social Sciences, and Humanities

"He was the best Chair that Modern Languages had. He is an internationally well recognized scholar, and true poet. He continues to represent WSU community after his retirement. He continues to publish. He established a fund for students, and he really new his faculty talents. He is a writer in English and Spanish, and the quality of his works speaks by itself."

David Lee Garrison was born in Bremerton, Washington. He earned his PhD from the Johns Hopkins University, taught Spanish and Portuguese at Wright State University from 1979 to 2009, and is now retired. He has published translations of Spanish poets from Lope de Vega to Gloria Fuertes, including collections of Nobel Laureate Vicente Aleixandre (with Willis Barnstone), José Bergamín, and Pedro Salinas.

Garrison's poems have appeared widely in journals such as Connecticut Review, Poem, and Rattle, and also in several anthologies. Two poems from his book, Sweeping the Cemetery, were read by Garrison Keillor on The Writer's Almanac, and one was included in Keillor's Good Poems, American Places. The title poem from his book, Playing Bach in the D. C. Metro, was featured by Ted Kooser on his website, American Life in Poetry.

Garrison has stated about his work, "My main goal is to communicate, so I create poems that are not hard to understand with a first reading. My poems emerge mainly from memories, as I fashion bits and pieces of my past into fictionalized configurations in a search for something more meaningful, more transcendent than my personal experience."

Corey Andrews, in a review of Sweeping the Cemetery in Ohioana Quarterly, states that Garrison's poetry "explores the nature of reflection and memory, probing the boundaries that separate the living and the dead. The honorable tradition of poetic memento mori is directly addressed by Garrison, serving not only as the title of one of his poems but also as the collection's thematic core."

David Lee Garrison lives in Dayton, Ohio, with his wife, Suzanne Kelly-Garrison, a writer and professor of law at Wright State University.

Kelli Zaytoun

Best Overall Faculty Member, Women's Studies
Best Researcher/Scholar, Women's Studies
Best Teacher, Women's Studies
Most Helpful to Students, Women's Studies

Research interests: identity and narrative, multi-ethnic American literature, feminist theory, memoirs, Gloria Anzaldúa

Education History: B.A., 1986, M.S., 1988, Ph.D., 2003 Miami University

XAVIER UNIVERSITY

Diane Ceo-DiFrancesco

Best Researcher/Scholar, Liberal Arts, Social Sciences, and Humanities

Margaret McDiarmid

Best Teacher, French

Jo Ann Recker

Best Researcher/Scholar, French

YALE UNIVERSITY

Teresa Carballal

Best Overall Faculty Member

YOUNGSTOWN STATE UNIVERSITY

Qi Jiang

Best Overall Faculty Member, Sociology

Scott Leonard

Best Teacher, Liberal Arts, Social Sciences, and Humanities
Best Researcher/Scholar, Liberal Arts, Social Sciences, and Humanities

Index

C

D

G

H

I

J

K

S

T

Y

Z

Lightning Source UK Ltd.
Milton Keynes UK
UKOW06f2202301115

263809UK00003B/32/P